Infosys Science Foundation Series

Infosys Science Foundation Series in Mathematical Sciences

The *Infosys Science Foundation Series in Mathematical Sciences,* a Scopus-indexed book series, is a sub-series of the *Infosys Science Foundation Series*. This sub-series focuses on high-quality content in the domain of mathematical sciences and various disciplines of mathematics, statistics, bio-mathematics, financial mathematics, applied mathematics, operations research, applied statistics and computer science. All content published in the sub-series are written, edited, or vetted by the laureates or jury members of the Infosys Prize. With this series, Springer and the Infosys Science Foundation hope to provide readers with monographs, handbooks, professional books and textbooks of the highest academic quality on current topics in relevant disciplines. Literature in this sub-series will appeal to a wide audience of researchers, students, educators, and professionals across mathematics, applied mathematics, statistics and computer science disciplines.

More information about this subseries at http://www.springer.com/series/13817

Ramji Lal

Algebra 4

Lie Algebras, Chevalley Groups, and their Representations

Springer

Ramji Lal
University of Allahabad
Prayagraj, Uttar Pradesh, India

ISSN 2363-6149 ISSN 2363-6157 (electronic)
Infosys Science Foundation Series
ISSN 2364-4036 ISSN 2364-4044 (electronic)
Infosys Science Foundation Series in Mathematical Sciences
ISBN 978-981-16-0477-5 ISBN 978-981-16-0475-1 (eBook)
https://doi.org/10.1007/978-981-16-0475-1

This Springer imprint is published by the registered company Springer Nature Singapore Pte Ltd.
The registered company address is: 152 Beach Road, #21-01/04 Gateway East, Singapore 189721, Singapore

Dedicated to the memory of
my younger sister
(Late) Smt Pushpa (Malti),
who left us at the age of 30.

Preface

The present volume, Algebra 4, in this series of books on Algebra, centers around the study of Lie algebras, Chevalley groups, and their representation theory. Lie groups and Lie algebras are very intrinsically related. The origin of Lie groups and Lie algebras lies in the study of geometric spaces with the very crucial observation that a geometric space is determined by the group of its continuous symmetries. Lie groups and Lie algebras play a very fundamental role in Physics also.

The main concerns in the book are the following:

1. The structure theory and the classification of semi-simple Lie algebras over $\mathbb{C}$ through root space decomposition, root systems, and Dynkin diagrams.
2. The representation theory of semi-simple Lie algebras including the theorem of Harish-Chandra and the theorems of Ado and Iwasava.
3. Chevalley groups including the twisted finite simple groups of Lie types.
4. The representation theory of Chevalley groups including the Steinberg characters, Principal and Discrete series representations, and an introduction to the Deligne–Lusztig characters.

The book can act as a text for graduate and advanced graduate students specializing in the field.

There is no prerequisite essential for the book except for some basics in algebra (as in Algebra 1 and Algebra 2) together with some amount of calculus and topology. An attempt to follow the logical ordering has been made throughout the book.

My teacher: (Late) Prof. B. L. Sharma; my colleagues at the University of Allahabad; my friends: Prof. Satyadeo, Prof. S. S. Khare, Prof. H. K. Mukherji, and Dr. H. S. Tripathi; my students: Prof. R. P. Shukla, Prof. Shivdatt, Dr. Brajesh Kumar Sharma, Mr. Swapnil Srivastava, Dr. Akhilesh Yadav, Dr. Vivek Jain, Dr. Vipul Kakkar, and Dr. Laxmikant; and above all the mathematics students of Allahabad University had always been a motivating force for me to write a series of books on various topics in Algebra. Without their continuous insistence, it would have not come in the present form. I wish to express my warmest thanks to all of them.

Harish-Chandra Research Institute Allahabad has always been a great source for me to learn more and more mathematics. I wish to express my deep sense of appreciation and thanks to HRI for providing me with all the infrastructural facilities to write these volumes.

Last but not least, I wish to express my thanks to my wife Veena Srivastava who had always been helpful in this endeavor.

In spite of all the care, some mistakes and misprints might have crept and escaped my attention. I shall be grateful to any such attention. Criticisms and suggestions for the improvement of the book will be appreciated and gratefully acknowledged.

Prayagraj, India Ramji Lal
September 2020

Contents

About the Author

Ramji Lal is Adjunct Professor at the Harish-Chandra Research Institute (HRI), Allahabad, India. He started his research career at the Tata Institute of Fundamental Research (TIFR), Mumbai, and served the University of Allahabad in different capacities for over 43 years, as a professor, head of the department and the coordinator of the DSA program. He was associated with HRI, where he initiated a postgraduate program in mathematics and coordinated the Nurture Program of National Board for Higher Mathematics (NBHM) from 1996 to 2000. After his retirement from the University of Allahabad, he was an advisor cum adjunct professor at the Indian Institute of Information Technology (IIIT), Allahabad, for over three years. His areas of interest include group theory, algebraic K-theory and representation theory

Notations

$[,]$	Lie product or bracket product
L	a Lie algebra (usually)
$C_L(S)$	centralizer of S
$N_L(V)$	normalizer of V
β_n	a bracket arrangement of weight n
$Z(L)$	center of L
$gl(V)(gl(n,F))$	general linear Lie algebra
$sl(V)(sl(n,F))$	special linear Lie algebra
$o(f)(o(2m+1,F)$	orthogonal Lie algebra
$sp(2m,F)$	symplectic Lie algebra
$d(n,F)$	Lie algebra of diagonal matrices
$t(n,F)$	Lie algebra of upper triangular matrices
$\nu(n,F)$	Lie algebra of strictly upper triangular matrices
$Der(L)$	Lie algebra of derivations
$Ider(L)$	Lie algebra of inner derivations
ad	adjoint representation
$A \succ B$	semi − direct product of A and B
$T_e(G)$	Tangent space of G at e
$L(G)$	Lie algebra of an algebraic (Lie) group G
Ad	adjoint representation of an algebraic group on its Lie algebra
$T(V)$	tensor algebra on V
$S(V)$	symmetric algebra on V
$(U(L),j_L)$	universal enveloping algebra of L
$R(L)$	radical of L
$x_s(x_n)$	semi − simple(nilpotent) part of x
κ	Killing form
c_ρ	Casimir element associated with ρ
$\Gamma(\rho)(\Gamma(V))$	set of weights of a repn(module) ρ(V)
Φ	root system
P_α	hyperplane determined by a root α

σ_α	reflection determined by α
$W(\Phi)$	the Weyl group of Φ
Δ	a basis of a root system Φ
$B(\Delta)$	standard Borel subalgebra associated with a basis δ
$\Lambda_{univ}(\Phi)(\Lambda_r)$	universal weight (root) lattice associated with Φ
$\mathbb{Z}(\lambda)$	universal standard cyclic module of highest weight λ
$V(\lambda)$	standard cyclic simple module with highest weight λ
Γ_λ	set of weights of $V(\lambda)$
Ch_λ	formal character of λ
Ch_V	formal character of the module V
c_{univ}	universal Casimir element
χ_λ	the character afforded by $\mathbb{Z}(\lambda)$
$L(K)$	Chevalley algebra of complex semi – simple Lie algebra over K
$G(V,K)$	Chevalley group associated with a L-module V and a field K
G_{univ}	universal Chevalley group
G_{adj}	adjoint Chevalley group
G_π	principal parabolic subgroup determined by $\pi \subseteq \Delta$
$St_{G(V,F_q)}$	Steinberg character of G
$R_{T,\Theta}$	Deligne – Lusztig generalized character

Chapter 1
Lie Algebras

The concept and the theory of Lie algebras originated and took momentum from the Lie theory of continuous groups. Locally, a Lie group is essentially a Lie algebra. To every Lie group (complex or real), there is an associated Lie algebra. Structurally, Lie subgroups and normal Lie subgroups of a Lie group associate faithfully with the Lie subalgebras, and Lie ideals of the Lie algebra associated with the Lie group. The isomorphism between Lie algebras corresponds to the local isomorphism between the corresponding Lie groups. Indeed, the category of simply connected Lie groups is equivalent to the category of Lie algebras. The theory of Lie algebras is indispensable in the theory of Lie groups.

In another development, Magnus (refer to the excellent book entitled "Combinatorial Group Theory" by Magnus, Karrass, and Solitar) initiated the use of Lie algebras in the study of discrete groups given in terms of presentations. The Lie algebras over fields of positive characteristics have been very effectively and successfully used in dealing with the restricted Burnside problem: "Is there a free object in the category $B(n, r)$ of finite groups generated by n elements having the exponent dividing r?". The problem was solved by E. Zelmonov in 1991 for which he got the Fields Medal in 1994.

In the present chapter, we develop the basic language of Lie algebras including universal enveloping algebras (PBW theorem), free Lie algebras, solvable, nilpotent, and semi-simple Lie algebras. We also establish the theorem of Weyl about the complete reducibility of representations of semi-simple Lie algebras. A field is usually be denoted by F or also sometimes by K.

1.1 Definitions and Examples

Definition 1.1.1 A **Lie algebra** over a field F is a vector space L over F together with a binary operation [,] on L such that the following conditions hold:

R. Lal, *Algebra 4*, Infosys Science Foundation Series,
https://doi.org/10.1007/978-981-16-0475-1_1

1. [,] is bi-linear in the sense that

$$[\alpha x + \beta y, z] = \alpha[x, z] + \beta[y, z],$$

and

$$[x, \alpha y + \beta z] = \alpha[x, y] + \beta[x, z]$$

for all $x, y, z \in L$ and $\alpha, \beta \in F$, where $[x, y]$ denotes the image of (x, y) under [,].
2. [,] is an alternating map in the sense that $[x, x] = 0$ for all $x \in L$.
3. $[x, [y, z]] + [y, [z, x]] + [z, [x, y]] = 0$ for all $x, y, z \in L$.

$[x, y]$ is called the **Lie product** of x and y. The third identity is termed as the **Jacobi identity**.

We also say that $(L, [,])$ is a Lie algebra over F, or [,] is a Lie algebra structure on the vector space L over F.

Proposition 1.1.2 *Let $(L, [,])$ be a Lie algebra over F. Then $[x, y] = -[y, x]$ for all $x, y \in L$. Conversely, let L be a vector space over a field F of characteristic different from 2 and [,] be a bilinear product on L such that conditions 1 and 3 of the Definition 1.1 together with the condition*
(2') $[x, y] = -[y, x]$ hold for all $x, y \in L$.
Then $(L, [,])$ is a Lie algebra.

Proof Suppose that $(L, [,])$ is a Lie algebra. Then by condition 2, $[x + y, x + y] = 0$ for all $x, y \in L$. Using condition 1 and condition 2 again, we see that $[x, y] + [y, x] = 0$. Conversely, suppose that [,] is bi-linear and $[x, y] = -[y, x]$ for all $x, y \in L$. In particular, $[x, x] = -[x, x]$ for all $x \in V$. Since the characteristic of F is different from 2, $[x, x] = 0$ for all $x \in L$. ♯

Proposition 1.1.3 *A Lie algebra structure determines, and is determined uniquely by, a vector space homomorphism ϕ from $L \bigwedge L$ to L satisfying the conditions*

$$\phi(x \wedge \phi(y \wedge z)) + \phi(y \wedge \phi(z \wedge x)) + \phi(z \wedge \phi(x \wedge y)) = 0$$

for all $x, y, z \in L$, where $L \bigwedge L$ denotes the exterior square of L.

Proof Let $(L, [,])$ be a Lie algebra. Since [,] is an alternating map, from the universal property of the exterior square, there is a unique vector space homomorphism ϕ from $L \bigwedge L$ to L given by $\phi(x \wedge y) = [x, y]$. The identity

$$\phi(x \wedge \phi(y \wedge z)) + \phi(y \wedge \phi(z \wedge x)) + \phi(z \wedge \phi(x \wedge y)) = 0$$

is a consequence of the Jacobi identity. Conversely, if such a homomorphism ϕ exists, then we have the Lie product [,] on L given by $[x, y] = \phi(x \wedge y)$. ♯

Let $(L, [,])$ be a Lie algebra over a field F. A subspace V of L is said to be a **Lie subalgebra** of L if $[x, y] \in V$ for all $x, y \in V$. Evidently, V is also a Lie algebra with respect to the induced Lie product at its own right. A subalgebra V of L is said be a **Lie ideal** if $[x, v] \in V$ for all $x \in L$ and $v \in V$. Evidently, a subspace V of L is an ideal of L if and only if $[v, x] \in V$ for all $v \in V$ and $x \in L$. If V is an ideal of L, then it can be easily observed that the quotient space L/V is a Lie algebra with respect to the Lie product $[,]$ given by $[x + V, y + V] = [x, y] + V$. This Lie algebra is called the **quotient Lie algebra** of L modulo V. A Lie algebra L having no nonzero proper ideal is called a **simple Lie algebra**.

Let $(L, [,])$ and $(L', [,]')$ be Lie algebras over a field F. A linear transformation f from L to L' is called **Lie algebra homomorphism** if $f([x, y]) = [f(x), f(y)]'$ for all $x, y \in L$. This gives us a category LA_F of Lie algebras over F. We have two forgetful functors from LA_F: One from LA_F to the category of vector spaces, and the second from LA_F to the category SET of sets. The question of the existence and the construction of adjoints to these functors will be discussed in detail in the following section. As usual, the image of a Lie subalgebra under a Lie homomorphism is a Lie subalgebra, whereas the image of a Lie ideal need not be a Lie ideal. The inverse image of a Lie subalgebra under a Lie homomorphism is a Lie subalgebra and also the inverse image of a Lie ideal under a Lie homomorphism is a Lie ideal. In particular, the kernel $Ker\ f = f^{-1}(\{0\})$ of a Lie homomorphism is a Lie ideal. Further, as usual the correspondence theorem, isomorphism theorems, and the Jordan-Holder theorem follow for Lie algebras.

Let L be a Lie algebra over a field F. Let A and B be ideals of L. Let $[A, B]$ denote the subspace generated by the set $\{[a, b] \mid a \in A \ and\ b \in B\}$. By the Jacobi identity

$$[x, [a, b]] = -[a, [b, x]] - [b, [x, a]]$$

for all $x \in L$, $a \in A$ and $b \in B$. It follows that $[x, [a, b]] \in [A, B]$ for all $x \in L$, $a \in A$ and $b \in B$. This shows that $[A, B]$ is also an ideal of L. In particular, $[L, L]$ is also an ideal of L. The ideal $[L, L]$ is called the **commutator** or the **derived subalgebra** of L. A Lie algebra L is said to be abelian if $[L, L] = \{0\}$. Evidently, $[L, L]$ is the smallest ideal of L by which if we factor, we get an abelian Lie algebra. $L/[L, L]$ is the largest quotient of L which is abelian. As in case of groups, we term $L/[L, L]$ as an abelianizer of L, and it is denoted by L_{ab}. Any homomorphism f from L to an abelian Lie algebra factors through L_{ab}. A Lie algebra L is said to be **perfect** if $[L, L] = L$. Evidently, every simple Lie algebra is perfect.

Let L be a Lie algebra over a field, and S be a subset of L. Let $C_L(S)$ denote the subset $\{x \in L \mid [x, s] = 0\ \forall s \in S\}$. The Jacobi identity ensures that $C_L(S)$ is a Lie subalgebra of L. This Lie subalgebra is called the **centralizer** of S in L. If $S = \{a\}$, then we denote the centralizer of S by $C_L(a)$. If A is an ideal of L, then $[z, [x, a]] = -[x, [a, z]] - [a, [z, x]] = 0$ for all $z, x \in L$ and $a \in C_L(A)$. This shows that the centralizer of an ideal is an ideal. In particular $C_L([L, L])$ is an ideal. The centralizer $C_L(L) = \{x \in L \mid [x, y] = 0\ \forall y \in L\}$ is called the **center** of L, and it is denoted by $Z(L)$. Let V be a Lie subalgebra of L. The set $N_L(V) = \{x \in L \mid [x, V] \subseteq V\}$ is a subalgebra of L which is called the **normalizer of V** in L. Evidently, $N_L(V)$ is characterized by the fact that it is the largest subalgebra of L in which V is an ideal. Note that a simple Lie algebra L is centerless in the sense that $Z(L) = \{0\}$.

Let L be a Lie algebra over a field F. Evidently, the intersection of a family of Lie subalgebras/ideals of L is a Lie subalgebra/ideal of L. Let X be a subset of L. The smallest Lie subalgebra of L containing X (the intersection of all Lie subalgebras of L containing X) is called the Lie subalgebra generated by X and it is denoted by $< X >$. Similarly, we can talk of the ideal generated by a subset of L.

In non-associative structures, the bracket arrangements are important tools. To describe the form of the elements of the subalgebra $< X >$ generated by X in terms of elements of X, we introduce the concept of **bracket arrangements**. For each $n \in \mathbb{N}\bigcup\{0\}$, we introduce the set B_n of elements called the bracket arrangements of weight n. This we do by the induction on n. $B_0 = \{\emptyset\}$, $B_1 = \{(\star)\}$, $B_2 = \{((\star)(\star))\}$, $B_3 = \{(((\star)(\star))(\star)), ((\star)((\star)(\star)))\}$. Assume that the set B_r has already been defined for all $r < n$. For each $r, s \leq n-1, r+s = n$, let $B_{r,s}$ denote the set $\{(\beta\gamma) \mid \beta \in B_r \ and\ \gamma \in B_s\}$. Define $B_n = \bigcup_{r+s=n} B_{r,s}$. Each $\star$ is called a place holder. This defines bracket arrangements of different weights. Given a sequence $x_1, x_2, \cdots x_m, x_{m+1}, \cdots$ of elements of a Lie algebra L, for each bracket arrangement β_n of weight n, we define the element $\beta_n(x_1, x_2, \cdots, x_n)$ of L. This, again, we do by induction on n as follows. Define $\beta_1(x_1) = x_1$. Assume that $\beta_r(x_1, x_2, \cdots, x_r)$ has been defined for all $r < n$. Suppose that $\beta_n = (\beta_r, \beta_s)$, where β_r is a bracket arrangement of weight r and β_s is that of weight s. Then we put $\beta_n(x_1, x_2, \cdots, x_n) = [\beta_r(x_1, x_2, \cdots, x_r), \beta_s(x_{r+1}, x_{r+2}, \cdots, x_{r+s})]$. The elements of the type $\beta_n(x_1, x_2, \cdots, x_n)$ are called the bracket arrangements of the sequence $x_1, x_2, \cdots x_m, x_{m+1}, \cdots$ of elements of L. Let X be a subset of L. Let $\beta(X)$ denote the set of all bracket arrangements corresponding to all sequences in X. Clearly, the Lie subalgebra $< X >$ of L generated by X is precisely the set of all linear combinations of members of $\beta(X)$. The reader is asked to describe the form of the elements of the ideal generated by X.

Lie Algebras of Low Dimensions

Example 1.1.4 In this example, we describe the isomorphism classes of Lie algebras of dimension at the most 2. Evidently, there is a unique one-dimensional Lie algebra which is abelian. Suppose that L is a non-abelian two-dimensional Lie algebra. Let $\{x, y\}$ be a basis of L. Then $[L, L]$ is the subspace generated by $[x, y] = z \neq 0$. Thus, $[L, L]$ is a one-dimensional subspace generated by $\{z\}$. Let $\{z, u\}$ be a basis of L. Then $[z, u] = \alpha z$ for some $\alpha \neq 0$. Taking $v = \alpha^{-1}u$, we see that L is the Lie algebra generated by $\{z, v\}$ subject to the relation $[z, v] = z$. An arbitrary element of L is of the form $\alpha z + \beta v$ and $[\alpha z + \beta v, \gamma z + \delta v] = (\alpha\delta - \beta\gamma)z$. Thus, up to isomorphism, there is only one non-abelian Lie algebra of dimension 2. Further, if $\alpha z + \beta v \in Z(L)$, then $-\beta z = [\alpha z + \beta v, z] = 0$ and $\alpha z = [\alpha z + \beta v, v] = 0$. This implies that $\alpha = 0 = \beta$. It follows that $Z(L) = \{0\}$. Evidently, no Lie algebra of dimension less than 3 is perfect.

Example 1.1.5 In this example, we classify all Lie algebras of dimension 3. Let L be a Lie algebra of dimension 3.

Case (1). $[L, L] = \{0\}$. In this case, L is abelian and there is only one such Lie algebra up to isomorphism.

Case (2). $Dim[L, L] = 1$ and $[L, L] \subseteq Z(L)$. Suppose that $[L, L] = Fx_1$, where $x_1 \neq 0$. Embed $\{x_1\}$ into a basis $\{x_1, x_2, x_3\}$ of L. Evidently, $[x_2, x_3] = \alpha x_1$ for some $\alpha \neq 0$. We may choose x_2 so that $\alpha = 1$. Since $x_1 \in Z(L)$, the Lie product in L is uniquely given by

$$[\alpha_1 x_1 + \alpha_2 x_2 + \alpha_3 x_3, \beta_1 x_1 + \beta_2 x_2 + \beta_3 x_3] = (\alpha_2\beta_3 - \alpha_3\beta_2)x_1.$$

It can be easily seen that the Jacobi identity is satisfied. Thus, in this case also we have a unique Lie algebra up to isomorphism given as above.

Case(3). $Dim\ [L, L] = 1$ and $[L, L] \not\subseteq Z(L)$. Suppose that $[L, L] = Fx_1$, where $x_1 \neq 0$. Since $[L, L] \not\subseteq Z(L)$, there is a nonzero element x_2 of L such that $[x_1, x_2] \neq 0$. We may take x_2 such that $[x_1, x_2] = x_1$. Evidently, $\{x_1, x_2\}$ is linearly independent. Embed it into a basis $\{x_1, x_2, x_3\}$ of L. Suppose that $[x_1, x_3] = \beta x_1$. If $\beta \neq 0$, we may choose x_1 so that $[x_1, x_3] = x_1$. Note that the relation $[x_1, x_2] = x_1$ will not change. We can replace x_3 by $x_3 - x_2$ and then $[x_1, x_3] = 0$. Suppose that $[x_2, x_3] = \gamma x_1$. If $\gamma \neq 0$, we may further modify x_3 by taking it to be $\gamma^{-1}x_3$ so that $[x_2, x_3] = x_1$. Observe that the remaining relations remain the same. Again replace x_3 by $x_1 + x_3$. Then $[x_2, x_3]$ becomes 0. Still note that the remaining relations remain the same. Also note that all the time $\{x_1, x_2, x_3\}$ remains a basis. We get a Lie algebra structure on L given by

$$[\alpha_1 x_1 + \alpha_2 x_2 + \alpha_3 x_3, \beta_1 x_1 + \beta_2 x_2 + \beta_3 x_3] = (\alpha_1\beta_2 - \alpha_2\beta_1)x_1.$$

It can be easily seen that the Jacobi identity is satisfied. Thus, in this case also we have a unique Lie algebra up to isomorphism described as above.

Case (4). $Dim\ [L, L] = 2$. We first show that $[L, L]$ is abelian. Suppose the contrary. Then as in Example 1.1.4, there is a basis $\{x_1, x_2\}$ of $[L, L]$ with $[x_1, x_2] = x_2$. In particular, $Z([L, L]) = \{0\}$. Consider the centralizer $C_L([L, L])$ of $[L, L]$ which is an ideal of L. Evidently, $[L, L] \bigcap C_L([L, L]) = \{0\}$. Let x be an element of L. Suppose that $[x, x_1] = \alpha x_1 + \beta x_2$ and $[x, x_2] = \gamma x_1 + \delta x_2$. Then using the Jacobi identity,

$$\gamma x_1 + \delta x_2 = [x, x_2] = [x, [x_1, x_2]] = -[x_1, [x_2, x]] - [x_2, [x, x_1] = (\alpha + \delta)x_2.$$

This means that $\alpha = 0 = \gamma$. Thus, $[x, x_1] = \beta x_2$ and $[x, x_2] = \delta x_2$. Hence

$$[x - \delta x_1 + \beta x_2, x_1] = 0 = [x - \delta x_1 + \beta x_2, x_2].$$

This shows that $x - \delta x_1 + \beta x_2$ belongs to $C_L([L, L])$. Consequently, $L = [L, L] \oplus C_L([L, L])$. Since $C_L([L, L])$ is a one-dimensional abelian Lie algebra, $[L, L] = [[L, L], [L, L]] = 0$. This is a contradiction. Thus, $[L, L]$ is abelian. Let $\{x_1, x_2, x_3\}$ be a basis of L, where $\{x_1, x_2\}$ is a basis of $[L, L]$. Suppose that $[x_3, x_1] = \alpha x_1 + \beta x_2$ and $[x_3, x_2] = \gamma x_1 + \delta x_2$. Since $[L, L]$ is abelian and of dimension 2, $\{[x_3, x_1], [x_3, x_2]\}$ is also a basis of $[L, L]$. This gives us a non-singular 2×2 matrix

$$A = \begin{bmatrix} \alpha & \beta \\ \gamma & \delta \end{bmatrix}.$$

Thus, for a Lie algebra L of dimension 3 for which the dimension of $[L, L]$ is 2, a choice of basis $\{x_1, x_2, x_3\}$ of L with $\{x_1, x_2\}$ as a basis of $[L, L]$ determines a non-singular 2×2 matrix A described as above. Conversely, given a non-singular 2×2 matrix A, take a vector space L with a set $\{x_1, x_2, x_3\}$ as a basis and define the product $[,]$ by

$$[\lambda_1 x_1 + \lambda_2 x_2 + \lambda_3 x_3, \mu_1 x_1 + \mu_2 x_2 + \mu_3 x_3] =$$
$$(-\lambda_1\mu_3 + \lambda_3\mu_1)(\alpha x_1 + \beta x_2) + (\lambda_3\mu_2 - \lambda_2\mu_3)(\gamma x_1 + \delta x_2).$$

It can be easily observed that $(L, [,])$ is a Lie algebra with $[L, L]$ being of dimension 2, and also the matrix associated with L with respect to the basis $\{x_1, x_2, x_3\}$ is the given matrix A. However, different matrices may determine isomorphic Lie algebras. We try to classify it faithfully. If we fix an element $x_3 \neq 0$ outside $[L, L]$ and change the basis $\{x_1, x_2\}$ of $[L, L]$ to a basis $\{x_1', x_2'\}$, then the matrix A changes to a matrix A' which is similar to A. However, if we fix a basis $\{x_1, x_2\}$ of L and change x_3 to x_3' so that $\{x_1, x_2, x_3'\}$ becomes a basis of L, then $x_3' = \lambda x_3 + x$, for some nonzero element λ in F and $x \in [L, L]$. Since $[L, L]$ is abelian, $[x_3', x_1] = \lambda(\alpha x_1 + \beta x_2)$ and $[x_3', x_2] = \lambda(\gamma x_1 + \delta x_2)$. The matrix A changes to the matrix λA. It follows that every Lie algebra L of dimension 3 in which $[L, L]$ is of dimension 2 determines, and is determined uniquely up to isomorphism by, a conjugacy class of Collineation group of 2×2 matrices. If F is an algebraically closed field, then using the Jordan theorem, we see that a conjugacy class of Collineation group of 2×2 matrices contains one and only one member from the class

$$\left\{\begin{bmatrix} 1 & 0 \\ 0 & \alpha \end{bmatrix} \mid \alpha \neq 0\right\} \bigcup \left\{\begin{bmatrix} 1 & \beta \\ 0 & 1 \end{bmatrix} \mid \beta \neq 0\right\}$$

of matrices. Thus, in this case, we have infinitely many Lie algebras L having a basis $\{x_1, x_2, x_3\}$ satisfying one and only one of the following two types of relations:

1. $[x_1, x_2] = 0$, $[x_3, x_1] = x_1$, $[x_3, x_2] = \alpha x_2$ or

2. $[x_1, x_2] = 0$, $[x_3, x_1] = x_1 + \beta x_2$, *and* $[x_3, x_2] = x_2$.

Case (5). *Dim* $[L, L] = 3$ or equivalently $[L, L] = L$. We describe and classify such Lie algebras up to isomorphisms. Let $PL(3, F)$ denote the set of isomorphism classes of perfect Lie algebras of dimension 3 over a field F. The isomorphism class determined by L will be denoted by $[L]$. A non-singular symmetric matrix A is said to be multiplicatively cogredient to a non-singular symmetric matrix B if there is a nonzero scalar λ and a non-singular matrix P such that $A = \lambda PBP^t$. Thus, if every element of the field F has a square root (for example $\mathbb{C}$), then the multiplicative cogredience is the same as the usual congruence. Let $\hat{S}(3, F)$ denote the set of multiplicatively cogredient classes of 3×3 non-singular symmetric matrices with entries in F. The cogredient class determined by the matrix A will be denoted by $[A]$. We exhibit a natural bijective correspondence between $PL(3, F)$ to $\hat{S}(3, F)$. Let L be a three-dimensional perfect Lie algebra, and let $\{x_1, x_2, x_3\}$ be a basis of L. Put $y_1 = [x_2, x_3]$, $y_2 = [x_3, x_1]$, and $y_3 = [x_1, x_2]$. Evidently, $\{y_1, y_2, y_3\}$ generates $[L, L] = L$, and as such it is also a basis of L. Suppose that $y_j = \sum_{i=1}^{3} a_{ij}x_i$. Then $A = [a_{ij}]$ is a non-singular matrix. We term A as the matrix of the basis $\{y_1, y_2, y_3\}$ with respect to the basis $\{x_1, x_2, x_3\}$. The Jacobi identity in L is equivalent to the identity

$$[x_1, y_1] + [x_2, y_2] + [x_3, y_3] = 0.$$

This is further equivalent to

$$[x_1, \sum_{i=1}^{3} a_{i1}x_i] + [x_2, \sum_{i=1}^{3} a_{i2}x_i] + [x_3, \sum_{i=1}^{3} a_{i3}x_i] = 0.$$

Using the bi-linear and alternating property of the product [,], we obtain the identity

$$(a_{21} - a_{12})y_3 + (a_{13} - a_{31})y_2 + (a_{32} - a_{23})y_1 = 0.$$

This shows that $A = [a_{ij}]$ is a symmetric 3×3 matrix. Thus, every perfect Lie algebra of dimension 3 together with a choice $\{x_1, x_2, x_3\}$ of a basis of L determines a unique non-singular symmetric matrix A described as above. Conversely, suppose that we are given a 3×3 symmetric matrix $A = [a_{ij}]$. Let L be a vector space with a basis $\{x_1, x_2, x_3\}$ consisting of 3 elements. Define a product [,] on L by

$$\left[\sum_{i=1}^{3} \alpha_i x_i, \quad \sum_{i=1}^{3} \beta_i x_i\right] = (\alpha_1\beta_2 - \alpha_2\beta_1)\sum_{i=1}^{3} a_{i3}x_i + (-\alpha_1\beta_3 - \alpha_3 \beta_1)\sum_{i=1}^{3} a_{i2}x_i + (\alpha_2\beta_3 - \alpha_3\beta_2)\sum_{i=1}^{3} a_{i1}x_i.$$

The fact that A is a non-singular symmetric matrix implies that L is a perfect Lie algebra of dimension 3 and the basis $\{x_1, x_2, x_3\}$, in turn, determines back the given matrix A. To classify these Lie algebras, we analyze the effect of the change of basis of L on the matrix A. Let $\{u_1, u_2, u_3\}$ be another basis with the associated basis $\{v_1, v_2, v_3\}$, where $v_1 = [u_2, u_3]$, $v_2 = [u_3, u_1]$, and $v_3 = [u_1, u_2]$. Let $B = [b_{ij}]$ be the associated non-singular symmetric matrix. Then $v_j = \sum_{i=1}^{3} b_{ij}u_i$. We relate A and B. Suppose that $u_j = \sum_{i=1}^{3} c_{ij}x_i$. Clearly, $C = [c_{ij}]$ is non-singular matrix. Now,

$$\begin{aligned}
v_1 &= [u_2, u_3] \\
&= [\textstyle\sum_{i=1}^{3} c_{i2}x_i, \sum_{i=1}^{3} c_{i3}x_i] \\
&= (c_{22}c_{33} - c_{32}c_{23})y_1 + (-c_{12}c_{33} + c_{32}c_{13})y_2 + (c_{12}c_{23} - c_{22}c_{12})y_3 \\
&= \textstyle\sum_{i=1}^{3} d_{i1}y_i,
\end{aligned}$$

where $[d_{i1}]$ represent the the first column of $(C^t)^{adj} = (C^{adj})^t = (Det\ C)(C^t)^{-1}$, where C^{adj} denotes the adjoint of C. Similarly, for each j, $v_j = \sum_{i=1}^{3} d_{ij}y_i$, where

$[d_{ij}]$ represent the jth column of adC^t. We have the following transformations among the bases of L:

$$U \overset{B^{-1}}{\to} V \overset{(C^t)^{adj}}{\to} Y \overset{A}{\to} X = U \overset{C}{\to} X,$$

where U denotes the basis $\{u_1, u_2, u_3\}$, V denotes the basis $\{v_1, v_2, v_3\}$, Y represents the basis $\{y_1, y_2, y_3\}$, and X represents the basis $\{x_1, x_2, x_3\}$. This shows that $A\,(C^t)^{adj}B^{-1} = C$. Since $(C^t)^{adj} = (Det\ C)(C^t)^{-1}$,

$$A = (Det\ C)^{-1}CBC^t,$$

where C is a non-singular matrix. This shows that different choice of bases for L give rise to cogredient matrices. More generally, if f is an isomorphism from a perfect Lie algebra L of dimension 3 with a basis $\{x_1, x_2, x_3\}$ to another perfect Lie algebra L' of dimension 3 with a basis $\{x_1', x_2', x_3'\}$, then the associated matrices A and A' are cogredient to each other. Consequently, we get a map η from $PL(3, F)$ to $\hat{S}(3, F)$ which is given by $\eta([L]) = [A]$, where A is the non-singular symmetric matrix associated with L with respect to a basis of L. Further, suppose that we are given a 3×3 symmetric matrix $A = [a_{ij}]$. Let L be a vector space with a basis $\{x_1, x_2, x_3\}$ consisting of 3 elements. Define a product $[,]$ on L by

$$\left[\sum_{i=1}^{3}\alpha_i x_i, \sum_{i=1}^{3}\beta_i x_i\right] = (\alpha_1\beta_2 - \alpha_2\beta_1)\sum_{i=1}^{3}a_{i3}x_i + (-\alpha_1\beta_3 + \alpha_3\beta_1)\sum_{i=1}^{3}a_{i2}x_i + (\alpha_2\beta_3 - \alpha_3\beta_2)\sum_{i=1}^{3}a_{i1}x_i.$$

The fact that A is a non-singular symmetric matrix implies that L is a perfect Lie algebra of dimension 3 and basis $\{x_1, x_2, x_3\}$, in turn, determines back the given matrix A. Thus, the map η is surjective. Next, let A and B be non-singular symmetric matrices which are cogredient. Let L be a Lie algebra together with a basis $\{x_1, x_2, x_3\}$ which is associated with the matrix A and let L' be a Lie algebra with a basis $\{x_1', x_2', x_3'\}$ which is associated with the matrix B. Then there is a nonzero scalar λ and a non-singular matrix P such that $A = \lambda PBP^t$. Put $P = \mu Q$. Then $A = \lambda\mu^2 QBQ^t$. We wish to choose μ so that $\lambda\mu^2 = (Det\ Q)^{-1}$. This is possible, since $(Det\ Q)^{-1} = \mu^3(Det\ P)^{-1}$. Thus, we have a non-singular matrix Q such that $A = (Det\ Q)^{-1}QBQ^t$. It follows from the above discussion that the linear transformation from L to L' having Q as the matrix representation with respect to the basis $\{x_1, x_2, x_3\}$ of L and the basis $\{x_1', x_2', x_3'\}$ of L' is an isomorphism from L to L'. This shows that the map η is a bijective correspondence.

Now suppose that the field F is of characteristic different from 2. Then every symmetric matrix with entries in F is congruent to a diagonal matrix (see Theorem 5.6.15, Algebra 2). Consequently, every equivalence class in $\hat{S}(3, K)$ contains a matrix of the form $Diag(\alpha, \beta, 1), \alpha \neq 0 \neq \beta$. Thus, every perfect three-dimensional Lie algebra over a field of characteristic different from 2 is isomorphic to a Lie algebra

$L_{\alpha,\beta}$ having a basis $\{x_1, x_2, x_3\}$ such that $[x_1, x_2] = x_3$, $[x_2, x_3] = \alpha x_1$, and $[x_3, x_1] = \beta x_2$. Note that $L_{\alpha,\beta}$ may be isomorphic to $L_{\gamma,\delta}$ even if $\{\alpha, \beta\} \neq \{\gamma, \delta\}$.

If the field F is the field $\mathbb{R}$ of real numbers, then it follows (see Theorem 5.6.22, Algebra 2) that every equivalence class contains one and only one of the following two matrices $Diag(1, 1, 1)$ or $Diag(-1, 1, 1)$. Thus, there are only two different perfect Lie algebras of dimension 3 over the field $\mathbb{R}$ of real numbers, and they are given by the set $\{[x_2, x_3] = x_1, [x_3, x_1] = x_2, [x_1, x_2] = x_3\}$ of relations or else it is given by the set $\{[x_2, x_3] = -x_1, [x_3, x_1] = x_2, [x_1, x_2] = x_3\}$ of relations. Identify them.

Suppose that the field F is the field $\mathbb{C}$ of complex numbers. Then every non-singular symmetric 3×3 matrix is congruent to the identity matrix. Hence there is only one perfect Lie algebra L of dimension 3 over the field $\mathbb{C}$ of complex numbers. L has a basis $\{x_1, x_2, x_3\}$ with relations $[x_2, x_3] = x_1$, $[x_3, x_1] = x_2$, $[x_1, x_2] = x_3$. Indeed, this Lie algebra has a nice representation as the Lie algebra $sl(2, \mathbb{C})$ consisting of 2×2 matrices with entries in $\mathbb{C}$ and with trace 0. The Lie product is given by $[x, y] = xy - yx$ (see Example 1.1.7).

A cyclic Lie algebra is a Lie algebra which is generated by a single element. Evidently, a nontrivial cyclic Lie algebra is an abelian algebra on a one-dimensional space. A Lie algebra L has no proper subalgebras if and only if L is a cyclic Lie algebra. A Lie algebra L is said to be a **simple Lie algebra** if it has no proper ideals. Evidently, an abelian Lie algebra is simple if and only if it is cyclic. As in the case of finite groups, the problem of classification of finite-dimensional Lie algebras reduces to the following two problems:

1. Classify all finite-dimensional simple Lie algebras up to isomorphisms.
2. Given a pair of (A, B) of Lie algebras, to classify Lie algebras L having an ideal A' isomorphic to A as Lie algebra such that L/A' isomorphic to B.

These problems will be discussed in due course.

Proposition 1.1.6 *$L/Z(L)$ cannot be a nontrivial cyclic Lie algebra.*

Proof Suppose that $L/Z(L) = < x + Z(L) >$ is a cyclic Lie algebra. Then any element of L is of the form $\alpha x + u$, where $u \in Z(L)$. Evidently, $[\alpha x + u, \beta x + v] = 0$ for all $\alpha, \beta \in K$ and $u, v \in Z(L)$. This implies that $L = Z(L)$. ♯

Example 1.1.7 Let L be a non-abelian Lie algebra of dimension 3. It follows from the above proposition that $Z(L) = \{0\}$ or else $Z(L)$ is of dimension 1. Suppose that L is perfect. Then every quotient of L is perfect. Since no Lie algebra of dimension less than 3 is perfect, it follows that L has no proper ideals. Thus, every perfect Lie algebra of dimension 3 is simple. Let $sl(2, F)$ denote the vector space of 2×2 matrices with entries in the field F having 0 trace. Then $sl(2, F)$ is a Lie algebra with respect to the Lie product given by $[A, B] = AB - BA$. Evidently, $sl(2, F)$ is three dimensional with $\{e_{12}, e_{21}, h\}$ as a basis, where

$$e_{12} = \begin{bmatrix} 0 & 1 \\ 0 & 0 \end{bmatrix}, \; e_{21} = \begin{bmatrix} 0 & 0 \\ 1 & 0 \end{bmatrix}, \; and\ h = \begin{bmatrix} 1 & 0 \\ 0 & -1 \end{bmatrix}.$$

It can be easily observed that $[e_{12}, e_{21}] = h$, $[h, e_{12}] = 2e_{12}$, *and* $[h, e_{21}] = -2e_{21}$. Thus, if the characteristic of F is different from 2, then $sl(2, F)$ is perfect. Consequently, $sl(2, F)$ is simple provided that the characteristic of F is different from 2.

Next, suppose that $Z(L)$ is of dimension 1 generated by $\{z\}$. Let $\{x, y, z\}$ be a basis of L. $L/Z(L)$ is a two-dimensional Lie algebra generated by $\{x + Z(L), y + Z(L)\}$. If $L/Z(L)$ is abelian, then $[L, L] \subseteq Z(L)$, and, there is a unique (up to isomorphism) non-abelian Lie algebra L of dimension 3 as described in Case 2 of Example 1.1.5. Now, suppose that $L/Z(L)$ is non-abelian. Then $[L, L] \nsubseteq Z(L)$. If dim $[L, L] = 1$, then we have a unique Lie algebra L as described in Case 3 of Example 1.1.5. Finally, if dim $[L, L] = 2$, then as described in Case 4 of Example 1.1.5, $[L, L]$ is abelian and there are two types of Lie algebras.

Now, we list some more important examples.

Example 1.1.8 Consider the usual vector space $\mathbb{R}^3$ over $\mathbb{R}$. The vector product $\times$ in $\mathbb{R}^3$ is an alternating product which satisfies the Jacobi identity. As such, $(\mathbb{R}^3, \times)$ is a three-dimensional real Lie algebra. All nontrivial proper subalgebras of $(\mathbb{R}^3, \times)$ are one dimensional. However, it has no nontrivial proper ideals. Thus, $(\mathbb{R}^3, \times)$ is a simple Lie algebra over $\mathbb{R}$.

Example 1.1.9 Let L be a three-dimensional vector space over F with $\{x, y, z\}$ as a basis. Then there is a unique Lie product [,] on L subject to $[x, y] = z$, $[x, z] = y$ and $[y, z] = 0$. Observe that the abelian subalgebra A of L generated by $\{y, z\}$ is an ideal such that L/A is also abelian.

Example 1.1.10 Let A be an associative algebra over a field F, for example, a polynomial algebra in non-commuting variables or a group algebra $F(G)$ over a group G. Define a product [,] on A given by $[a, b] = ab - ba$. It is easily observed that $(A, [,])$ is a Lie algebra. This Lie algebra will be denoted by A_L, and it will be termed as the Lie algebra associated with the associative algebra A. As such, we have a functor from the category ASS_F of associative algebras over F to the category LA_F of Lie algebras over the field F. The adjoint to this functor is an important functor which is extremely useful in the representation theory and the structure theory of Lie algebras. This adjoint functor will be discussed in detail in the next section. This also provides several examples of Lie algebras.

Example 1.1.11 Let $(L, [,])$ be a finite-dimensional Lie algebra over F. Let $\{x_1, x_2, \cdots, x_n\}$ be an ordered basis of L. This gives us n skew symmetric matrices $M^1 = [a^1_{ij}]$, $M^2 = [a^2_{ij}], \cdots, M^n = [a^n_{ij}]$ given by

$$[x_i, x_j] = \sum\nolimits_{k=1}^{n} a^k_{ij} x_k.$$

Further, the Jacobi identity induces the following identity among the entries of the matrices M^k, $1 \leq k \leq n$:

$$\sum_{k=1}^{n}(a_{ij}^{k}a_{kl}^{p} + a_{jl}^{k}a_{ki}^{p} + a_{li}^{k}a_{kj}^{p}) = 0$$

for all i, j, k, l, p. Conversely, given the set of skew symmetric matrices $M^k = [a_{ij}^k]$ satisfying the above condition, there is a unique Lie algebra structure [,] on F^n given by

$$[e_i, e_j] = \sum_{k=1}^{n} a_{ij}^k e_k,$$

where $\{e_1, e_2, \cdots, e_n\}$ is the standard basis of F^n. The entries a_{ij}^k are called the **structure constants** of the Lie algebra L associated with the basis $\{x_1, x_2, \cdots, x_n\}$. Describe the effect of change of the base on the structure constants.

Example 1.1.12 Let V be a vector space over a field F. Then $End_F(V)$ is an associative algebra over F. The associated Lie algebra is denoted by $gl(V)$. Thus, $gl(V) = End_F(V)$ and the Lie product [,] is given by $[A, B] = AB - BA$ for all $A, B \in gl(V)$. The Lie algebra $(gl(V), [,])$ is called the **general linear Lie algebra** on V. Any Lie subalgebra of $gl(V)$ is called a **linear Lie algebra**. Indeed, it is a fact (Ado and Iwasava theorems: Theorems 3.1.23 and 3.1.25) that every finite-dimensional Lie algebra is a linear Lie algebra over a vector space of finite dimension. If V is of dimension n, then $gl(V)$ is of dimension n^2. Fixing a basis of $gl(V)$, $gl(V)$ can be identified with the Lie algebra $gl(n, F)$ of $n \times n$ matrices with entries in F. Clearly, $\{e_{ij} \mid 1 \leq i \leq n, 1 \leq j \leq n\}$ is a basis of $gl(n, F)$, where e_{ij} is the matrix all of whose entries are 0 except the ij entry which is 1. Evidently, $e_{ij}e_{kl} = \delta_{jk}e_{il}$, where δ_{jk} is the Kronecker delta. Thus,

$$[e_{ij}, e_{kl}] = \delta_{jk}e_{il} - \delta_{li}e_{kj}$$

for all i, j, k, l. This means that the structure constants of $gl(n, F)$ with respect to the standard basis $\{e_{ij} \mid 1 \leq i \leq n, 1 \leq j \leq n\}$ of $gl(n, F)$ are members of the prime field in F.

Next, we introduce some important families of Lie subalgebras of $gl(V) \approx gl(n, F)$ termed as **classical linear Lie algebras**.

Example 1.1.13 The family A_n. Let V be a vector space of dimension $n + 1$ over a field F. Let $sl(V) \approx sl(n + 1, F)$ denote the set of linear endomorphisms of V which are of trace 0. Evidently, $sl(V)$ is a vector subspace of $gl(V)$ which is of dimension $(n + 1)^2 - 1$. Since $Tr(AB - BA) = 0$, it follows that $[A, B] \in sl(V)$ for all $A, B \in sl(V)$. This shows that $sl(V)$ is a Lie subalgebra of $gl(V)$. Observe that $sl(V)$ is not an associative subalgebra of the associative algebra $gl(V)$. The Lie subalgebra $sl(V) \approx sl(n + 1, F)$ is called a **special linear Lie algebra**. Evidently, $\{e_{ij} \mid i \neq j\} \bigcup \{e_{ii} - e_{i+1i+1} \mid 1 \leq i \leq n + 1\}$ is a basis of $sl(V)$. This basis is called the standard basis of $sl(V)$. Determine the structure constants.

Example 1.1.14 The family B_n. Let f be a nondegenerate symmetric bi-linear form on a vector space V of odd dimension $m = 2n + 1$ over a field F. Let $o(f)$ denote the set of all endomorphisms A of V such that $f(A(v), w) = -f(v, A(w))$ for all $v, w \in V$. Evidently, $o(f)$ is a subspace of $gl(V)$. If $A, B \in o(f)$, then

$$f((AB - BA)(v), w) = f(A(B(v)), w) - f((B(A(v)), w) = -f(v, (AB - BA)(w))$$

for all $v, w \in V$. Hence $[A, B] \in o(f)$ for all $A, B \in o(f)$. Thus, $o(f)$ is a Lie subalgebra of $gl(V)$. This Lie algebra is called the **orthogonal Lie algebra** associated with f. We try to represent it in matrix form. Since f is nondegenerate, there is a basis (see Algebra 2, Sect. 5.6) $\{v_1, v_2, \cdots, v_{2n+1}\}$ such that the matrix $M(f) = [f(v_i, v_j]$ of f with respect to this basis is given by

$$M(f) = \begin{bmatrix} 1 & 0_{1\times n} & 0_{1\times n} \\ 0_{n\times 1} & 0_{n\times n} & I_n \\ 0_{n\times 1} & I_n & 0_{n\times n} \end{bmatrix},$$

where $0_{r\times s}$ denotes the $r \times s$ zero matrix and I_n is the $n \times n$ identity matrix. Evidently, $A \in o(f)$ if and only if $M(f)M(A) = -M(A)^t M(f)$. Suppose that the matrix $M(A)$ of A with respect to the above basis is expressed as

$$M(A) = \begin{bmatrix} a & \overline{\alpha} & \overline{\beta} \\ \overline{\gamma}^t & P & Q \\ \overline{\delta}^t & R & S \end{bmatrix},$$

where $a \in F, \overline{\alpha}, \overline{\beta}, \overline{\gamma}, \overline{\delta}$ are row vectors in F^n, whereas P, Q, R, and S are members of $gl(n, F)$. The condition $M(f)M(A) = -M(A)^t M(f)$ implies that $a = 0$, $\overline{\alpha} = -\overline{\gamma}$, $\overline{\beta} = -\overline{\delta}$, R and Q are skew symmetric matrices, whereas $P^t = -S$. This, in turn implies that the trace of each member of $o(f)$ is 0, and $o(f)$ is a Lie subalgebra of $sl(V)$. The Lie algebra $o(f)$ is called an **orthogonal Lie algebra** associated with f. The matrices of the form $M(A)$ described above form a Lie algebra under the Lie product of matrices. This Lie algebra is isomorphic to $o(f)$, and it is denoted by $o(2n + 1, F)$. Evidently, the dimension of $o(2n + 1, F)$ is $2n^2 + n$. Determine a standard basis of $o(2n + 1, F)$ as in the above example. Determine the structure constants with respect to this basis. Check that they also belong to the prime field contained in F.

Example 1.1.15 The family C_n. Let f be a nondegenerate alternating bi-linear form on a vector space V of dimension $m = 2n$ over a field F. Observe that there is no degenerate alternating form on odd-dimensional spaces unless the field is of characteristic 2. Let $sp(f)$ denote the set of all endomorphisms A of V such that $f(A(v), w) = -f(v, A(w))$ for all $v, w \in V$. Evidently, $sp(f)$ is a subspace of $gl(V)$. As in the above example, $sp(f)$ is a Lie subalgebra of $gl(V)$. This Lie algebra is called the **symplectic Lie algebra** associated with f. We try to represent it in matrix form. Since f is nondegenerate, there is a basis (see Algebra 2, Sect. 5.6) $\{v_1, v_2, \cdots, v_{2n}\}$ such that the matrix $M(f) = [f(v_i, v_j)]$ of f with respect to this basis is given by

$$M(f) = \begin{bmatrix} 0_n & I_n \\ -I_n & 0_n \end{bmatrix}.$$

Evidently, $A \in sp(f)$ if and only if $M(f)M(A) = -M(A)^t M(f)$. Suppose that the matrix $M(A)$ of A with respect to the above basis is expressed as

$$M(A) = \begin{bmatrix} P & Q \\ R & S \end{bmatrix},$$

where P, Q, R, and S are members of $gl(n, F)$. The condition $M(f)M(A) = -M(A)^t M(f)$ implies that R and Q are symmetric matrices, whereas $P^t = -S$. This, in turn implies that the trace of each member of $sp(f)$ is 0, and $sp(f)$ is a Lie subalgebra of $sl(V)$. The set of matrices of the form $M(A)$ described above is denoted by $sp(2n, F)$ and it is a Lie algebra with respect to the Lie product of matrices. This Lie algebra is isomorphic to $sp(f)$. Evidently, the dimension of $sp(2n, F)$ is $2n^2 + n$. Determine a standard basis of $sp(2n, F)$ and also the structure constants with respect to this basis. Observe that they belong to the prime field of F.

Example 1.1.16 The family D_n. In Example 1.1.14, we described the orthogonal Lie algebras on odd-dimensional spaces. In this example, we describe orthogonal Lie algebras on even-dimensional spaces. Let f be a symmetric nondegenerate bi-linear form on a vector space V of even dimension $2n$. There is a basis $\{v_1, v_2, \cdots, v_{2n}\}$ of V such that the matrix $M(f)$ of f with respect to this basis is

$$M(f) = \begin{bmatrix} 0_n & I_n \\ I_n & 0_n \end{bmatrix}.$$

As in Example 1.1.14, we have a Lie subalgebra $o(f)$ of $sl(2n, F)$ which is isomorphic to the Lie algebra $o(2n, F) = \{A \mid M(f)A = -A^t M(f)\}$ of matrices. As in previous examples, it can be seen that the dimension of $o(2n, F)$ is $2n^2 - n$. The reader may determine the standard basis, and also the structure constants.

Apart from these family of linear Lie algebras, there are other important linear Lie subalgebras of $gl(n, F)$ which are important for the structure theory of Lie algebras.

Example 1.1.17 The set $d(n, F)$ of all diagonal matrices in $gl(n, F)$ form an abelian Lie subalgebra of $gl(n, F)$ which is called a total subalgebra of $gl(n, F)$. The set $t(n, F)$ of upper triangular matrices in $gl(n, F)$ is also a Lie subalgebra of $gl(n, F)$ which is termed as a Borel subalgebra of $gl(n, F)$. The set $n(n, F)$ of strict upper triangular matrices (diagonal entries 0) also form a Lie subalgebra of $gl(n, F)$. It is easy to observe that $[t(n, F), t(n, F)] = \{[A, B] \mid A, B \in t(n, F)\} = n(n, F)$. Find the dimension of each of these subalgebras.

Derivations and Lie Algebras

Let A be an algebra over a field F which may be non-associative. More explicitly, A is a vector space over a field F together with a bi-linear product on A denoted by juxtaposition. A F-linear transformation d from A to A is called a **derivation** on A if

$$d(ab) = d(a)b + ad(b)$$

for all $a, b \in A$. Let $Der(A)$ denote the set of all derivations on A. Evidently, $Der(A)$ is a subspace of $End_F(A)$. The composite of two derivations need not be a derivation (give an example). However, $[d, d']] = dd' - d'd$ can easily be seen to be a derivation. Thus, $Der(A)$ is a Lie subalgebra of $gl(A)$. In particular, if L is a Lie algebra over F, a derivation d on L is, by the definition, a linear transformation from L to L such that

$$d([v, w]) = [d(v), w] + [v, d(w)]$$

for all $v, w \in L$. For each $v \in L$, the map $ad(v)$ from L to L given by $ad(v)(w) = [v, w]$ is clearly a linear transformation. Further,

$$ad(v)([w, u]) = [v, [w, u]] = [ad(v)(w), u] + [w, ad(v)(u)]$$

for all $u, v, w \in L$, thanks to the Jacobi identity. Hence for each $v \in L$, $ad(v)$ is a derivation on L. Such a derivation is called an **inner derivation** of L determined by the element v. Other derivations are called the **outer derivations**. In turn, we get a map ad from L to $Der(L)$ which associates with each $v \in L$ the inner derivation $ad(v)$. Evidently, ad is a linear transformation. Further,

$$[ad(v), ad(w)](u) = ad(v)(ad(w)(u)) - ad(w)(ad(v)(u)) =$$
$$[v, [w, u]] - [w, [v, u]] = [[v, w], u] = ad([v, w])(u)$$

for all $u, v, w \in L$, thanks to the Jacobi identity. Hence ad is a Lie homomorphism from L to $Der(L)$. This is called the **adjoint representation** of L. The image of ad is precisely the Lie algebra of inner derivations of L, and it is denoted by $ad(L)$ or $Ider(L)$. The kernel $Ker\ ad = \{v \in L \mid [v, w] = 0\ \forall w \in L\}$ of ad is the center $Z(L)$ of the Lie algebra L. By the fundamental theorem of homomorphism

$$L/Z(L) \approx Ider(L) = ad(L).$$

Let $d \in Der(L)$ and $v \in L$. Then

$$[d, ad(v)](w) = d(ad(v)(w)) - ad(v)(d(w)) = d([v, w]) - [v, d(w)] =$$
$$[d(v), w] + [v, d(w)] - [v, d(w)] = [d(v), w] = ad(d(v))(w)$$

for all $v, w \in L$. This shows that $Ider(L)$ is an ideal of $Der(L)$. The quotient Lie algebra $Der(L)/Ider(L)$ is called the Lie algebra of outer derivations of L which will be denoted by $Oder(L)$. We get a short exact sequence

$$0 \longrightarrow Ider(L) \stackrel{i}{\to} Der(L) \stackrel{\nu}{\to} Oder(L) \longrightarrow 0,$$

and also an exact sequence

$$0 \longrightarrow Z(L) \stackrel{i}{\to} L \stackrel{ad}{\to} Der(L) \stackrel{\nu}{\to} Oder(L) \longrightarrow 0$$

of Lie algebras.

Semi-Direct Product and Split Extension

As in the case of groups, we introduce the notion of semi-direct product and split extensions in the category of Lie algebras. Here in this case, the derivation algebras play the role of automorphism groups. Thus, let A and B be Lie algebras over a field F. Let σ be a Lie algebra homomorphism from B to the derivation algebra $Der(A)$. Consider the vector space $L = A \times B$. Define the product [,] on L by

$$[(a, b), (a', b')] = ([a, a'] + \sigma(b')(a) - \sigma(b)(a'), [b, b'])$$

for all $a, a' \in A$ and $b, b' \in B$. It is a straightforward verification to show that L is a Lie algebra. The first inclusion map i_1 from A to L given by $i_1(a) = (a, o)$ is a monomorphism whereas the second projection map p_2 from L to B given by $p_2((a, b)) = b$ is an epimorphism. Further, $A \times \{0\}$ is the kernel of p_2 and so it is an ideal of L. We get a short exact sequence

$$0 \longrightarrow A \stackrel{i_1}{\to} L \stackrel{p_2}{\to} B \longrightarrow 0.$$

Evidently, the second inclusion i_2 from B to L is a splitting of the short exact sequence. Note that $L = A' + B'$, where $A' = A \times \{0\}$ and $B' = \{0\} \times B$, A' is an ideal isomorphic to A, and B' is a subalgebra isomorphic to B. Further $A' \bigcap B' = \{0\}$. We say that L is an external semi-direct product of A with B, and it is said to be an internal semi-direct product of A' with B'. Note that if σ is trivial homomorphism, then the semi-direct product is the direct product. Suppose that L is the internal semi-direct product of its ideal A by its subalgebra B. Then $L = A + B$ and $A \bigcap B = \{0\}$. An element x of L is uniquely expressible as $x = a + b$, where $a \in A$ and $b \in B$. For each element b of B, we have a derivation $\sigma(b)$ of A given by $\sigma(a)(b) = [a, b]$. It can easily be seen that σ is a homomorphism from B to $Der(A)$. Further, the map $(a, b) \mapsto a + b$ from the external semi-direct product $A \times B$ to the internal semi-direct product L is easily seen to be an isomorphism. Thus, an internal semi-direct product is isomorphic to an external semi-direct product. The external semi-direct product of A with B relative to a homomorphism σ is denoted by $A \succ_\sigma B$. If L is a Lie algebra having an ideal A and a subalgebra B such that $L = A + B$ and $A \bigcap B = \{0\}$, then we simply denote it by $A \succ B$.

Linear Algebraic Groups and Linear Lie Algebras

The discussions to follow will be useful in the representation theory. Recall that an algebraic variety G together with a group structure on G is called an algebraic group or a group variety if the group operations ($(a, b) \mapsto ab,\ a \mapsto a^{-1}$) are morphisms between the corresponding varieties. Let F be an algebraically closed field. Consider the general linear group $GL(n, F)$ of non-singular $n \times n$ matrices with entries in F. Then $GL(n, F)$ can be identified with the affine algebraic subset $\{(A, (Det(A))^{-1}) \mid A \in GL(n, F)\} = V(Det([X_{ij}])X - 1)$ of F^{n^2+1}. Thus, $GL(n, F)$ is an algebraic variety. Since the matrix multiplication and inversion on $GL(n, F)$ can be treated as polynomial maps on the entries of the matrices, the operations on $GL(n, F)$ are morphisms of varieties. Consequently, $GL(n, F)$ is an algebraic group. This algebraic group is called a general linear algebraic group over F. A closed (with respect to the Zariski topology) subgroup G of $GL(n, F)$ is also an algebraic group which is termed as a linear algebraic group over F. Let $G \subseteq GL(n, F)$ be a connected linear algebraic group, where F is an algebraically closed field. Let $\Gamma(G)$ denote the coordinate ring of G which consists of polynomial maps on G. Then $\Gamma(G)$ is an algebra over F. For each $x \in G$, the right multiplication R_x from G to G given by $R_x(g) = gx$ is an isomorphism of G considered as a variety. Thus, for each $x \in G$, we have a map $\rho(x)$ from $\Gamma(G)$ to itself given by $\rho(x)(f)(g) = f(gx)$. Evidently, $\rho(x) \in GL(\Gamma(G)) = Aut_F(\Gamma(G))$ for each $x \in G$, and also $\rho(xy) = \rho(x)\rho(y)$ for all $x, y \in G$. This gives us homomorphism ρ from G to $GL(\Gamma(G)) = Aut_F(\Gamma(G))$, and correspondingly, $\Gamma(G)$ is a right $F(G)$-module, where $F(G)$ denotes the group algebra over F. Now, consider the Lie subalgebra $Der(\Gamma(G))$ of $gl(\Gamma(G))$ consisting of derivations on $\Gamma(G)$. A member of $Der(\Gamma(G))$ need not be a G-homomorphism from $\Gamma(G)$ to $\Gamma(G)$. A derivation d on $\Gamma(G)$ is a $F(G)$-homomorphism if $do\rho(x) = \rho(x)od$ for all $x \in G$. Let $(Der(\Gamma(G)))^G = End_{F(G)}(\Gamma(G)) \bigcap Der(\Gamma(G))$ denote the set of all derivations on $\Gamma(G)$ which are $F(G)$-homomorphisms. Evidently, $End_{F(G)}(\Gamma(G))$ is a Lie subalgebra of $gl(\Gamma(G))$. Hence $(Der(\Gamma(G)))^G$ is a Lie subalgebra of $gl(\Gamma(G))$.

Treat the field F as a $\Gamma(G)$-module by putting $f\alpha = f(e)\alpha = \alpha f(e) = \alpha f,\ f \in \Gamma(G),\ and\ \alpha \in F$. This $\Gamma(G)$-module F is denoted by F_e. A $\Gamma(G)$-derivation from $\Gamma(G)$ to F_e is called a point derivation of $\Gamma(G)$ at e. Recall (Definition 4.3.29, Algebra 3) that the tangent space $T_e(G)$ of G at e is the space $Der(\Gamma(G), F_e)$ of point derivations of $\Gamma(G)$ at e. More explicitly, $T_e(G)$ is the F vector space of all maps d from $\Gamma(G)$ to F satisfying the condition

$$d(fg) = d(f)g(e) + f(e)d(g)$$

for all $f, g \in \Gamma(G)$, where e denotes the identity of G.

Proposition 1.1.18 *We have a natural F-isomorphism χ from $(Der(\Gamma(G)))^G$ to $Der(\Gamma(G), F_e)$ given by $\chi(d)(f) = d(f)(e)$.*

Proof Let d be a member of $(Der(\Gamma(G)))^G$. Then the map $\chi(d)$ from $\Gamma(G)$ to F given by $\chi(d)(f) = d(f)(e)$ is a point derivation of $\Gamma(G)$ at e, since $\chi(d)(ff') = d(ff')(e) = (d(f)f' + fd(f'))(e) = d(f)(e)f'(e) + f(e)d(f')(e) = \chi(d)(f)f'(e) + f(e)\chi(d)(f')$ for all $f, f' \in \Gamma(G)$. This gives us a natural linear map χ from $(Der(\Gamma(G)))^G$ to $Der(\Gamma(G), F_e)$ defined by $\chi(d)(f) = d(f)(e)$. Suppose that $\chi(d) = \chi(d')$, where $d, d' \in (Der(\Gamma(G)))^G$. Then $d(f)(e) =$

$d'(f)(e)$ for all $f \in \Gamma(G)$. Since d and d' are G-module endomorphisms of $\Gamma(G)$, $d(f)(x) = xd(f(e)) = xd'(f)(e) = d'(f)(x)$ for all $x \in G$. This shows that $d(f) = d'(f)$ for all $f \in \Gamma(G)$. It follows that χ is injective. Next, let ϕ be a member of $Der(\Gamma(G), F_e)$. Then $\phi(ff') = \phi(f)f' + f\phi(f')$ for all $f, f' \in \Gamma(G)$. Define a map d from $\Gamma(G)$ to itself by putting $d(f)(x) = \phi(\rho(x)(f)) = \phi(f \cdot x)$. Using the fact that ϕ is a point derivation of $\Gamma(G)$ at e, it is easily observed that d is a derivation of $\Gamma(G)$ which is a G-module homomorphism. Evidently, $\chi(d) = \phi$. ♯.

The vector space isomorphism χ introduced above induces a Lie algebra structure on $T_e(G)$. $T_e(G)$ with this Lie algebra structure is called the **Lie algebra of G**, and it is denoted by $L(G)$. Let ϕ be an algebraic homomorphism from an algebraic group G to an algebraic group G'. Define a map $d\phi_e$ from $L(G) = T_e(G) = Der(\Gamma(G), F_e)$ to $L(G') = T_e(G') = Der(\Gamma(G'), F_e)$ by putting $d\phi_e(d)(f) = d(fo\phi)$. It can be easily seen that $d\phi_e$ is a Lie algebra homomorphism. There is no loss in adopting the notation $d\phi$ for $d\phi_e$. If ϕ is a homomorphism from G to G' and ψ is a homomorphism from G' to G'', then $d(\psi o\phi) = d(\psi)od(\phi)$. This gives us a functor L from the category ALG of connected algebraic groups over an algebraically closed field F to the category LA_F of Lie algebras over F which associates with each connected algebraic group G the Lie algebra $L(G)$ of G, and with each homomorphism ϕ from G to G', the Lie algebra homomorphism $L(\phi) = d\phi$.

For each $g \in G$, let i_g denote the inner automorphism of G determined by g. Then i_g is an algebraic automorphism of G, and di_g is a Lie automorphism of $L(G)$. This automorphism is denoted by $Ad(g)$. We have a map Ad from G to the group $Aut(L(G))$ of the automorphisms of the Lie algebra $L(G)$ of G. It can be observed that $Ad(gh) = Ad(g)oAd(h)$. Thus, Ad is a representation of G on its Lie algebra $L(G)$. This representation is called the **adjoint representation** of the algebraic group G. We shall have occasions to discuss the adjoint representations.

Example 1.1.19 Consider the general linear algebraic group $GL(n, F) = V(\{Det\ [X_{ij}]X - 1\}) \subseteq A_F^{n^2+1}$. The coordinate ring $\Gamma(GL(n, F))$ of $GL(n, F)$ is the F-algebra

$$\frac{F[X_{11}, X_{12}, \cdots, X_{1n}, X_{21}, X_{22}, \cdots, X_{2n}, \cdots, X_{nn}, X]}{I(GL(n, F))}.$$

Evidently, the map ϕ from $\Gamma(GL(n, F))$ to the polynomial ring

$$F[X_{11}, X_{12}, \cdots, X_{1n}, X_{21}, X_{22}, \cdots, X_{2n}, \cdots, X_{nn}, Det\ [X_{ij}]^{-1}]$$

defined by

$$\phi(f[X_{11}, X_{12}, \cdots, X_{1n}, X_{21}, X_{22}, \cdots, X_{2n}, \cdots, X_{nn}, X] + I(GL(n, F))) = f[X_{11}, X_{12}, \cdots, X_{1n}, X_{21}, X_{22}, \cdots, X_{2n}, \cdots, X_{nn}, Det\ [X_{ij}]^{-1}]$$

is a natural isomorphism. The map η from $Der(\Gamma(GL(n, F)), F_I)$ to $gl(n, F)$ is defined by $\eta(d) = [a_{ij}]$, where $d(X_{ij}) = a_{ij}$ can be easily seen to be a Lie algebra isomorphism. Thus, $L(GL(n, F)) = gl(n, F)$.

Next, consider the special linear group $SL(n, F) = V(\{Det\,[X_{ij}] - 1\})$. Clearly,

$$\Gamma(SL(n, F)) = \frac{F[X_{11}, X_{12}, \cdots, X_{1n}, X_{21}, X_{22}, \cdots, X_{2n}]}{I(SL(n, F))}.$$

Let $\overline{X_{ij}}$ denote the coset $X_{ij} + I(SL(n, F))$. Let $d \in Der(\Gamma(SL(n, F)), F_I)$. Put $d(\overline{X_{ij}}) = \alpha^d_{ij} \in F$. Since d is a derivation

$$0 = d(1) = d(Det\,[\overline{X_{ij}}]) = D(Det\,[X_{ij}])(I) = \sum_i D(X_{ii})(I) = \sum_i d(X_{ii}) = \sum_i \alpha^d_{ii},$$

where D is the derivation of $F[X_{11}, X_{12}, \cdots, X_{1n}, X_{21}, X_{22}, \cdots, X_{2n}]$ associated with d. This shows that the matrix $[\alpha^d_{ij}]$ is of trace 0. The map η from $Der(\Gamma(SL(n, F)), F_I)$ to $sl(n, F)$ given by $\eta(d) = [\alpha^d_{ij}]$ is easily seen to be a Lie algebra isomorphism. Thus, $L(SL(n, F)) = sl(n, F)$.

Lie Groups and Lie Algebras

Recall that a map f from an open subset U of $\mathbb{R}^n$ to $\mathbb{R}$ is a C^r-map if for each $k \leq r$ and an n-tuple $(t_1, t_2, \cdots, t_n)$ of nonnegative integers with $t_1 + t_2 + \cdots + t_n = k$, the partial derivative

$$\frac{\partial^k f}{\partial^{t_1} x_1 \partial^{t_2} x_2 \cdots \partial^{t_n} x_n}$$

exists and it is continuous on U. It is said to be a C^∞-map if it is a C^r-map for each $r \geq 0$. A map ϕ from an open subset U of $\mathbb{R}^n$ to $\mathbb{R}^m$ can be expressed as $\phi = (\phi^1, \phi^2, \cdots, \phi^m)$, where $\phi^i = p_i o \phi$ are maps from U to R. We say that ϕ is a C^∞-map if each ϕ^i is a C^∞-map.

A Hausdorff topological space M (usually second countable) is called a **topological manifold** or a **locally Euclidean space** if there is a nonnegative integer n such that every point $p \in M$ has an open neighborhood U_p which is homeomorphic to $\mathbb{R}^n$. By the invariance of the domain (see Corollary 3.2.3, Algebra 3), n is uniquely determined and it is called the dimension of the manifold. Thus, every nonempty open subset U of $\mathbb{R}^n$ is a manifold of dimension n. Since $S^n - \{p\}$ is homeomorphic to $\mathbb{R}^n$ for each $p \in S^n$, S^n is a manifold of dimension n.

Let M be a manifold of dimension n. A family $\Sigma = \{(U_\alpha, h_\alpha) \mid \alpha \in \Lambda\}$ is called a C^r (C^∞)-differential atlas on M if the following hold:

(i) $\{(U_\alpha \mid \alpha \in \Lambda\}$ is an open cover of M.
(ii) For each $\alpha \in \Lambda$, h_α is a homeomorphism from U_α to $\mathbb{R}^n$.
(iii) For each pair $\alpha, \beta \in \Lambda$, the restriction map $h_\alpha o h_\beta^{-1}|_{h_\beta(U_\alpha \bigcap U_\beta)}$ from $h_\beta(U_\alpha \bigcap U_\beta)$ to $h_\alpha(U_\alpha \bigcap U_\beta)$ is a $C^r(C^\infty)$-map.

A C^r (C^∞)-atlas Σ is said to compatible with a C^r (C^∞)-atlas Σ' if $\Sigma \bigcup \Sigma'$ is also a C^r (C^∞)-atlas. We have an obvious partial ordering on the set of all C^r (C^∞)-atlases. A maximal C^r (C^∞)-atlas Σ on M is called C^r (C^∞)**- differential structure** on M. If M admits a C^r (C^∞)-atlas Σ, then the union of all atlases compatible with Σ is a C^r (C^∞)-differential structure containing Σ. Thus, every C^r (C^∞)-atlas Σ is contained in a unique C^r (C^∞)-differential structure. A manifold may admit several differential structures. Indeed, Milnor showed the existence of 28 distinct differential structures on S^7. A $C^r(C^\infty)$-manifold is a pair (M, Σ), where M is a manifold and Σ is a $C^r(C^\infty)$ differential structure. We shall be interested in C^∞-manifolds.

Example 1.1.20 1. If U is an open subset of $\mathbb{R}^n$, then U is union of a countable family $\{B_n \mid n \in \mathbb{N}\}$ of open balls in $\mathbb{R}^n$. For each n, we have a homeomorphism h_n from B_n to $\mathbb{R}^n$. Clearly, $\{(B_n, h_n) \mid n \in \mathbb{N}\}$ is a C^∞-atlas on U which determines a unique C^∞-differential structure Σ on U.

2. $GL(n, \mathbb{R})$, being an open subset of $M_n(\mathbb{R})$, is a C^∞-manifold as described above. Similarly, $GL(n, \mathbb{C})$ being an open subset of $\mathbb{C}^{n^2} \approx \mathbb{R}^{(2n)^2}$ is a C^∞-manifold of dimension $4n^2$. It can be shown (see Chap. 5) that all closed subgroups of $GL(n, \mathbb{C})$ are C^∞-manifolds.

3. Consider $S^m \subseteq \mathbb{R}^{m+1}$. Let U_1 denote the open subset $S^m - \{p_n\}$ and U_2 the open subset $S^m - \{p_s\}$, where $p_n = (0, 0, \cdots, 0, 1)$ is the north pole and $p_s = (0, 0, \cdots, 0, -1)$ is the south pole of S^m. Define a map h_1 from U_1 to $\mathbb{R}^n$ by putting $h_1(x_1, x_2, \cdots, x_{m+1}) = (\frac{x_1}{1-x_{m+1}}, \frac{x_2}{1-x_{m+1}}, \cdots, \frac{x_m}{1-x_{m+1}})$ and a map h_2 from U_2 to $\mathbb{R}^n$ by putting $h_2(x_1, x_2, \cdots, x_{m+1}) = (\frac{x_1}{1+x_{m+1}}, \frac{x_2}{1+x_{m+1}}, \cdots, \frac{x_m}{1+x_{m+1}})$. It can be easily observed that $\{(U_1, h_1), (U_2, h_2)\}$ is a C^∞-atlas which determines a unique C^∞-manifold structure on S^n.

Let (M_1, Σ_1) and $M_2, \Sigma_2)$ be C^∞-manifolds of dimensions n and m, respectively. Clearly, $M_1 \times M_2$ is a manifold of dimension $n + m$. Further, $\Sigma_1 \times \Sigma_2$ is a C^∞-atlas which determines a unique C^∞-differential structure Σ. The C^∞-manifold $(M_1 \times M_2, \Sigma)$ is called the product of (M_1, Σ_1) and (M_2, Σ_2).

Let (M, Σ) be a C^∞-manifold and U be an open subset of M. We say that a map f from U to $\mathbb{R}$ is a C^∞-function if $foh_\alpha^{-1}|_{h_\alpha(U_\alpha \bigcap U)}$ is a C^∞ function for each $(U_\alpha, h_\alpha) \in \Sigma$. This is also equivalent to say that $fok_\beta^{-1}|_{k_\beta(V_\beta \bigcap U)}$ is a C^∞ function for each $(V_\beta, k_\beta) \in \Sigma'$, where Σ' is a C^∞-atlas defining the differential structure Σ. Let (M, Σ) and (N, Σ') be two C^∞-manifolds. A continuous map f from M to N is called a C^∞-map if for any open subset V of N and a C^∞-map h from V to $\mathbb{R}$, hof is a C^∞-map on $f^{-1}(V)$. This is equivalent to say that $h_\alpha ofok_\beta^{-1}$ is a C^∞-map for each $(U_\alpha, h_\alpha) \in \Sigma$ and $(V_\beta, k_\beta) \in \Sigma'$. Clearly, composites of C^∞-maps are C^∞-maps. Thus, we have the category of C^∞-manifolds.

Let $C^\infty(M)$ denote the set of all C^∞-functions from M to $\mathbb{R}$. Then $C^\infty(M)$ is a commutative algebra over $\mathbb{R}$. A derivation on $C^\infty(M)$ is called a **vector field** on M. The set $Der(C^\infty(M))$ of all vector fields is a vector space over $\mathbb{R}$ with respect to the obvious operations. If $d, d' \in Der(C^\infty(M))$, then it is easily seen that $[d, d'] = dd' - d'd$ is also a member of $Der(C^\infty(M))$. In turn, $Der(C^\infty(M))$ is a Lie subalgebra of $gl(C^\infty(M))$ over $\mathbb{R}$.

Let (M, Σ) be a C^∞-manifold and $p \in M$. Consider the set $\Omega_p = \{(U, \phi) \mid$ *U is an open set containing p and ϕ is C^∞ map on U*$\}$. Define a relation $\approx$ on Ω_p by putting $(U, \phi) \approx (V, \psi)$ if there is an open subset W, $p \in W \subseteq U \bigcap V$ such that $\phi = \psi$ on W. It is clear that $\approx$ is an equivalence relation. Let $C^\infty(p) = \{\overline{(U, \phi)} \mid (U, \phi) \in \Omega_p\}$ denote the quotient set Ω_p modulo

$\approx$. $C^\infty(p)$ is called the set of germs of the C^∞ functions defined at p. Suppose that $\overline{(U, \phi)} = \overline{(U', \phi')}$ and $\overline{(V, \psi)} = \overline{(V', \psi')}$. It is easily observed that $\overline{(U \bigcap V, \phi + \psi)} = \overline{(U' \bigcap V', \phi' + \psi')}$, $\overline{(U \bigcap V, \phi \cdot \psi)} = \overline{(U' \bigcap V', \phi' \cdot \psi')}$, and $\overline{(U, a\phi)} = \overline{(U', a\phi')}$, $a \in \mathbb{R}$. This makes $C^\infty(p)$ a commutative $\mathbb{R}$-algebra. Further, $\mathbb{R}$ is a $C^\infty(p)$-module with the external product given by $\overline{(U, \phi)} \cdot a = \phi(p)a = a \cdot \phi(p) = a \cdot \overline{(U, \phi)}$. The **tangent space** $T_p(M)$ of M at p is defined to be the space $Der(C^\infty(p), \mathbb{R})$ of all point derivations of $C^\infty(p)$ at p.

If M is a complex manifold, then we can talk of an analytic atlas and an analytical structure on M together with other related concepts by replacing C^∞- maps with analytic maps.

A group G together with C^∞-structure on G is called a **Real Lie Group** if all the group operations are C^∞-maps. For example, $GL(n, \mathbb{C})$, $GL(n, \mathbb{R})$, $SL(n, \mathbb{C})$, and $SU(n)$ are all examples of real Lie groups. Similarly, A group G together with an analytic-structure on G is called a **Complex Lie Group** if all the group operations are analytic-maps. For example, $GL(n, \mathbb{C})$, $SL(n, \mathbb{C})$ are all examples of complex Lie groups. Note that $SU(n)$ is not a complex Lie group (why?).

Let G be a real Lie group. Then for each $g \in G$, the left multiplication L_g and the right multiplication R_g are bijective C^∞-maps whose inverses are also C^∞-maps on G. For each $g \in G$, we have a map $\rho(g)$ from $C^\infty(G)$ to $C^\infty(G)$ defined by $(\rho(g)(f))(x) = f(xg)$. Evidently, $\rho(g) \in GL(C^\infty(G)) = Aut_{\mathbb{R}}(C^\infty(G))$ for each $g \in G$, and also $\rho(xy) = \rho(x)\rho(y)$ for all $x, y \in G$. This gives us a homomorphism ρ from G to $GL(C^\infty(G)) = Aut_{\mathbb{R}}(C^\infty(G))$, and correspondingly $C^\infty(G)$ is a right $\mathbb{R}(G)$-module, where $\mathbb{R}(G)$ denotes the group algebra over $\mathbb{R}$. Now, consider the Lie subalgebra $Der(\mathbb{C}^\infty(G))$ of $gl(C^\infty(G))$ consisting of derivations on $C^\infty(G)$. A member of $Der(C^\infty(G))$ need not be a G-homomorphism from $C^\infty(G)$ to $C^\infty(G)$. A derivation d on $C^\infty(G)$ is a $\mathbb{R}(G)$-homomorphism if and only if $do\rho(x) = \rho(x)od$ for all $x \in G$. Let $(Der(C^\infty(G)))^G = End_{\mathbb{R}(G)}(C^\infty(G)) \bigcap Der(C^\infty(G))$ denote the set of all derivations on $C^\infty(G)$ which are $\mathbb{R}(G)$-homomorphisms. Evidently, $End_{\mathbb{R}(G)}(C^\infty(G))$ is a Lie subalgebra of $gl(C^\infty(G))$. Hence $(Der(C^\infty(G)))^G$ is a Lie subalgebra of $gl(C^\infty(G))$. The members of $(Der(C^\infty(G)))^G$ are called the G-invariant vector fields. Each $f \in C^\infty(G)$ determines an element $\overline{(M, f)}$ of $C^\infty_e(G)$ which we denote by f itself. As in Proposition 1.1.18, we obtain a natural $\mathbb{R}$-isomorphism χ from $(Der(C^\infty(G)))^G$ to $T_e(G) = Der(C^\infty(e), \mathbb{R})$ by putting $\chi(d)(f) = d(f)(e)$. It follows that $T_e(G)$ is a real Lie algebra. This Lie algebra is called the Lie algebra of G and it is denoted by $L(G)$. We can do the same for complex Lie groups. In Chap. 5, we shall describe Lie algebras of linear Lie groups.

Remark 1.1.21 Hilbert's fifth problem posed by Hilbert in 1900 was to see if a Locally Euclidean group is a Lie group. Gleason and Montgomery settled it by proving that every Locally Euclidean Group (a group together with a manifold structure such that the group operations are continuous) is a Lie group.

Let L be a real Lie algebra. Then $L_{\mathbb{C}} = L \otimes_{\mathbb{R}} \mathbb{C}$ is a Lie algebra over $\mathbb{C}$ with obvious Lie product. This Lie algebra is called the complexification of L. Show that the complexification of $su(n)$ is $GL(n, \mathbb{C})$ which is also the complexification of $gl(n, \mathbb{R})$.

Exercises

1.1.1. Assuming that the characteristic of F is 0, show that all classical Lie algebras in the families A_n, B_n, C_n, and D_n are perfect and center-less.
1.1.2. Determine the centers of all the Lie algebras discussed so far.
1.1.3. Show that for any Lie algebra L, $ad(L)$ cannot be one dimensional.
1.1.4. Show that $sl(3, F)$ is simple if and only if the characteristic of F is different from 3.
1.1.5. Let $T \in gl(n, F)$ which has all its eigenvalues in F and all are different. Show that $ad(T)$ is diagonalizable.
1.1.6. Show that the derived algebra of $gl(n, F)$ is $sl(n, F)$.
1.1.7. Describe the derived algebras of $d(n, F)$, $t(n, F)$, and $n(n, F)$. What are their normalizers in $gl(n, F)$.
1.1.8. Describe the Lie algebras of the algebraic groups $D(n, F)$, $T(n, F)$, and $U(n, F)$, where $D(n, F)$ is the group of non-singular diagonal matrices, $T(n, F)$ is the group of non-singular upper triangular matrices, and $U(n, F)$ is the group of uni-upper triangular matrices.
1.1.9. Describe all three-dimensional Lie algebras over $\mathbb{Z}_3$.
1.1.10. Show that $(\mathbb{R}^3, \times)$ is a Lie algebra over $\mathbb{R}$. Is it simple? Describe its complexification.
1.1.11. Describe all three-dimensional Lie algebras over $\mathbb{Q}$. What are their complexifications?
1.1.12. Establish the Leibnitz rule

$$\delta^n(xy) = \sum_{i=0}^{n} {}^nC_r\delta^r(x)\delta^{n-r}(y)$$

for a derivation δ on a Lie algebra over a field of characteristic 0.
1.1.13. Let δ be a nilpotent derivation of an algebra L over a field of characteristic 0. Suppose that $\delta^n = 0$. Show that

$$exp(\delta) = \sum_{r=0}^{n-1} \frac{\delta^r}{r!}$$

is an automorphism of L.
1.1.14. Describe the Lie subalgebra of $sl(n, F)$ generated by $\{e_{12}, e_{21}\}$. Show that the ideal of $sl(n, F)$ generated by $\{e_{12}\}$ is $sl(n, F)$.
1.1.15. Let b_n denote the number of bracket arrangements of weight n. Consider the power series $b(t) = \sum_{r=0}^{\infty} b_r t^r$ in $\mathbb{Z}[[t]]$. Show that $b(t)^2 - b(t) + t$ with $b(0) = 0$. Deduce that $b(t) = \frac{1 - \sqrt{1-4t}}{2}$. Expanding this in power series show that $b_n = \frac{1}{n-1}\, {}^{2n-2}C_n$.
1.1.16. Show that the Lie algebra L over $\mathbb{C}$ having a basis $\{x_1, x_2, x_3\}$ subject to the relation $[x_2, x_3] = x_1$, $[x_3, x_1] = x_2$, *and* $[x_1, x_2] = x_3$ is isomorphic to $sl(2, \mathbb{C})$.
1.1.17. Let us call a Lie algebra L to be a **complete Lie algebra** if $Z(L) = \{0\}$ and every derivation is an inner derivation. Let A be an ideal of L which

is complete as an algebra. Show that there is an ideal B of L such that $L = A \oplus B$.

1.1.18. Show that every non-abelian Lie algebra of dimension 2 is complete.

1.2 Universal Enveloping Algebras: PBW Theorem

This section is devoted to introducing and studying some universal objects in the category of Lie algebras such as universal enveloping algebras, free Lie algebras, and also to describing Lie algebras through presentations. We also establish the Poincaré–Birkhoff–Witt (PBW) theorem together with some of its consequences. Every group G is isomorphic to a subgroup of $GL(V)$ for some vector space V (consequence of Cayley's theorem). We shall show that every Lie algebra is isomorphic to a Lie subalgebra of $gl(V)$ for some vector space V (not necessarily finite dimensional). In case of groups, every finite group is isomorphic to a subgroup of a matrix group over a field F. The corresponding analogue for the Lie algebras are the theorems of Ado and of Iwasawa (Theorems 3.1.23 and 3.1.25) which assert that a finite-dimensional Lie algebra over F is isomorphic to a Lie subalgebra of $gl(n, F)$ for some n.

Let V be a vector space (not necessarily finite dimensional) over an arbitrary field F. Recall (Sect. 7, Algebra 2) the tensor algebra $T(V)$ on V. Thus, as a vector space, $T(V) = \oplus\sum_{n=0}^{\infty} \otimes^n V$, where $\otimes^0 V = F$, $\otimes^1 V = V$, and $\otimes^n V = \underbrace{V \otimes V \otimes \cdots \otimes V}_{n}$. For each $m, n \geq 0$, we have a bi-linear product η_{nm} from $\otimes^n V \times \otimes^m V$ to $\otimes^{n+m} V$ given by

$$\eta_{nm}(x_1 \otimes x_2 \otimes \cdots \otimes x_n, x_1 \otimes x_2 \otimes \cdots \otimes x_m) = x_1 \otimes x_2 \otimes \cdots \otimes x_n \otimes x_1 \otimes x_2 \otimes \cdots \otimes x_m.$$

These partial products can be extended by linearity to a unique associative bi-linear product on $T(V)$. This makes $T(V)$ an associative F-algebra. The algebra $T(V)$ is called the **tensor algebra** of V. We have the tautological injective linear map i_V from V to $\otimes^1 V \subseteq T(V)$ given by $i_V(v) = v$. The pair $(T(V), i_V)$ has the following universal property: $(\star)$ If ϕ is a linear map from V to an associative algebra A, then there is a unique algebra homomorphism $\overline{\phi}$ from $T(V)$ to A such that $\overline{\phi} o i_V = \phi$. Further, if (B, j) is a pair which also satisfies the universal property $(\star)$, then there is a unique algebra homomorphism $\overline{j}$ from $T(V)$ to B, and a unique algebra homomorphism $\overline{i_V}$ from B to $T(V)$ such that $\overline{j} o i_V = j$ and $\overline{i_V} o j = i_V$. Thus, $\overline{i_V} o \overline{j} o i_V = i_V = I_{T(V)} o i_V$. From the universal property of $(T(V), i_V)$, $\overline{i_V} o \overline{j} = I_{T(V)}$. Similarly, $\overline{j} o \overline{i_V} = I_B$. This shows that $(T(V), i_V)$ is the unique pair up to an isomorphism which satisfies the universal property $(\star)$. Consequently, if f is a linear map from a vector space V to a vector space W, then there is a unique algebra homomorphism $T(f)$ from $T(V)$ to $T(W)$ such that $T(f) o i_V = i_W o f$. Thus, T defines a functor from the category of vector spaces over F to the category of associative algebras over F which is adjoint to the forgetful functor from the category of associative algebras over F to the category of vector spaces over F.

$T(V)$ has a simpler description as a polynomial algebra: Let B be a basis of V and let S be a set of symbols in bijective correspondence with B. Let $\hat{F}[S]$ denote the polynomial algebra over F in **non-commuting indeterminates** belonging to S. We have a linear map i from V to $\hat{F}[S]$ induced by the bijective correspondence from the basis B to the set S of indeterminates. It is an easy observation that the pair $(\hat{F}[S], i)$ also satisfies the universal property $\star$. In fact, $\hat{F}[S]$ is a free associative algebra on the set S, and V is the free F-module on S.

Let I be the ideal of the tensor algebra $T(V)$ generated by $\{x \otimes y - y \otimes x \mid x, y \in V\}$. The algebra $T(V)/I$ is called the **symmetric algebra** of V and it is denoted by $S(V)$. The natural quotient map from $T(V)$ to $S(V)$ will be denoted by π. Clearly, $S(V)$ is an associative and commutative algebra over F. Further, $I(V) = \oplus \sum_{n=2}^{\infty}(I \bigcap \otimes^n V)$ (note that $I \bigcap F = \{0\} = I \bigcap \otimes^1 V$). Thus, $S(V) = \oplus \sum_{n=0}^{\infty} S^m(V)$, where $S^0(V) = F$, $S^1(V) = V$, and $S^m(V) = \otimes^m V/(I \bigcap \otimes^m V)$, $m \geq 2$. $S^m(V)$ is called the mth **symmetric power** of V. We have the tautological linear map i_V from V to $S(V)$. The pair $(S(V), i_V)$ has the following universal property: ($\star\star$) If (B, j) is a pair, where B is an associative and commutative algebra and j is a linear map from V to B, then there is a unique homomorphism $\overline{j}$ from $S(V)$ to B such that $\overline{j}oi = j$. As in the case of tensor algebra, the pair $(S(V), i_V)$ is unique up to isomorphism which satisfies the universal property ($\star\star$). Evidently, S defines a functor from the category of vector spaces to the category of commutative algebras over F which is adjoint to the forgetful functor. The symmetric algebra of V has a simpler description as polynomial algebra $F[S]$ in the set S of commuting variables, where S is in bijective correspondence with a basis of V.

Now, we introduce the adjoint functor to the functor from the category of associative algebras to the category of Lie algebras which associates with each associative algebra A the associated Lie algebra A_L (see Example 1.1.10).

Definition 1.2.1 A **universal enveloping algebra** of a Lie algebra L over a field F is a pair $(U(L), j_L)$, where $U(L)$ is an associative algebra over F and j_L is a Lie algebra homomorphism from L to the associated Lie algebra $U(L)_L$ such that it satisfies the following universal property: ($\star\star\star$) If (B, j) is also a pair, where B is an associative algebra and j is a Lie algebra homomorphism from L to B_L, then there is a unique algebra homomorphism $\overline{j}$ from $U(L)$ to B such that $\overline{j}oj_L = j$.

As usual, the pair $(U(L), j_L)$ is unique up to a natural isomorphism.

We show the existence of a universal enveloping algebra by constructing it. Let $T(L)$ denote the tensor algebra of L considered as a vector space. Let J denote the ideal of $T(L)$ generated by $\{x \otimes y - y \otimes x - [x, y] \mid x, y \in L\}$ (note that $(x \otimes y - y \otimes x)$ belongs to the component $\otimes^2 L$ of $T(L)$ and $[x, y]$ belongs to the component L of $T(L)$). Take $U(L)$ to be $T(L)/J$ and $j_L = \nu o i_L$, where ν is the quotient map from $T(L)$ to $T(L)/J$. Let B be an associative algebra and j be a linear map from L to B such that $j([x, y]) = j(x)j(y) - j(y)j(x)$ for all $x, y \in L$. From the universal property of $(T(L), i_L)$, there is a unique algebra homomorphism ϕ from

$T(L)$ to B such that $\phi o i_L = j$. Since $j([x, y]) = j(x)j(y) - j(y)j(x)$, it follows that $\phi(x \otimes y - y \otimes x - [x, y]) = 0$. Hence ϕ induces a unique homomorphism ψ from $U(L)$ to B such that $\psi o j_L = j$. This shows that the pair $(U(L), j_L)$ is universal enveloping algebra of L.

Let η be a homomorphism from a Lie algebra L to a Lie algebra L'. Then $j_{L'} o \eta$ is a Lie algebra homomorphism from L to $U(L')_L$. From the universal property of the universal enveloping algebra, we have a unique algebra homomorphism $U(\eta)$ from $U(L)$ to $U(L')$ such that $j_{L'} o \eta = U(\eta) o j_L$. This defines a functor U from the category of Lie algebras to the category of associative algebras. The functor U is adjoint to the functor from the category of associative algebras to the category of Lie algebras which associates with each associative algebra A the corresponding Lie algebra A_L.

Example 1.2.2 If L is an abelian Lie algebra, then $[x, y] = 0$ for all $x, y \in L$. Thus, the ideal J is the same as I. Hence in this case, $U(L) = S(L)$. If L is a non-abelian Lie algebra of dimension 2, then there is a basis $\{x, y\}$ of L such that $[x, y] = x$. In this case, $U(L)$ is isomorphic to $\hat{F}[X, Y]/J$, where $\hat{F}[X, Y]$ is the polynomial ring in non-commuting variables X and Y, and J is the principal ideal of $\hat{K}[X, Y]$ generated by $XY - YX - X = X(Y - 1) - YX$. Observe that in both the cases j_L is injective. Also observe that in this case $\{X^r Y^s + J \mid r \geq 0, s \geq 0\}$ forms a basis of $U(L)$.

Theorem 1.2.3 *Let L be a Lie algebra. Then the universal enveloping algebra $(U(L), j_L)$ of L satisfies the following properties:*

1. *The image $j_L(L)$ generates $U(L)$ as an associative algebra.*
2. *Let A be an ideal of L. Let $\hat{A}$ denote the ideal of $U(L)$ which is generated by $j_L(A)$. Then the pair $(U(L)/\hat{A}, \hat{j_L})$ is the universal enveloping algebra of L/A, where $\hat{j_L}$ is given by $\hat{j_L}(x + A) = j_L(x) + \hat{A}$.*
3. *There is a unique anti-automorphism ρ of $U(L)$ such that $\rho(j_L(x)) = -j_L(x)$ for each $x \in L$. Further $\rho^2 = I_{U(L)}$.*
4. *We have a unique algebra homomorphism Δ from $U(L)$ to the algebra $U(L) \otimes_F U(L)$ (note that $U(L) \otimes_F U(L)$ is an associative algebra) such that*

$$\Delta(j_L(x)) = j_L(x) \otimes 1 + 1 \otimes j_L(x)$$

   for each $x \in L$. The homomorphism Δ is called the ***diagonal homomorphism****.*
5. *For each derivation d on L, we have a unique derivation $\hat{d}$ on $U(L)$ such that $\hat{d} o j_L = j_L o d$.*

Proof 1. Since $T(L)$ is generated by $i_L(L)$ and $j_L = \nu o i_L$, it follows that $U(L)$ is generated by $j_L(L)$.

2. Evidently, the map $\hat{j_L}$ from L/A to $U(L)/\hat{A}$ given by $\hat{j_L}(x + A) = j_L(x) + \hat{A}$ is a linear map. Also
$\hat{j_L}([x + A, y + A])$
$= \hat{j_L}([x, y] + A)$

$$\begin{aligned}
&= j_L([x, y]) + \hat{A} \\
&= (j_L(x)j_L(y) - j_L(y)j_L(x)) + \hat{A} \\
&= [j_L(x) + \hat{A}, j_L(y) + \hat{A}] \\
&= [\hat{j_L}(x + A), \hat{j_L}(y + A)].
\end{aligned}$$

This shows that $\hat{j_L}$ is a Lie algebra homomorphism from L/A to $(U(L)/\hat{A})_L$. Next, let B be an associative algebra and let μ be a Lie algebra homomorphism from L/A to B_L. Then μ induces a Lie algebra homomorphism ν from L to B_L given by $\nu(x) = \mu(x + A)$. Since $(U(L), j_L)$ is a universal enveloping algebra of L, we have a unique Lie algebra homomorphism $\hat{\mu}$ from $U(L)$ to B such that $\hat{\mu}oj_L = \nu$. Hence $(\hat{\mu}oj_L)(x) = \nu(x) = \mu(x + A) = 0$ whenever $x \in A$. Hence $\hat{\mu}(j_L(A)) = 0$. Consequently, $\hat{\mu}(\hat{A}) = 0$. Thus, $\hat{\mu}$ induces a unique homomorphism $\overline{\mu}$ from $U(L)/\hat{A}$ to B such that $\overline{\mu}o\hat{j_L} = \mu$. It follows that $(U(L)/\hat{A}, \hat{j_L})$ is the universal enveloping algebra of L/A.

3. Consider the universal enveloping algebra $(U(L), j_L)$ of L. $U(L)$ is also an associative algebra with respect to the new product $\star$ given by $u \star v = v \cdot u$, where $\cdot$ is the product in $U(L)$. Let us denote this new associative Lie algebra by $U(L)'$. Let $[,]'$ denote the associated Lie algebra structure. Consider the map η from L to $U(L)$ given by $\eta(x) = -j_L(x)$. Then $\eta([x, y]) = -j_L[x, y] = [j_L(y), j_L(x)] = [\eta(y), \eta(x)] = [\eta(x), \eta(y)]'$. Thus, η is a Lie algebra homomorphism from L to $(U(L)')_L$. From the universal property of $(U(L), j_L)$, we get a homomorphism ρ from $U(L)$ to $U(L)'$ such that $\rho o J_L = \eta = -j_L$. Evidently, ρ is the anti-automorphism of $U(L)$ such that $\rho^2 = I_L$.

4. Consider the associative algebra $U(L) \otimes_F U(L)$ and the map ϕ from L to $U(L) \otimes_K U(L)$ given by $\phi(x) = J_L(x) \otimes 1 + 1 \otimes j_L(x)$. Clearly, ϕ is a linear map. Further,

$$\begin{aligned}
\phi([x, y]) = j_L([x, y]) \otimes 1 + 1 \otimes j_L([x, y]) = (j_L(x)j_L(y) - j_L(y)j_L(x)) \otimes \\
1 + 1 \otimes (j_L(x)j_L(y) - j_L(y)j_L(x)) = \phi(x)\phi(y) - \phi(y)\phi(x).
\end{aligned}$$

This means that ϕ is a Lie algebra homomorphism from L to $(U(L) \otimes_F U(L))_L$. From the universal property of $(U(L), j_L)$, we have a unique algebra homomorphism Δ with the required property.

5. Let d be a derivation of the Lie algebra L. Consider the algebra $M_2(U(L))$ of 2×2 matrices with entries in the universal enveloping algebra $U(L)$ of L. Consider the map ϕ from L to $M_2(U(L))$ which is given by

$$\phi(x) = \begin{bmatrix} j_L(x) & j_L(d(x)) \\ 0 & j_L(x) \end{bmatrix}.$$

Clearly ϕ is a linear map. Also using the fact that d is a derivation and j_L is a linear map, it can be easily seen that $\phi[x, y] = \phi(x)\phi(y) - \phi(y)\phi(x)$. Hence from the universal property of $(U(L), j_L)$, there is a unique homomorphism $\overline{\phi}$ from $U(L)$ to $M_2(U(L))$ such that $\overline{\phi}oj_L = \phi$. Since $j_L(L)$ generates $U(L)$ and since the set of upper triangular matrices with same diagonal entries is closed under the matrix product, we have a unique map $\hat{d}$ from $U(L)$ to itself such that for $u \in U(L)$,

$$\overline{\phi}(u) = \begin{bmatrix} u & \hat{d}(u) \\ 0 & u \end{bmatrix}.$$

The fact that $\overline{\phi}$ is a homomorphism implies that $\hat{d}$ is a derivation of $U(L)$. Clearly, $\hat{d}oj_L = j_Lod$. ♯

Our next aim is to determine the structure of $U(L)$ (PBW theorem). We have the filtration

$$T^0(L) \subseteq T^1(L) \subseteq T^2(L) \subseteq \cdots\cdots T^n(L) \subseteq T^{n+1}(L) \subseteq \cdots$$

of $T(L)$ considered as a vector space, where $T^m(L) = \oplus\sum_{i=0}^{m} \otimes^i L$. Consequently, we get a filtration

$$U^0(L) \subseteq U^1(L) \subseteq U^2(L) \subseteq \cdots\cdots U^n(L) \subseteq U^{n+1}(L) \subseteq \cdots$$

of $U(L)$ considered as a vector space, where $U^m(L) = \nu(T^m)$, ν being the quotient map from $T(L)$ to $U(L)$. Evidently, $U^p(L)U^q(L) \subseteq U^{p+q}(L)$. Let $\Omega^m(L)$ denote the vector space $U^m(L)/U^{m-1}(L)$. Put $\Omega(L) = \oplus\sum_{m=0}^{\infty} \Omega^m(L)$. The product in $U(L)$ induces a bi-linear product from $\Omega^p(L) \times \Omega^q(L)$ to $\Omega^{p+q}(L)$.. These bi-linear products can be extended by linearity to an associative product on $\Omega(L)$ which makes $\Omega(L)$ an associative algebra. For each m, we have a linear map μ_m from $\otimes^m L$ to $\Omega^m(L)$ given by $\mu_m(x) = \nu(x) + U^{m-1}(L)$. Given any $u \in U^m(L)$, there is an element $x = x_o + x_1 + \cdots + x_m$ of $T^m(L)$ such that $\nu(x) = u$, where $x_i \in \otimes^i L$ for each i. Evidently, $\mu_m(x_m) = \nu(x_m) + U^{m-1}(L) = \nu(x) + U^{m-1}(L) = u + U^{m-1}(L)$. This shows that μ_m is a surjective linear map from $\otimes^m(L)$ to $\Omega^m(L)$. This gives us a unique linear map μ from $T(L)$ to $\Omega(L)$ subject to $\mu|_{\otimes^m L} = \mu_m$ for each m. If $x \in \otimes^p(L)$ and $y \in \otimes^q(L)$, then $xy \in \otimes^{p+q}(L)$ and $\mu(xy) = \mu_{p+q}(xy) = \nu(xy) + U^{p+q-1}(L) = \nu(x)\nu(y) + U^{p+q-1}(L) = (\nu(x) + U^{p-1}(L))(\nu(y) + U^{q-1}(L)) = \mu(x)\mu(y)$. This shows that μ is a surjective algebra homomorphism from $T(L)$ to $\Omega(L)$. Consider the element $x \otimes y - y \otimes x \in \otimes^2 L$. Then $\nu(x \otimes y - y \otimes x) \in \Omega^2(L)$. Also $x \otimes y - y \otimes x - [x, y] \in J$ for all $x, y \in L$. Further, $\mu(x \otimes y - y \otimes x) = \mu_2(x \otimes y - y \otimes x) + U_1(L) = U_1(L)$, since $x \otimes y - y \otimes x = [x, y] \in U_1(L)$. It follows that $x \otimes y - y \otimes x \in ker\ \mu$ for each pair $x, y \in L$. Hence $I \subseteq \ker\ \mu$. In turn, μ induces a unique surjective algebra homomorphism $\overline{\mu}$ from the symmetric algebra $S(L)$ of L to $\Omega(L)$ such that $\overline{\mu}o\pi = \mu$, where π is the natural quotient map from $T(L)$ to $S(L)$.

Theorem 1.2.4 (Poincaré–Birkhoff–Witt) *The algebra homomorphism $\overline{\mu}$ from $S(L)$ to $\Omega(L)$ is an isomorphism.*

Before proving the theorem, we derive some of its important consequences as corollaries.

Corollary 1.2.5 *Let W be a subspace of $\otimes^m L$ such that the quotient map η_m from $\otimes^m L$ to $S^m(L) = \otimes^m L/(I \bigcap \otimes^m L)$ restricted to W is an isomorphism. Then $U^m(L) = \nu(W) \oplus U^{m-1}(L)$.*

Proof Under the hypothesis of the corollary, it is clear that $\otimes^m L = W \oplus (I \bigcap \otimes^m L)$. Thus, it follows from the PBW theorem that $\mu_m|_W$ is an isomorphism from W to $\Omega^m(L) = U^m(L)/U^{m-1}(L)$. Consequently, $U^m(L) = \nu(W) \oplus U^{m-1}(L)$. ♯

Corollary 1.2.6 *The linear map j_L from L to $U(L)$ is an injective linear map which is also a Lie homomorphism from L to $U(L)_L$.*

Proof The map η_1 is a tautological isomorphism from $L = \otimes^1 L$ to $S^1(L)$. The result follows from the above corollary, since $U_0 = \{0\}$. ♯

Let L be a Lie algebra. Let $\{x_\lambda \mid \lambda \in \Lambda\}$ be a basis of L, where $(\Lambda, \leq)$ is a well-ordered index set. Such a basis exists because of the well-ordering theorem. For each member $\Sigma = (\lambda_1, \lambda_2, \cdots, \lambda_m) \in \Lambda^m$, let x_Σ denote the element $x_{\lambda_1} \otimes x_{\lambda_2} \otimes \cdots \otimes x_{\lambda_m}$ of $\otimes^m L$. We put $x_\emptyset = 1$ for $\emptyset \in \Lambda^0$, and for $(\lambda) \in \Lambda^1$, we denote $x_{(\lambda)}$ by x_λ. Evidently, $X_m = \{x_\Sigma \mid \Sigma \in \Lambda^m\}$ is a basis of $\otimes^m L$. Next, let Λ_m denote the subset $\{(\lambda_1, \lambda_2, \cdots, \lambda_m) \in \Lambda^m \mid \lambda_1 \leq \lambda_2 \leq \cdots \leq \lambda_m\}$ of Λ^m. For $m = 0$, $\Lambda_0 = \{\emptyset\}$ and for $m = 1$, $\Lambda_1 = \Lambda^1$. For each m and for each ordered m-tuple $\Gamma = (\lambda_1, \lambda_2, \cdots, \lambda_m) \in \Lambda_m$, the element $x_{\lambda_1} \otimes x_{\lambda_2} \otimes \cdots \otimes x_{\lambda_m} + (I \bigcap \otimes^m L)$ of $S^m(L)$ is denoted by $\overline{x_\Gamma}$. We put $\overline{x_{(\lambda)}} = \overline{x_\lambda}$. Thus, $\overline{x_\Sigma} = \overline{x_{\lambda_1}}\,\overline{x_{\lambda_2}} \cdots \overline{x_{\lambda_m}}$. By the convention, $\overline{x_\emptyset} = 1$. More generally, the image of the element $x \in L$ under the embedding i_L of L to $S_1(L) \subseteq S(L)$ is denoted by $\overline{x}$.

Let W_m be the subspace of $\otimes^m L$ generated by $\{x_\Gamma \mid \Gamma \in \Lambda_m\}$. Since $\{\overline{x_\Gamma} \mid \Gamma \in \Lambda_m\}$ is a basis of $S^m(L)$, the quotient map $\eta_m|_{W_m}$ from W_m to $S^m(L)$ is an isomorphism. It follows from Corollary 1.2.5 that $U^m(L) = \nu(W_m) \oplus U^{m-1}(L)$ and $\nu(X_m) = \{x_\Gamma + J \mid \Gamma \in \Lambda_m\}$ is a basis of $\nu(W_m)$. Since

$$U^0(L) \subseteq U^1(L) \subseteq U^2(L) \subseteq \cdots \subseteq U^n(L) \subseteq U^{n+1}(L) \subseteq \cdots$$

is a filtration of $U(L)$, we obtain the following corollary:

Corollary 1.2.7 $\bigcup_{m=1}^{\infty} \nu(X_m) \bigcup \{1\}$ *is a basis of $U(L)$. More explicitly, $\{x_{\lambda_1} \otimes x_{\lambda_2} \otimes \cdots \otimes x_{\lambda_m} + J \mid \lambda_1 \leq \lambda_2 \leq \cdots \leq \lambda_m\} \bigcup \{1\}$ is a basis of $U(L)$.* ♯

The basis of $U(L)$ described in the above corollary is termed as the PBW basis of $U(L)$.

Corollary 1.2.8 *Let M be a Lie subalgebra of a Lie algebra L. Then $U(i)$ is injective algebra homomorphism from $U(M)$ to $U(L)$, where i is the inclusion homomorphism. Further, $U(L)$ is a free $U(M)$-module.*

Proof Let $X = \{x_\lambda \mid \lambda \in \Lambda\}$ be a basis of M, where $(\Lambda, \leq)$ is a well-ordered set. Embed it into a basis $\{x_\lambda \mid \lambda \in \Lambda'\}$, where $\Lambda \subseteq \Lambda'$. By the well-ordering theorem, we have a well order on $\Lambda' - \Lambda$. In turn, we get a well order $\leq'$ on Λ' such that $\leq'|_\Lambda = \Lambda$ and $\lambda <' \mu$ for all $\lambda \in \Lambda$ and $\mu \in \Lambda' - \Lambda$. From the above corollary,

$(\bigcup_{m=1}^{\infty}\nu(X'_m))\bigcup\{1\}$ is a basis of $U(L)$ whereas $(\bigcup_{m=1}^{\infty}\nu(X_m))\bigcup\{1\}$ is a basis of $U(M)$. Evidently, $\nu(X_m)\subseteq\nu(X'_m)$. This shows that $U(M)$ can be realized as a subalgebra of $U(L)$ and $U(L)$ is a free $U(M)$-module with basis $\bigcup_{m=1}^{\infty}(\nu(X'_m)-\nu(X_m))\bigcup\{1\}$. ♯

Corollary 1.2.9 *If L is a Lie algebra, then $U(L)$ is without zero divisors.*

Proof Since $S(L)\approx\Omega(L)$ and the symmetric algebra $S(L)$, being isomorphic to the polynomial algebra, is without zero divisors, it follows that the algebra $\Omega(L)$ is without zero divisors. We show that $U(L)$ is without zero divisors. Let x, y be members of $U(L)$ such that $xy = 0$ with $y\neq 0$. Then there is the smallest $r\geq 1$ such that $y\in U^r(L)$ but $y\notin U^{r-1}(L)$. Suppose that $x\in U^m(L)$. Then $(x+U^{m-1}(L))(y+U^{r-1}(L)) = 0$ in $\Omega(L)$. Since $\Omega(L)$ is without zero divisors and $y+U^{r-1}(L)$ is a nonzero element, $x+U^{m-1}(L)$ is zero in $\Omega^m(L)$. Hence $x\in U^{m-1}(L)$. Proceeding inductively, we see that $x\in U^0(L) = F$. Since $xy = 0$, $x = 0$. ♯

Consider the universal enveloping algebra $(U(L), j_L)$ of L. For each $u\in U(L)$, we have the left multiplication map l_u from $U(L)$ to $U(L)$ given by $l_u(v) = uv$. Clearly, $l_u\in End_K(U(L)) = gl(U(L))$. The map ϕ from $U(L)$ to $End_K(U(L))$ defined by $\phi(u) = l_u$ is easily seen to be an algebra homomorphism. Since $U(L)$ has no nonzero divisors, ϕ is injective. Consequently, $\phi o j_L$ is an injective Lie algebra homomorphism from L to $gl(U(L))_L$. We have established the following analogue of Cayley's theorem for Lie algebras.

Corollary 1.2.10 *Every Lie algebra is isomorphic to a Lie subalgebra of $gl(V)$ for some vector space V. In the language of the representation theory, every Lie algebra has a faithful representation.* ♯

Observe that even if L is finite dimensional, the algebra $U(L)$ need not be finite dimensional. However, the theorem of Ado–Iwasava asserts that every finite-dimensional Lie algebra is the Lie subalgebra of $gl(V)$ for some finite-dimensional space V. Ado proved it for the characteristic 0 case, and Iwasava proved it for the positive characteristic. The proofs of these results will follow in Chap. 3.

We still postpone the proof of the PBW theorem and use it further to introduce the concept of free Lie algebras and that of the presentations of Lie algebras.

Definition 1.2.11 Let X be a set. A pair $(L(X), i_X)$, where $L(X)$ is a Lie algebra over F and i_X is a map from X to $L(X)$, is called a **free Lie algebra** on X if it satisfies the following universal property: For any pair (B, j), where B is a Lie algebra over F and j is a map from X to B, there is a unique Lie algebra homomorphism $\overline{j}$ from $L(X)$ to B such that $\overline{j} o i_X = j$.

As usual, the pair $(L(X), i_X)$ is unique up to isomorphism. For the existence, let V be a vector space over F with X as a basis. Consider the tensor algebra $(T(V), i_V)$ on V. Let $L(X)$ denote the Lie subalgebra of the Lie algebra $(T(V))_L$ generated by $i_V(X)$. Observe that $(T(V), i_X)$ is the free associative algebra on X, where $i_X = i_V|_X$. We show that $(L(X), i_X)$ is a free Lie algebra on X. Let M be a Lie algebra and

j be a map from X to M. Then $j_M o j$ is a map from X to the universal enveloping algebra $U(M)$ of M. Since $T(V)$ is free associative algebra on X, we get an algebra homomorphism ϕ from $T(V)$ to $U(M)$ such that $\phi o i_X = j_M o j$. In turn, ϕ is also a Lie algebra homomorphism from $T(V)_L$ to $U(M)_L$. Since $j_M(M)$ is a Lie subalgebra of $U(M)_L$ containing $(j_M o j)(X)$ and $L(X)$ is the Lie subalgebra of $T(V)_L$ generated by $i_X(X)$, $\eta = (j_M)^{-1} o \phi|_{L(X)}$ is a Lie algebra homomorphism from $L(X)$ to M such that $\eta o i_X = j$. This ensures that $(L(X), i_X)$ is a free Lie algebra on X. The members of $L(X)$ are called the Lie elements in X.

In particular, if X is a set, W is a vector space over F, and $\cdot$ is a map from $X \times W$ to W such that $x \cdot (\alpha w + \beta w') = \alpha(x \cdot w) + \beta(x \cdot w')$ for all $w, w' \in W$ and $\alpha, \beta \in F$, then we have a unique $L(X)$-module structure on W subject to $i_X(x) \cdot w = x \cdot w$.

Presentations of Lie Algebras

Let F be a fixed field. A **presentation** of a Lie algebra is a pair $< X; R >$, where X is a set and R is a set of Lie elements in X. Let L be a Lie algebra. A presentation $< X; R >$ together with a map ϕ from X to L is called a presentation of L if the induced homomorphism $\overline{\phi}$ from $L(X)$ to L is surjective and the kernel of $\overline{\phi}$ is the ideal of $L(X)$ generated by R. The members of R are called the defining relations of the presentation. Thus, $< X; \emptyset >$ together with the map i_X is a presentation of the free Lie algebra $L(X)$ on X. The presentation $< \{x, y\}; \{x \otimes y - y \otimes x - x\} >$ is a presentation of a non-abelian two-dimensional Lie algebra. As in the case of the presentation theory of groups, one can have the analogue of the word problem and also the isomorphism problems in the theory of presentations of Lie algebras. In general, a finite presentation may represent an infinite-dimensional Lie algebra. One of the most important aims of the book is to have a faithful classification of finite-dimensional semi-simple Lie algebras over an algebraically closed characteristic 0 in terms of presentations.

We state without proof the following analogue of the Nielsen–Schreier theorem.

Theorem 1.2.12 *(Širšov–Witt Theorem) Every subalgebra of a free Lie algebra is free.* $\sharp$

Proof of the PBW Theorem

Now, we proceed for a proof of the PBW theorem. We establish a few lemmas for this purpose.

We have a filtration

$$S_0(L) \subseteq S_1(L) \subseteq S_2(L) \subseteq \cdots S_m(L) \subseteq S_{m+1}(L) \subseteq \cdots$$

of $S(L)$, where $S_m(L) = \oplus \sum_{i=0}^{m} S^i(L)$. Thus, $S_0(L) = F$ and $S_1(L) = F \oplus S^1(L)$, where $S^1(L) = \overline{L} = \{\overline{x} \mid x \in L\}$ is isomorphic to L as a vector space.

Lemma 1.2.13 *Given a Lie algebra L, there is a unique sequence* $\{f_m \in Hom(L \otimes_F S_m(L), S(L)) \mid m \in \mathbb{N} \bigcup \{0\}\}$ *of linear maps satisfying the following conditions:*

1. $f_m|_{L \otimes_F S_{m-1}(L)} = f_{m-1}$ *for each m.*
2. $f_m(x_\lambda \otimes \overline{x_\Gamma}) = \overline{x_\lambda}\,\overline{x_\Gamma}$, *where* $\Gamma = (\lambda_1, \lambda_2, \cdots, \lambda_m) \in \Lambda_m$ *and* $\lambda \leq \lambda_1$.

3. *If* $\Gamma \in \Lambda_r$, $r \leq m$, *then* $f_m(x_\lambda \otimes \overline{x_\Gamma}) = \overline{x_\lambda}\,\overline{x_\Gamma} + u_r$ *for some* $u_r \in S_r(L)$.
4. *If* $\Gamma \in \Lambda_{m-1}$, *then* $f_m([x_\lambda, x_\mu] \otimes \overline{x_\Gamma}) = f_m(x_\lambda \otimes (f_m(x_\mu \otimes \overline{x_\Gamma}))) - f_m(x_\mu \otimes (f_m(x_\lambda \otimes \overline{x_\Gamma})))$.

Proof We construct the sequence $\{f_m \mid m \geq 0\}$ by the induction on m. The uniqueness is also established by the induction. Indeed, for each m, we construct the finite sequence $\{f_r \mid r \leq m\}$ satisfying the conditions of the lemma. For $m = 0$, $S_0(L) = F$, $\Lambda_0 = \{\emptyset\}$, $\overline{x_\emptyset} = 1$. We have the unique linear map f_0 from $L \otimes S_0(L)$ to $S(L)$ such that $f_0(x_\lambda \otimes \overline{x_\emptyset}) = \overline{x_\lambda}\,\overline{x_\emptyset} = \overline{x_\lambda}$. Indeed, it is given by $f_0(x \otimes \alpha) = \alpha\overline{x}$. Conditions 3 and 4 are vacuously satisfied. Uniqueness is evident because of condition 2. For more clarity in the arguments, we construct f_1 also. We need to define $f_1(x_\lambda \otimes \overline{x_\mu})$ for all $\lambda, \mu \in \Lambda$ which we can further extend by linearity. Condition 2 forces us to define $f_1(x_\lambda \otimes \overline{x_\mu}) = \overline{x_\lambda}\,\overline{x_\mu}$ whenever $\lambda \leq \mu$ and also $f_1(x_\lambda \otimes 1) = f_1(x_\lambda \otimes \overline{x_\emptyset}) = \overline{x_\lambda}\,\overline{x_\emptyset} = \overline{x_\lambda}$. Thus, $f_1|_{L \otimes S_0(L)} = f_0$. Suppose that $\mu < \lambda$. Condition 3 requires the existence of a $u_1 \in S_1(L)$ such that $f_1(x_\lambda \otimes \overline{x_\mu}) = \overline{x_\lambda x_\mu} + u_1$. We use condition 4 to determine u_1. By 2 and 4,

$$f_1(x_\lambda \otimes \overline{x_\mu}) = f_1(x_\lambda \otimes f_1(x_\mu \otimes \overline{x_\emptyset})) = f_1(x_\mu \otimes f_1(x_\lambda \otimes \overline{x_\emptyset}) + f_1([x_\lambda, x_\mu] \otimes \overline{x_\emptyset}) =$$
$$f_1(x_\mu \otimes \overline{x_\lambda}) + \overline{[x_\lambda, x_\mu]} = \overline{x_\mu}\,\overline{x_\lambda} + \overline{[x_\lambda, x_\mu]} = \overline{x_\mu}\,\overline{x_\lambda} + \overline{x_\lambda}\,\overline{x_\mu} - \overline{x_\mu}\,\overline{x_\lambda} = \overline{x_\lambda}\,\overline{x_\mu}.$$

Thus, $u_1 = 0$ and $f_1(x_\lambda \otimes \overline{x_\mu}) = \overline{x_\lambda}\,\overline{x_\mu}$ for all $\lambda, \mu \in \Lambda$. This completes the construction of the sequence $\{f_0, f_1\}$.

Assume that the sequence $\{f_0, f_1, \cdots, f_m\}, m \geq 1$ with the required conditions has already been constructed. We construct f_{m+1}. Condition 1 forces us to define $f_{m+1}(x \otimes u) = f_r(x \otimes u)$ whenever $u \in S_r(L), r \leq m$. Thus, it is sufficient to define $f_{m+1}(x_\lambda \otimes \overline{x_\Gamma})$ for all $\Gamma = (\lambda_1, \lambda_2, \cdots, \lambda_{m+1}\} \in \Lambda_{m+1}$ and then extend it to a linear map from $L \otimes S_{m+1}(L)$ to $S(L)$. If $\lambda \leq \lambda_1$, then condition 2 forces us to put $f_{m+1}(x_\lambda \otimes \overline{x_\Gamma}) = \overline{x_\lambda}\,\overline{x_\Gamma}$. In general to ensure condition 3, $f_{m+1}(x_\lambda \otimes \overline{x_\Gamma})$ should be $\overline{x_\lambda}\,\overline{x_\Gamma} + u_{m+1}$ for some $u_{m+1} \in S_{m+1}(L)$ to be determined. Suppose that $\lambda_1 < \lambda$. Let us denote the element $(\lambda_2, \lambda_3, \cdots, \lambda_{m+1})$ of Λ_m by Γ'. Since $\overline{x_{\Gamma'}} \in S_m(L)$, $f_{m+1}(x_{\lambda_1} \otimes \overline{x_{\Gamma'}}) = f_m(x_{\lambda_1} \otimes \overline{x_{\Gamma'}}) = \overline{x_{\lambda_1}}\,\overline{x_{\Gamma'}} = \overline{x_\Gamma}$. Again, since $\overline{x_{\Gamma'}} \in S_m(L)$, by the induction hypothesis, there is a unique $u_m \in S_m(L)$ such that $f_{m+1}(x_\lambda \otimes \overline{x_{\Gamma'}}) = f_m(x_\lambda \otimes \overline{x_{\Gamma'}}) = \overline{x_\lambda}\,\overline{x_{\Gamma'}} + u_m$. Condition 4 forces us to put

$$f_{m+1}(x_\lambda \otimes \overline{x_\Gamma}) = \overline{x_{\lambda_1}}\,\overline{x_\lambda}\,\overline{x_{\Gamma'}} + f_m(x_{\lambda_1} \otimes u_m) + f_m([x_\lambda, x_{\lambda_1}] \otimes \overline{x_{\Gamma'}}).$$

This completes the forced construction of f_{m+1}. Conditions 1, 2, and 3 hold by the construction. If $\Gamma' = (\lambda_2, \lambda_3, \cdots \lambda_{m+1}) \in \Lambda_m$ and $\mu = \lambda_1 < \lambda$ then also condition 4 follows from the construction of f_{m+1} as above. Since $[x_\mu, x_\lambda] = -[x_\lambda, x_\mu]$, condition 4 also holds when $\lambda < \mu \leq \lambda_2$. Since $[x_\lambda, x_\lambda] = 0$, condition 4 also holds for $\lambda = \mu$. Suppose that $\lambda_2 < \lambda$ and also $\lambda_2 < \mu$. Denote $(\lambda_3, \lambda_4, \cdots \lambda_{m+1}) \in \Lambda_{m-1}$ by Γ''. Since $\lambda_2 \leq \lambda_i$ for all $i \geq 3$, by the induction assumption,

$$f_{m+1}(x_{\lambda_2} \otimes \overline{x_{\Gamma''}}) = f_{m-1}(x_{\lambda_2} \otimes \overline{x_{\Gamma''}}) = \overline{x_{\lambda_2}}\,\overline{x_{\Gamma''}} = \overline{x_{\Gamma'}}.$$

Again using the induction hypothesis,

$$f_{m+1}(x_\mu \otimes \overline{x_{\Gamma'}}) = f_m(x_\mu \otimes f_m(x_{\lambda_2} \otimes \overline{x_{\Gamma''}})) =$$
$$f_m(x_{\lambda_2} \otimes f_m(x_\mu \otimes \overline{x_{\Gamma''}})) + f_m([x_\mu, x_{\lambda_2}] \otimes \overline{x_{\Gamma''}}) \cdots \quad (1.1)$$

Thus, since $\lambda_2 < \mu$,

$$\begin{aligned} & f_{m+1}(x_\lambda \otimes f_{m+1}(x_\mu \otimes \overline{x_{\Gamma'}})) \\ &= f_{m+1}(x_\lambda \otimes (f_m(x_{\lambda_2} \otimes f_m(x_\mu \otimes \overline{x_{\Gamma''}})) + f_m([x_\mu, x_{\lambda_2}] \otimes \overline{x_{\Gamma''}}))) \\ &= f_{m+1}(x_\lambda \otimes (f_m(x_{\lambda_2} \otimes f_m(x_\mu \otimes \overline{x_{\Gamma''}})))) + f_{m+1}(x_\lambda \otimes (f_m([x_\mu, x_{\lambda_2}] \otimes \overline{x_{\Gamma''}}))) \\ &= f_{m+1}(x_{\lambda_2} \otimes f_m(x_\lambda \otimes f_m(x_\mu \otimes \overline{x_{\Gamma''}}))) + f_m([x_\lambda, x_{\lambda_2}] \otimes f_m(x_\mu \otimes \overline{x_{\Gamma''}})) + \\ & f_m([x_\mu, x_{\lambda_2}] \otimes f_m(x_\lambda \otimes \overline{x_{\Gamma''}})) + f_m([x_\lambda, [x_\mu, x_{\lambda_2}]] \otimes \overline{x_{\Gamma''}}) \cdots \end{aligned} \quad (1.2)$$

Since $\lambda_2 < \lambda$, we may interchange λ and μ to get the equation

$$\begin{aligned} & f_{m+1}(x_\mu \otimes f_{m+1}(x_\lambda \otimes \overline{x_{\Gamma'}})) \\ &= f_{m+1}(x_{\lambda_2} \otimes f_m(x_\mu \otimes f_m(x_\lambda \otimes \overline{x_{\Gamma''}}))) + f_m([x_\mu, x_{\lambda_2}] \otimes f_m(x_\lambda \otimes \overline{x_{\Gamma''}})) + \\ & f_m([x_\lambda, x_{\lambda_2}] \otimes f_m(x_\mu \otimes \overline{x_{\Gamma''}})) + f_m([x_\mu, [x_\lambda, x_{\lambda_2}]] \otimes \overline{x_{\Gamma''}}) \cdots \end{aligned} \quad (1.3)$$

Subtracting 3 from 2, applying condition 4 for f_m again, and using the Jacobi identity for L, we arrive at

$$f_{m+1}(x_\lambda \otimes f_{m+1}(x_\mu \otimes \overline{x_{\Gamma'}})) - f_{m+1}(x_\mu \otimes f_{m+1}(x_\lambda \otimes \overline{x_{\Gamma'}})) = f_m([x_\lambda, x_\mu] \otimes$$
$$f_m(x_{\lambda_2} \overline{x_{\Gamma''}})) + f_m(([x_{\lambda_2}, [x_\lambda, x_\mu]] + [x_\lambda, [x_\mu, x_{\lambda_2}]] + [x_\mu, [x_{\lambda_2}, x_\lambda]]) \otimes \overline{x_{\Gamma''}}) =$$
$$f_m([x_\lambda, x_\mu] \otimes f_m(x_{\lambda_2} \overline{x_{\Gamma''}})).$$

Since $\lambda_2 \leq \lambda_i$ for all $i \geq 3$, $f_m(x_{\lambda_2} \otimes \overline{x_{\Gamma''}}) = \overline{x_{\lambda_2}}\, \overline{x_{\Gamma''}} = \overline{x_{\Gamma'}}$. Thus,

$$f_{m+1}(x_\lambda \otimes f_{m+1}(x_\mu \otimes \overline{x_{\Gamma'}})) - f_{m+1}(x_\mu \otimes f_{m+1}(x_\lambda \otimes \overline{x_{\Gamma'}})) = f_{m+1}([x_\lambda, x_\mu] \otimes \overline{x_{\Gamma'}}).$$

This verifies condition 4 also for f_{m+1}. The proof of the lemma is complete. ♯

As a consequence of the above lemma, we have the following:

Lemma 1.2.14 *We have a representation ρ of L on $S(L)$ such that the following hold:*
For each $\Gamma = (\lambda_1, \lambda_2, \cdots, \lambda_m)$ in Λ_m,

$$\rho(x_\lambda)(\overline{x_\Gamma}) = \overline{x_\lambda}\, \overline{x_\Lambda}$$

whenever $\lambda \leq \lambda_1$, and in general

$$\rho(x_\lambda)(\overline{x_\Gamma}) = \overline{x_\lambda}\,\overline{x_\Lambda} + u_m$$

for some $u_m \in S_m(L)$.

Proof Let $x \in L$ and $u \in S(L)$. Then there is a $m \geq 0$ such that $u \in S_m(L)$. If $u \in S_m(L)$ and also $u \in S_n(L)$ with $m \leq n$, then from Lemma 1.2.13, $f_n(x \otimes u) = f_m(x \otimes u)$. Define $\rho(x)(u) = f_m(x \otimes u)$. This gives us a linear map ρ from L to $End_F(S(L))$. Condition 4 of Lemma 1.2.13 ensures that it is a Lie algebra homomorphism from L to $(End_F(S(L)))_L = gl(S(L))_L$. It is also clear that ρ satisfies the required condition. ♯

Lemma 1.2.15 *Let $x = x_0 + x_1 + \cdots + x_m$ be a member of $T_m(L) = \oplus\sum_{r=0}^{m} \otimes^r L$, where $x_r \in \otimes^r L$, $r \leq m$. Suppose that $x \in J$, where J is the kernel of the natural quotient map ν from $T(L)$ to $U(L)$. Then $x_m \in I$, where I is the kernel of the natural quotient map π from $T(L)$ to $S(L)$.*

Proof For each $\Gamma = (\lambda_1, \lambda_2, \cdots, \lambda_m) \in \Lambda^m$, let x_Γ denote the member $x_{\lambda_1} \otimes x_{\lambda_2} \otimes \cdots \otimes x_{\lambda_m}$ of $T(L)$. Evidently, $\pi(x_\Sigma)$ is the monomial $\overline{x_{\lambda_1}}\,\overline{x_{\lambda_2}} \cdots \overline{x_{\lambda_m}}$ in $\overline{x_{\lambda_i}}$ (note that $S(L)$ is commutative polynomial algebra in $\{\overline{x_\lambda} \mid \lambda \in \Lambda\}$). Clearly, $\{x_\Gamma \mid \Gamma \in \Lambda^m\}$ is a basis of $\otimes^m L$. From Lemma 1.2.14, we have a Lie algebra homomorphism ρ from L to $gl(S(L))_L = (End_F(S(L)))_L$ satisfying the conditions described in Lemma 1.2.14. Again, from the universal property of $U(L)$, we have a unique algebra homomorphism $\overline{\rho}$ from $U(L) = T(L)/J$ to $End_F(S(L))$ such that $\overline{\rho} o j_L = \rho$. In turn, we obtain an algebra homomorphism $\hat{\rho}$ from $T(L)$ to $End_F(S(L))$ such that $\hat{\rho} o i_L = \rho$. It follows that $J \subseteq Ker\ \hat{\rho}$. By Lemma 1.2.14, $\rho(x_\lambda)(1) = \rho(x_\lambda)(\overline{x_\emptyset}) = \overline{x_\lambda}$ and $\rho(x_\lambda)(\overline{x_\Gamma}) = \overline{x_\lambda}\,\overline{x_\Gamma} + u_m$ for some $u_m \in S_m(L)$. It follows that $\rho(x)(1)$ is a linear combination of monomials in $\overline{x_\lambda}$ with the highest m degree terms representing $\rho(x_m)$. Since $x \in J \subseteq Ker\ \hat{\rho}$, $\rho(x) = 0$. In turn, $\rho(x)(1) = 0$. Consequently, $\pi(x_m) = 0$. This shows that $x_m \in I$. ♯

Proof of the PBW Theorem. We need to show that $\overline{\mu}$ from $S(L)$ to $\Omega(L)$ is an isomorphism. Recall the discussion before the statement of the PBW theorem. We have already seen that $\overline{\mu}$ is a surjective homomorphism. By the definition, $\overline{\mu} o \pi = \mu$, where μ is the surjective homomorphism from $T(L)$ to $\Omega(L)$. Let x be a member of $T(L)$ such that $\mu(x) = 0$. We need to show that $x \in I$. Clearly, $x = x_0 + x_1 + \cdots + x_m$ belong to $T_m(L)$ for some m, where $x_i \in \otimes^i(L)$ for each $i \leq m$. Recall that $\mu(x) = \mu_m(x_m) = \nu(x_m) + U_{m-1}(L)$. Hence $\nu(x_m) \in U_{m-1}(L)$. Consequently, there is a member $x'_m \in T_{m-1}(L)$ such that $\nu(x_m) = \nu(x'_m)$. In turn, $x_m - x'_m$ belongs to J. From the previous lemma, $x_m \in I$. Hence $\mu(x) = \mu_m(x_m) = 0$. This shows that $x \in I$. ♯

Exercises

1.2.1. Suppose that two Lie algebras L and L' are isomorphic. Show that $U(L)$ is isomorphic to $U(L')$. Is the converse true? Support.
1.2.2. Describe the structure of $U(sl(2, \mathbb{C}))$.

1.2.3. Suppose that L is a finitely generated Lie algebra. Show that $U(L)$ is a Noetherian ring. What about the converse? Hint: Use the Hilbert basis theorem.

1.2.4. Let L be a free Lie algebra. Show that $U(L)$ is a free associative algebra.

1.2.5. Characterize Lie algebras for which $U(L)$ is a semi-simple ring.

1.2.6. Describe the free Lie algebra on a singleton.

1.2.7. Let $L(X)$ be a free Lie algebra on X over a field of characteristic 0. Suppose that X contains m elements. Show that $Dim\ (L(X) \bigcap U(L)^n) = \frac{1}{n}\sum_{d|n}\mu(d)m^{\frac{n}{d}}$, where μ is the Möbius function.

In the following exercises (1.2.8-1.2.12), we give another description of free Lie algebras and also of presentations.

1.2.8. Let $X = \{x_\lambda \mid \lambda \in \Lambda\}$ be a set and $\Omega(X)$ denote the set of all bracket arrangements in X of different weights with 1 representing the empty bracket arrangement. Define a product $\cdot$ in $\Omega(X)$ by taking 1 as the identity and by putting

$$\beta^r(x_{\lambda_1}, x_{\lambda_2}, \cdots, x_{\lambda_r}) \cdot \beta^s(x_{\delta_1}, x_{\delta_2}, \cdots, x_{\delta_s}) = \beta^{r+s}(x_{\lambda_1}, x_{\lambda_2}, \cdots, x_{\lambda_r}, x_{\delta_1}, x_{\delta_2}, \cdots, x_{\delta_s}),$$

where $\beta^{r+s} = (\beta_r\beta_s)$, $\lambda_i, \delta_j \in \Lambda$. Show that $\Omega(X)$ is a free magma with identity on X.

1.2.9. Let $F(\Omega(X))$ denote the vector space over F with $\Omega(X)$ as a basis. Extend the multiplication in $\Omega(X)$ to $F(\Omega(X))$ by bi-linearity and then $F(\Omega(X))$ becomes an algebra over F with identity (recall that an algebra over F is a vector space with a bi-linear product). Show that $F(\Omega(X))$ is a free algebra (with identity) on X.

1.2.10. An ideal of $F(\Omega(X))$ is a subalgebra of $F(\Omega(X))$ which is invariant under left and right multiplications by members of $\Omega(X)$. Describe the ideal of $F(\Omega(X))$ generated by a single element of $F(\Omega(X))$.

1.2.11. Let A be the ideal of $F(\Omega(X))$ generated by the set

$$\Sigma = \{u \cdot u \mid u \in F(\Omega(X))\} \bigcup \{u \cdot (v \cdot w) + v \cdot (w \cdot u) + w \cdot (u \cdot v) \mid u, v, w \in F(\Omega(X))\}.$$

Show that the quotient $F(\Omega(X))/A$ together with the map i from X to $F(\Omega(X))/A$ given by $i(x) = x + A$ is a free Lie algebra on X. The members of Σ are called the trivial relations.

1.2.12. Consider a pair $< X; R >$, where X is a set and $R \subset F(\Omega(X))$. The pair $< X; R >$ together with a map f from X to a Lie algebra L is called a presentation of L if the induced homomorphism $\overline{f}$ from $F(\Omega(X))$ to L is a surjective algebra homomorphism whose kernel is the ideal generated by $A \bigcup R$. Thus, $< X, \emptyset >$ is a presentation of the free Lie algebra on X. The members of R are called the relators of the presentation. Describe the Lie algebra having the presentation $< X; R >$, where $X = \{x, y\}$ and $R = \{(xy) - x\}$.

1.3 Solvable and Nilpotent Lie Algebras

The terminology of solvability and nilpotency in Lie algebra has been borrowed from the corresponding concepts of solvability and nilpotency in groups. Indeed, there is an intrinsic relationship between solvability/nilpotency of groups and the solvability/nilpotency of corresponding Lie algebras. We develop the theory in an analogous manner.

Definition 1.3.1 Let L be a Lie algebra. Define the ideals L^n of L inductively as follows. Put $L^0 = L$, $L^1 = [L, L]$. Assuming that L^n has already been defined, define $L^{n+1} = [L^n, L^n]$. Evidently, each L^n is an ideal of L such that $L^{n+1} \subseteq L^n$ and L^n/L^{n+1} is an abelian Lie algebra. Indeed, each L^n is a fully invariant subalgebra of L in the sense that $f(L^n) \subseteq L^n$ for all endomorphism f of L. The series

$$L = L^0 \triangleright L^1 \triangleright \cdots \triangleright L^n \triangleright L^{n+1} \triangleright \cdots$$

is called the **derived series** of the Lie algebra L. L^n is called the nth term of the derived series. A Lie algebra L is said to be a **solvable** Lie algebra if $L^n = \{0\}$ for some n. If $L^n = \{0\}$ but $L^{n-1} \neq \{0\}$, then we say that L is solvable of derived length n.

Thus, a nontrivial abelian Lie algebra is solvable of derived length 1. A non-abelian Lie algebra of dimension 2 is solvable of derived length 2. If the characteristic of F is 2, then $sl(2, F)$ is solvable. However, if the characteristic of F is different from 2, then $sl(2, F)$ is not solvable. The classical Lie algebras A_n, B_n, C_n, and D_n are non-solvable. The Lie algebra $t(n, F)$ is solvable of derived length n, and $n(n, F)$ is solvable of derived length $n-1$. Indeed, $t(n, F)^r$ is the Lie algebra of $n \times n$ matrices $[a_{ij}]$ for which $a_{ij} = 0$ whenever $j < i + r$.

Proposition 1.3.2 *A nontrivial solvable Lie algebra L is simple if and only if L is a cyclic Lie algebra.*

Proof Let L be a nontrivial solvable Lie algebra. Then $[L, L] \neq L$. Suppose that L is simple, then $[L, L]$ is $\{0\}$. Consequently, every subspace of L is an ideal. Since L is simple, it has no nontrivial proper subspace. This means that L is one dimensional. ♯

Corollary 1.3.3 *If A is a maximal ideal of a nontrivial solvable Lie algebra L, then A is of co-dimension 1.*

Proof A is a maximal ideal of a solvable Lie algebra L if and only if L/A is simple and solvable. Consequently, the dimension of L/A is 1. ♯

The proof of the following proposition is easy, and it is similar to the proof in the case of groups. As such, it is left as an exercise.

Proposition 1.3.4 *(i) A Lie algebra L is solvable if and only if there is a finite chain*

$$L = L_0 \triangleright L_1 \triangleright \cdots \triangleright L_{n-1} \triangleright \{0\}$$

of subalgebras of L terminating to $\{0\}$, where L_{i+1} is an ideal of L_i and L_i/L_{i+1} is abelian for each $i \leq n-1$.

(ii) Subalgebras and homomorphic images of solvable Lie algebras are solvable. In particular, the quotient of a solvable Lie algebra is solvable.

(iii) If A is an ideal of L such that A and L/A are solvable, then L is solvable. In particular, L is solvable if and only if $ad(L)$ is solvable.

(iv) If A and B are solvable ideals of L, then $A + B$ is also a solvable ideal of L. In particular, every finite-dimensional Lie algebra contains the largest solvable ideal. ♯

The largest solvable ideal of L is called the **radical of L**, and it is denoted by $Rad(L)$. A nontrivial finite-dimensional Lie algebra L is called a **semi-simple Lie algebra** if the radical $Rad(L)$ of L is trivial. Thus, a non-abelian simple Lie algebra is semi-simple. If L is not solvable, then $L/Rad(L)$ is semi-simple. We have a short exact sequence

$$0 \longrightarrow Rad(L) \stackrel{i}{\to} L \stackrel{\nu}{\to} L/Rad(L) \longrightarrow 0.$$

We shall see (Theorem 1.5.16) that L is a semi-direct product $Rad(L) \succ L/Rad(L)$. As such, the structure theory of Lie algebras depends on the structure theory of solvable Lie algebras, and even more importantly on the structure theory of semi-simple Lie algebras.

Theorem 1.3.5 *Let L be a solvable Lie subalgebra of $gl(V)$, where V is a nonzero finite-dimensional vector space over an algebraically closed field F of characteristic 0. Then there is a nonzero vector $v \in V$ which is an eigenvector of each member of L.*

Proof The proof is by the induction of $dim\ L$. If $dim\ L = 0$, then there is nothing to do. Assume the induction hypothesis. Let A be a maximal ideal of L. From Corollary 1.3.3, A is of co-dimension 1 in L. If $A = \{0\}$, then $L = Fx$ is one dimensional, where x is a nonzero endomorphism of V. Since F is algebraically closed, there is an eigenvector v_0 of x. Evidently, v_0 is the eigenvector of each member of L, and we are done. Suppose that $A \neq \{0\}$, and $L = A \oplus Fx_0$, where x_0 is a nonzero member of L. From the induction hypothesis, there is a nonzero vector $v \in V$ such that v is an eigenvector of each member $x \in A$. We get a linear functional λ on A such that $x(v) = \lambda(x)v$ for all $x \in A$. Let $V_{\lambda(x)}$ denote $\lambda(x)$-eigenspace of x in V. Evidently, $v \in V_{\lambda(x)}$ for all $x \in A$. Consider the subspace $W = \bigcap_{x \in A} V_{\lambda(x)}$ of V. Clearly, $v \in W$, and so $W \neq \{0\}$. Suppose that we are able to show that W is invariant under all endomorphisms in L. Then, in particular, W is invariant under the endomorphism x_0. Since F is algebraically closed, there is an eigenvector v_0 of x_0 in W. Since $L = A \oplus Fx_0$, it follows that v_0 is an eigenvector of all members of L.

Now, we show that W is invariant under all transformations in L. Let $w \in W$ and $y \in L$. We need to show that $y(w) \in W = \bigcap_{x\in A} V_{\lambda(x)}$. Let $x \in A$. Then $x(y(w)) = y(x(w)) - [x, y](w) = y(\lambda(x)w) - [x, y](w) = \lambda(x)y(w) - \lambda([x, y])w$ (note that $[x, y] \in A$). Thus, it is sufficient to show that $\lambda([x, y]) = 0$ for all $y \in L$ and $x \in A$. Let n be the smallest positive integer such that the set $\{w, y(w), y^2(w), \cdots, y^{n-1}(w)\}$ is linearly independent. For each $z \in A$, $z(w) = \lambda(z)w$, $z(y(w)) = y(z(w)) - [y, z](w) = \lambda(z)y(w) - \lambda([y, z])w$. Proceeding inductively, we can show that $z(y^i(w))$ is a linear combination of $\{w, y(w), y^2(w), \cdots, y^i(w)\}$. Again, for each i, $0 \leq i \leq n-1$, We show that

$$x(y^i(w)) = \lambda(x)y^i(w) + \sum_{j=0}^{i-1} \alpha^i_j y^j(w)$$

for some $\alpha^i_j \in F$. The proof is by the induction on i. It is evident for $i = 0$. Assume it to be true for i. Now,

$$x(y^{i+1}(w)) = x(y(y^i)(w)) = y(x(y^i(w))) - [y, x](y^i(w)) =$$
$$y(\lambda(x)y^i(w) + \sum_{j=0}^{i-1} \alpha^i_j y^j(w)) - [y, x](y^i(w)) =$$
$$\lambda(x)y^{i+1}(w) + \sum_{j=0}^{i-1} \alpha^i_j y^{j+1}(w)) - [y, x](y^i(w)).$$

Since $[y, x] \in A$, the result for $i + 1$ follows from the previous observation. It follows that the subspace U of V generated by $\{w, y(w), y^2(w), \cdots, y^{n-1}(w)\}$ is invariant under x and the matrix representation of x on U is lower triangular with all diagonal entries $\lambda(x)$. Thus, the trace of x on U is $n\lambda(x)$. Since U is invariant under x as well as y, it is invariant under $[x, y]$. As such, the trace of $[x, y]$ on U is zero. But already the trace of $[x, y]$ is $n\lambda([x, y])$. Since the field is of characteristic 0, it follows that $\lambda([x, y]) = 0$ for all $x \in A$ and $y \in L$. ♯

A **Flag** on a finite-dimensional vector space V of dimension n is a chain

$$\{0\} \subseteq V_1 \subseteq V_2 \subseteq \cdots V_{n-1} \subseteq V_n = V$$

of subspaces of V such that dimension of V_i is i for each i. The above flag is said to be invariant under an endomorphism x in $gl(V)$ if $x(V_i) \subseteq V_i$ for all i.

Corollary 1.3.6 *(Theorem of Lie) Let L be a solvable Lie subalgebra of $gl(V)$, where V is a finite-dimensional vector space over an algebraically closed field. Then there is a flag of V which is invariant under all members of L.*

Proof The proof is by the induction on the dimension of V. If $dim\ V = 1$, then there is nothing to do. Assume that the result is true for vector spaces of dimension n. Let V be a vector space of dimension $n + 1$ and L is a solvable Lie subalgebra of $gl(V)$. From the above theorem, there is a common eigenvector v_0 of elements of L. Evidently, the subspace V_1 generated by v_0 is invariant under all elements of L. Consider the vector space $V' = V/V_1$. Each $x \in L$ determines an element $\hat{x} \in gl(V')$ given by $\hat{x}(v + V_1) = x(v) + V_1$. The map ϕ from L to $gl(V')$ defined by $\phi(x) = \hat{x}$ is easily seen to be a Lie algebra homomorphism. Evidently, $\phi(L)$ is a solvable Lie subalgebra of $gl(V')$. By the induction hypothesis, there is a flag

$$V_1/V_1 \subseteq V_2/V_1 \subseteq \cdots \subseteq V_n/V_1 \subseteq V' = V/V_1$$

of V' which is invariant under $\phi(L)$. Clearly,

$$\{0\} \subseteq V_1 \subseteq V_2 \subseteq \cdots V_n \subseteq V_{n+1} = V$$

is a flag of V which is invariant under L. ♯

Corollary 1.3.7 *Let L be a finite-dimensional solvable Lie algebra over an algebraically closed field F of characteristic* 0. *Then there is a chain*

$$\{0\} \subseteq L_1 \subseteq L_2 \subseteq \cdots L_{n-1} \subseteq L_n = L$$

of ideals of L such that $dim\ L_i = i$ for each i.

Proof We have the Lie homomorphism ad from L to $ad(L) \subseteq gl(L)$. Evidently, $ad(L)$ is solvable. From the previous corollary, there is a flag

$$\{0\} \subseteq L_1 \subseteq L_2 \subseteq \cdots L_{n-1} \subseteq L_n = L$$

of subspaces of L which is invariant under $ad(L)$. This means that each L_i is also an ideal of L. ♯

Recall that $t(n, F)$ is a solvable Lie algebra of derived length n. We have the following corollary.

Corollary 1.3.8 *Let V be an n-dimensional vector space over an algebraically closed field F of characteristic 0. Then every solvable Lie subalgebra L of $gl(V)$ is isomorphic to a Lie subalgebra of $t(n, F)$. Indeed, from Ado's theorem, it further follows that a Lie algebra is solvable if and only if it is isomorphic to a Lie subalgebra of $t(n, F)$ for some n.*

Proof Let L be a solvable Lie subalgebra of $gl(V)$, where V is an n-dimensional vector space over an algebraically closed field K. From Corollary 1.3.6, there is a flag

$$\{0\} \subseteq V_1 \subseteq V_2 \subseteq \cdots V_{n-1} \subseteq V_n = V$$

of V which is invariant under the members of L. Select a basis $\{v_1, v_2, \cdots, v_n\}$ of V such that $\{v_1, v_2, \cdots, v_i\}$ is a basis of V_i. Evidently, the matrix representation of each member of L with respect to this basis is a member of $t(n, K)$. The result follows. ♯

Corollary 1.3.9 *Let L be a solvable Lie subalgebra of $gl(V)$, where V is a finite-dimensional vector space over an algebraically closed field of characteristic 0. Suppose that the dimension of L is n. Let $x \in [L, L]$. Then $ad(x)$ is nilpotent endomorphism. Indeed, $ad(x)^n = 0$.*

Proof From Corollary 1.3.7, there is a chain

$$\{0\} \subseteq L_1 \subseteq L_2 \subseteq \cdots L_{n-1} \subseteq L_n = L$$

of ideals of L such that $dim\ L_i = i$ for each i. We have a basis $\{x_1, x_2, \cdots, x_n\}$ of L such that $\{x_1, x_2, \cdots, x_i\}$ is a basis of L_i. Evidently, the matrix representation of each member of $ad(L)$ with respect to this basis lies in $t(n, F)$. As such, the matrix representation of each member of $ad(L)$ lies in $\nu(n, F)$. Since $A^n = 0$ for each $A \in \nu(n, F)$, it follows that $ad(x)^n = 0$ for all $x \in [L, L]$. ♯

Definition 1.3.10 Let L be a Lie algebra over a field F. Define ideals L_n of L inductively as follows. Define $L_0 = L$. Assuming that ideal L_n has already been defined, define $L_{n+1} = [L, L_n]$. We get a chain

$$L = L_0 \supseteq L_1 \supseteq L_2 \supseteq \cdots \supseteq L_n \supseteq L_{n+1} \supseteq \cdots$$

of ideals. This chain is called the **lower central series** of L. L_n is called the nth term of the lower central series of L. Similarly, we define ideals $Z_n(L)$ of L inductively as follows. Define $Z_0(L) = \{0\}$. Assuming that $Z_n(L)$ has already been defined, define $Z_{n+1}(L)$ by the property $Z_{n+1}(L)/Z_n(L) = Z(L/Z_n(L))$. Evidently, $Z_1(L) = Z(L)$ the center of L. This gives us an ascending chain

$$\{0\} \subseteq Z_1(L) \subseteq Z_2(L) \subseteq \cdots \subseteq Z_n(L) \subseteq Z_{n+1}(L) \cdots$$

of ideals of L. This is called the **upper central series** of L. Inductively, it follows that the terms of the lower central series of L are fully invariant.

Proposition 1.3.11 *The lower central series of L terminates to $\{0\}$ at the nth step if and only if the upper central series terminates to L at the nth step.*

Proof Suppose that the lower central series of L terminates to $\{0\}$ at the nth step. Then $L_n = \{0\}$. By the induction on i, we show that $L_{n-i} \subseteq Z_i(L)$ for each i. Evidently, $L_n \subseteq Z_0(L)$. Assume the result for i. We need to show that $L_{n-i-1} \subseteq Z_{i+1}(L)$. Let $a \in L_{n-i-1}$. Then for each $x \in L$, $[x, a] \in L_{n-i}$. By the induction assumption, $[x, a] \in Z_i(L)$ for each $x \in L$. This means that $a + Z_i(L) \in Z(L/Z_i(L))$. Hence $a \in Z_{i+1}(L)$. It follows that $L_{n-i} \subseteq Z_i(L)$ for each i. In particular, $L_0 = L_{n-n} \subseteq Z_n(L)$. This shows that $Z_n(L) = L$. Similarly, assuming that the upper central series terminates to L at the nth step, we can show that the lower central series of L terminates to $\{0\}$ at the nth step. ♯

Definition 1.3.12 A Lie algebra L is said to be a **nilpotent Lie algebra** if its lower central series terminates to $\{0\}$ after finitely many steps or equivalently, its upper central series terminates to L after finitely many steps. We say that it is nilpotent of class n if $L_n = \{0\}$ but $L_{n-1} \neq \{0\}$.

Evidently, every nilpotent Lie algebra is solvable, whereas $t(n, F)$ is solvable but it is not nilpotent. Indeed, $[t(n, F), t(n, F)] = \nu(n, F)$, and $[t(n, F), \nu(n, F)] = \nu(n, F)$. Clearly, Lie subalgebra and homomorphic images of nilpotent Lie algebras are nilpotent. The following proposition is the immediate consequence of Proposition 1.3.11.

Proposition 1.3.13 *(i) If L is a nontrivial nilpotent Lie algebra, then $Z(L) \neq \{0\}$. (ii) A Lie algebra L is nilpotent if and only if $L/Z(L) \approx ad(L)$ is nilpotent.* ♯

If L is nilpotent of index n, then $ad(x)^n(y) = 0$ for all $x, y \in L$. This means that the endomorphism $ad(x)$ is nilpotent for all $x \in L$. We shall show that the converse of this is also true.

Theorem 1.3.14 *Let L be a Lie subalgebra of $gl(V)$, where V is a finite-dimensional vector space. Suppose that all the members of L are nilpotent endomorphisms V. Then there is a common eigenvector v_0 of members of L corresponding to the eigenvalue 0. More explicitly, there is a nonzero vector $v_0 \in V$ such that $x(v_0) = 0$ for all $x \in L$.*

Proof The proof is by the induction on $dim\ L$. If $dim\ L = 0$, then there is nothing to do. If $dim\ L = 1$, then $L = Fx$ for some nonzero member x of L. Since x is nilpotent, 0 is an eigenvalue of x. Hence there is a nonzero vector v_0 of V such that $x(v_0) = 0$. Evidently, v_0 is an eigenvector of each member of L corresponding to the eigenvalue 0. Assume that the result is true for all those Lie algebras whose dimensions are less than $dim\ L$. Let A be a proper nontrivial Lie subalgebra of L. Then $\{ad(x) \mid x \in A\}$ is a Lie subalgebra of $End(L) = gl(L)$ of dimension less than that of L. Further, $ad(x)(y) = [x, y] = xy - yx$ for all $x \in A,\ y \in L$. As such $ad(x) = (l_x - r_x)|_L$, where l_x is the left multiplication by x in $End(V)$ and r_x is the right multiplication by x. Since x is nilpotent, l_x and r_x are nilpotent. Clearly, l_x and r_x commute with each other. This means that $l_x - r_x$ is nilpotent endomorphism of $End(End(V))$. Hence $ad(x)$ is nilpotent. Indeed, for each $x \in A$, $ad(x)$ induces an endomorphism $\widehat{ad}(x)$ of $gl(L/A)$ given by $\widehat{ad}(x)(y + A) = [x, y] + A$ which is nilpotent. Thus, $\{\widehat{ad}(x) \mid x \in A\}$ is a Lie subalgebra of $gl(L/A)$ consisting of nilpotent endomorphisms whose dimension is less than that of L. By the induction hypothesis, there is a member $y \in L - A$ such that $[x, y] \in A$ for all $x \in A$. This means that A is properly contained in the normalizer $N_L(A)$ of A. Thus, every maximal Lie subalgebra of L is an ideal. Consequently, every maximal Lie subalgebra A of L is of co-dimension 1. Let M be a maximal Lie subalgebra of L. Then $L = M + Fx$ for some nonzero member x of L. Also $[x, M] \subseteq M$. Consider the subspace $W = \{v \in V \mid y(v) = 0\ \forall y \in M\}$ of V. By the induction hypothesis, $W \neq \{0\}$. Further, $y(x(w)) = x(y(w)) - [x, y](w) = 0$ for all $w \in W$ and $y \in M$. This means that x restricted to W is an endomorphism of W which, of course, is nilpotent. Hence there is a member v_0 of W such that $x(v_0) = 0$. Evidently, v_0 is a common eigenvector of each member of L corresponding to the eigenvalue 0. ♯

Corollary 1.3.15 *Under the hypothesis of the above theorem, there is a flag*

$$\{0\} \subseteq V_1 \subseteq V_2 \subseteq \cdots \subseteq V_{n-1} \subseteq V_n = V$$

which is stable under the endomorphisms in L, and even more, $x(V_i) \subseteq V_{i-1}$ for each i. Consequently, there is a basis of V with respect to which the matrix representation of each member in L is strictly lower triangular.

Proof The result follows by the induction on the dimension of V, thanks to the above theorem. ♯

Theorem 1.3.16 *(Engel) Let L be a Lie algebra which is ad-nilpotent in the sense that $ad(x)$ is nilpotent for each $x \in L$. Then L is nilpotent.*

Proof The proof is by the induction on the $dim\ L$. If $dim\ L$ is 0 or 1, then there is nothing to do. Assume that the result is true for all those Lie algebras whose dimensions are less than that of L. Clearly, $ad(L)$ is a Lie subalgebra of $gl(L)$. By the hypothesis, each member of $ad(L)$ are nilpotent endomorphisms of L. From Theorem 1.3.14, there is a nonzero member y of L such that $ad(x)(y) = [x, y] = 0$ for all $x \in L$. This means that $y \in Z(L)$, and $Z(L) \neq \{0\}$. Hence $L/Z(L) \approx ad(L)$ is of dimension less than that of L. Clearly, $L/Z(L)$ is ad-nilpotent. By the induction hypothesis, $L/Z(L)$ is nilpotent. By Proposition 1.3.13, L is nilpotent. ♯

Corollary 1.3.17 *Let L be a nilpotent Lie algebra, and A be a nontrivial ideal of L. Then $A \bigcap Z(L) \neq \{0\}$.*

Proof $\{ad(x)|_A \mid x \in L\}$ is a Lie subalgebra of $gl(A)$ consisting of nilpotent endomorphisms. From Theorem 1.3.14, there is a nonzero element $y \in A$ such that $ad(x)(y) = 0$ for all $x \in L$. This means that $y \in A \bigcap Z(L)$. ♯

Corollary 1.3.18 *A Lie algebra L is solvable if and only if $[L, L]$ is nilpotent.*

Proof If $[L, L]$ is nilpotent, then $[L, L]$ and $L/[L, L]$ both are solvable. Hence L is solvable. Conversely, suppose that L is solvable. From Corollary 1.3.9, $ad(x)$ is nilpotent for each $x \in [L, L]$. This means that $[L, L]$ is ad-nilpotent. The result follows from the theorem of Engel. ♯

Remark 1.3.19 Note that in the case of groups, $[G, G]$ need not be nilpotent even if G is solvable. For example, S_4 is solvable, but $[S_4, S_4] = A_4$ is not nilpotent.

Our next aim is to establish Trace criteria for the solvability of a linear Lie algebra.

We first recall the Jordan–Chevalley decomposition theorem (cf. Sect. 6.3, Algebra 2) for this purpose.

Let V be a finite-dimensional vector space over an algebraically closed field (not necessarily of characteristic 0). An element $x \in gl(V) = End(V)$ is called a semi-simple element if it is diagonalizable in the sense that the matrix representation of x with respect to a suitable basis of V is a diagonal matrix. A necessary and sufficient condition for x to be diagonalizable is that all the roots of the minimal polynomial of x are distinct. An element $x \in gl(V)$ is nilpotent if $x^m = 0$ for some m. The Jordan–Chevalley decomposition theorem asserts that any element $x \in gl(V)$ is uniquely expressible as

$$x = x_s + x_n$$

such that the following hold:

(i) x_s is semi-simple and x_n is nilpotent.

(ii) x_s and x_n commute with each other.

Further, it also follows that

(iii) $x_s = p(x)$ and also $x_n = q(x)$, where $p(X)$ and $q(X)$ are polynomials without constant terms.

x_s is called the **semi-simple part** and x_n is called the **nilpotent part** of x.

Proposition 1.3.20 *Let V be a finite-dimensional vector space over a field F. If x is a nilpotent element of $End(V) = gl(V)$, then $ad(x)$ is nilpotent in $End(End(V))$. Again if x is a semi-simple element of $End(V)$, then $ad(x)$ is a semi-simple element of $End(End(V))$. Further, suppose that F is algebraically closed and $x = x_s + x_n$ is the Jordan–Chevalley decomposition of x in $End(V)$. Then $ad(x) = ad(x_s) + ad(x_n)$ is the Jordan–Chevalley decomposition of $ad(x)$ in $End(End(V))$.*

Proof Let x be a semi-simple element in $End(V)$. Then there is a basis $\{v_1, v_2, \cdots, v_n\}$ of V and scalars $\{\lambda_1, \lambda_2, \cdots, \lambda_n\}$ such that $x(v_i) = \lambda_i v_i$ for all i. We have a corresponding basis $\{\hat{v_{ij}} \mid 1 \leq i, j \leq n\}$ of $End(V)$, where $\hat{v_{ij}}$ is the member of $End(V)$ given by $\hat{v_{ij}}(v_j) = v_i$ and $\hat{v_{ij}}(v_k) = 0$ for $k \neq j$. Clearly,

$$ad(x)(\hat{v_{ij}})(v_l) = x(\hat{v_{ij}}(v_l)) - \hat{v_{ij}}(x(v_l)) = \delta_{jl}(\lambda_i - \lambda_j)v_i,$$

where δ_{lj} is the Kronecker delta. This shows that $ad(x)(\hat{v_{ij}}) = (\lambda_i - \lambda_j)\hat{v_{ij}}$. This shows that $ad(x)$ is also semi-simple.

Next, suppose that x is a nilpotent element of $gl(V)$. Then $x^m = 0$ for some m. Now, $ad(x)(y) = xy - yx$. Thus, $ad(x) = l_x - r_x$, where l_x is the left multiplication by x and r_x is the right multiplication by x in $End(V)$. Clearly, l_x and r_x commute and $l_x^m = 0 = r_x^m$. Using the binomial theorem, $ad(x)^{2m+1} = (l_x - r_x)^{2m+1} = 0$.

Finally, suppose that $x = x_s + x_n$ is the Jordan–Chevalley decomposition of x. Then $ad(x) = ad(x_s) + ad(x_n)$, where $ad(x_s)$ is a semi-simple element in $End(End(V))$ and $ad(x_n)$ is a nilpotent element in $End(End(V))$. Also $[ad(x_s), ad(x_n)] = ad([x_s, x_n]) = 0$, since x_s and x_n commute. This means that $ad(x_s)$ and $ad(x_n)$ also commute. It follows that $ad(x) = ad(x_s) + ad(x_n)$ is the Jordan–Chevalley decomposition of $ad(x)$. ♯

Proposition 1.3.21 *Let d be a derivation on an algebra A (not necessarily associative) over a field F (not necessarily algebraically closed). Then for any $\lambda, \mu \in F$,*

$$(d - (\lambda + \mu)I)^n(xy) = \sum_{r=0}^{n} {}^nC_r(d - \lambda I)^r(x)(d - \mu I)^{n-r}(y)$$

for all $x, y \in A$.

Proof The proof is by the induction on n. Clearly,

$$(d-(\lambda+\mu)I)(xy) = d(xy) - \lambda(xy) - \mu(xy) =$$
$$d(x)y + xd(y) - \lambda(xy) - \mu(xy) = x(d-\mu I)(y) + (d-\lambda I)(x)y$$

for all $x, y \in A$. Thus, the identity holds for $n = 1$. Assume that the identity holds for n. Then
$(d - (\lambda+\mu)I)^{n+1}(xy)$
$= (d - (\lambda+\mu)I)^n((d - (\lambda+\mu)I)(xy))$
$= (d - (\lambda+\mu)I)^n(x(d-\mu I)(y) + (d-\lambda I)(x)y)$
$= (d - (\lambda+\mu)I)^n(x(d-\mu I)(y)) + (d - (\lambda+\mu)I)^n((d-\lambda I)(x)y)$
$= \sum_{r=0}^{n} {}^nC_r(d-\lambda I)^r(x)(d-\mu I)^{n-r}((d-\mu I)(y)) + \sum_{r=0}^{n} {}^nC_r(d-\lambda I)^r$
$((d-\lambda I)(x))(d-\mu I)^{n-r}(y)$
$= \sum_{r=0}^{n+1} {}^{n+1}C_r(d-\lambda I)^r(x)(d-\mu I)^{n+1-r}(y)$
for all $x, y \in A$. This completes the proof of the identity. ♯

Proposition 1.3.22 *Let A be an algebra (not necessarily associative) over an algebraically closed field F. Let d be a derivation on A. Let* $d = d_s + d_n$ *be the Jordan–Chevalley decomposition of d considered as an element of* $End_F(A)$. *Then* d_s *and* d_n *are also derivations of A.*

Proof It is sufficient to show that d_s is a derivation of A. For each $\lambda \in F$, let A_λ denote the λ-eigenspace of d. More explicitly, $A_\lambda = \{x \in A \mid (d-\lambda I)^m(x) = 0 \; for \; some \; m\}$. A_λ may be trivial also. Then as a vector space $A = \oplus\sum_{\lambda\in F} A_\lambda$. Note that eigenvalues of d are the same as those of d_s. From the above proposition, it follows that $A_\lambda A_\mu \subseteq A_{\lambda+\mu}$. Thus, if $x \in A_\lambda$ and $y \in A_\mu$, then $d_s(xy) = (\lambda+\mu)xy = d_s(x)y + xd_s(y)$. Since $A = \oplus\sum_{\lambda\in K} A_\lambda$, $d_s(xy) = d_s(x)y + xd_s(y)$ for all $x, y \in A$. This shows that d_s is also a derivation of A. ♯

Corollary 1.3.23 *If A is an algebra (not necessarily associative) over an algebraically closed field F, then the Lie algebra Der(A) of derivations on A contains the semi-simple and nilpotent parts of each of its elements considered as elements of* $End_F(A)$. ♯

Theorem 1.3.24 (Cartan's Trace Criteria) *Let L be a Lie subalgebra of gl(V), where V is a finite-dimensional vector space over an algebraically closed field of characteristic 0. Suppose that* $Tr(xy) = 0$ *for all* $x \in [L, L]$ *and* $y \in L$. *Then L is solvable.*

Before proving the theorem, we establish a lemma for this purpose.

Lemma 1.3.25 *Let x be a member of gl(V), where V is a finite-dimensional vector space over an algebraically closed field F of characteristic 0. Suppose that there are two subspaces W and U of V such that* $[x, W] \subseteq U$, *and* $Tr(xy) = 0$ *whenever* $[y, W] \subseteq U$. *Then x is nilpotent.*

Proof Let x be a member of $gl(V)$ satisfying the hypothesis of the lemma. Suppose that $x = x_s + x_n$ is the Jordan–Chevalley decomposition of x. We need to show that $x_s = 0$. Let $\{v_1, v_2, \cdots, v_m\}$ be a basis of V consisting of the eigenvectors of x_s corresponding to the eigenvalues $\lambda_1, \lambda_2, \cdots, \lambda_m$, respectively. We are required to show that each λ_i is 0. Suppose not. Consider F as the vector space over its prime field $\mathbb{Q}$. Let K be the $\mathbb{Q}$-subspace of F generated by the subset $\{\lambda_1, \lambda_2, \cdots, \lambda_m\}$ of F. Then $K \neq \{0\}$. Let f be a nonzero linear functional on K. Then $f(\lambda_{i_0}) \neq 0$ for at least one i_0. As in the proof of Proposition 1.3.20, we have the basis $\{\hat{v_{ij}} \mid 1 \leq i, j \leq m\}$ of $End(V)$ consisting of the eigenvectors of $ad(x_s)$ corresponding to the eigenvalues $\lambda_i - \lambda_j,\ 1 \leq i, j \leq m$. Let y be another semi-simple element of $gl(V)$ having the eigenvectors v_i with corresponding eigenvalues $f(\lambda_i)$. Then again as in the proof of Proposition 1.3.20, $\hat{v_{ij}}$ are the eigenvectors of the semi-simple element $ad(y)$ of $End(V)$ corresponding to the eigenvalues $f(\lambda_i) - f(\lambda_j)$. Since f is a linear functional, $\lambda_i - \lambda_j = \lambda_k - \lambda_l$ implies that $f(\lambda_i) - f(\lambda_j) = f(\lambda_k) - f(\lambda_l)$. Thus, using the Lagrange interpolation, we can find a polynomial $\phi(X)$ such that $\phi(\lambda_i - \lambda_j) = f(\lambda_i) - f(\lambda_j)$. Since $\phi(ad(x_s))$ and $ad(y)$ are semi-simple elements in $End(End(V))$ with the same eigenvalues corresponding to the eigenvectors $\hat{v_{ij}}$, $\phi(ad(x_s)) = ad(y)$. Since $ad(x_s)$ is a semi-simple part of $ad(x)$ (Proposition 1.3.20), there is a polynomial $p(X)$ without constant term such that $p(ad(x)) = ad(x_s)$. In turn, we have the polynomial $\psi(X) = \phi(p(X))$ without constant term such that $\psi(ad(x)) = ad(y)$. Since $ad(x)(W) \subseteq U$, it follows that $ad(y)(W) \subseteq U$. By the hypothesis, $Tr(xy) = 0$. Thus, $\sum_{i=1}^{m} \lambda_i f(\lambda_i) = 0$. Since f is a linear functional on K, $f(\sum_{i=1}^{m} \lambda_i f(\lambda_i)) = \sum_{i=1}^{m}(f(\lambda_i))^2 = 0$. This is a contradiction, since $f(\lambda_{i_0}) \neq 0$. ♯

Proof of Theorem 1.3.24. From Corollary 1.3.18 and the fact that every nilpotent Lie algebra is solvable, it is sufficient to show that $[L, L]$ is a nilpotent Lie algebra. In turn, by Engel's theorem, it is sufficient to show that $[L, L]$ is ad-nilpotent. More so, it is sufficient to show that the members of $[L, L]$ are nilpotent endomorphisms in $End(V)$. Take $W = L$ and $U = [L, L]$ in the above lemma. Consider the element $[x, y] \in [L, L],\ x, y \in L$ and the element $z \in gl(V)$ with the property that $[z, L] \subseteq [L, L]$. Now, $Tr([x, y]z) = Tr((xy - yx)z) = Tr(xyz) - Tr(yxz) = Tr(xyz) - Tr(xzy) = Tr(x[y, z]) = Tr([y, z]x)$ for all $x, y, z \in gl(V)$. Thus, if $[z, L] \subseteq L$, then by the hypothesis of the theorem, for all $x, y \in L$, $Tr([y, z]x) = 0$. In turn, $Tr([x, y]z) = 0$ for all $x, y \in L$ and $z \in gl(V)$ for which $[z, L] \subseteq [L, L]$. Since any element of $[L, L]$ is a linear combination of members of the type $[x, y],\ x, y \in L$, it follows that $Tr(uz) = 0$ for all $z \in gl(V)$ for which $[z, L] \subseteq [L, L]$ and $u \in [L, L]$. It follows from the lemma that u is nilpotent for all $u \in [L, L]$. ♯

Corollary 1.3.26 *Let L be a finite-dimensional Lie algebra over an algebraically closed field F of characteristic 0 such that* $Tr(ad(x)ad(y)) = 0$ *for all* $x, y \in L$. *Then L is solvable.*

Proof Consider the subalgebra $ad(L)$ of $gl(L)$. Under the hypothesis, it follows from Cartan's criteria that $ad(L)$ is solvable. Thus, $L/Z(L)$ is solvable. Since $Z(L)$ is solvable, it follows that L is solvable. ♯

Exercises

1.3.1. Show that every Lie algebra of dimension 2 is a solvable Lie algebra. Give an example of a two-dimensional Lie algebra which is not nilpotent.
1.3.2. Show that L is solvable(nilpotent) if and only if $ad(L)$ is solvable (nilpotent).
1.3.3. Let L be a finite-dimensional nilpotent Lie algebra. Show that there is a natural number n such that $N_L^n(A) = L$ for all Lie subalgebra A of L. What about the converse?
1.3.4. Show that sum of two nilpotent ideals is a nilpotent ideal. Deduce that every finite-dimensional Lie algebra has the unique largest nilpotent ideal.
1.3.5. Let L be a nilpotent Lie algebra of finite dimension. Show that it has an ideal of co-dimension 1.
1.3.6. Let L be a finite-dimensional Lie algebra over an algebraically closed field F of characteristic 0. Let A be an ideal of L such that L/A is nilpotent and $ad(x)|_A$ is nilpotent for all $x \in L$. Show that L is nilpotent.
1.3.7. Show by means of an example that Lie's theorem is not true for fields of positive characteristics. However, if the dimension of V is less than the characteristic of the field, then show that it is true.
1.3.8. Check if the Lie algebra $(\mathbb{R}^3, \times)$ over $\mathbb{R}$ is solvable.
1.3.9. Show that every nilpotent Lie algebra has an outer derivation. What about solvable Lie algebras?
1.3.10. Show that the radical of each classical Linear Lie algebra described in Sect. 1 is the center of the corresponding Lie algebra.

1.4 Semi-Simple Lie Algebras

This section is devoted to the basic structure theory including the basic representation theory of semi-simple Lie algebras. All fields will be assumed to be algebraically closed fields of characteristic 0 unless stated otherwise. Recall that a finite-dimensional Lie algebra L is semi-simple if its radical is trivial. As for solvable Lie algebras, we give a trace criteria for a Lie algebra to be semi-simple.

Definition 1.4.1 Let L be a Lie algebra over F. The symmetric bi-linear form κ on L defined by $\kappa(x, y) = Tr(ad(x)ad(y))$ is called the **Killing form** on L.

Thus, if L is an abelian Lie algebra, then $ad(x) = 0$ for each x and so κ is trivial on L. If L is a non-abelian two-dimensional Lie algebra, then there is a basis $\{x, y\}$ of L such that $[x, y] = x$. It can be easily checked that the matrix of κ with respect to the basis $\{x, y\}$ is

$$\kappa = \begin{bmatrix} 0 & 0 \\ 0 & 1 \end{bmatrix}.$$

Recall the Lie algebra $sl(2, F)$ in Example 1.1.7. It has a basis $\{e_{12}, e_{21}, h\}$ and it satisfies the relations $[e_{12}, e_{21}] = h,\ [h, e_{12}] = 2e_{12}$ and $[h, e_{21}] = -2e_{21}$. It can be easily checked that the matrix of the Killing form κ with respect to this basis is given by

$$\kappa = \begin{bmatrix} 0 & 0 & 4 \\ 0 & 4 & 0 \\ 8 & 0 & 0 \end{bmatrix}.$$

Recall that a symmetric bi-linear form f on a vector space V is called a nondegenerate bi-linear form if $f(x, y) = 0$ for all $y \in V$ implies that $x = 0$. This is equivalent to say that the matrix of f with respect to a basis is non-singular. Thus, the Killing form κ on $sl(2, F)$ is nondegenerate if and only if the characteristic of F is different from 2.

Proposition 1.4.2 *The Killing form κ on a Lie algebra L is associative in the sense that $\kappa([x, y], z) = \kappa(x, [y, z])$ for all $x, y, z \in L$.*

Proof $\kappa([x, y], z) = Tr(ad([x, y])ad(z)) = Tr([ad(x), ad(y)]ad(z)) = Tr(ad(x)ad(y)ad(z) - ad(y)ad(x)ad(z)) = Tr(ad(x)ad(y)ad(z) - ad(x)ad(z)ad(y)) = Tr(ad(x)ad([y, z])) = \kappa(x, [y, z])$ for all $x, y, z \in L$. ♯

Proposition 1.4.3 *Let L be a finite-dimensional Lie algebra over F. Let A be an ideal of L. Then the restriction of the Killing form κ of L to the ideal A is in fact the Killing form on A.*

Proof Let $\{x_1, x_2, \cdots, x_r\}$ be a basis of A. Extend it to a basis $\{x_1, x_2, \cdots, x_r, x_{r+1}, \cdots, x_n\}$ of L. Let $x, y \in A$. Since A is an ideal, $ad(x)ad(y)$ maps L into A. Thus, the trace of $(ad(x)ad(y))$ considered as a linear transformation is the same as the trace of $ad(x)|_A ad(y)|_A$. The result follows. ♯

The following result describes a semi-simple Lie algebra in terms of its Killing form.

Theorem 1.4.4 *Let L be a nontrivial Lie algebra. Then L is semi-simple if and only if its Killing form κ is nondegenerate.*

Proof Suppose that L is semi-simple. Then the radical $R(L)$ of L is $\{0\}$. Consider the subset $A = \{x \in L \mid \kappa(x, y) = Tr(ad(x)ad(y)) = 0 \ \forall y \in L\}$. Since κ is bi-linear, A is a subspace of L. Let $x \in A$ and $y \in L$. Then $\kappa(x, z) = 0$ for all $z \in L$. From the associative property of κ, $\kappa([x, y], z) = \kappa(x, [y, z]) = 0$ for all $z \in L$. This shows that $[x, y] \in A$ for all $y \in L$. Hence A is an ideal of L. In particular, $Tr(ad(x)ad(y)) = 0$ for all $x \in [A, A]$ and $y \in L$. By Cartan's criteria, $ad(A)$ is solvable. Hence A is solvable. Since L is semi-simple, $A = \{0\}$. This shows that κ is nondegenerate.

Conversely, suppose that κ is nondegenerate. Then the ideal A given above is trivial. Let B be an abelian ideal of L. Then $ad(x)ad(y)$ is a linear transformation from L to B for all $x \in B$. Since B is abelian, $(ad(x)ad(y))^2 = 0$ for all $x \in B$ and $y \in L$. This means that all the eigenvalues of $ad(x)ad(y)$ are 0. In particular, $\kappa(x, y) = 0$ for all $x \in B$ and $y \in L$. This shows that $B \subseteq A = \{0\}$. It follows that the radical of L is $\{0\}$. ♯

Remark 1.4.5 In the proof of the converse part of the above theorem, we have not used the fact that characteristic F is 0. Hence the nondegeneracy of the Killing form implies the semi-simplicity for Lie algebras over an algebraically closed field of arbitrary characteristic. Observe that in the proof of the direct part, we have used Cartan's criteria which requires that the field is of characteristic 0.

Representations of Lie Algebras

Let L be a Lie algebra over a field F. Let V be a vector space over F. A Lie homomorphism ρ from L to the Lie algebra $gl(V)$ is called a **representation** of L on V. If V is finite dimensional, then we term it as a finite-dimensional representation of L. The dimension of V is called the degree of the representation ρ. It is also convenient to view representations of L as L-modules. A module over a Lie algebra L over a field F is a vector space V over F together with an external operation $\cdot$ from $L \times V$ to V (the image of (x, v) under $\cdot$ is denoted by $x \cdot v$ or simply by the juxtaposition xv) such that the following hold:

(i) $(\alpha x + \beta y) \cdot v = \alpha(x \cdot v) + \beta(y \cdot v)$,
(ii) $x \cdot (\alpha v + \beta w) = \alpha(x \cdot v) + \beta(x \cdot w)$, and
(iii) $[x, y] \cdot v = x \cdot (y \cdot v) - y \cdot (x \cdot v)$ for all $x, y \in L$; $v, w \in V$, *and* $\alpha, \beta \in F$.
A module over L is also termed as L-module. Thus, a representation ρ of L on V determines and is determined uniquely by an L-module structure on V by putting

$$\rho(x)(v) \equiv x \cdot v,$$

$x \in L,\ v \in V$. Without any loss, representations of a Lie algebra L and L-modules can be treated as synonyms. Every representation has the associated module and every module has the associated representation. Evidently, the vector space part of the Lie algebra L is an L-module with respect to the Lie product. More explicitly, if we put $x \cdot y = [x, y]$, then the bi-linearity of the Lie product and the Jacobi identity ensure that L is a module over itself. The representation associated with the L-module structure on L is precisely the adjoint representation ad.

On the pattern of modules over rings, we can develop the language of modules over a Lie algebra. We recall it quickly. Let V and W be L-modules. A linear transformation ϕ from V to W is called an L-homomorphism if $\phi(x \cdot v) = x \cdot \phi(v)$ for all $x \in L$ and $v \in V$. If ρ is the representation associated with the L-module V, and η is the representation associated with the L-module W, then ϕ is an L-homomorphism from V to W if and only if $\eta(x)o\phi = \phi o\rho(x)$ for all $x \in L$. Such a ϕ is also termed as an intertwining operator from the representation ρ to the representation η. If ϕ is also an isomorphism, then $\eta(x) = \phi o\rho(x)o\phi^{-1}$ for all $x \in L$, and we say that ρ and η are equivalent representations. We have the category L-Mod of L-modules which is isomorphic to the category of representations of L. An isomorphism in this category are bijective homomorphisms, a monomorphism is an injective homomorphism, and an epimorphism is simply a surjective homomorphism.

Let V be an L-module. A subspace W of V is called an L-submodule if $x \cdot w \in W$ for all $x \in L$ and $w \in W$. Thus, W is also an L-module in its own right.

Consequently, a representation η of L on W is a subrepresentation of a representation ρ on V if W is a subspace of V and $\eta(x) = \rho(x)|_W$ for all $x \in L$. Evidently, the submodules of L considered as a module over itself are precisely the ideals of L. If W is an L-submodule of V, then the quotient space V/W is also an L-module with respect to the external operation $\cdot$ given by $x \cdot (v + W) = (x \cdot v) + W$. This module is called the quotient module. The quotient map ν from V to V/W is an epimorphism in L-Mod. An L-module V is called a **simple** L-module if it is neither zero dimensional nor one dimensional and it has no nontrivial proper L-submodules (note that a one-dimensional module has no proper submodules). A simple L-module is also termed as an irreducible L-module. A representation associated with a simple L-module is called an irreducible representation of L. The correspondence theorem, the isomorphism theorems, and the Jordan–Holder theorem for finite-dimensional modules can be established by imitating the proofs of the corresponding results in the theory of modules over rings.

If V_1 and V_2 are L-modules, then we have an L-module structure on the direct sum $V_1 \oplus V_2$ given by $x \cdot (v_1 \oplus v_2) = (x \cdot v_1) \oplus (x \cdot v_2)$. This also defines the direct sum of two representations. An L-submodule W of an L-module V is called a direct summand if there is an L-submodule U of V such that $L = W \oplus U$.

We construct different representations of L (L-modules) in terms of given representations of L (L-modules). Let ρ be a representation of L on V and η be a representation of L on W. We have a map $Hom(\rho, \eta)$ from L to $gl(Hom(V, W))$ given by

$$Hom(\rho, \eta)(x)(f) = \eta(x)of - fo\rho(x), \ x \in L, \ f \in Hom(V, W).$$

Clearly, $Hom(\rho, \eta)$ is a linear map. Now,

$$Hom(\rho, \eta)([x, y])(f) = \eta([x, y])of - fo\rho([x, y]) = \\ (\eta(x)o\eta(y) - \eta(y)o\eta(x))of - fo(\rho(x)o\rho(y) - \rho(y)o\rho(x)),$$

for all $x, y \in L$ and $f \in Hom(V, W)$. Also,

$$[Hom(\rho, \eta)(x), Hom(\rho, \eta)(y)](f) = (Hom(\rho, \eta)(x)oHom(\rho, \eta)(y) - \\ Hom(\rho, \eta)(y)oHom(\rho, \eta)(x))of = \eta(x)o(\eta(y)of - fo\rho(y)) - (\eta(y)of - \\ fo\rho(y))o\rho(x) - \eta(y)o(\eta(x)of - fo\rho(x)) + (\eta(x)of) - fo\rho(x))o\rho(y) = \\ (\eta(x)o\eta(y) - \eta(y)o\eta(x))of - fo(\rho(x)o\rho(y) - \rho(y)o\rho(x)) = \\ Hom(\rho, \eta)([x, y])(f)$$

for all $x, y \in L$ and $f \in Hom(V, W)$. Thus,

$$Hom(\rho, \eta)[x, y] = [Hom(\rho, \eta)(x), Hom(\rho, \eta)(y)]$$

for all $x, y \in L$. This shows that $Hom(\rho, \eta)$ is a representation of L. In the language of modules, $Hom(V, W)$ is an L-module with respect to the external product $\cdot$

given by $(x \cdot f)(v) = x \cdot f(v) - f(x \cdot v)$, $x \in L$ and $v \in V$. In the language of category theory, Hom defines a bi-functor from the category of representations of L (of L-modules) to itself. Further, given a representation ρ of L on V, we have the representation $\rho^\star$ of L on $Hom(V, K) = V^\star$ given by $\rho^\star(x)(f) = -fo\rho(x)$. This representation is called the **dual** representation of ρ.

Let ρ and η be representations of L on V and W, respectively. Consider the tensor product $V \otimes_F W$. Define a map $\rho \otimes \eta$ from L to $gl(V \otimes_F W)$ by

$$(\rho \otimes \eta)(x)(v \otimes w) = \rho(x)(v) \otimes w + v \otimes \eta(w).$$

Evidently, $\rho \otimes \eta$ is a linear map from L to $gl(V \otimes_F W)$. Now,

$$\begin{aligned}(\rho \otimes \eta)([x, y])(v \otimes w) &= \rho([x, y])(v) \otimes w + v \otimes \eta([x, y])(w) = \\ (\rho(x)\rho(y) - \rho(y)\rho(x))(v) \otimes w &+ v \otimes (\eta(x)\eta(y) - \eta(y)\eta(x))(w) = \\ ((\rho \otimes \eta)(x)(\rho \otimes \eta)(y) &- (\rho \otimes \eta)(x)(\rho \otimes \eta)(y))(v \otimes w)\end{aligned}$$

for all $x, y \in L$, $v \in V$ and $w \in W$. Thus,

$$(\rho \otimes \eta)([x, y]) = [(\rho \otimes \eta)(x), (\rho \otimes \eta)(y)]$$

for all $x, y \in L$. This shows that $\rho \otimes \eta$ is a representation of L. The representation $\rho \otimes \eta$ is called the **tensor product** of ρ and η.

As in group representations, the following lemma of Schur is very crucial and important in the representation theory of Lie algebras.

Lemma 1.4.6 *(Schur)(i) Let V and W be simple L-modules. Then any L-homomorphism from V to W is either a zero homomorphism or an isomorphism.*

(ii) Let L be a Lie algebra over an algebraically closed field K and V be a simple L-module. Then any L-endomorphism of V is the multiplication by a scalar. As such, $End_L(V)$ is a field isomorphic to F.

Proof (i) Let f be an L-homomorphism from V to W, where V and W are simple L-modules. Then $ker\ f$ is an L-submodule of V. If $ker\ f = V$, then f is zero homomorphism. Suppose that $ker\ f \neq V$. Since V is simple, $ker\ f = \{0\}$ and so, f is injective. Further, $f(V)$ is an L-submodule of W and since W is also simple, $f(V) = W$. This shows that f is an isomorphism.

(ii) Let f be a member of $End_L(V)$, where V is a simple L-module. Since F is an algebraically closed field, f has an eigenvalue λ. Let V_λ denote the λ-eigensubspace of V. Then $V_\lambda \neq \{0\}$. If $v \in V_\lambda$, then $f(x \cdot v) = x \cdot f(v) = x \cdot \lambda v = \lambda(x \cdot v)$. Hence $x \cdot v \in V_\lambda$ for all $x \in L$ and $v \in V_\lambda$. Since V is a simple L-module, $V_\lambda = V$. The last assertion is immediate. ♯

Proposition 1.4.7 *Let L be a Lie algebra over a field F, and V be a finite-dimensional L-module. Then the following two conditions are equivalent.*

1. Every submodule of V is a direct summand.

2. *V is the direct sum of simple submodules.*

Proof ($1 \Longrightarrow 2$) Assume 1. Let W be a submodule of V. Let U be a submodule of W. Then there is a submodule U' of V such that $V = U \oplus U'$. Clearly, $W = U \oplus (W \bigcap U')$. Thus, every submodule of V also satisfies condition 1. The proof of 2 is by the induction on the dimension of V. If the dimension of V is 0 or 1, then there is nothing to do. Assume that condition 2 holds for all L-modules satisfying 1 whose dimensions are less than the dimension of V, and V satisfies 1. If V is simple, then there is nothing to do. If not, there is a nonzero proper L-submodule W of V. From 1, there is a submodule W' of V such that $V = W \oplus W'$. Evidently, the dimensions of W and W' are less than the dimension of V. Further, from our earlier observation, W and W' both satisfy 1. By the induction hypothesis, W and W' are both direct sums of simple modules. Hence V is the direct sum of simple submodules.

($2 \Longrightarrow 1$) Assume 2. Suppose that $V = V_1 \oplus V_2 \oplus \cdots \oplus V_r$, where V_i is simple for each i. Let W be a nontrivial proper submodule of V. There is i_1, $1 \leq i_1 \leq r$ such that $W \bigcap V_{i_1} = \{0\}$ for otherwise $V_i \bigcap W = V_i$ and $V_i \subseteq W$ for each i. If $V = W + V_{i_1} = W \oplus V_{i_1}$, we are done. If not, there is $i_2 \neq i_1$ such that $(W \oplus V_{i_1}) \bigcap V_{i_2} = \{0\}$. But then $W + V_{i_1} + V_{i_2} = W \oplus V_{i_1} \oplus V_{i_2}$. If $V = W \oplus V_{i_1} \oplus V_{i_2}$, we are done. Proceed inductively. This process stops after finitely many steps giving a submodule W' such that $V = W \oplus W'$. ♯

Definition 1.4.8 A finite-dimensional L-module V is said to be **completely reducible** or **semi-simple** if it satisfies any of the above two (and hence both) conditions.

Remark 1.4.9 Observe (see Algebra 2, Theorem 9.1.5) that for modules over rings, the equivalence of the two conditions holds without any restriction of finiteness on the module. Can we imitate the proof for modules over rings to relax the condition of finite dimensionality of V?

It is clear from Proposition 1.4.7 that submodules, quotient modules, and homomorphic images of semi-simple modules are semi-simple modules. A finite-dimensional L-module V is semi-simple if and only if every short exact sequence of L-modules of the type

$$0 \longrightarrow U \stackrel{\alpha}{\to} V \stackrel{\beta}{\to} W \longrightarrow 0$$

splits.

Theorem 1.4.10 *Let L be a semi-simple Lie algebra. Then L considered as L-module is semi-simple. There exist simple ideals $A_1, A_2, \cdots, A_m$ of L which are simple as Lie algebras such that $L = A_1 \oplus A_2 \oplus, \cdots, \oplus A_r$. Also $\{A_1, A_2, \cdots, A_m\}$ is the set of all simple ideals of L. Further, the restrictions of the Killing form of L on A_i is the Killing form of A_i for each i.*

Proof We consider the semi-simple Lie algebra L as an L-module. Let A be a submodule of L. Then A is an ideal of L. Consider $A^{\perp} = \{x \in L \mid \kappa(x, y) =$

$0 \forall y \in A\}$, where κ is the Killing form on L. The associativity of κ ensures that $A^{\perp}$ is also an ideal. Cartan's criteria implies that $A \bigcap A^{\perp}$ is a solvable ideal of L. Since L is semi-simple, it follows that $A \bigcap A^{\perp} = \{0\}$. Suppose that $A + A^{\perp} \neq L$. Let $\{x_1, x_2, \cdots, x_r, x_{r+1}, \cdots, x_n\}$ be a basis of L with $\{x_1, x_2, \cdots, x_r\}$ a basis of $A + A^{\perp}$, $n > r$. Since κ is nondegenerate, the matrix of κ with respect to the above basis is non-singular. Hence $x_{r+1} \in (A + A^{\perp})^{\perp} \subseteq A \bigcap A^{\perp}$. This is a contradiction. Thus, $A + A^{\perp} = L$, and hence $L = A \oplus A^{\perp}$. It follows from Proposition 1.4.7 that L is a semi-simple module considered as a module over itself. In turn, there exist simple ideals $A_1, A_2, \cdots, A_r$ of L which are simple as Lie algebras such that $L = A_1 \oplus A_2 \oplus \cdots \oplus A_r$. Let A be a nonzero simple ideal of L. Then $[A, L]$ is an ideal of A and also of L. Since $Z(L) = \{0\}$, $[A, L] \neq \{0\}$. Since A is supposed to be simple, $[A, L] = A$. Hence $A = [A, A_1] \oplus [A, A_2] \oplus \cdots \oplus [A, A_m]$. Since A is simple, $[A, A_i] = A$ for a unique i. Again, since $[A, A_i] \subseteq A_i$, it follows that $A = A_i$ for a unique i. The rest of the assertion follows from Proposition 1.4.3. ♯

The following is an immediate corollary of the above theorem and the preceding discussions.

Corollary 1.4.11 *Every ideal and also every homomorphic image of a semi-simple Lie algebra is semi-simple. A semi-simple Lie algebra L is perfect in the sense that $[L, L] = L$. All ideals of a semi-simple Lie algebra L are the direct sums of some simple ideals of L.* ♯

Theorem 1.4.12 *Every derivation of a semi-simple Lie algebra is the inner derivation. Every semi-simple Lie algebra L is isomorphic to the Lie algebra of its derivations.*

Proof Let L be a semi-simple Lie algebra. Then $Z(L) = \{0\}$ and so L is isomorphic to $ad(L)$. Thus, the second assertion of the theorem follows from the first. We prove the first assertion. We have already observed that $ad(L)$ is an ideal of the Lie algebra $Der(L)$ of derivations of L. Let κ_D denote the Killing form on $Der(L)$. Also from Proposition 1.4.3, it follows that the restriction of κ_D to $ad(L)$ is the Killing form κ on $ad(L)$. Since L is a semi-simple Lie algebra, $ad(L)$ (being isomorphic to L) is also semi-simple. Hence $\kappa_D|_{ad(L)}$ is nondegenerate on $ad(L)$. Let U denote the ideal $ad(L)^{\perp} = \{d \in Der(L) \mid \kappa_D(u, d) = 0 \forall u \in ad(L)\}$. Since κ is nondegenerate on $ad(L)$, $ad(L) \bigcap U = \{0\}$. Since $ad(L)$ and U are ideals of $Der(L)$, $[ad(L), U] = \{0\}$. Now, $d[x, y] = [d(x), y] + [x, d(y)]$ for all $x, y \in L$. Hence $ad(d(x))(y) = [d, ad(x)](y) = 0$ for all $d \in U, x \in L$, and $y \in L$. This means that $ad(d(x)) = 0$ for all $d \in U$. Since ad is injective, $d(x) = 0$ for all $d \in U$ and $x \in L$. It follows that $U = \{0\}$. Consequently, $ad(L) = Der(L)$. ♯

Thus, unlike groups, every simple Lie algebra is complete. In particular, derivation algebra $DerL$ is complete for any semi-simple Lie algebra.

Let L be a semi-simple Lie algebra. It follows from the above theorem that $ad(L) = Der(L)$, and ad is an isomorphism from L to $der(L)$. Let $ad(x) = (ad(x))_s + (ad(x))_n$ denote the Jordan decomposition of $ad(x)$ considered as an element of $End_F(L)$. It follows from Proposition 1.3.22 that $(ad(x))_s$ and $(ad(x))_n$ are also members of $ad(L)$. Since ad is injective, there is a unique element x_s and a unique element x_n in L such that $ad(x_s) = (ad(x))_s$ and $ad(x_n) = (ad(x))_n$. Again, $ad(x_s + x_n) = ad(x_s) + ad(x_n) = (ad(x))_s + (ad(x))_n = ad(x)$. It follows that $x = x_s + x_n$. Further, $ad([x_s, x_n]) = [ad(x_s), ad(x_n)] = [(ad(x))_s, (ad(x))_n] = 0$. It follows that $[x_s, x_n] = 0$. This decomposition of x is termed as the **abstract Jordan decomposition** of x. The element x_s is called the semi-simple part and x_n is called the nilpotent part of x.

Remark 1.4.13 If $L \subseteq gl(V)$ is a semi-simple linear Lie algebra, then an element $x \in L$ has Jordan decomposition considered as an element of $End_F(V)$. At this point , there is no reason to believe that this decomposition agrees with the abstract Jordan decomposition. However, we shall see soon that it is so. For a particular case when $L = sl(V)$, it can be seen easily as follows. Let x be a member of $sl(V)$. Suppose that $x = x_s + x_n$ is the Jordan decomposition of x considered as an element of $End_F(V)$. Since x_n is nilpotent, $Tr(x_n) = 0$. Hence $x_n \in sl(V)$. In turn, x_s also belongs to $sl(V)$. Since x_s is semi-simple in $gl(V)$, $ad_{gl(V)}(x_s)$ is also semi-simple in $End(End_F(V))$. Similarly, $ad_{gl(V)}(x_n)$ is nilpotent in $End(End_F(V))$. Consequently, $ad(x_s)$ is semi-simple and $ad(x_n)$ is nilpotent in $ad(sl(V))$. Also $[ad(x_s), ad(x_n)] = ad([x_s, x_n]) = 0$. Thus, indeed, $x = x_s + x_n$ is also the abstract Jordan decomposition of x.

Casimir Element

Let ρ be a representation of a Lie algebra L on V. We have an associated symmetric bi-linear form β_ρ on L given by $\beta_\rho(x, y) = Tr(\rho(x)\rho(y))$. Thus, the Killing form κ on a Lie algebra is β_{ad}. As usual, β_ρ is also associative. In turn, $N_\rho = \{x \in L \mid \beta_\rho(x, y) = 0 \ for\ all\ y \in L\}$ is an ideal of L. It follows from Cartan's criteria that $\rho(N_\rho)$ is solvable. If ρ is a faithful representation, then N_ρ is also solvable. In particular, if L is semi-simple and ρ is faithful, then N_ρ is trivial, and so the associated bi-linear form β_ρ is nondegenerate. Indeed, the nondegeneracy of the Killing form on a semi-simple Lie algebra is a particular case of the above observation.

Let L be a semi-simple Lie algebra, ρ be a faithful representation of L on V, and β_ρ the associated nondegenerate bi-linear form as described above. For each $x \in L$, we have a map χ_x from V to F given by $\chi_x(y) = \beta_\rho(y, x)$. Evidently, $\chi_x \in V^\star$, where $V^\star$ is the dual space of V. The map χ from V to $V^\star$ given by $\chi(x) = \chi_x$ is a vector space homomorphism. Since β_ρ is nondegenerate, χ is injective. Since V and $V^\star$ are of the same dimension, χ is an isomorphism. Let $\{x_1, x_2, \cdots, x_n\}$ be a basis of V, and let $\{f_1, f_2, \cdots, f_n\}$ be the dual basis of $\{x_1, x_2, \cdots, x_n\}$. From our above discussion, we get a basis $\{y_1, y_2, \cdots, y_n\}$ of V such that $f_i = \chi(y_i) = \chi_{y_i}$ for all i. More explicitly, $\beta_\rho(x_i, y_j) = \delta_{ij}$, where δ is the Kronecker delta.

Proposition 1.4.14 *Let ρ be a faithful representation of a semi-simple Lie algebra on V. Let $\{x_1, x_2, \cdots, x_n\}$ be a basis of L and $\{y_1, y_2, \cdots, y_n\}$ be the dual basis with respect to the bi-linear form β_ρ. Let $[a_{ij}]$ be the matrix of $ad(x)$ with respect to the basis $\{x_1, x_2, \cdots, x_n\}$. Then the matrix of $ad(x)$ with respect to the dual basis $\{y_1, y_2, \cdots, y_n\}$ is $-[a_{ij}]^t$.*

Proof We have $[x, x_i] = \sum_{j=1}^{n} a_{ji}x_j$. Suppose that $[b_{ij}]$ is the matrix representation of $ad(x)$ with respect to the dual basis $\{y_1, y_2, \cdots, y_n\}$. Then $[x, y_i] = \sum_{j=1}^{n} b_{ji}y_j$. Now

$$a_{jk} = \sum_{i=1}^{n} a_{ik}\beta_\rho(x_i, y_j) = \beta_\rho(\sum_{i=1}^{n} a_{ik}x_i, y_j) = \beta_\rho([x, x_k], y_j) = \beta_\rho(-[x_k, x], y_j) = \beta_\rho(-x_k, [x, y_j]) = -a_{kj}. \sharp$$

Definition 1.4.15 The endomorphism $c_\rho = \sum_{i=1}^{n} \rho(x_i)\rho(y_i)$ of V is called the **Casimir element** of the representation ρ relative to the basis $\{x_1, x_2, \cdots, x_n\}$.

Proposition 1.4.16 *Let L be a semi-simple Lie algebra of dimension n. Let ρ be a faithful representation of L on V. Then $Tr(c_\rho) = n$.*

Proof By the definition,

$$Tr(c_\rho) = Tr(\sum_{i=1}^{n} \rho(x_i)\rho(y_i)) = \sum_{i=1}^{n} Tr(\rho(x_i)\rho(y_i)) = \sum_{i=1}^{n} \beta_\rho(x_i, y_i) = n. \sharp$$

Proposition 1.4.17 *Let ρ be a faithful representation of a semi-simple Lie algebra L on V. Then $[\rho(L), c_\rho] = 0$. More explicitly, c_ρ commutes with $\rho(x)$ for all $x \in L$.*

Proof We have $[\phi, \psi\eta] = [\phi, \psi]\eta + \psi[\phi, \eta]$ for all ϕ, ψ and $\eta \in gl(V)$. Thus,

$$[\rho(x), c_\rho] = \sum_{i=1}^{n} [\rho(x), \rho(x_i)]\rho(y_i) + \sum_{i=1}^{n} \rho(x_i)[\rho(x), \rho(y_i)] = \sum_{i=1}^{n}\sum_{k=1}^{n} a_{ki}\rho(x_k)\rho(y_i) + \sum_{i=1}^{n}\sum_{k=1}^{n} b_{kj}\rho(x_i)\rho(y_k) = 0,$$

thanks to Proposition 1.4.14. $\sharp$

The following corollary is immediate.

Corollary 1.4.18 *Under the hypothesis of the above proposition, a Casimir element $c_\rho \in End_L(V)$.* $\sharp$

Corollary 1.4.19 *Let L be a semi-simple Lie algebra and ρ be an irreducible representation of L on V. Then, c_ρ is the multiplication by $\frac{dim\ L}{dim\ V}$.*

Proof Since V is a simple L-module and c_ρ is an L-endomorphism of V, by the Schur lemma, c_ρ is the multiplication by a scalar. Since $Tr(c_\rho) = dim\ L$ (Proposition 1.4.16), the result follows. $\sharp$

Example 1.4.20 Let ρ denote the standard inclusion representation of $sl(2, F)$ on $V = F^2$. Take the standard ordered basis $\{e_{12}, h, e_{21}\}$ of $sl(2, F)$. The dual basis of this standard basis can be easily seen to be $\{e_{21}, \frac{h}{2}, e_{12}\}$. Thus, the Casimir element c_ρ of the representation is given by

$$c_\rho = e_{12}e_{21} + \frac{h^2}{2} + e_{21}e_{12} = \begin{bmatrix} \frac{3}{2} & 0 \\ 0 & \frac{3}{2} \end{bmatrix}.$$

The trace of c_ρ is $3 = dim\, sl(2, F)$. Evidently, c_ρ is the multiplication by $\frac{3}{2}$. Observe that the module F^2 associated with this representation is a simple L-module.

Recall (Theorem 9.1.10, Algebra 2) that a ring R is semi-simple if and only if every module over R is semi-simple (completely reducible). In the case of Lie algebras, we have already shown that a semi-simple Lie algebra L is a completely reducible module over itself (Theorem 1.4.10). The converse is true here also for Lie algebras. However, the proof of the converse for Lie algebras is a little subtle. In the case of rings, the proof is an easy consequence of the fact that a free R-module is the direct sum of several copies of R considered as an R-module, and a homomorphic image of a semi-simple module is semi-simple.

Theorem 1.4.21 (Weyl) *Let L be a semi-simple Lie algebra. Then every finite-dimensional L-module is completely reducible. In the language of the representation theory, every finite-dimensional representation of a semi-simple Lie algebra is the direct sum of irreducible representations.*

Proof Let L be a semi-simple Lie algebra. Let ρ be a representation of L on V. Since L is perfect, and $[gl(V), gl(V)] = sl(V)$, it follows that $\rho(L) \subseteq sl(V)$. Thus, $Tr(\rho(x)) = 0$ for all $x \in L$. In particular, if V is a one-dimensional L-module, then $x \cdot v = 0$ for all $x \in L$ and $v \in V$.

Let ρ be a representation of a semi-simple Lie algebra L on V. Let A be the kernel of ρ. Then A is an ideal of L and by Theorem 1.4.10, $L = A \oplus B$. B is also a semi-simple Lie algebra, and $\rho|_B$ is a faithful representation. Further, L-submodules of V are precisely B - submodules of V. As such, without any loss of generality, we may assume that the representation ρ is faithful.

We prove the theorem in several steps:

Step 1. Every irreducible submodule W of V of co-dimension 1 is a direct summand.

Proof of Step 1. Let

$$0 \longrightarrow W \overset{i}{\rightarrow} V \overset{\nu}{\rightarrow} V/W \longrightarrow 0$$

be an exact sequence of L-modules, where V is a faithful L-module and W is a simple submodule of co-dimension 1. By Corollary 1.4.18, the Casimir operator $c_\rho = \sum_{i=1}^{n} \rho(x_i)\rho(y_i)$ is an L-endomorphism of V. Hence $c_\rho(W)$ is an L-submodule of V. If $c_\rho(W) = \{0\}$, then since L acts trivially on V/W, $Tr(c_\rho) = 0$. This is a contradiction, since $Tr(c_\rho) = Dim\, L \neq 0$. Thus, $c_\rho(W) \neq \{0\}$. Since W is a simple submodule, $(c_\rho(W) \bigcap W)$ is $\{0\}$ or it is W. If $(c_\rho(W) \bigcap W)$ is $\{0\}$, then $V = W \oplus c_\rho(W)$ and we are done. If not, then $W \subseteq c_\rho(W)$. By the dimension consideration, $c_\rho(W) = W$. Since L acts trivially on one-dimensional modules, c_ρ induces trivial endomorphism $\hat{c_\rho}$ on V/W. Clearly, $Tr(c_\rho) = Tr(c_\rho|_W) + Tr(\hat{c_\rho}) = Tr(c_\rho|_W)$. Since $Tr(c_\rho) = Dim\ L \neq 0$, $Tr(c_\rho|_W) \neq 0$. Further, since W is an irreducible

L-module, $c_\rho|_W$ is the multiplication by a nonzero member of F. Evidently, $ker\ c_\rho$ is a one-dimensional L-submodule and $W \bigcap ker\ c_\rho = \{0\}$. This shows that $V = W \oplus ker\ c_\rho$. This proves step 1.

Step 2. Every submodule W of V of co-dimension 1 is a direct summand.

Proof of Step 2. The proof is by the induction on the dimension of W. If the dimension of W is zero, then there is nothing to do. Assume that the assertion is true for all submodules of co-dimensions 1 whose dimensions are less than the dimension of W. Let

$$0 \longrightarrow W \xrightarrow{i} V \xrightarrow{\nu} V/W \longrightarrow 0$$

be an exact sequence of L-modules, where V is a faithful L-module and W is a submodule of co-dimension 1. If W is simple, then the result follows from step 1. Suppose that W is not simple. Let W' be a nonzero proper submodule of W. We have the exact sequence

$$0 \longrightarrow W/W' \xrightarrow{i} V/W' \xrightarrow{\nu} V/W \longrightarrow 0.$$

By the induction hypothesis, there is a one-dimensional submodule U/W' of V/W' such that $V/W' = W/W' \oplus U/W'$. We get a short exact sequence

$$0 \longrightarrow W' \xrightarrow{i} U \xrightarrow{\nu} U/W' \longrightarrow 0$$

of L-modules. By the induction hypothesis, there is a one-dimensional submodule P of U such that $U = P \oplus W'$. Also $V/W' = W/W' \oplus U/W'$. Since $P \not\subseteq W$ and $W \bigcap P = \{0\}$, $V = W \oplus P$. This proves step 2.

Proof of the Theorem. Let W be a submodule of V. We need to show that W is a direct summand of V. We have the short exact sequence

$$0 \longrightarrow W \xrightarrow{i} V \xrightarrow{\nu} V/W \longrightarrow 0$$

of L-modules. We have the L-module $Hom(V, W)$, where $x \cdot f$ is given by $(x \cdot f)(v) = x \cdot f(v) - f(x \cdot v)$. Similarly, $Hom(W, W)$ is also an L-module. We have the restriction vector space homomorphism $i^\star$ given by $i^\star(f) = f|_W$. Clearly, $i^\star$ is an L-homomorphism. For each $a \in F$, let f_a denote the multiplication by the scalar a on W. Then the subspace $\hat{F} = \{f_a \mid a \in F\}$ of $Hom(W, W)$ consisting of scalar multiplications on W can be easily seen to be an L-submodule of $Hom(W, W)$. Indeed, $L \cdot \hat{F}$ is $0 \in Hom(V, W)$. We have the short exact sequence

$$0 \longrightarrow ker\ i^\star \xrightarrow{i} (i^\star)^{-1}(\hat{F}) \xrightarrow{\nu} (i^\star)^{-1}(\hat{F})/ker\ i^\star \longrightarrow 0$$

of L-modules. Since $\hat{F}$ is one dimensional, $ker\ i^\star$ is of co-dimension 1. From step 2, $(i^\star)^{-1}(\hat{F}) = ker\ i^\star \oplus < g >$, where $< g >$ is the one-dimensional submodule of $(i^\star)^{-1}(\hat{F})$ generated by g. Since $g \notin ker\ i^\star$, $i^\star(g) = g|_W = f_a$, where $a \neq 0$.

Replace g by $h = a^{-1}g$. Then $h|_W = I_W$. Since $L \cdot \hat{F} = \{0\}$, $(x \cdot h)(v) = x \cdot (h(v)) - h(x \cdot v) = 0$ for all $x \in L$ and $v \in V$. This shows that h is an L-homomorphism from V to W such that $h|_W = I_W$. This shows that W is a direct summand of V. ♯

Weyl's theorem is the most fundamental result initiating the representation theory of semi-simple Lie algebras. As the first application of the theorem, we establish that under any representation of a semi-simple Lie algebra, Jordan decompositions are preserved.

Theorem 1.4.22 *Let L be a semi-simple linear Lie algebra contained in $gl(V)$, where V is a finite-dimensional vector space over an algebraically closed field. Then L contains semi-simple and nilpotent parts of each of its elements considered as elements of $End_F(V)$. Consequently, abstract and usual Jordan decompositions of members of L are the same.*

Proof Let $x \in L$. Let $x = x_s + x_n$ represent the usual Jordan decomposition of x in $End_F(V)$. We have to show that x_s and x_n both belong to L. By Proposition 1.3.20, $ad_{gl(V)}(x) = ad_{gl(V)}(x_s) + ad_{gl(V)}(x_n)$ is the Jordan decomposition of $ad_{gl(V)}(x)$ in $End_F(End_F(V))$. Since $ad_{gl(V)}(x)(L) \subseteq L$ and $ad_{gl(V)}(x_s)$ and $ad_{gl(V)}(x_n)$ are polynomials of $ad_{gl(V)}(x)(L)$ without constant terms, it follows that $ad_{gl(V)}(x_s)(L) \subseteq L$ and also $ad_{gl(V)}(x_n)(L) \subseteq L$. This means that x_s and x_n both belong to the normalizer $N_{gl(V)}(L)$ of L in $gl(V)$ for each $x \in L$. For each L-submodule W of V, consider the subspace $\hat{W} = \{y \in gl(V) \mid y(W) \subseteq W \text{ and } Tr(y|_W) = 0\}$. Since L is semi-simple, $[L, L] = L$ and consequently $L \subseteq \hat{W}$ for all L-submodules W of V. It is also clear from the earlier observation that x_s and x_n belong to $\hat{W}$ for each L-submodule W of V. Denote the intersection $N_{gl(V)}(L) \bigcap (\bigcap_{W \in S(V)} \hat{W})$ by $\hat{L}$, where $S(V)$ denotes the set of all L-submodules of V. Evidently, $\hat{L}$ is a Lie subalgebra of $gl(V)$ containing L as an ideal such that x_s and x_n both belong to $\hat{L}$ for each $x \in L$. It is sufficient to show that $\hat{L} = L$. Clearly, $\hat{L}$ is a finite-dimensional L-module with L as an L-submodule. Since L is semi-simple, by the theorem of Weyl, there is an L-submodule U of $\hat{L}$ such that $\hat{L} = L \oplus U$. Since $[L, \hat{L}] \subseteq L$, it follows that $[L, U] = \{0\}$. We show that $U = \{0\}$. Let $u \in U$. Since $[L, u] = \{0\}$, u represents an L-endomorphism of V. Let W be a simple submodule of V. Then by the Schur lemma, u acts on W by a scalar multiplication. Also, since $u \in \hat{W}$, $Tr(u|_W) = 0$. This means that the endomorphism u restricted to W is zero for all simple L-submodules of V. Since V is the direct sum of simple submodules, it follows that $u = 0$. Hence $U = \{0\}$, and $\hat{L} = L$. This completes the proof of the first part of the theorem. The rest of the assertion of the theorem also follows. ♯

Corollary 1.4.23 *Let ρ be a finite-dimensional representation of a semi-simple Lie algebra L on V. Let $x = x_s + x_n$ be the abstract Jordan decomposition of x in L. Then $\rho(x) = \rho(x_s) + \rho(x_n)$ is the usual Jordan decomposition of $\rho(x)$ in $End_F(V)$.*

Proof Since $ad(x_s)$ is semi-simple in $End_F(L)$, L is generated by the set of eigenvectors of $ad(x_s)$. Hence $\rho(L)$ is also generated by the eigenvectors $ad_{\rho(L)}(\rho(x_s))$. This shows $ad_{\rho(L)}(\rho(x_s))$ is semi-simple. Again since $ad(x_n)$ is nilpotent, $ad_{\rho(L)}(\rho(x_n))$ is nilpotent. This shows that $\rho(x) = \rho(x_n) + \rho(x_s)$ is the abstract Jordan decomposition of $\rho(x)$ in $\rho(L)$. It follows from the above theorem that it is the usual Jordan decomposition of $\rho(x)$ in $End_F(V)$. ♯

Representations of $sl(2, F)$

Let F be an algebraically closed field of characteristic 0. Recall that the Lie algebra $sl(2, F)$ is a simple Lie algebra having the standard basis $\{e_{12}, e_{21}, h\}$ satisfying the relations

$$[h, e_{12}] = 2e_{12}, \ [h, e_{21}] = -2e_{21}, \ [e_{12}, e_{21}] = h.$$

For simplicity, we denote e_{12} by x and e_{21} by y. Let ρ be a representation of $sl(2, F)$ on a finite-dimensional vector space V. Clearly, h is a semi-simple element while x, y are nilpotent elements in $sl(2, F)$. From Corollary 1.4.23, it follows that $\rho(h)$ is semi-simple in $gl(V)$, whereas $\rho(x)$ and $\rho(y)$ are nilpotent in $gl(V)$. Let Γ denote the set of all eigenvalues of the transformation $\rho(h)$. The members of Γ are also called the **weights** of the representation ρ. For a weight λ, the λ-eigenspace $V_\lambda = \{v \in V \mid \rho(h)(v) = \lambda v\}$ of $\rho(h)$ is called the **weight space** of the representation associated with λ. Evidently, $V_\lambda \neq \{0\}$. Since $\rho(h)$ is semi-simple, $V = \oplus \sum_{\lambda \in \Gamma} V_\lambda$. Evidently, Γ is finite.

Proposition 1.4.24 *Let $\lambda \in \Gamma$ and $v \in V_\lambda$. Then $\rho(x)(v) \in V_{\lambda+2}$. Further, $\rho(y)(v) \in V_{\lambda-2}$.*

Proof $\rho(h)(\rho(x)(v)) = \rho([h, x])(v) + \rho(x)(\rho(h)(v)) = 2\rho(x)(v) + \lambda\rho(x)(v)v = (\lambda + 2)\rho(x)(v)$. Thus, $\rho(x)(v) \in V_{\lambda+2}$. Similarly, the second assertion follows. ♯

Corollary 1.4.25 *There is a $\lambda \in \Gamma$ such that $\lambda + 2 \notin \Gamma$ and in this case $\rho(x)(v) = 0$ for all $v \in V_\lambda$. Similarly, there is a $\mu \in \Gamma$ such that $\mu - 2 \notin \Gamma$ and in this case $\rho(y)(v) = 0$ for all $v \in V_\mu$.*

Proof Since Γ is nonempty and finite, there is a $\lambda \in \Gamma$ such that $\lambda + 2 \notin \Gamma$. It is evident that $\rho(x)(v) = 0$ for all $v \in V_\lambda$. The remaining assertion follows, similarly. ♯

A weight $\lambda \in \Gamma$ is called a **maximal weight** if $\lambda + 2 \notin \Gamma$. An eigenvector corresponding to a maximal weight is called a **maximal vector**.

Proposition 1.4.26 *Let ρ be an irreducible representation of $sl(2, F)$ on a finite-dimensional vector space V. Let λ be a maximal weight of the representation and v_0 a maximal vector corresponding to the maximal weight λ. For each $r \geq 0$, put $v_r = \frac{1}{r!}(\rho(y))^r(v_0)$. Then*
(i) $\rho(h)(v_r) = (\lambda - 2r)v_r$,
(ii) $\rho(y)(v_r) = (r + 1)v_{r+1}$, and
(iii) $\rho(x)(v_r) = (\lambda - r + 1)v_{r-1}$
for all $r \geq 0$. We put $v_{-1} = 0$ in (iii).

Proof (i). We prove it by the induction on r. For $r = 0$, $\rho(h)(v_0) = \lambda v_0$ by the choice. Assume that $\rho(h)(v_r) = (\lambda - 2r)v_r$. By Proposition 1.4.24, $\rho(h)(\rho(y)(v_r)) = (\lambda - 2r - 2)\rho(y)(v_r)$. Hence $\rho(h)(v_{r+1}) = \rho(h)(\frac{1}{r+1}\rho(y)(v_r)) = \frac{1}{r+1}\rho(h)(\rho(y)(v_r)) = \frac{1}{r+1}(\lambda - 2r - 2)\rho(y)(v_r) = (\lambda - 2(r+1))v_{r+1}$.

(ii). Evident from the definition.

(iii). By the induction, it can be easily seen that

$$\rho(y)(\rho(x)(v_r)) = (\lambda - r + 1)\rho(y)(v_{r-1})$$

for each r. Now,

$$\begin{aligned}
&\rho(x)(v_r)\\
&= \tfrac{1}{r}\rho(x)(\rho(y)(v_{r-1}))\\
&= \tfrac{1}{r}(\rho([x, y])(v_{r-1}) + \rho(y)(\rho(x)(v_{r-1})))\\
&= \tfrac{1}{r}(\rho(h)(v_{r-1}) + (\lambda - r + 2)\rho(y)(v_{r-2}))\\
&= \tfrac{1}{r}((\lambda - 2r + 2)v_{r-1} + (r - 1)(\lambda - r + 2)\rho(y)(v_{r-1}))\\
&= (\lambda - r + 1)v_{r-1}. \sharp
\end{aligned}$$

Corollary 1.4.27 *The set $\{v_r \mid v_r \neq 0, r \geq 0\}$ forms a linearly independent subset of V. Consequently, there is a nonnegative integer n such that $v_n \neq 0$ but $v_{n+1} = 0$.*

Proof The result follows from the fact that v_r are eigenvectors of h corresponding to distinct eigenvalues of h. Finally, since V is of finite dimension, the rest of the assertion is clear. $\sharp$

Theorem 1.4.28 *Let ρ be an irreducible representation of $sl(2, F)$ on a vector space V of dimension $m + 1$, $m \geq 0$. Then the following hold:*

1. *The set Γ of weights of the representation is given by*

$$\Gamma = \{m - 2r \mid 0 \leq r \leq m\} = [-m, m] \bigcap m + 2\mathbb{Z}.$$

2. *Each weight space V_λ, $\lambda \in \Gamma$ is one dimensional.*
3. *There is a basis $\{v_0, v_1, \cdots, v_m\}$ of V with $v_r \in V_{m-2r}$ such that*

(i) $\rho(h)(v_r) = (m - 2r)v_r$,
(ii) $\rho(y)(v_r) = (r + 1)v_{r+1}$, *and*
(iii) $\rho(x)(v_r) = (m - r + 1)v_{r-1}$.

Proof From Corollary 1.4.27, there is a linearly independent subset $\{v_0, v_1, \cdots, v_n\}$ of V satisfying conditions (i), (ii), and (iii) of Proposition 1.4.26. Let W be the subspace of V generated by $\{v_0, v_1, \cdots, v_n\}$. Again conditions (i), (ii), and (iii) of Proposition 1.4.26 imply that $\rho(z)(W) \subseteq W$ for all $z \in sl(2, F)$. Since ρ is irreducible, $W = V$, and hence $n = m$. For $r = m + 1$, condition (iii) of Proposition 1.2.26 implies that $0 = (\lambda - m)v_m$. Since $v_m \neq 0$, the maximal weight $\lambda = m = (dim\ V - 1)$. The theorem follows from Proposition 1.4.26. $\sharp$

Corollary 1.4.29 *For each* $n = m + 1, m \geq 0$, *there is a unique irreducible representation of* $sl(2, F)$ *of degree* n *up to equivalence.* ♯

Proof The uniqueness follows from the above theorem. For existence, let V be a vector space having a basis $\{v_0, v_1, \cdots, v_m\}$. We have linear transformations $\rho(h)$, $\rho(x)$, and $\rho(y)$ on V given by (i), (ii), and (iii) of Theorem 1.4.28. In turn, we get a linear map ρ from $sl(2, F)$ to $gl(V)$. It can be seen that $\rho([h, x]) = [\rho(h), \rho(x)]$, $\rho([h, y]) = [\rho(h), \rho(y)]$, and $\rho([x, y]) = [\rho(x), \rho(y)]$. Hence ρ is a representation which is irreducible. ♯

Corollary 1.4.30 *Let* ρ *be a finite-dimensional representation (not necessarily irreducible) of* $sl(2, F)$ *on* V. *Then all the eigenvalues of* $\rho(h)$ *are integers. If* r *is an eigenvalue of* $\rho(h)$, *then* $-r$ *is also an eigenvalue. Indeed, the pair* $(r, -r)$ *appears equal times. Let* V_0 *denote the eigenspace of* $\rho(h)$ *corresponding to the eigenvalue* 0, *and* V_1 *denotes that corresponding to the eigenvalue 1. Then* ρ *is the direct sum of* $Dim\ V_0 + Dim\ V_1$ *irreducible representations.*

Proof By the theorem of Weyl, ρ is the direct sum of irreducible representations. The eigenvalues of $\rho(h)$ are precisely the eigenvalues of $\rho(h)$ restricted to the irreducible components. The first assertion is immediate from the above theorem. Let $\{u_1, u_2, \cdots, u_s\}$ be a basis of V_0 and $\{v_1, v_2, \cdots, v_t\}$ be a basis of V_1. It is clear from the Theorem 1.4.28 that each u_i determines a unique irreducible component of ρ and distinct u_i determine different irreducible components. Similarly, each v_j determines a unique irreducible component of ρ and distinct v_j determine different irreducible components. The result follows. ♯

Exercises

1.4.1. Compute the matrix of the Killing form of each of the Lie algebras described so far with respect to their standard bases.
1.4.2. Compute the Casimir operator of the adjoint representation of $sl(2, F)$ and $sl(3, F)$ with respect to their standard bases. Find their traces also.
1.4.3. Show that every irreducible representation of a solvable Lie algebra is one dimensional.
1.4.4. Call a Lie algebra L to be a **reductive Lie algebra** if $Rad(L) = Z(L)$. Show that a semi-simple Lie algebra is reductive. Give an example of a reductive Lie algebra which is not semi-simple.
1.4.5. Show that a Lie algebra is reductive if and only if L considered as $ad(L)$-module is completely reducible.
1.4.6. Let L be a semi-simple Lie algebra. Describe the structure of the ring $End_L(L)$ if possible.
1.4.7. Let L be a simple Lie algebra. Let β be a symmetric associative nondegenerate bi-linear form on L. Use Schur lemma to show that there is a scalar $a \in F^\star$ such that $\beta = a\kappa$, where κ is the Killing form on L.
1.4.8. Determine the matrix of trace form Tr on $sl(n, F)$ given by $Tr(x, y) = Tr(xy)$ with respect to the standard basis of $sl(n, F)$. Show that it is symmetric associative and nondegenerate. Assuming that $sl(n, F)$ is a simple Lie algebra, show that $\kappa = 2nTr$.

1.5 Extensions of Lie Algebras and Co-homology

In this section, we introduce the co-homology of Lie algebras, describe low-dimensional co-homologies, and discuss the Schreier theory of extensions of Lie algebras. There is a close analogy between the co-homology theory of groups and the co-homology theory of Lie algebras.

Let L be a Lie algebra over a field K. Recall that L-modules are precisely $U(L)$-modules. We have bi-functors $EXT^n_{U(L)}(-,-)$ and $Tor^{U(L)}_n(-,-)$ from the category of L $(U(L))$-modules to the category of groups. These functors are also denoted by $EXT^n_L(-,-)$ and $Tor^L_n(-,-)$, respectively. Treat K as a trivial L-module. Let A be an L-module. Then $EXT^n_L(K, A)$ is denoted by $H^n(L, A)$ and it is called the n_{th} co-homology of L with a coefficient in A. Further, $Tor^{U(L)}_n$ is denoted by $H_n(L, A)$ and it is called the n_{th} homology of L with coefficient in A. Thus, we need to construct a convenient projective $U(L)$-resolution of the trivial L-module K ($x \cdot a = 0$ for all $x \in L$ and $a \in K$) for this purpose.

The trivial module structure on K induces an algebra homomorphism ϵ from $U(L)$ to $gl(K) \approx K$. This homomorphism is called the augmentation map. The $Ker\epsilon$ is called the augmentation ideal and it is denoted by $I(L)$. Since K is the trivial L-module, it follows that $I(L)$ is the ideal of $U(L)$ generated by $j_L(L)$. We have a free presentation

$$0 \longrightarrow I(L) \longrightarrow U(L) \stackrel{\epsilon}{\to} K \longrightarrow 0$$

of the trivial module K.

Proposition 1.5.1 $H^0(L, A) = \{a \in A \mid xa = 0 \text{ for all } x \in L\}$. ♯

Proof By the definition, $H^0(L, A) = Ext^0_{U(L)}(K, A) \approx Hom_{U(L)}(K, A)$. Now, a $U(L)$-homomorphism f from $U(L)$ to A determines and is uniquely determined by the image $f(1) = a \in A$ of 1 such that $0 = f(0) = f(j_L(x) \cdot 1) = j_L(x) \cdot f(1) = x \cdot a$ for all $x \in L$. Thus, $H^0(L, A) \approx \{a \in A \mid xa = 0 \text{ for all } x \in L\}$. ♯

Proposition 1.5.2 *$H^1(L, A) \approx Der(L, A)/Ider(L, A)$ and if A is a trivial L-module, $H^1(L, A) \approx Hom_K(L_{ab}, A)$.* ♯

Proof Consider the short exact sequence

$$0 \longrightarrow I(L) \longrightarrow U(L) \stackrel{\epsilon}{\to} K \longrightarrow 0.$$

We have the long exact sequence (see Corollary 2.1.17, Algebra 3)

$$0 \longrightarrow Ext^0_{U(L)}(K, A) \stackrel{\epsilon^\star}{\to} Ext^0_{U(L)}(U(L), A) \stackrel{i^\star}{\to} Ext^0_{U(L)}(I(L), A) \stackrel{\partial}{\to} \\ Ext^1_{U(L)}(K, A) \stackrel{\epsilon^\star}{\to} Ext^1_{U(L)}(U(L), A) \stackrel{i^\star}{\to} \cdots .$$

Since $U(L)$ is a free $U(L)$-module, $Ext^1_{U(L)}(U(L), A) = \{0\}$. Also $Ext^0_{U(L)}(-,-) = Hom_{U(L)}(-,-)$. Consequently, we get the exact sequence

$$0 \longrightarrow Hom_{U(L)}(K, A) \stackrel{\epsilon^\star}{\to} Hom_{U(L)}(U(L), A) \stackrel{i^\star}{\to} Hom_{U(L)}(I(L), A) \stackrel{\partial}{\to} Ext^1_{U(L)}(K, A) \longrightarrow 0.$$

Thus,

$$H^1(L, A) = Ext^1_{U(L)}(K, A) \approx \frac{Hom_{U(L)}(I(L), A)}{Ker\partial} = \frac{Hom_{U(L)}(I(L), A)}{image\ i^\star}.$$

Now, let us interpret the members of $Hom_{U(L)}(I(L), A)$ and of $image\ i^\star$. Let $f \in Hom_{U(L)}(I(L), A)$. Define a map D_f from L to A by putting $D_f(x) = f(j_L(x))$. The fact that f is a $U(L)$-homomorphism implies that $D_f \in Der(L, A)$. The map D from $Hom_{U(L)}(I(L), A)$ to $Der(L, A)$ given by $D(f) = D_f$ can be easily seen to be a vector space isomorphism. Further, if f is the restriction to $I(L)$ of a homomorphism from $U(L)$ to A, then it is determined uniquely by an element $a = f(1)$ of a such that $f(x) = j_L(x) \cdot a = x \cdot a$ and D_f is an inner derivation D_a determined by a. The rest is an easy verification. ♯

Now, we shall describe a convenient projective resolution of the trivial L-module K for the purpose of computing co-homologies. Let $E(L)$ denote the the exterior algebra (see Algebra 2, Sect. 7.3) of L considered as a vector space. Thus,

$$E(L) = \oplus\sum_{r=0}^{\infty} \bigwedge^r(L),$$

where $\bigwedge^r L$ denote the r_{th} exterior power of L. In particular, $\bigwedge^0 L = K, \bigwedge^1 L = L$. If $\{x_1, x_2, \cdots, x_n\}$ is a basis of L, then $\{x_{i_1} \wedge x_{i_2} \wedge \cdots \wedge x_{i_r} \mid 1 \leq i_1 < i_2 < \cdots < i_r \leq n\}$ is a basis of $\bigwedge^r L$. In particular, $Dim \bigwedge^r L = {}^nC_r$ and $DimE(V) = 2^n$. The product $\cdot$ from $\bigwedge^r L \times \bigwedge^s L$ to $\bigwedge^{r+s} L$ defined by

$$(x_{i_1} \wedge x_{i_2} \wedge \cdots \wedge x_{i_r}) \cdot (x_{j_1} \wedge x_{j_2} \wedge \cdots \wedge x_{j_s}) = x_{i_1} \wedge x_{i_2} \wedge \cdots \wedge x_{i_r} \wedge x_{j_1} \wedge x_{j_2} \wedge \cdots \wedge x_{j_s}$$

can be extended to the whole of $E(L)$ by linearity which makes $E(L)$ an alternating algebra. Let X_n denote the $U(L)$-module $U(L) \otimes_K \bigwedge^r L$. Clearly, X_n is a free $U(L)$-module of rank ${}^nC_r$, since $\bigwedge^r L$ is a vector space of dimension ${}^nC_r$. Without any confusion, we can denote $1 \otimes (x_{i_1} \wedge x_{i_2} \wedge \cdots \wedge x_{i_r}), 1 \leq i_1 < i_2 < \cdots < i_r \leq n$ by $x_{i_1} \wedge x_{i_2} \wedge \cdots \wedge x_{i_r}$. Thus, $\{x_{i_1} \wedge x_{i_2} \wedge \cdots \wedge x_{i_r}), 1 \leq i_1 < i_2 < \cdots < i_r \leq n\}$ is a free $U(L)$-basis of X_n. We have a unique $U(L)$-homomorphism d_r from X_r to X_{r-1} subject to

$d_r(x_{i_1} \wedge x_{i_2} \wedge \cdots \wedge x_{i_r}) = \sum_{j=1}^{r}(-1)^{j+1} j_L(x_{i_j})(x_{i_1} \wedge x_{i_2} \wedge \cdots \hat{x_{i_j}} \wedge x_{i_{j+1}} \cdots \wedge x_{i_r}) + \sum_{1 \leq k < l \leq r}(-1)^{k+l}[x_{i_k}, x_{i_l}] \wedge x_{i_1} \wedge \cdots \wedge \hat{x_{i_k}} \wedge \cdots \wedge \hat{x_{i_l}} \wedge \cdots x_{i_r}.$

As usual, one can establish the following proposition:

Proposition 1.5.3 *The chain*

$$\hat{X} = 0 \longrightarrow X_n \xrightarrow{d_n} X_{n-1} \xrightarrow{d_{n-1}} \cdots \xrightarrow{d_2} X_1 \xrightarrow{d_1} X_0 \xrightarrow{\epsilon} K \longrightarrow 0$$

is a $U(L)$-free resolution of the trivial module K, where $n = DimL$. ♯

Thus, $H^n(L, A)$ is the n_{th} co-homology of the cochain complex $Hom_{U(L)}(\hat{X}, A)$.

Corollary 1.5.4 *If n is the dimension of L, then $H^{n+1}(L, A) = \{0\}$.* ♯

Our next aim is develop the theory of Schreier Extensions and interpret the second and third co-homologies in terms of the extensions.

A short exact sequence

$$E \equiv 0 \longrightarrow A \xrightarrow{\alpha} L \xrightarrow{\beta} B \longrightarrow 0$$

of Lie algebras is called an extension of A by B. Let t be a vector space splitting. More explicitly, t is a vector space homomorphism such that $\beta o t = I_B$. The map t need not be a Lie algebra homomorphism. If t is a Lie algebra homomorphism, then it is termed as a splitting and the extension is termed as a split extension. For example,

$$0 \longrightarrow sl(n, F) \xrightarrow{i} gl(n, F) \xrightarrow{Tr} F \longrightarrow 0$$

is a split exact sequence, where Tr is the trace map. Note that it is not a direct sum extension. An extension E is called a central extension if $\alpha(A) \subseteq Z(L)$. Thus, a central split extension is a direct sum extension.

Let *EXT* denote the category whose objects are short exact sequences

$$0 \longrightarrow A \xrightarrow{\alpha} L \xrightarrow{\beta} B \longrightarrow 0$$

of Lie algebras, and a morphism between two extensions E_1 and E_2 given by the short exact sequences

$$0 \longrightarrow A_1 \xrightarrow{\alpha_1} L_1 \xrightarrow{\beta_1} B_1 \longrightarrow 0$$

and

$$0 \longrightarrow A_2 \xrightarrow{\alpha_2} L_2 \xrightarrow{\beta_2} B_2 \longrightarrow 0$$

is a triple (λ, μ, ν), where λ is a homomorphism from A_1 to A_2, μ is a homomorphism from L_1 to L_2, and ν is a homomorphism from B_1 to B_2 such that the following diagram is commutative:

$$\begin{array}{ccccccccc}
0 & \longrightarrow & A_1 & \xrightarrow{\alpha_1} & L_1 & \xrightarrow{\beta_1} & B_1 & \longrightarrow & 0 \\
 & & \Big\downarrow{\scriptstyle \lambda} & & \Big\downarrow{\scriptstyle \mu} & & \Big\downarrow{\scriptstyle \nu} & & \\
0 & \longrightarrow & A_2 & \xrightarrow{\alpha_2} & L_2 & \xrightarrow{\beta_2} & B_2 & \longrightarrow & 0
\end{array}$$

The category *EXT* is called the category of **Schreier extensions** of Lie algebras. The isomorphisms in this category are called the **equivalences of extensions**. Using the five lemmas, it follows that a morphism (λ, μ, ν) is an equivalence if and only if λ and ν are isomorphisms.

Let

$$E \equiv 0 \longrightarrow A \stackrel{\alpha}{\to} L \stackrel{\beta}{\to} B \longrightarrow 0 \tag{1.1}$$

be an extension of A by B. Since β is surjective, there is a vector space homomorphism t (not necessarily a Lie homomorphism) from B to L (called a **section** or a **transversal**) such that $\beta o t = I_B$ (note that we are using the axiom of choice). $\alpha(A) = ker\beta$ is an ideal of L. Thus, for each $x \in B$ and $a \in A$, $[t(x), \alpha(a)]$ belongs to $\alpha(A)$. Since α is injective, there is a unique element $\sigma_x^t(a)$ in A depending on x and a such that

$$[t(x), \alpha(a)] = \alpha(\sigma_x^t(a)) \tag{1.2}$$

This gives us a map σ_x^t from A to A given by (1.2). It can be easily seen that σ_x^t is a vector space endomorphism of A. Further,
$\alpha(\sigma_x^t([a, b]))$
$= [t(x), \alpha([a, b])]$
$= [t(x), [\alpha(a), \alpha(b)]]$
$= [[t(x), \alpha(a)], \alpha(b)] + [\alpha(a), [t(x), \alpha(b)]]$ (by the Jacobi identity)
$= \alpha([\sigma_x^t(a), b] - [\sigma_x^t(b), a])$.
Since α is injective,

$$\sigma_x^t([a, b]) = [\sigma_x^t(a), b] + [a, \sigma_x^t(b)] \tag{1.3}$$

This shows that σ_x^t is a derivation of A.. Thus, the extension E together with a choice of the section t gives us a map σ^t from B to $Der(A)$. Since t is a vector space homomorphism, σ^t is also a vector space homomorphism.

As observed earlier, t need not be a Lie homomorphism. However, since β is a Lie algebra homomorphism,

$$\beta(t([x, y]) = [x, y] = [\beta(t(x)), \beta(t(y))] = \beta([t(x), t(y)])$$

for all $x, y \in B$. Hence, there is a unique element $f^t(x, y) \in A$ such that

$$\alpha(f^t(x, y)) = [t(x), t(y)] - t([x, y]) \tag{1.4}$$

for all $x, y \in B$. This gives us a map f^t from $B \times B$ to A which is given by (1.4). Since t is a vector space homomorphism,

$$\alpha(f^t(ax + by, z)) = [t(ax + by), t(z)] - t([ax + by, z]) = \alpha(af^t(x, z) + bf^t(y, z))$$

for all $x, y, z \in B$. Thus,

$$f^t(ax + by, z) = af^t(x, z) + bf^t(y, z) \tag{1.5}$$

for all $x, y, z \in B$ and $a, b \in F$. Similarly,

$$f^t(x, by + cz) = bf^t(x, y) + cf^t(x, z) \tag{1.6}$$

for all $x, y, z \in B$ and $b, c \in F$. Also since $[t(x), t(x)] = 0 = t([x, x])$, we have

$$f^t(x, x) = 0 \tag{1.7}$$

for all $x \in B$. Next,

$$[[t(x), t(y)], t(z)] + [[t(y), t(z)], t(x)] + [[t(z), t(x)], t(y)] = 0$$

implies that

$$[t([x, y]) + \alpha(f^t(x, y)), t(z)] + [t([y, z]) + \alpha(f^t(y, z)), t(x)] + [t([z, x])+ \alpha(f^t(z, x)), t(y)] = 0$$

or equivalently,

$$t([[x, y], z]) + t([[y, z], x]) + t([[z, x], y]) + \alpha(f^t([x, y], z)) - \alpha(\sigma_z^t(f^t(x, y)))+ \alpha(f^t([y, z], x)) - \alpha(\sigma_x^t(f^t(y, z))) + \alpha(f^t([z, x], y)) - \alpha(\sigma_y^t(f^t(z, z))) = 0.$$

Since t is a vector space homomorphism, using the Jacobi identity and the fact that α is injective, we obtain

$$f^t([x, y], z) + f^t([y, z], x) + f^t([z, x], y) - \sigma_z^t(f^t(x, y)) - \sigma_x^t(f^t(y, z))- \sigma_y^t(f^t(z, x)) = 0 \tag{1.8}$$

Similarly, the Jacobi identity

$$[[t(x), t(y)], \alpha(a)] + [[t(y), \alpha(a)], t(x)] + [[\alpha(a), t(x)], t(y)] = 0$$

implies that

$$\sigma^t_{[x,y]}(a) = \sigma^t_x\sigma^t_y(a) - \sigma^t_y\sigma^t_x(a) - [f^t(x, y), a] \tag{1.9}$$

for all $x, y \in B$ and $a \in F$. We are prompted to have the following definition:

Definition 1.5.5 A quadruple (B, A, σ, f), where A and B are Lie algebras over a field F, σ is a linear transformation from B to $Der(A)$, and f is an alternating map from $B \times B$ to A, is called a **Schreier factor system** if it satisfies the equations (1.3), (1.8), and (1.9) with σ^t and f^t replaced by σ and f, respectively.

Thus, given an extension E of A by B together with a section t, we obtain a factor system (B, A, σ^t, f^t). Conversely, let (B, A, σ, f) be a factor system. Consider $L = A \times B$. L is a vector space with respect to the coordinate-wise operation. Define the bracket operation [,] by putting

$$[(a, x), (b, y)] = ([a, b] - \sigma_y(a) + \sigma_x(b) + f(x, y), [x, y]).$$

Using the defining conditions for the factor system, it can be easily verified that L is a Lie algebra and

$$0 \longrightarrow A \stackrel{\alpha}{\to} L \stackrel{\beta}{\to} B \longrightarrow 0$$

is an extension of A by B, where $\alpha(a) = (a, 0)$ and $\beta((a, x)) = x$. The map t from B to L given by $t(x) = (0, x)$ is a section such that $(B, A, \sigma, f) = (B, A, \sigma^t, f^t)$.

Let

$$E_1 \equiv 0 \longrightarrow A_1 \stackrel{\alpha_1}{\to} L_1 \stackrel{\beta_1}{\to} B_1 \longrightarrow 0$$

and

$$E_2 \equiv 0 \longrightarrow A_2 \stackrel{\alpha_2}{\to} L_2 \stackrel{\beta_2}{\to} B_2 \longrightarrow 0$$

be extensions in EXT together with sections t_1 and t_2 of E_1 and E_2, respectively. Let (λ, μ, ν) be a morphism from E_1 to E_2. Using the commutativity of the relevant diagram,

$$\beta_2\mu t_1(x) = \nu\beta_1 t_1(x) = \nu(x) = \beta_2 t_2\nu(x)$$

for all $x \in B_1$. Consequently, there is a unique element $g(x) \in A_2$ such that

$$\mu t_1(x) = t_2\nu(x) + \alpha_2 g(x) \tag{1.10}$$

for all $x \in B_1$. This defines a map g from B_1 to A_2. Since μ, ν, t_1, t_2 are vector space homomorphisms and α_2 is an injective homomorphism, it follows that g is a vector space homomorphism. Now,

$$\begin{aligned}&\mu([t_1(x), t_1(y)])\\ &= \mu(t_1([x, y]) + \alpha_1(f^{t_1}(x, y)))\\ &= t_2(\nu([x, y])) + \alpha_2(g([x, y]) + \alpha_2(\lambda(f^{t_1}(x, y)))\end{aligned} \tag{1.11}$$

On the other hand, since μ is a Lie algebra homomorphism,

$$\begin{aligned}&\mu([t_1(x), t_1(y)])\\ &= [\mu(t_1(x)), \mu(t_1(y))]\\ &= [t_2(\nu(x)) + \alpha_2(g(x)), t_2(\nu(y)) + \alpha_2(g(y))]\\ &= t_2(\nu([x, y])) + \alpha^2(f^{t_2}(\nu(x), \nu(y))) - \alpha_2(\sigma^{t_2}_{\nu(y)}(g(x))) +\\ &\alpha_2(\sigma^{t_2}_{\nu(x)}(g(y))) + \alpha_2([g(x), g(y)])\end{aligned} \tag{1.12}$$

Comparing (1.11) and (1.12), we get

$$\begin{aligned}f^{t_2}(\nu(x), \nu(y)) - \sigma^{t_2}_{\nu(y)}(g(x)) + \sigma^{t_2}_{\nu(x)}(g(y)) + [g(x), g(y)] =\\ \lambda(f^{t_1}(x, y)) + g([x, y])\end{aligned} \tag{1.13}$$

for all $x, y \in B_1$. Further,
$\alpha_2(\lambda(\sigma^{t_1}_x(a)))$
$+ \mu(\alpha_1(\sigma^{t_1}_x(a)))$
$= [\mu(t_1(x)), \mu(\alpha_1(a))]$
$= [t_2(\nu(x)) + \alpha_2(g(x)), \alpha_2(\lambda(a))]$
$= \alpha_2(\sigma^{t_2}_{\nu(x)}(\lambda(a))) + \alpha_2([g(x), \lambda(a)])$.
Since α_2 is an injective homomorphism,

$$\lambda(\sigma^{t_1}_x(a)) = \sigma^{t_2}_{\nu(x)}(\lambda(a)) + [g(x), \lambda(a)] \tag{1.14}$$

for all $x \in B_1$ and $a \in A$. Thus, a morphism (λ, μ, ν) between extensions E_1 and E_2 together with choices of sections t_1 and t_2 of the corresponding extensions induces a linear map g from B_1 to A_2 such that the triple (ν, g, λ) satisfies (1.13) and (1.14), and it may be viewed as a morphism from the factor system $(B_1, A_1, \sigma^{t_1}, f^{t_1})$ to $(B_2, A_2, \sigma^{t_2}, f^{t_2})$.

Let $(\lambda_1, \mu_1, \nu_1)$ be a morphism from an extension

$$E_1 \equiv 0 \longrightarrow A_1 \stackrel{\alpha_1}{\to} L_1 \stackrel{\beta_1}{\to} B_1 \longrightarrow 0$$

to an extension

$$E_2 \equiv 0 \longrightarrow A_2 \stackrel{\alpha_2}{\to} L_2 \stackrel{\beta_2}{\to} B_2 \longrightarrow 0,$$

and $(\lambda_2, \mu_2, \nu_2)$ be that from the extension E_2 to

$$E_3 \equiv 0 \longrightarrow A_3 \stackrel{\alpha_3}{\rightarrow} L_3 \stackrel{\beta_3}{\rightarrow} B_3 \longrightarrow 1.$$

Let t_1, t_2 and t_3 be corresponding choices of the sections. Then as in (1.10),

$$\mu_1(t_1(x)) = t_2(\nu_1(x)) + \alpha_2(g_1(x))$$

and

$$\mu_2(t_2(u)) = t_3(\nu_2(u)) + \alpha_3(g_2(u)),$$

where g_1 is the uniquely determined linear map from B_1 to A_2, and g_2 is that from B_2 to A_3. In turn,

$$\mu_2(\mu_1(t_1(x))) = \mu_2(t_2(\nu_1(x)) + \alpha_2(g_1(x))) = \mu_2(t_2(\nu_1(x))) + \mu_2(\alpha_2(g_1(x))) =$$
$$t_3(\nu_2(\nu_1(x))) + \alpha_3(g_2(\nu_1(x))) + \alpha_3(\lambda_2(g_1(x))) = t_3((\nu_2 o \nu_1)(x)) + \alpha_3(g_3(x)),$$

where $g_3(x) = g_2(\nu_1(x)) + \lambda_2(g_1(x))$. It follows that the composition $(\lambda_2 \circ \lambda_1, \mu_2 \circ \mu_1, \nu_2 \circ \nu_1)$ induces the triple $(\nu_2 \circ \nu_1, g_3, \lambda_2 \circ \lambda_1)$, where $g_3(x) = g_2(\nu_1(x)) + \lambda_2(g_1(x))$ for each $x \in B_1$.

Prompted by the above discussion, we introduce the category **FACS** whose objects are factor systems, and a morphism from $(B_1, A_1, \sigma^1, f^1)$ to $(B_2, A_2, \sigma^2, f^2)$ is a triple (ν, g, λ), where ν is a homomorphism from B_1 to B_2, λ a homomorphism from A_1 to A_2, and g a linear map from B_1 to A_2 satisfying (1.13) and (1.14) with f^{t_1} replaced by f^1, f^{t_2} replaced by f^2, σ^{t_1} replaced by σ^1, and σ^{t_2} replaced by σ^2.

The following theorem is the consequence of the above discussion.

Theorem 1.5.6 *Let t_E be a choice of a section of the extension E of a Lie algebra by another Lie algebra (such a choice function t exists because of the axiom of choice). Then the association Fac which associates with each extension E the factor system $Fac(E, t_E)$ is an equivalence between the category EXT of extensions to the category* **FACS** *of factor systems.* ♯

Fix a pair A and B of Lie algebras. We try to describe the equivalence classes of extensions of A by B. Let L be an extension of A by B given by the exact sequence

$$E \equiv 0 \longrightarrow A \stackrel{\alpha}{\rightarrow} L \stackrel{\beta}{\rightarrow} B \longrightarrow 0.$$

Let (λ, μ, ν) be an equivalence from this extension to another extension L' of A' by B' given by the exact sequence

$$E' \equiv 0 \longrightarrow A' \stackrel{\alpha'}{\rightarrow} L' \stackrel{\beta'}{\rightarrow} B' \longrightarrow 0$$

Then μ is an isomorphism and it follows that L' is also an extension of A by B given by the exact sequence

$$E'' \equiv 0 \longrightarrow A \overset{\alpha' \circ \lambda}{\rightarrow} L' \overset{\nu^{-1} \circ \beta'}{\rightarrow} B \longrightarrow 0$$

such that (I_A, μ, I_B) is an equivalence from E to E''. Also, $(\lambda, I_{L'}, \nu)$ is an equivalence from E'' to E'.

As such, there is no loss of generality in restricting the concept of equivalence on the class $E(A, B)$ of all extensions of A by B by saying that two extensions

$$E_1 \equiv 0 \longrightarrow A \overset{\alpha_1}{\rightarrow} L_1 \overset{\beta_1}{\rightarrow} B \longrightarrow 0$$

and

$$E_2 \equiv 0 \longrightarrow A \overset{\alpha_2}{\rightarrow} L_2 \overset{\beta_2}{\rightarrow} B \longrightarrow 0$$

in $E(A, B)$ are equivalent if there is an isomorphism ϕ from L_1 to L_2 such that the diagram

$$\begin{array}{ccccccccc}
0 & \longrightarrow & A & \overset{\alpha_1}{\longrightarrow} & L_1 & \overset{\beta_1}{\longrightarrow} & B & \longrightarrow & 0 \\
 & & \downarrow {\scriptstyle I_A} & & \downarrow {\scriptstyle \phi} & & \downarrow {\scriptstyle I_B} & & \\
0 & \longrightarrow & A & \underset{\alpha_2}{\longrightarrow} & L_2 & \underset{\beta_2}{\longrightarrow} & B & \longrightarrow & 0
\end{array}$$

is commutative. Indeed, for any extension E in *EXT* which is equivalent to a member E' of $E(A, B)$, there is a member E'' of $E(A, B)$ such that E is equivalent to E'' in the category *EXT* and E'' in $E(A, B)$ is an equivalent E' in the sense described above.

Let

$$E \equiv 0 \longrightarrow A \overset{\alpha}{\rightarrow} L \overset{\beta}{\rightarrow} B \longrightarrow 0$$

be an extension of A by B. Let t and t' be sections of the extension. Then there is a linear map g from B to A given by

$$t'(x) = t(x) + \alpha(g(x)).$$

Now,

$$\alpha(\sigma_x^{t'}(a)) = [t'(x), \alpha(a)] = [t(x) + \alpha(g(x)), \alpha(a)] = \alpha(\sigma_x^t(a) + [g(x), a])$$

for all $x \in B$ and $a \in A$. Thus,

$$\sigma_x^{t'}(a) = \sigma_x^t(a) + ad(g(x))(a)$$

for all $x \in B$ and $a \in A$. Thus, the linear map σ^t from B to $Der(A)$ depends on t (note that it is not a Lie homomorphism). It follows from the above equation that the map $x \mapsto \sigma^t_x IDer(A)$ from B to the Lie algebra $OutDer(A)$ of outer derivations of A is independent of the choice of the section t. Thus, for each extension E of A by B, we have a map Ψ_E from B to $OutDer(A)$ given by $\Psi_E(x) = \sigma^t_x IDer A$. Evidently, Ψ_E is a linear map. It follows from Eq. (1.9) that Ψ_E is a Lie algebra homomorphism. If an extension E_1 of A by B is equivalent to an extension E_2 of A by B, then it can be easily seen that $\Psi_{E_1} = \Psi_{E_2}$.

Definition 1.5.7 A homomorphism from B to $OutDer(A) = Der(A)/IDer(A)$ is called a **coupling** or an **abstract kernel** of B to A.

We have established the following theorem.

Theorem 1.5.8 *Let* Ext (A, B) *denote the set of all equivalence classes of extensions in E(A, B). Then there is a natural map* Ψ *from* Ext (A, B) *to the set* $Hom(B, OutDer(A))$ *of all abstract kernels (couplings) of B to A given by* $\Psi([E]) = \Psi_E$ *as defined above.*♯

The map Ψ described in the above theorem is called the **abstract kernel** map.

The abstract kernel map Ψ need not be injective. In other words, two nonequivalent extensions of A by B may induce the same abstract kernels of B to A (give an example). We shall see that the map Ψ may not be surjective also. Indeed, we have two basic problems in the theory of extensions of Lie algebras.

1. To determine the abstract kernels $\eta \in Hom(B, OutDer(A))$ which are realizable from an extension E of A by B in the sense that $\Psi([E]) = \eta$.
2. Given an abstract kernel $\eta \in Hom(B, OutDer(A))$ which is realizable from an extension, to determine and classify all extensions E up to equivalence such that $\Psi(E) = \eta$. Such abstract kernels are call couplings.

Theorem 1.5.9 *Let A be a Lie algebra with* $Z(A) = \{0\}$. *Then the map* Ψ *from* Ext (A, B) *to the set* $Hom(B, OutDer(A))$ *is bijective. More explicitly, every abstract kernel* η *of B to A determines and is determined uniquely by an equivalence class of extensions in* Ext (A, B).

Proof Let $\eta \in Hom(B, OutDer(A))$ be an abstract kernel of B to A. Consider the Pull Back Diagram

$$\begin{array}{ccc} L & \xrightarrow{p_2} & B \\ {\scriptstyle p_1}\downarrow & & \downarrow{\scriptstyle \eta} \\ Der(A) & \xrightarrow{\nu} & OutDer(A) \end{array}$$

More explicitly, L is the Lie subalgebra of the direct product $Der(A) \times B$ given by $L = \{(\sigma, x) \mid \sigma \in Der(A) \text{ and } \sigma IDer(A) = \eta(x)\}$, p_1 the first projection, and

p_2 the second projection. Clearly, p_2 is a surjective homomorphism from L to B. The $kerp_2 = \{(\sigma, 0) \mid \sigma IDer(A) = \eta(0) = IDer(A)\} = IDer(A) \times \{0\}$. Since the center $Z(A)$ of A is trivial, the map α from A to L defined by $\alpha(a) = (i_a, 0)$ (i_a denotes the inner derivation determined by a) is an injective homomorphism with $image\alpha = kerp_2$. This gives an extension E of A by B given by the exact sequence

$$E \equiv 0 \longrightarrow A \stackrel{\alpha}{\to} L \stackrel{p_2}{\to} B \longrightarrow 0.$$

Using the axiom of choice, there is a map ξ from B to $Der(A)$ such that $\xi(x)IDer(A) = \eta(x)$. This determines a section t of the extension E given by $t(x) = (\xi(x), x)$. Recall that the abstract kernel $\Psi(E)$ associated with the extension E is given by $\Psi(E)(x) = \sigma_x^t IDer(A)$, where σ_x^t is given by

$$[t(x), \alpha(a)] = \alpha(\sigma_x^t(a)).$$

Now,

$$\alpha(\sigma_x^t(a)) = [t(x), \alpha(a)] = [(\xi(x), x), (i_a, 0)] = ([\xi(x), i_a], 0) = (i_{\xi(x)(a)}, 0) = \alpha(\xi(x)(a)).$$

Thus, $\sigma_x^t(a) = \xi(x)(a)$ for all $a \in A$. In turn, $\sigma_x^t = \xi(x)$. By the definition, $\Psi(E)(x) = \sigma_x^t IDer(A) = \xi(x)IDer(A) = \eta(x)$ for all $x \in B$. This shows that $\Psi(E) = \eta$ and so Ψ is surjective.

To prove the injectivity, suppose that $\Psi(E_1) = \Psi(E_2)$, where E_1 and E_2 are extensions of A by B given by

$$E_1 \equiv 0 \longrightarrow A \stackrel{\alpha_1}{\to} L_1 \stackrel{\beta_1}{\to} B \longrightarrow 0$$

and

$$E_2 \equiv 0 \longrightarrow A \stackrel{\alpha_2}{\to} L_2 \stackrel{\beta_2}{\to} B \longrightarrow 0.$$

Let t_1 be a section of E_1 with a corresponding factor system $(B, A, \sigma^{t_1}, f^{t_1})$, and t_2 be a section of E_2 with the corresponding factor system $(B, A, \sigma^{t_2}, f^{t_2})$. Under our assumption,

$$\sigma_x^{t_1} IDer(A) = \Psi(E_1)(x) = \Psi(E_2)(x) = \sigma_x^{t_2} IDer(A)$$

for all $x \in B$, where $\sigma_x^{t_1}$ and $\sigma_x^{t_2}$ are given by the equations

$$[t_1(x), \alpha_1(a)] = \alpha_1(\sigma_x^{t_1}(a))$$

and

$$[t_2(x), \alpha_2(a)] = \alpha_2(\sigma_x^{t_2}(a)).$$

Since $\sigma_x^{t_1} IDer(A) = \sigma_x^{t_2} IDer(A)$ for all $x \in B$, and since A is center-less, there is a unique linear map g from B to A such that

$$\sigma_x^{t_1} = i_{g(x)} \circ \sigma_x^{t_2} \cdots \quad (1.15)$$

for all $x \in B$. Again by 9, we have

$$\sigma_{[x,y]}^{t_1}(a) = \sigma_x^{t_1}\sigma_y^{t_1}(a) - \sigma_y^{t_1}\sigma_x^{t_1}(a) - [f^{t_1}(x, y), a] \quad (1.16)$$

and

$$\sigma_{[x,y]}^{t_2}(a) = \sigma_x^{t_2}\sigma_y^{t_2}(a) - \sigma_y^{t_2}\sigma_x^{t_2}(a) - [f^{t_2}(x, y), a] \quad (1.17)$$

for all $x, y \in B$ and $a \in A$. Using (1.15), (1.16), (1.17), and the fact that $Z(A) = \{0\}$, it can be shown that (1.13) holds with $\nu = I_B$ and $\lambda = I_A$. It follows that (I_A, g, I_B) is an equivalence between the factor system $(B, A, \sigma^{t_1}, f^{t_1})$ and $(B, A, \sigma^{t_2}, f^{t_2})$. By Theorem 1.5.6, it follows that E_1 and E_2 are equivalent. ♯

Indeed, the proof of the above theorem establishes the following more general result.

Proposition 1.5.10 *Let E_1 and E_2 be extensions of A by B with $\Psi(E_1) = \Psi(E_2)$. Then the following induced extensions E_1' and E_2' of $A/Z(A)$ by B given below are equivalent:*

$$E_1' \equiv 0 \longrightarrow A/Z(A) \stackrel{\overline{\alpha_1}}{\to} L_1/\alpha_1(Z(A)) \stackrel{\overline{\beta_1}}{\to} B \longrightarrow 0,$$

$$E_2' \equiv 0 \longrightarrow A/Z(A) \stackrel{\overline{\alpha_2}}{\to} L_2/\alpha_2(Z(A)) \stackrel{\overline{\beta_2}}{\to} B \longrightarrow 1. \sharp$$

So far we described the extensions of Lie algebras with trivial centers. Let us consider the other extreme case when the center of the Lie algebra is the Lie algebra itself. More explicitly, we describe the extensions of abelian Lie algebras. Let A be an abelian Lie algebra. Then the derivation algebra $Der(A)$ is a subalgebra of the endomorphism algebra $End(A)$ and $IderA = \{0\}$. The Lie algebra $OutDer(A)$ is naturally identified with $Der(A)$. An abstract kernel of B to A is a homomorphism σ from B to $Der(A)$. We discuss the following problem:
Problem: Let A be an abelian Lie algebra. Classify all extensions of A by B (up to equivalence) with the given abstract kernel σ.

Let us denote by $EXT_\sigma(A, B)$ the set of equivalence classes of extensions of A by B with the given abstract kernel σ. We have at least one such extension, viz., the semi-direct product extension of A by B associated with the homomorphism σ. Clearly, a factor system associated with the split extension is (B, A, σ, f_0), where f_0 is trivial in the sense that $f_0(x, y) = 0$ for all $x, y \in K$. Let $Z^2_\sigma(B, A)$ denote the set of factor systems (B, A, σ, f) associated with the abstract kernel σ. Indeed, a factor system in $Z^2_\sigma(B, A)$ determines, and it is uniquely determined by, the corresponding map f which satisfies the equations (8) and (9) with f^t replaced by f and σ^t_x replaced by $\sigma(x)$. By the abuse of language, we shall call such an f as a factor system in $Z^2_\sigma(B, A)$. f is also called a 2-co-cycle associated with (B, A, σ). It is easily observed that $f + f' \in Z^2_\sigma(B, A)$, since A is abelian. Also $-f \in Z^2_\sigma(B, A)$ for all $f \in Z^2_\sigma(B, A)$. Thus, $Z^2_\sigma(B, A)$ is an abelian group with respect to the operation defined above. f_0 is the identity of the group. Let $B^2_\sigma(B, A)$ denote the set of factor systems which are equivalent to the trivial factor system f_0. More precisely, from (1.13), $f \in B^2_\sigma(B, A)$ if and only if there is a vector space homomorphism g from B to A such that $f(x, y) = g([x, y]) - \sigma(x)(g(y)) + \sigma(y)(g(x))$. Note that for any vector space homomorphism g, the map ∂g from $B \times B$ to A defined by $\partial g(x, y) = g([x, y]) - \sigma(x)(g(y)) + \sigma(y)(g(x))$ is a factor system. The members of $B^2_\sigma(B, A)$ are called the 2-co-boundaries associated with (B, A, σ). The quotient group $Z^2_\sigma(B, A)/B^2_\sigma(B, A)$ is called the **second co - homology group** associated with (B, A, σ), and it is denoted by $H^2_\sigma(B, A)$.

Theorem 1.5.11 *Let A be an abelian Lie algebra, and B be a Lie algebra. Let σ be an abstract kernel of B to A. Then, there is a natural bijective correspondence Γ between the set $EXT_\sigma(A, B)$ of equivalence classes of extensions of A by B with the given abstract kernel σ and the second co-homology group $H^2_\sigma(B, A)$.*

Proof Let E be an extension of A by B with the abstract kernel σ. Let t be a section of the extension, and (B, A, σ, f^t) be the corresponding factor system. Then $f^t \in Z^2_\sigma(B, A)$. Let E' be another equivalent extension of A by B, and t' be a section of the extension E'. Let $(B, A, \sigma, f^{t'})$ be the corresponding factor system. Then (see the equation 13) there is a vector space homomorphism g from B to A such that $f^t(x, y) + g([x, y]) = \sigma(x)(g(y)) - \sigma(y)(g(x)) + f^{t'}(x, y)$ (note that A is abelian). This shows that $f^t + B^2_\sigma(B, A) = f^{t'} + B^2_\sigma(B, A)$. Thus, the association $(E, t) \mapsto f^t$ induces a map Γ from $EXT_\sigma(A, B)$ to $H^2_\sigma(B, A)$ given by $\Gamma([E]) = f^t + B^2_\sigma(B, A)$, where t is a section of E. Let $f \in Z^2_\sigma(B, A)$. Then by the discussion following Definition 1.5.5, there is an extension E of A by B, and a section t such that $f^t = f$. This shows that Γ is surjective. Let E_1 and E_2 be extensions of A by B with sections t_1 and t_2 and the abstract kernel σ such that $\Gamma([E_1]) = \Gamma([E_2])$. Then $f^{t_1} + B^2_\sigma(B, A) = f^{t_2} + B^2_\sigma(B, A)$. Hence there exists a vector space homomorphism g from B to A such that $f^{t_1}(x, y) + g([x, y]) = \sigma(x)(g(y)) - \sigma(y)(g(x)) + [g(x), g(y)] + f^{t_2}(x, y)$ (note that $[g(x), g(y)] = 0$). It follows that the factor system f^{t_1} is equivalent f^{t_2}. Hence the corresponding extensions E_1 and E_2 are equivalent. ♯

Let A be a Lie algebra (not necessarily abelian), and B be another Lie algebra. Let $\psi : B \longmapsto OutDer(A) = Der(A)/IDer(A)$ be an abstract kernel. If $d \in Der(A)$ and $a \in Z(A)$, then $[d(a), b] = d([a, b]) - [a, d(b)] = 0$ for all $b \in A$. This means that $d(a) \in Z(A)$. In turn, we get a homomorphism $\chi : Der(A) \longmapsto Der(Z(A))$ given by $\chi(d) = d|_{(Z(A)}$. Let $\sigma : B \longmapsto Der(A)$ be a vector space homomorphism which is the lifting of ψ in the sense that $\nu o \sigma = \psi$, where ν is the quotient map from $Der(A)$ to $OutDer(A)$. Since ψ is a homomorphism, $\sigma([x, y])IDer(A) = [\sigma(x), \sigma(y)]IDer(A)$. Hence there is a map f from $B \times B$ to A such that $[\sigma(x), \sigma(y)] = i_{f(x,y)}\sigma([x, y])$ (recall that i_a denote the inner derivation determined by a). It follows that $(\chi o\sigma)([x, y]) = [(\chi o\sigma)(x), (\chi o\sigma)(y)]$ $for\ all\ x, y \in B$. This means that $\chi o\sigma$ is a homomorphism from B to $Der(Z(A)) = End(Z(A))$. Let τ be another lifting of ψ. Then $\sigma(x)IDer(A) = \tau(x)IDer(A)$ $for\ all\ x \in B$. Hence there is a vector space homomorphism g from B to A such that $\sigma(x) = i_{g(x)}\tau(x)$ for all $x \in B$. But then $(\chi o\sigma(x)) = (\chi o\tau(x))$ for all $x \in B$. Thus, $\chi o\sigma$ depends only on ψ and not on any particular lifting σ. In turn, χ induces a map $\overline{\chi}$ from the set $Hom(B,\ OutDer(A))$ of abstract kernels from B to A to the set $Hom(B,\ Z(A))$ of abstract kernel from B to $Z(A)$ given by $\overline{\chi}(\psi) = \chi o\sigma$, where σ is a lifting of ψ.

Proposition 1.5.12 *Let*

$$E \equiv 0 \longrightarrow A \overset{\alpha}{\rightarrow} L \overset{\beta}{\rightarrow} B \longrightarrow 0$$

and

$$E' \equiv 0 \longrightarrow A \overset{\alpha'}{\rightarrow} L' \overset{\beta'}{\rightarrow} B \longrightarrow 0$$

be extensions of A by B such that $\psi_E = \psi_{E'} = \psi$. *Then there is a section t of E and a section t′ of E′ such that* $\sigma^t = \sigma^{t'} = \overline{\chi}(\psi)$, *and* $-f^t(x, y) + f^{t'}(x, y) \in Z(A)$ $for\ all\ x, y \in B$. *Then the map h from* $B \times B$ *to* $Z(A)$ *defined by* $h(x, y) = -f^t(x, y) + f^{t'}(x, y)$ *is a 2-co-cycle in* $Z^2_{\overline{\chi}(\psi)}(B, Z(A))$.

Proof Let t be a section of E, and s be a section of E'. Since $\psi_E = \psi_{E'}$, $\sigma^t(x)IDer(A) = \sigma^s(x)IDer(A)$ for all $x \in B$. This means that there is a linear map g from B to A such that $\sigma^t(x) = i_{g(x)}\sigma^s(x)$ $for\ all\ x \in B$. The map t' from B to L' given by $t'(x) = g(x) + s(x)$ is also a linear section of E'. Further, $\sigma^{t'}(x) = i_{g(x)}\sigma^s(x) = \sigma^t(x)$ for all x in B. This shows that $\sigma^t = \sigma^{t'}$. Now $f^t(x, y) = [t(x), t(y)] - t([x, y])$ and $f^{t'}(x, y) = [t'(x), t'(y)] - t'([x, y])$. Hence $i_{f^t(x,y)} = [\sigma^t(x), \sigma^t(y)] - (\sigma^t([x, y])) = [\sigma^{t'}(x), \sigma^{t'}(y)] - (\sigma^{t'}([x, y])) = i_{f^{t'}(x,y)}$ for all $x, y \in B$. Thus, $i_{-f^t(x,y) + f^{t'}(x,y)} = 0$. This shows that $-f^t(x, y) + f^{t'}(x, y) \in Z(A)$ for all $x, y \in B$. Put $h(x, y) = -f^t(x, y) + f^{t'}(x, y)$. It is straightforward to verify that $h \in Z^2_{\overline{\chi}(\psi)}(B, Z(A))$. ♯

Theorem 1.5.13 *Let* $\psi : B \longmapsto OutDer(A)$ *be an abstract kernel from B to A which is realizable by an extension of A by B. Then the second co-homology group* $H^2_{\overline{\chi}(\psi)}(B, Z(A))$ *acts sharply transitively on the set* $EXT_\psi(A, B)$ *of equivalence classes of extensions of A by B associated with the abstract kernel* ψ.

Proof Let E be an extension of A by B which realizes the abstract kernel ψ, and let t be a section of E. Let (B, A, σ^t, f^t) be the corresponding factor system. Then $\psi(x) = \sigma^t_x IDer(A)$ $for\ all\ x \in B$. Let $h \in Z^2_{\overline{\chi}(\psi)}(B, Z(A))$. It is easily seen that $(B, A, \sigma^t, f^t + h)$ is again a factor system. Let $E \star h$ denote the corresponding extension. Clearly, $E \star h$ also realizes ψ. Let h' be another 2-co-cycle in

$Z^2_{\overline{\chi}(\psi)}(B, Z(A))$ such that the co-homology class $[h] = h + B^2_{\overline{\chi}(\psi)}(B, Z(A)) = [h'] = h' + B^2_{\overline{\chi}(\psi)}(B, Z(A))$ in $H^2_{\overline{\chi}(\psi)}(B, Z(A))$. Then there is a linear map $g : B \longmapsto Z(A) \subseteq A$ such that $h'(x, y) = \partial g(x, y) + h(x, y)$ for all $x, y \in B$. Clearly $f^t + h' = f^t + h + \partial g$. Hence $(B, A, \sigma^t, f^t + h)$ is equivalent to $(B, A, \sigma^t, f^t + h')$. This shows that $[E \star h] = [E \star h']$. Let E and E' be equivalent extensions of A by B which realize ψ and $h \in Z^2_{\overline{\chi}(\psi)}(B, Z(A))$. By Theorem 1.5.6, we have sections t and t' of E and E', respectively, such that (B, A, σ^t, f^t) is equivalent to $(B, A, \sigma^{t'}, f^{t'})$. Hence, there is a linear map g from B to A such that $f^t(x, y) = \partial g(x, y) + f^{t'}(x, y)$ for all $x, y \in B$. Clearly, $(B, A, \sigma^t, f^t + h)$ is equivalent to $(B, A, \sigma^{t'}, f^{t'} + h)$, and so $[E \star h] = [E' \star h]$. Thus, we get an action $\star$ of $H^2_{\overline{\chi}(\psi)}(B, Z(A)$ on $EXT_\psi(A, B)$ given by $[E] \star [h] = [E \star h]$. We show that this action is sharply transitive. Let E and E' be extensions realizing the abstract kernel ψ. By the Proposition 1.5.12, there is a section t of E, and there is a section t' of E' such that $\sigma^t = \sigma^{t'} = \overline{\chi}(\psi)$, and the map h from $B \times B$ to $Z(A)$ defined by $h(x, y) = -f^t(x, y) + f^{t'}(x, y)$ is a 2-co-cycle in $Z^2_{\overline{\chi}(\psi)}(B, Z(A))$. Clearly, $[E] \star [h] = [E']$. This shows that the action $\star$ is transitive. Next, suppose that $[E] \star [h] = [E]$. Then there is a section t of E such that the factor system (B, A, σ^t, f^t) is equivalent to $(B, A, \sigma^t, f^t + h)$. Hence there is a linear map g from B to A such that $f^t(x, y) + h(x, y) = \partial g(x, y) + + f^t(x, y)$ for all $x, y \in B$ and also $i_{g(x)} o \sigma^t_x(h) = \sigma^t_x(h)$ for all $x \in B$ and $h \in A$. Since σ^t_x is a derivation of A, it follows that $g(x) \in Z(A)$ for all $x \in B$. Thus, $h(x, y) = \sigma^t_x(g(y)) - g([x, y]) + \sigma^t_y(g(x))$ for all $x, y \in B$. This shows that $h = \partial g$, where g is a linear map from B to $Z(A)$. It follows that $[h] = 0$. This completes the proof of the fact that the action $\star$ is sharply transitive. ♯

Corollary 1.5.14 *There is a bijective correspondence between $EXT_\psi(A, B)$ to $H^2_{\overline{\chi}(\psi)}(B, Z(A))$ provided there is an extension of A by B which realizes ψ.* ♯

We state the lemma of Whitehead and prove the Radical splitting theorem due to Levi:

Lemma 1.5.15 *If L is a semi-simple Lie algebra and A is a finite-dimensional L-module, then $H^2(L, A) = \{0\} = H^1(L, A)$.* ♯

Theorem 1.5.16 (Levi–Malcev) *Let L be a finite-dimensional Lie algebra. Then the following exact sequence is a split exact sequence:*

$$0 \longrightarrow R(L) \xrightarrow{i} L \xrightarrow{\nu} L/R(L) \longrightarrow 0,$$

where $R(L)$ is the solvable radical of L.

Proof The proof is by the induction on the derived length of $R(L)$. Suppose that $R(L)$ is abelian. Since $L/R(L)$ is semi-simple, by the Whitehead lemma, $H^2(L/R(L), A) = 0$. By Theorem 1.5.11, the result follows. Assume that the result holds for all Lie algebras for which the derived length is n. Let L be a Lie algebra such that the derived length of $R(L)$ is $n + 1$. We have the exact sequence

$$0 \longrightarrow R(L)/[R(L), R(L)] \xrightarrow{i} L/[R(L), R(L)] \longrightarrow L/R(L) \longrightarrow 0.$$

Clearly, $R(L)/[R(L), R(L)]$ is abelian and it is the radical of $L/[R(L), R(L)]$. It follows from our earlier argument that the above exact sequence splits. Let t be an splitting. Then $t(L/R(L)) = L'/[R(L), R(L)]$ for some subalgebra L' of L. Again, we have an exact sequence

$$0 \longrightarrow [R(L), R(L)] \xrightarrow{i} L' \xrightarrow{\nu} L'/[R(L), R(L)] \longrightarrow 0,$$

where $[R(L), R(L)]$ is the radical of L' of derived length n. By the induction hypothesis, the above sequence splits. If s is the splitting, then st can be realized as the splitting of

$$0 \longrightarrow R(L) \xrightarrow{i} L \longrightarrow L/R(L) \longrightarrow 0.\sharp$$

Exercises

1.5.1. Show that if L is a free Lie algebra, then $H^n(L, A) = \{0\}$ for all L-modules A and $n \geq 2$.

1.5.2. Let

$$0 \longrightarrow A \xrightarrow{\alpha} L \xrightarrow{\beta} B \longrightarrow 0$$

be an exact sequence. As in the case of groups (see Algebra 2, Chap. 10), establish the five-term exact sequence

$$H_2(L, V) \longrightarrow H_2(B, V) \longrightarrow A_{ab} \otimes V \longrightarrow H_1(L, V) \longrightarrow H_1(B, V) \longrightarrow 0.$$

1.5.3. Let L be a Lie algebra and

$$0 \longrightarrow R \xrightarrow{i} F \xrightarrow{f} L \longrightarrow 0$$

be a presentation of the Lie algebra, where F is a free Lie algebra. Note (Theorem 1.2.12) that R is also a free Lie algebra. Show that

$$E = 0 \longrightarrow R/[R, F] \xrightarrow{\hat{i}} F/[R, F] \xrightarrow{\hat{f}} L \longrightarrow 0$$

is a free central extension of L in the sense that given any central extension

$$E' = 0 \longrightarrow A \xrightarrow{\alpha} B \xrightarrow{\beta} L' \longrightarrow 0$$

and a homomorphism η from L to L', there is a morphism (ρ, τ, η) from E to E'.

1.5.4*. Let

$$0 \longrightarrow R \xrightarrow{i} F \xrightarrow{f} L \longrightarrow 0$$

and

$$0 \longrightarrow R' \stackrel{i}{\to} F' \stackrel{f}{\to} L \longrightarrow 0$$

be two free presentations of L. Show that $(R \bigcap [F, F])/[R, F] \approx (R' \bigcap [F', F'])/[R', F]'$. The unique (up to isomorphism) group $(R \bigcap [F, F])/[R, F]$ is called the Schur Multiplier of L. Suppose further that L is of finite dimension. Show that $[F, F]/[R, F]$ is of finite dimension.

1.5.5. Describe the Schur multiplier of the non-abelian Lie algebra of dimension 2.

Chapter 2
Semi-Simple Lie Algebras and Root Systems

The structure theory of semi-simple Lie algebras, Geometry of root systems, Dynkin diagrams, classification of semi-simple Lie algebras, Existence theorem, the Theorem of Serre, and the isomorphism theorem constitute the subject matter of this chapter.

2.1 Root Space Decomposition

This section is devoted to studying the structure of a semi-simple Lie algebra and describing it through root space decomposition.

A subalgebra T of a Lie algebra L is called a **Toral subalgebra** if $ad(x)$ is a semi-simple element of $gl(L)$ for all $x \in T$. If L is a nilpotent Lie algebra, then from the theorem of Engel, $ad(x)$ is nilpotent for all $x \in L$. Hence, a nilpotent Lie algebra has no nontrivial toral subalgebra. Suppose that L is a semi-simple Lie algebra. Then there is an element $x \in L$ such that $ad(x)$ is non-nilpotent. Suppose that $ad(x) = (ad(x))_s + (ad(x))_n = ad(x_s) + ad(x_n)$ is the Jordan decomposition of $ad(x)$ in $gl(L)$. Then the subalgebra $< x_s >$ generated by x_s is a nontrivial toral subalgebra. Thus, every semi-simple Lie algebra has a nontrivial toral subalgebra.

Proposition 2.1.1 *Every toral subalgebra T of a Lie algebra L is abelian.*

Proof We have to show that $ad(x)(y) = 0$ for all $x, y \in T$. We need to show that all the eigenvalues of $ad(x)|_T$ is zero. Suppose not. Then there is a $y \in T - \{0\}$ and $a \in F^\star$ such that $ad(x)(y) = [x, y] = ay$. But then $ad(y)(x) = -ay$. Thus, $ad(y)(x)$ is an eigenvector of $ad(y)$ corresponding to the eigenvalue 0. Since $ad(y)$ is semi-simple, x is a linear combination of eigenvectors of $ad(y)$. Consequently, applying $ad(y)$ to x, we obtain a linear combination eigenvectors of $ad(y)$ corresponding to nonzero eigenvalues, if there is any. This is a contradiction to the earlier observation. ♯

R. Lal, *Algebra 4*, Infosys Science Foundation Series,
https://doi.org/10.1007/978-981-16-0475-1_2

Evidently, a maximal toral subalgebra of a finite-dimensional Lie algebra exists. Let H be a maximal toral subalgebra of a semi-simple Lie algebra. Then H is abelian. Hence $\{ad(h) \mid h \in H\}$ is a family of pairwise commuting semi-simple endomorphisms of L. Consequently, L is the direct sum of common eigenspaces of $\{ad(h) \mid h \in H\}$. More explicitly, there is a set $\hat{H}$ of maps from H to F such that for all $\alpha \in \hat{H}$, $L_\alpha = \{x \in L \mid ad(h)(x) = \alpha(h)x\} \neq \{0\}$ and $L = \oplus \sum_{\alpha \in \hat{H}} L_\alpha$. Since $ad(h + k)(x) = (\alpha(h) + \alpha(k))(x)$ and $\alpha(ah)x = ad(ah)(x) = aad(h)(x) = a\alpha(h)x$, it follows that $\hat{H} \subseteq H^\star$, where $H^\star$ is a dual space of H. Evidently, $L_0 = C_L(H)$. Let Φ denote the set $\hat{H} - \{0\}$. Φ is called the **root system** of the semi-simple Lie algebra L associated with the maximal toral subalgebra H. The members of Φ are called **roots**. Thus, $L = C_L(H) \oplus (\oplus \sum_{\alpha \in \Phi} L_\alpha)$. This decomposition is termed as the **Cartan decomposition** or the **Root space decomposition** of L associated with the maximal toral subalgebra H (a maximal toral subalgebra is also called a **Cartan subalgebra**).

Example 2.1.2 Consider $sl(n, F)$. Evidently, $d(n, F)$ consisting of diagonal matrices of trace 0 is a toral subalgebra. Since no member of $sl(n, F)$ outside $d(n, F)$ commutes with all the members of $d(n, F)$, it follows that $d(n, F)$ is a maximal toral subalgebra.

For each $\alpha \in H^\star$, let L_α denote the subspace $\{x \in L \mid ad(h)(x) = \alpha(h)x\}$. Evidently, $L_\alpha = \{0\}$ whenever $\alpha \notin \{0\} \bigcup \Phi$.

Proposition 2.1.3 *$[L_\alpha, L_\beta] \subseteq L_{\alpha+\beta}$ for all $\alpha, \beta \in H^\star$. For $\alpha \in \Phi$, all members of L_α are nilpotent. Again, if $\alpha + \beta \neq 0$, then $\kappa(L_\alpha, L_\beta) = 0$, where κ is the Killing form on L.*

Proof Let $x \in L_\alpha$ and $y \in L_\beta$. By the Jacobi identity,
$ad(h)([x, y]) = [h, [x, y]]$
$= -[x, [y, h]] - [y, [h, x]]$
$= \beta(h)[x, y] + \alpha(h)[x, y]$
$= (\alpha(h) + \beta(h))[x, y]$.
This shows that $[x, y] \in L_{\alpha+\beta}$. Let $x \in L_\alpha$, where $\alpha \neq 0$. Since L is finite dimensional, there is a $n \in \mathbb{N}$ such that $L_{n\alpha+\beta} = \{0\}$ for all β. Using the first assertion, we observe that $ad(x)^n = 0$. Consequently, $ad(x)$ and so also the element x is nilpotent. Finally, suppose that $\alpha + \beta \neq 0$. Let $h \in H$ such that $(\alpha + \beta)(h) \neq 0$. Let $x \in L_\alpha$ and $y \in L_\beta$. Using the associativity of the bi-linear form κ,
$\alpha(h)\kappa(x, y)$
$= \kappa([h, x], y])$
$= -\kappa([x, h], y])$
$= -\kappa([x, [h, y])$
$= -\beta(h)\kappa(x, y)$.
Since $\alpha(h) + \beta(h) \neq 0$, $\kappa(x, y) = 0$. ♯

Corollary 2.1.4 *$\kappa|_{C_L(H)}$ is nondegenerate.*

Proof Let $h \in C_L(H)$ and $\kappa(h, k) = 0$ for all $k \in C_L(H)$. It follows from the above proposition that $\kappa(h, x) = 0$ for all $x \in \oplus\sum_{\alpha\in\Phi} L_\alpha$. It also follows that $\kappa(h, z) = 0$ for all $z \in L$. Since L is semi-simple, κ is nondegenerate and so $h = 0$. ♯

Theorem 2.1.5 *If H is a maximal toral subalgebra of a semi-simple Lie algebra L, then $C_L(H) = H$.*

Proof We divide the proof into several easy steps:

Step 1. The subalgebra $C_L(H)$ contains the semi-simple and nilpotent part of each of its elements: Let $x \in C_L(H)$. Then $ad(x)(h) = 0$ for all $h \in H$. Since $ad(x_s)$ and $ad(x_n)$ are polynomials in $ad(x)$ with non-constant terms, it follows that $ad(x_s)(h) = 0 = ad(x_n)(h)$ for each $h \in H$. This shows that $x_s, x_n \in C_L(H)$.

Step 2. H contains all semi-simple elements of $C_L(H)$: Suppose not. Let s be a semi-simple element of $C_L(H)$ which is not in H. Then $H \oplus Ks$ is an abelian Lie subalgebra of L containing only semi-simple elements. Consequently, $H \oplus Ks$ is a toral subalgebra containing H. This is a contradiction, since H is a maximal toral subalgebra of L.

Step 3. $\kappa|_H$ is nondegenerate: Let $h \in H$ such that $\kappa(h, k) = Tr(ad(h)ad(k)) = 0$ for all $k \in H$. We need to show that $h = 0$. Let x be a nilpotent element in $C_L(H)$. Then $ad(x)$ is nilpotent. Since $[x, h] = 0$, $ad(x)$ and $ad(h)$ commute pairwise. Hence $ad(x)ad(h)$ is nilpotent. Consequently, $\kappa(h, x) = Tr(ad(h)ad(x)) = 0$. Since every element of $C_L(H)$ is some semi-simple element and a nilpotent element of $C_L(H)$ and a semi-simple element of $C_L(H)$ is in H, an element y of $C_L(H)$ is expressed as $y = k + x$, where $k \in H$ and x is a nilpotent element of $C_L(H)$. It follows that $\kappa(h, y) = 0$ for all $y \in C_L(H)$. From Corollary 2.1.4, $h = 0$.

Step 4. $C_L(H)$ is a nilpotent subalgebra: Using the Engel theorem, it is sufficient to show that $ad(x)|_{C_L(H)}$ is nilpotent for all $x \in C_L(H)$. Let $x = x_s + x_n$ be the Jordan decomposition of x. Then $ad(x_s)$ and $ad(x_n)$ commute. It is sufficient to show that $ad(x_s)|_{C_L(H)}$ and $ad(x_n)|_{C_L(H)}$ are nilpotent. Already $ad(x_n)$ is nilpotent. Now, by step 2, $x_s \in H$. Hence $ad(x_s)|_{C_L(H)} = 0$. This shows that $ad(x)|_{C_L(H)}$ is nilpotent for all $x \in C_L(H)$.

Step 5. $H\bigcap[C_L(H), C_L(H)] = 0$. Let $x = \sum_{i=1}^{r}[x_i, y_i]$ be a member of $H\bigcap[C_L(H), C_L(H)]$, where $x_i, y_i \in C_L(H)$. Since κ is associative and $[h, x_i] = 0 = [h, y_i]$ for each i, $\kappa(h, [x_i, y_i]) = 0$ for each i. It follow that $\kappa(h, x) = 0$ for all $h \in H$. Since $\kappa|_H$ is nondegenerate, $x = 0$.

Step 6. $C_L(H)$ is an abelian Lie subalgebra: Suppose not. Then $[C_L(H), C_L(H)] \neq \{0\}$. Since $C_L(H)$ is nilpotent, there is a nonzero element $x \in Z(C_L(H))\bigcap [C_L(H), C_L(H)]$. The element x cannot be semi-simple for otherwise x will lie in H, a contradiction to step 5. Thus, the nilpotent part x_n is a nonzero element of $C_L(H)$. Since $ad(x_n)$ is a polynomial of $ad(x)$ without constant term, and $ad(x) = 0$ on $C_L(H)$, it follows that $ad(x_n)$ is zero on $C_L(H)$. This means that $x_n \in Z(C_L(H))$. In turn, it follows that $\kappa(x_n, y) = 0$ for all $y \in C_L(H)$. This is a contradiction, since $\kappa|_{C_L(H)}$ is nondegenerate.

Step 7. $C_L(H) = H$: Suppose not. Then there is a nonzero element $k \in C_L(H) - H$. From step 2, it follows that k is a nilpotent element of $C_L(H)$. Hence $ad(k)$ is nilpotent. Since $C_L(H)$ is abelian, it follows that $\kappa(x, k) = 0$ for all $x \in C_L(H)$. This is a contradiction, since $\kappa|_{C_L(H)}$ is nondegenerate. This completes the proof of the theorem. ♯

Since κ is nondegenerate on H, for each $\lambda \in H^\star$, there is a unique member $t_\lambda \in H$ such that $\kappa(t_\lambda, h) = \lambda(h)$ for all $h \in H$. In particular, we have the subset $\{t_\alpha \mid \alpha \in \Phi\}$ of nonzero elements of H which is in bijective correspondence with Φ and is such that $\alpha(h) = \kappa(t_\alpha, h)$ for all $h \in H$ and $\alpha \in \Phi$.

Theorem 2.1.6 *Let H be a maximal toral subalgebra of a semi-simple Lie algebra L. Let $\Phi \subseteq H^\star$ be the corresponding root system. Let $L = H \oplus \sum_{\alpha \in \Phi} L_\alpha$ be the corresponding root space decomposition. Then the following hold:*

(i) The root system Φ generates the space $H^\star$.
(ii) If $\alpha \in \Phi$, then $-\alpha \in \Phi$.
(iii) For each $\alpha \in \Phi$, $[L_\alpha, L_{-\alpha}]$ is one dimensional having $\{t_\alpha\}$ as a basis.
(iv) For each $\alpha \in \Phi$, $\kappa(t_\alpha, t_\alpha) = \alpha(t_\alpha) \neq 0$.
(v) For each $\alpha \in \Phi$ and a nonzero element $x_\alpha \in L_\alpha$, there is an element $y_\alpha \in L_{-\alpha}$ such that $[x_\alpha, y_\alpha] = h_\alpha = \frac{2t_\alpha}{\kappa(t_\alpha, t_\alpha)}$. Further, the Lie subalgebra generated by $\{x_\alpha, y_\alpha, h_\alpha\}$ is three-dimensional simple Lie algebra isomorphic to $sl(2, F)$ under the correspondence $x_\alpha \mapsto e_{12}$, $y_\alpha \mapsto e_{21}$, and $h_\alpha \mapsto e_{11} - e_{22}$.
(vi) For each $\alpha \in \Phi$, $-t_\alpha = t_{-\alpha}$.

Proof (i) Suppose that $< \Phi > \neq H^\star$. Then there is a nonzero member $h \in H$ such that $\alpha(h) = 0$ for all $\alpha \in \Phi$. This means that $[h, L_\alpha] = 0$ for all $\alpha \in \Phi$. Already $[h, H] = 0$. Hence $h \in Z(L)$. This is a contradiction, since L is semi-simple.

(ii) Let $\alpha \in \Phi$. Then $L_\alpha \neq \{0\}$. Suppose that $-\alpha \notin \Phi$. Then $L_{-\alpha} = \{0\}$. This means that $\kappa(L_\alpha, L_\beta) = 0$ for all $\beta \in \Phi$ (Proposition 2.1.3). Already $\kappa(L_\alpha, H) = 0$. This is a contradiction, since κ is nondegenerate.

(iii) Let $x \in L_\alpha$ and $y \in L_{-\alpha}$. Then $[x, y] \in L_0 = H$. Using the associativity of κ, for each $h \in H$,

$$\begin{aligned}
&\kappa(h, [x, y] - \kappa(x, y)t_\alpha) \\
&= \kappa(h, [x, y]) - \kappa(h, \kappa(x, y)t_\alpha) \\
&= \kappa([h, x], y) - \kappa(h, \kappa(x, y)t_\alpha) \\
&= \kappa(\alpha(h)x, y) - \kappa(h, \kappa(x, y)t_\alpha) \\
&= \alpha(h)\kappa(x, y) - \kappa(h, \kappa(x, y)t_\alpha) \\
&= \kappa(t_\alpha, h)\kappa(x, y) - \kappa(h, \kappa(x, y)t_\alpha) \\
&= 0.
\end{aligned}$$

Since κ is nondegenerate on H, $[x, y] = \kappa(x, y)t_\alpha$. Thus, $[L_\alpha, L_{-\alpha}]$ is generated by t_α. Since $[L_\alpha, L_{-\alpha}] \neq \{0\}$, $t_\alpha \neq 0$.

(iv) Suppose that $\kappa(t_\alpha, t_\alpha) = \alpha(t_\alpha) = 0$. Then $[t_\alpha, x] = 0 = [t_\alpha, y]$ for all $x \in L_\alpha$ and $y \in L_{-\alpha}$. Using (iii), we can find a $x \in L_\alpha$ and a $y \in L_{-\alpha}$ such that $[x, y] = t_\alpha$. It follows that the subspace V generated by $\{x, y, t_\alpha\}$ is a three-dimensional solvable subalgebra. Since L is semi-simple, the map $v \mapsto ad(v)$ is an injective homomorphism from V to $gl(L)$. We denote the image by $ad_L(V)$. By the theorem of Lie, $ad(w)$ is nilpotent for all $w \in [V, V]$. In particular, $ad(t_\alpha)$

is nilpotent. Already, $ad(t_\alpha)$ is semi-simple. Consequently, $ad(t_\alpha) = 0$. This means that $t_\alpha \in Z(L)$. This is impossible, since $t_\alpha \neq 0$.

(v) For each $\alpha \in \Phi$, and for each nonzero element $x_\alpha \in L_\alpha$, there is a member $y_\alpha \in L_{-\alpha}$ such that $\kappa(x_\alpha, y_\alpha) \neq 0$. Indeed, we can choose y_α so that $\kappa(x_\alpha, y_\alpha) = \frac{2}{\kappa(t_\alpha,t_\alpha)}$. Put $h_\alpha = \frac{2t_\alpha}{\kappa(t_\alpha,t_\alpha)}$. Then $[x_\alpha, y_\alpha] = \kappa(x_\alpha, y_\alpha)t_\alpha = h_\alpha$. Further, $[h_\alpha, x_\alpha] = \frac{2[x_\alpha,t_\alpha]}{\kappa(t_\alpha,t_\alpha)} = 2x_\alpha$, $[h_\alpha, y_\alpha] = -2y_\alpha$. This shows that the Lie subalgebra S_α of L generated by $\{x_\alpha, y_\alpha, h_\alpha\}$ is a three-dimensional simple Lie algebra isomorphic to $sl(2, F)$ under correspondence $x_\alpha \mapsto e_{12}$, $y_\alpha \mapsto e_{21}$, *and* $h_\alpha \mapsto e_{11} - e_{22}$.

(vi) Since $h_\alpha = [x_\alpha, y_\alpha] = -[y_\alpha, x_\alpha] = h_{-\alpha}$, $-t_\alpha = t_{-\alpha}$. ♯

Proposition 2.1.7 *Let L be a semi-simple Lie algebra with H as a maximal toral subalgebra. Let Φ be the associated root system. Then for $\alpha \in \Phi$, an integral multiple $i\alpha$ belongs to Φ if and only if $i = \pm 1$.*

Proof Let Φ be a root system. We have already seen (Theorem 2.1.6(ii)) that if $\alpha \in \Phi$, then $-\alpha \in \Phi$. Since L is finite dimensional, for $\alpha \in \Phi$, there is a positive integer $n \geq 1$ such that $L_{-k\alpha} = \{0\}$ for all $k \geq n + 1$. Consider the subspace $V = Fh_\alpha \oplus L_\alpha \oplus L_{-\alpha} \oplus \sum_{k=2}^{n} L_{-k\alpha}$ of L. Clearly, V is invariant under $ad(h)$ for all $h \in H$. Choose $x_\alpha \in L_\alpha, x_{-\alpha} \in L_{-\alpha}$ such that $[x_\alpha, x_{-\alpha}] = h_\alpha$. Now, $ad(h_\alpha)(z) = -k\alpha(h_\alpha)z$ for all $z \in L_{-k\alpha}$. Hence

$$Tr(ad(h_\alpha))|_V = \alpha(h_\alpha)(1 - DimL_{-\alpha} - 2DimL_{-2\alpha} - \cdots - nDimL_{-n\alpha}).$$

Since $h_\alpha = [x_\alpha, x_{-\alpha}]$, $Tr(ad(h_\alpha)) = 0$. Also $DimL_{-\alpha} = 1$. This shows that the $DimL_{-i\alpha} = 0$ for all $i \geq 2$. Hence $-i\alpha$ is a root if and only if $i = 1$. Similarly, $i\alpha$ is a root if and only if $i = 1$. ♯

Corollary 2.1.8 *Let Φ be a root system and $\alpha \in \Phi$. Then a scalar multiple $a\alpha$ of α is a root if and only if $a = \pm 1$.*

Proof It follows from the fact that the weights of representations of h_α are integers (Theorem 1.4.28). ♯

Proposition 2.1.9 *Let L be a semi-simple Lie algebra with H as a maximal toral subalgebra. Let Φ be the associated root system. Let $\alpha, \beta \in \Phi$, $\beta \neq \pm\alpha$. Let r be the largest nonnegative integer such that $L_{\beta-r\alpha} \neq \{0\}$ and q be the largest nonnegative integer such that $L_{\beta+q\alpha} \neq \{0\}$. Then the subspace $W = \oplus_{i=-r}^{q} L_{\beta+i\alpha}$ is a simple S_α-module with weights $\beta(h_\alpha) + 2i$, $-r \leq i \leq q$. Further, $\beta(h_\alpha)$ is an integer and $\beta(h_\alpha) = \frac{2\kappa(t_\beta,t_\alpha)}{\kappa(t_\alpha,t_\alpha)} = r - q$. Also $[L_\alpha, L_\beta] = L_{\alpha+\beta}$.*

Proof Since L is finite dimensional, there is the largest nonnegative integer r such that $L_{\beta-r\alpha} \neq \{0\}$ and there is the largest nonnegative integer q such that $L_{\beta+q\alpha} \neq \{0\}$. Take $W = \oplus \sum_{i=-r}^{q} L_{\beta+i\alpha}$. Evidently, W is a S_α-submodule and $L_{\beta+i\alpha}$ are weight spaces of the S_α-module W corresponding to the weights $\beta(h_\alpha) + 2i$. Since 0 and 1 both cannot be weights of W, W is an irreducible S_α-submodule (Corollary 1.4.30). Again by Theorem 1.4.28, $(\beta(h_\alpha) - 2r) = -(\beta(h_\alpha) + 2q)$. Thus, $\beta(h_\alpha) = r - q$.

Finally, for $\alpha, \beta, \alpha+\beta \in \Phi$, $[L_\alpha, L_\beta] \neq 0$ and since $L_{\alpha+\beta}$ is one dimensional, it follows that $[L_\alpha, L_\beta] = L_{\alpha+\beta}$. ♯

The integers $\beta(h_\alpha)$ are called **Cartan Integers**.

Theorem 2.1.10 *Let L, H, Φ be as above. Then κ induces positive definite inner product χ on $H^\star$ given by $\chi(\lambda, \mu) = \kappa(t_\lambda, t_\mu)$ such that the matrix of χ with any basis of $H^\star$ consisting of elements of Φ is a matrix with rational entries. For each pair $\alpha, \beta \in \Phi$, $\frac{2\chi(\alpha,\beta)}{\chi(\alpha,\alpha)}$ is an integer and $\beta - \frac{2\chi(\beta,\alpha)}{\chi(\alpha,\alpha)}\alpha$ is a member of Φ.*

Proof Since κ is nondegenerate on H and the map $\lambda \mapsto t_\lambda$ is isomorphism from $H^\star$ to H, χ is a nondegenerate symmetric bi-linear form. Let $\{\alpha_1, \alpha_2, \ldots, \alpha_l\}$ be a basis of $H^\star$ consisting of members of Φ. From Proposition 2.1.9, $\frac{2\chi(\alpha_j,\alpha_i)}{\chi(\alpha_i,\alpha_i)}$ are integers. Let $\beta \in \Phi$. Suppose that $\beta = \sum_{i=1}^{l} a_i\alpha_i$. We show that a_i are rational numbers. Now, $\chi(\beta, \alpha_j) = \sum_{i=1}^{l} a_i\chi(\alpha_i, \alpha_j)$ Multiply this equation by $\frac{2}{\chi(\alpha_j,\alpha_j)}$. This gives l equations with coefficient matrix $[\frac{2\chi(\alpha_i,\alpha_j)}{\chi(\alpha_i,\alpha_i)}]$ with entries in $\mathbb{Z}$, and the augmented matrix is also an integer matrix. Further, since χ is nondegenerate, the coefficient matrix is non-singular. It follows that all a_i are rational numbers. Next, for $\lambda \in H^\star$, $\chi(\lambda, \lambda) = \kappa(t_\lambda, t_\lambda) = Trace(ad(t_\lambda)ad(t_\lambda)) = \sum_{i=1}^{l} \alpha_i(t_\lambda)\alpha_i(t_\lambda)$ is always positive unless λ is zero. This shows that χ is positive definite. The rest of the assertions follow from Proposition 2.1.9. ♯

Corollary 2.1.11 *Let L, H, Φ be as above, where $Dim\ H = l$. Then we have naturally associated subset $\hat{\Phi}$ of $\mathbb{R}^l$ such that the following hold:*
(i) $\overline{0} \notin \hat{\Phi}$ and $\hat{\Phi}$ generate $\mathbb{R}^l$.
(ii) If $\alpha \in \hat{\Phi}$, then $n\alpha \in \hat{\Phi}$ if and only if $n = \pm 1$.
(iii) $\frac{2<\beta,\alpha>}{<\alpha,\alpha>} \in \mathbb{Z}$ for all $\alpha, \beta \in \hat{\Phi}$.
(vi) If $\alpha, \beta \in \hat{\Phi}$, then $\beta - \frac{2<\beta,\alpha>}{<\alpha,\alpha>}\alpha \in \hat{\Phi}$.
(v) Given any basis B of $\mathbb{R}^l$ consisting of members of $\hat{\Phi}$, $\hat{\Phi} \subseteq \hat{E}_{\mathbb{Q}}$, where $\hat{E}_{\mathbb{Q}}$ is the rational space generated by B.

Proof Let $E_{\mathbb{Q}}$ denote the rational subspace of $H^\star$ generated by Φ. Then χ is the positive definite inner product on $E_{\mathbb{Q}}$. χ can be extended to an inner product on the real space E generated by Φ by putting $\chi(\sum_{i=1}^{l} a_i\alpha_i, \sum_{i=1}^{l} b_i\alpha_i) = \sum_{i,j} a_i\chi(\alpha_i, \alpha_j)b_j$. It is a fact of linear algebra that there is an isometric isomorphism η from E to $\mathbb{R}^l$. Take $\hat{\phi} = \eta(\Phi)$. The rest is immediate. ♯

We shall denote $\hat{\Phi}$ also by Φ.
More concretely, we have the following definition:

Definition 2.1.12 A **root system** is a triple $(E, <,>, \Phi)$, where E is a finite-dimensional real vector space, $<,>$ is an inner product on E, and Φ is a finite subset of E such that the following hold: (i) $\overline{0} \notin \Phi$ and Φ generate E.
(ii) If $\alpha \in \Phi$, then $n\alpha \in \Phi$ if and only if $n = \pm 1$.
(iii) $\frac{2<\beta,\alpha>}{<\alpha,\alpha>} \in \mathbb{Z}$ for all $\alpha, \beta \in \Phi$.

(vi) If $\alpha, \beta \in \Phi$, then $\beta - \frac{2<\beta,\alpha>}{<\alpha,\alpha>}\alpha \in \Phi$.
(v) Given any basis B of E consisting of members of Φ, $\Phi \subseteq \hat{E}_{\mathbb{Q}}$, where $\hat{E}_{\mathbb{Q}}$ is the rational space generated by B.

Thus, to each pair (L, H), where L is a semi-simple Lie algebra and H a maximal toral subalgebra of L, we get a unique root system as described above. Indeed, the category of root systems with suitable morphisms is equivalent to the category of pairs (L, H) with obvious morphisms. We shall classify the irreducible root systems, which, in turn, will classify all finite-dimensional semi-simple Lie algebras over an algebraically closed field of characteristic 0.

Exercises

2.1.1. Determine a maximal toral subalgebra of each earlier example of semi-simple Lie algebras over $\mathbb{C}$ and also find the corresponding root systems.
2.1.2. Determine all maximal toral subalgebras of $sl(2, \mathbb{C})$ and also the corresponding root systems.
2.1.3. Does there exist a four-dimensional semi-simple algebra over $\mathbb{C}$? Support.

2.2 Root Systems

Let E be a real vector space of dimension l with a positive definite inner product $<, >$. A subspace of dimension $l - 1$ is called a **hyperplane** of E. Let P be a hyperplane of E. Then the orthogonal complement $P^{\perp}$ of P is a one-dimensional subspace. Thus, there is a nonzero vector (unique up to a scalar multiple) $\alpha \in E$ such that $P = \{x \in E \mid < x, \alpha >= 0\}$. Conversely, for any nonzero vector $\alpha \in E$, $P_{\alpha} = \{x \in E \mid < x, \alpha >= 0\}$ is a hyperplane. Evidently, $P_{\alpha} = P_{\beta}$ if and only if $\alpha = a\beta$ for some $a \neq 0$. If we fix an orientation of a hyperplane P, then there is a unique unit vector α compatible with the chosen orientation of P such that $P = P_{\alpha}$. A **reflection** σ in E about a hyperplane P is a linear transformation of E which fixes P element-wise and maps every vector $x \in P^{\perp}$ to $-x$. Evidently, reflection about a hyperplane is unique.

Proposition 2.2.1 *Let α be a nonzero vector. Then the reflection σ_{α} about the hyperplane P_{α} is given by $\sigma_{\alpha}(x) = x - \frac{2<x,\alpha>}{<\alpha,\alpha>}\alpha$.*

Proof Let σ be the reflection about P_{α}. Every element x is uniquely expressible as $x = y + a\alpha$, where $y \in P_{\alpha}$ and $a \in \mathbb{R}$. Clearly, $< x, \alpha >= a < \alpha, \alpha >$. Since σ is a reflection about P_{α}, $\sigma(x) = y - a\alpha = x - 2a\alpha = x - \frac{2<x,\alpha>}{<\alpha,\alpha>}\alpha = \sigma_{\alpha}(x)$. ♯

Corollary 2.2.2 *Every reflection is an orthogonal transformation of order 2 and it is of determinant -1.*

Proof Clearly, σ_{α} is a linear transformation of order 2. Next, $< \sigma_{\alpha}(x), \sigma_{\alpha}(y) >=$, $< x, y >$ for all $x, y \in E$. It is also clear that the matrix of σ_{α} with respect to an orthonormal basis of P_{α} extended to an orthonormal basis of E is $Diag(1, 1, \ldots, 1, -1)$. ♯

Proposition 2.2.3 *Let α and β be nonzero vectors in E such that the angle between α and β is θ. Then $\sigma_\alpha\sigma_\beta$ is a rotation through an angle 2θ.*

Proof Without any loss, we may take $E = \mathbb{R}^2$, $\alpha = e_1 = (1, 0)$, and $\beta = (cos\theta, sin\theta) = cos\theta e_1 + sin\theta e_2$. The matrix A of σ_α with respect to $\{e_1, e_2\}$ is given by

$$A = \begin{bmatrix} -1 & 0 \\ 0 & 1 \end{bmatrix}.$$

Similarly, the matrix B of σ_β with respect to this basis is given by

$$B = \begin{bmatrix} -cos2\theta & -sin2\theta \\ -sin2\theta & cos2\theta \end{bmatrix}.$$

The product AB is given by

$$B = \begin{bmatrix} cos2\theta & sin2\theta \\ -sin2\theta & cos2\theta \end{bmatrix}.$$

Evidently, this represents rotation through an angle 2θ. ♯

For convenience, we denote $\frac{2<\beta,\alpha>}{<\alpha,\alpha>}$ by $\prec \beta, \alpha \succ$. Observe that $\prec, \succ$ is linear only in the first coordinate and it is not symmetric. Clearly $\sigma_\alpha(\beta) = \beta - \prec \beta, \alpha \succ \alpha$ and if Φ is a root system, then $\prec \beta, \alpha \succ \in \mathbb{Z}$ for all $\alpha, \beta \in \Phi$.

Proposition 2.2.4 *Let Φ be a finite set of nonzero elements of E which generates E. Suppose that $\sigma_\gamma(\Phi) = \Phi$ for all $\gamma \in \Phi$. Let σ be a member of $GL(E)$ such that $\sigma(\Phi) = \Phi$. Suppose also that there is a hyperplane P together with a member $\alpha \in \Phi$ such that σ fixes each member of P and sends α to $-\alpha$. Then $\sigma = \sigma_\alpha$ and $P = P_\alpha$.*

Proof Put $\tau = \sigma\sigma_\alpha$. Since $\sigma_\alpha^{-1} = \sigma_\alpha$, it is sufficient to show that $\tau = I$. Evidently, $\tau(\Phi) = \Phi$ and $\tau(\alpha) = \alpha$. Thus, τ is an identity on $\mathbb{R}\alpha$. Since $\sigma(\alpha) = -\alpha$ and σ fixes P element-wise, $\alpha \notin P$ (observe that $\alpha \neq 0$). Thus, $E = P \oplus \mathbb{R}\alpha$ (vector space direct sum). Let $v \in E$. Then $v = w + a\alpha$ for some $w \in P$ and $a \in \mathbb{R}$. Now,

$$\sigma\sigma_\alpha(v) = \sigma(\sigma_\alpha(w) - a\alpha) = \sigma(w - \prec w, \alpha \succ \alpha - a\alpha) = v + \prec w, \alpha \succ \alpha.$$

This means that τ is a unipotent transformation having only 1 as the eigenvalue. Thus, the characteristic polynomial of τ is $(\lambda - 1)^l$ and hence the minimum polynomial $m_\tau(\lambda)$ of τ divides $(\lambda - 1)^l$. Next, since Φ is finite, for each $\beta \in \Phi$, there is a positive integer m_β such that $\tau^{m_\beta}(\beta) = \beta$. Taking m sufficiently large, we see that $\tau^m(\beta) = \beta$ for all $\beta \in \Phi$. Since Φ generates E, it follows that $\tau^m = I$. Therefore, the minimum polynomial $m_\tau(\lambda)$ of τ divides $\lambda^m - 1$ also. Since $g.c.d.$ of $(\lambda - 1)^l$ and $(\lambda^m - 1)$ is $\lambda - 1$, it follows that $m_\tau(\lambda) = \lambda - 1$. This means that $\tau = I$. ♯

Proposition 2.2.5 *Let Φ be a root system in E. Let $\sigma \in GL(E)$ such that $\sigma(\Phi) = \Phi$. Then $\sigma\sigma_\alpha\sigma^{-1} = \sigma_{\sigma(\alpha)}$ for all $\alpha \in \Phi$. Also $\prec \sigma(\alpha), \sigma(\beta) \succ = \prec \alpha, \beta \succ$ for all $\alpha, \beta \in \Phi$. Note that σ need not be an isometry.*

Proof Since $\sigma_\alpha(\beta) \in \Phi$ for all $\alpha, \beta \in \Phi$, $(\sigma\sigma_\alpha\sigma^{-1})(\sigma(\beta)) = \sigma(\sigma_\alpha(\beta))$ belongs to Φ for all $\alpha, \beta \in \Phi$. Since $\sigma(\Phi) = \Phi$, $(\sigma\sigma_\alpha\sigma^{-1})(\Phi) = \Phi$ for all $\alpha \in \Phi$. Suppose that $\sigma(\beta) \in \sigma(P_\alpha)$, where $\beta \in P_\alpha$, then $(\sigma\sigma_\alpha\sigma^{-1})(\sigma(\beta))=\sigma(\beta)$ and $(\sigma\sigma_\alpha\sigma^{-1})(\sigma(\alpha)) = -\sigma(\alpha)$. From Proposition 2.2.4, $\sigma\sigma_\alpha\sigma^{-1} = \sigma_{\sigma(\alpha)}$ for each α. Next, $\sigma(\beta)- \prec \sigma(\beta)$, $\sigma(\alpha) \succ \sigma(\alpha) = \sigma_{\sigma(\alpha)}(\sigma(\beta)) = (\sigma\sigma_\alpha\sigma^{-1})(\sigma(\beta)) = \sigma\sigma_\alpha(\beta) = \sigma(\beta- \prec \beta, \alpha \succ \alpha)$ $= \sigma(\beta)- \prec \beta, \alpha \succ \sigma(\alpha)$. This shows that $\prec \sigma(\beta), \sigma(\alpha) \succ = \prec \beta, \alpha \succ$ for all $\alpha, \beta \in \Phi$. ♯

This prompts to have the following definition:

Definition 2.2.6 Two root systems (E, Φ) and (E', Φ') are said to be isomorphic if there is a vector space isomorphism η from E to E' such that $\eta(\Phi) = \Phi'$ and $\prec \eta(\alpha), \eta(\beta) \succ = \prec \alpha, \beta \succ$ for all $\alpha, \beta \in \Phi$.

Note that an isomorphism need not be an isometry. However, if η is an isometry from E to E' and Φ is a root system in E, then η is an isomorphism from (E, Φ) to $(E', \eta(\Phi))$. Another extremely important invariant associated with a root system is that of a Weyl group. Let (E, Φ) be a root system. For each $\alpha \in \Phi$, σ_α can be viewed as a member of $Sym(\Phi)$. The subgroup $W(\Phi)$ of $Sym(\Phi)$ generated by $\{\sigma_\alpha \mid \alpha \in \Phi\}$ is called the **Weyl group** of Φ. Evidently, $W(\Phi)$ is finite. Each member of $W(\Phi)$ is an automorphism of (E, Φ). Thus, $W(\Phi)$ is a subgroup of $Aut(E, \Phi)$. It is also clear that if η is an isomorphism from (E, Φ) to (E', Φ'), then $\sigma_\alpha \mapsto \sigma_{\eta(\alpha)}$ induces isomorphisms between the corresponding Weyl groups.

Before having some examples, let us have some more crucial observations:

Let (E, Φ) be a root system. Let $\alpha, \beta \in \Phi$. Then

$$\prec \alpha, \beta \succ \prec \beta, \alpha \succ = 4\frac{< \alpha, \beta >< \beta, \alpha >}{|| \alpha ||^2 || \beta ||^2} = 4cos^2\theta$$

is a nonnegative integer, where θ is angle between α and β. Evidently, the possibilities for $cos^2\theta$ are 0, $\frac{1}{4}$, $\frac{1}{2}$, $\frac{3}{4}$, and 1. The corresponding angles are $\theta = \frac{\pi}{2}$; $\frac{\pi}{3}$, $\frac{2\pi}{3}$; $\frac{\pi}{4}$, $\frac{3\pi}{4}$; $\frac{\pi}{6}$, $\frac{5\pi}{6}$, $0, \pi$. If $cos\theta = \pm 1$, then $\beta = \pm\alpha$. Note that $\{\alpha, -\alpha\}$ is a root system which corresponds to $sl(2, F)$. Assume that α, β are non-proportional. Then $\alpha \neq \pm\beta$. Assume that $|| \beta || \geq || \alpha ||$. If $cos^2\theta = 0$, then $\theta = \frac{\pi}{2}$, $\prec \alpha, \beta \succ = 0 = \prec \beta, \alpha \succ$ and the ratio $\frac{||\beta||}{||\alpha||}$ can be anything positive. Suppose that $cos^2\theta = \frac{1}{4}$. Then $\theta = \frac{\pi}{3}$ or $\frac{2\pi}{3}$. Suppose that $\theta = \frac{\pi}{3}$. Then $\prec \beta, \alpha \succ \prec \beta, \alpha \succ = 1$. Since $\prec \beta, \alpha \succ$ and $\prec \alpha, \beta \succ$ are positive integers, $\prec \beta, \alpha \succ = 1 = \prec \alpha, \beta \succ$ whereas $\frac{||\beta||}{||\alpha||} = 1$. Similarly, when $\theta = \frac{2\pi}{3}$, $\prec \beta, \alpha \succ = -1 = \prec \alpha, \beta \succ$ whereas $\frac{||\beta||}{||\alpha||} = 1$. Next, if $cos^2\theta = \frac{1}{2}$, then $\theta = \frac{\pi}{4}$ or $\frac{3\pi}{4}$. Suppose that $\theta = \frac{\pi}{4}$. Then $\prec \alpha, \beta \succ = 1$, $\prec \beta, \alpha \succ = 2$, and $\frac{(||\beta||)^2}{(||\alpha||)^2} = 2$. Similarly, if $\theta = \frac{3\pi}{4}$, then $\prec \alpha, \beta \succ = -1$, $\prec \beta, \alpha \succ = -2$, and $\frac{(||\beta||)^2}{(||\alpha||)^2} = 2$. Finally, suppose that $cos^2\theta = \frac{3}{4}$. Then $\theta = \frac{\pi}{6}$ or it is $\frac{5\pi}{6}$. Suppose that $\theta = \frac{\pi}{6}$. Then it is easily seen that $\prec \alpha, \beta \succ = 1$, $\prec \beta, \alpha \succ = 3$, and $\frac{(||\beta||)^2}{(||\alpha||)^2} = 3$. Similarly, if $\theta = \frac{5\pi}{6}$, then $\prec \alpha, \beta \succ = -1$, $\prec \beta, \alpha \succ = -3$, and $\frac{(||\beta||)^2}{(||\alpha||)^2} = 3$.

Proposition 2.2.7 *Let $\alpha, \beta \in \Phi, \alpha \neq \pm\beta$. If $< \alpha, \beta >$ is greater than 0, then $\alpha - \beta \in \Phi$ and if $< \alpha, \beta >$ is less than 0, then $\alpha + \beta \in \Phi$.*

Proof It is clear from the discussion preceding the proposition that $< \alpha, \beta > > 0$ implies that $\prec \alpha, \beta \succ = 1$. Thus, $\alpha - \beta = \alpha - \prec \alpha, \beta \succ \beta = \sigma_{\beta}(\alpha)$ belongs to Φ. The rest of the assertions follows from the similar observations. ♯

Let $\alpha, \beta \in \Phi$. The subset $\{\beta + i\alpha \in \Phi \mid i \in \mathbb{Z}\}$ of Φ is called a – **string through fi.**

Proposition 2.2.8 *Let α and β be non-proportional roots in Φ. Let r be the largest integer such that $\beta - r\alpha \in \Phi$ and q be the largest integer such that $\beta + q\alpha \in \Phi$. Then the following hold:*

(i) The α-root string through β remains unbroken in the sense that $\beta + i\alpha \in \Phi$ for all i, $-r \leq i \leq q$.
(ii) σ_{α} reverses the string in the sense that $\sigma_{\alpha}(\beta + (q - i)\alpha) = \beta - (r + i\alpha)$ for all i. In particular, $\sigma_{\alpha}(\beta + q\alpha) = \beta - r\alpha$.
(iii) $r - q = \prec \beta, \alpha \succ$.
(iv) The length of the α-string through β is $r + q + 1 \leq 4$, where $\beta \neq \pm\alpha$. Further, $r \leq 3$ and $q \leq 3$.

Proof (i) Suppose that for some i, $-r \leq i \leq q$, $\beta + i\alpha \notin \Phi$. Since $\beta - r\alpha$ and $\beta + q\alpha$ belong to Φ, there exists a pair of integers j, k such that $-r \leq j < k \leq q$, $\beta + j\alpha \in \Phi$, $\beta + (j + 1)\alpha \notin \Phi$, $\beta + (k - 1)\alpha \notin \Phi$, and $\beta + k\alpha \in \Phi$. It follows from the previous proposition that $< \beta + j\alpha, \alpha > \geq 0$ and $< \beta + k\alpha, \alpha > \leq 0$. This is a contradiction, since $j < k$ and $< \alpha, \alpha >$ is greater than 0. This shows that the string remains unbroken.

(ii) The fact that α-string through β remains invariant under σ_{α} and the fact that σ_{α} reverses the string are the consequences of the definition of σ_{α} and the fact that $\prec \beta, \alpha \succ$ is an integer.

(iii) From (ii) $\sigma_{\alpha}(\beta + q\alpha) = (\beta + q\alpha) - \prec \beta + q\alpha, \alpha \succ \alpha = \beta - r\alpha$. Since $\prec \alpha, \alpha \succ = 2$, $r - q = \prec \beta, \alpha \succ$.

(iv) Evidently, length of the α-string through $\beta \neq \pm\alpha$ is $r + q + 1$, where r and q are as above. We have to show that the length of α-string through β is ≤ 4. Suppose not. Then we have a sub-string of length 5. After rearranging, we may suppose that it is $\beta - 2\alpha, \beta - \alpha, \beta, \beta + \alpha, \beta + 2\alpha$. Since $2\alpha = \beta + 2\alpha - \beta$ and $2(\beta + \alpha) = \beta + 2\alpha + \beta$ are not roots, it follows that β-string through $\beta + 2\alpha$ consists of just one term $\beta + 2\alpha$. Consequently, $< \beta + 2\alpha, \beta > = 0$. Similarly, we can see that $\beta - 2\alpha - \beta$ and $\beta - 2\alpha + \beta$ are not roots. In turn, $< \beta - 2\alpha, \beta > = 0$. This is a contradiction, since $< \beta, \beta > \neq 0$. The rest is evident. ♯

Geometrically, the following figures describe the root systems of rank 2:
1. The case $< \alpha, \beta > = 0$.

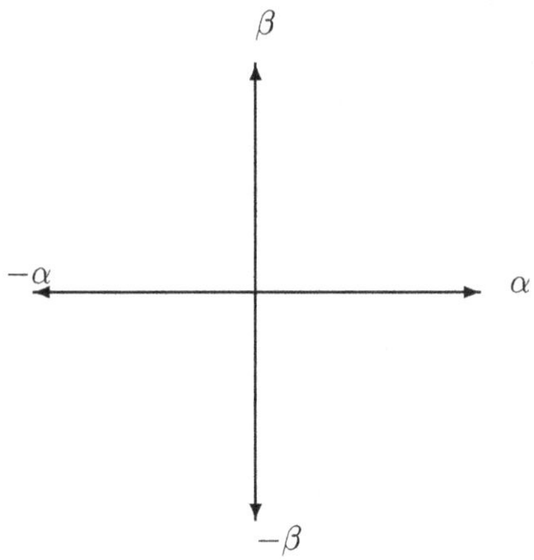

The root system $\Phi = \{\alpha, \beta, -\alpha, -\beta\}$ is associated with the Lie algebra $A_1 \times A_1$.

Suppose that $< \alpha, \beta > \neq 0$. Since $< \alpha, \beta > > 0$ implies that $< \alpha, -\beta > < 0$, we may assume that $< \alpha, \beta > < 0$.

2. Suppose that the angle θ between α and β is $\frac{2\pi}{3}$. Then $\frac{||\beta||}{||\alpha||} = 1$.

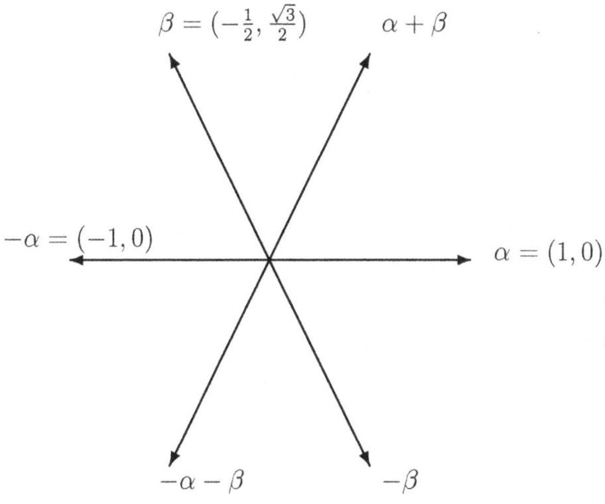

The root system $\Phi = \{\alpha, \beta, -\alpha, -\beta, \alpha + \beta, -\alpha - \beta\}$ is associated with the Lie algebra A_2.

3. Suppose that the angle θ between α and β is $\frac{3\pi}{4}$. Then $\frac{||\beta||}{||\alpha||} = \sqrt{2}$.

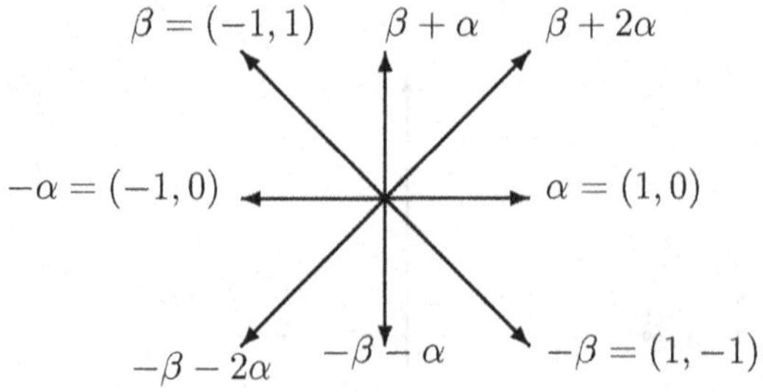

The root system $\Phi = \{\alpha, \beta, -\alpha, -\beta, \alpha + \beta, -\alpha - \beta, \beta + 2\alpha, -\beta - 2\alpha\}$ is associated with the Lie algebra B_2.

4. Suppose that the angle θ between α and β is $\frac{5\pi}{6}$. Then $\frac{||\beta||}{||\alpha||} = \sqrt{3}$.

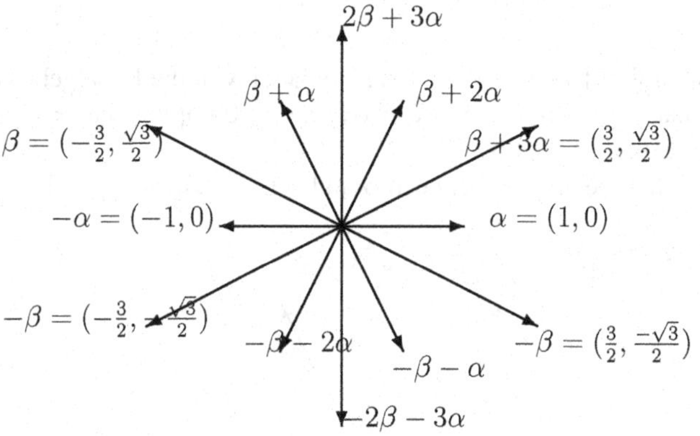

The root system $\Phi = \{\alpha, \beta, -\alpha, -\beta, \beta + \alpha, -\beta - \alpha, \beta + 2\alpha, -\beta - 2\alpha, \beta + 3\alpha, -\beta - 3\alpha, 2\beta + 3\alpha, -2\beta - 3\alpha\}$ is associated with a the Lie algebra G_2 to be described later.

Proposition 2.2.9 *Let Φ be a root system of rank l in E. Then $\Phi^\nu = \{\alpha^\nu = \frac{2\alpha}{<\alpha,\alpha>} \mid \alpha \in \Phi\}$ is also a root system of rank l such that $W(\Phi) \approx W(\Phi^\nu)$. Also, $\prec \alpha^\nu, \beta^\nu \succ = \prec \beta, \alpha \succ$.*

Proof Since Φ generates E, Φ^ν also generates E. Let $\alpha^\nu \in \Phi^\nu$, where $\alpha \in \Phi$. Suppose that $a\alpha^\nu = a\frac{2\alpha}{<\alpha,\alpha>}$ belongs to Φ^ν. Then there is a $\beta \in \Phi$ such that $\frac{2a\alpha}{<\alpha,\alpha>} = \beta^\nu = \frac{2\beta}{<\beta,\beta>}$. This means that α and β are proportional roots. Hence $\beta = \pm\alpha$. Consequently, $a = \pm 1$. Now, $< \beta^\nu, \alpha^\nu > = \frac{4<\beta,\alpha>}{<\beta,\beta><\alpha,\alpha>}$. Consequently,

$$\prec \beta^\nu, \alpha^\nu \succ = 2\frac{\frac{4<\beta,\alpha>}{<\beta,\beta><\alpha,\alpha>}}{\frac{4<\alpha,\alpha>}{<\alpha,\alpha><\alpha,\alpha>}} = \frac{2 < \beta, \alpha >}{< \beta, \beta >} = \prec \alpha, \beta \succ .$$

Further, it may be verified that $\sigma_{\alpha^\nu}(\beta^\nu) = (\sigma_\alpha(\beta))^\nu$. Finally, $\sigma_\alpha \mapsto \sigma_{\alpha^\nu}$ induces an isomorphism from $W(\Phi)$ to $W(\Phi^\nu)$. ♯

Using Proposition 2.2.3, we observe that the Weyl groups of $A_1 \times A_1$, A_2, B_2, and G_2 are the dihedral groups of order 4, 6, 8, and 12, respectively.

Proposition 2.2.10 *Let Φ be a root system and $\alpha, \beta \in \Phi, \alpha \neq \pm\beta$. Suppose that the α-string through β is $\beta - r\alpha, \beta - (r-1)\alpha, \ldots, \beta + q\alpha$ and β-string through α is $\alpha - r'\beta, \alpha - (r'-1)\beta, \ldots, \alpha + q'\beta$. Then*

(i) $\prec \beta, \alpha \succ = r - q$,
(ii) $\frac{r-q}{<\beta,\beta>} = \frac{r'-q'}{<\alpha,\alpha>}$,
(iii) $\frac{q(r+1)}{<\beta,\beta>} = \frac{q'(r'+1)}{<\alpha,\alpha>}$, *and*
(iv) $\frac{r-q}{q(r+1)} = \frac{r'-q'}{q'(r'+1)}$.

Proof Suppose that $< \alpha, \beta >= 0$. Then it follows that $r = 0 = r'$ and $q = 0' = q'$. All the identities follow trivially. Suppose that the angle between α and β is $\frac{\pi}{3}$. Looking at A_2, we see that $r = -1 = r'$, $q = 0 = q'$. Again the result follows. Suppose that the angle between α and β is $\frac{2\pi}{3}$. Looking at A_2 again, we see that $r = 0 = r'$, $q = 1 = q'$. The result follows. Similarly, looking at B_2 and G_2, we establish the result in the rest of the cases. ♯

Let Φ be a root system in E. Then Φ generates E. If B is a basis of E consisting of members of Φ, then, as already seen, every member of Φ is a rational linear combination of B. However, we look for a special type of basis.

Definition 2.2.11 A subset Δ of Φ which is a basis of E is called the **basis of the root system** Φ if every element of Φ is either a nonnegative integral linear combination or it is a nonpositive integral linear combination of members of Δ. A member α of Φ is called positive/negative with respect to Δ (in symbol $\alpha > / < 0$) if α is a nonnegative/nonpositive linear combination of members of Δ. It further induces a partial order $\geq$ on Φ. Explicitly, "$\alpha \geq \beta$" if and only if $\alpha = \beta$ or else $\alpha - \beta > 0$. The members of Δ are called **simple roots**. If $\alpha = \sum_{\gamma \in \Delta} a_\gamma \gamma$, then $\sum_{\gamma \in \Delta} a_\gamma$ is called the **height of the root** α with respect to Δ.

Proposition 2.2.12 *Let Δ be a basis of a root system Φ in E. Then the angles between distinct members of Δ are obtuse angles.*

Proof Suppose the contrary. Let $\alpha, \beta \in \Delta, \alpha \neq \beta$ be such that $< \alpha, \beta >$ is positive. From Proposition 2.2.7, $\alpha - \beta \in \Phi$. This contradicts the fact that a member of Φ is either a positive linear combination of members of Δ or it is a negative linear combination of members of Δ. ♯

We shall show the existence of a basis of a root system by constructing all bases. Let Φ be a root system in E. Since no infinite vector space can be expressed as a finite union of proper subspaces, $E - \bigcup_{\alpha \in \Phi} P_\alpha \neq \emptyset$. Let γ be a member of $E - \bigcup_{\alpha \in \Phi} P_\alpha$. Let $\Phi^+(\gamma)$ denote the subset $\{\alpha \in \Phi \mid < \alpha, \gamma > \ is\ positive\}$. Evidently, Φ is the disjoint union of $\Phi^+(\gamma)$ and $-\Phi^+(\gamma)$. Let us call an element $\alpha \in \Phi^+(\gamma)$ to be **indecomposable** if it cannot be expressed as the sum of two members of $\Phi^+(\gamma)$. Further, $\Delta(\gamma)$ will denote the set of all indecomposable elements in $\Phi^+(\gamma)$.

Theorem 2.2.13 *For each $\gamma \in E - \bigcup_{\alpha \in \Phi} P_\alpha$, the set $\Delta(\gamma)$ of indecomposable elements of $\Phi^+(\gamma)$ forms a basis of Φ.*

Proof We first show that every member of Φ is a nonnegative integral linear combination or it is a nonpositive integral linear combination of members of $\Delta(\gamma)$. It is sufficient to show that every member of $\Phi^+(\gamma)$ is a nonnegative integral linear combination of members of $\Delta(\gamma)$. Suppose the contrary. Choose a $\alpha \in \Phi^+(\gamma)$ with the smallest $< \alpha, \gamma >$ which is not a nonnegative integral linear combination of members of $\Delta(\gamma)$. Evidently, $\alpha \notin \Delta(\gamma)$. Suppose that $\alpha = \alpha_1 + \alpha_2$, where $\alpha_1, \alpha_2 \in \Phi^+(\gamma)$. Clearly, $< \alpha_i, \gamma >$ is less than $< \alpha, \gamma >$. Hence α_1 and α_2 are nonnegative integral linear combinations of members of $\Delta(\gamma)$. In turn, α is a nonnegative integral linear combination of members of $\Delta(\gamma)$. This is a contradiction.

Next, we show that a distinct pair of members of $\Delta(\gamma)$ form obtuse angles. Suppose not. Then there exists a pair $\alpha, \beta \in \Delta(\gamma)$, $\alpha \neq \beta$ such that $< \alpha, \beta >$ is greater than 0. Consequently, $\alpha - \beta \in \Phi$. Evidently, $\alpha \neq -\beta$. Thus, $\alpha - \beta \in \Phi^+(\gamma)$ or $\alpha - \beta \in -\Phi^+(\gamma)$. If $\alpha - \beta \in \Phi^+(\gamma)$, then $\alpha = \alpha - \beta + \beta$ becomes reducible and if $\beta - \alpha \in \Phi^+(\gamma)$, then $\beta = \beta - \alpha + \alpha$ becomes reducible. This is a contradiction.

Finally, we show that $\Delta(\gamma)$ is linearly independent. Suppose that $\sum_{\alpha \in \Delta(\gamma)} a_\alpha \alpha = 0$. We need to show that $A = \{\alpha \in \Delta(\gamma) \mid a_\alpha \neq 0\} = \emptyset$. Suppose not. Then $B = \{\alpha \in \Delta(\gamma) \mid a_\alpha > 0\} \neq \emptyset \neq C = \{\alpha \in \Delta(\gamma) \mid a_\alpha < 0\}$ and A is a disjoint union of B and C. Clearly, $b = \sum_{\mu \in B} a_\mu \mu = \sum_{\nu \in C} a_\nu \nu = c \neq 0$. From the earlier step, $< b, b >= \sum_{\mu, \nu \in B} a_\mu a_\nu < 0$. This is a contradiction. ♯

Let E be a finite-dimensional real inner product space. Let $\alpha \in E$. Then $P_\alpha^+ = \{\beta \in E \mid < \beta, \alpha > > 0\}$ is called the **positive open half space** associated with the vector α. Evidently, E is a disjoint union $P_\alpha^+ \bigcup P_\alpha \bigcup P_\alpha^-$. For example, $P_{e_1}^+$ in $\mathbb{R}^2$ is the upper half plane.

Proposition 2.2.14 *Let B be a vector space basis of a finite-dimensional real inner product space E. Then $\bigcap_{\alpha \in B} P_\alpha^+ \neq \emptyset$.*

Proof For each $\alpha \in B$, let U_α denote the subspace of E generated by $B - \{\alpha\}$. Let $\hat{\alpha}$ denote the projection of α on the orthogonal complement $U_\alpha^\perp$ of U_α. Clearly, $< \alpha, \hat{\alpha} >$ is greater than 0. Again, if $\beta \neq \alpha$ and $\beta \in B$, then $\alpha \in U_\beta$. Consequently, $< \alpha, \hat{\beta} >= 0$. Take $\gamma = \sum_{\alpha \in B} \hat{\alpha}$. Then it is clear from the above discussion that $< \gamma, \alpha >$ is greater than 0 for all $\alpha \in B$. This means that $\gamma \in \bigcap_{\alpha \in B} P_\alpha^+$. ♯

Proposition 2.2.15 *Given a basis Δ of Φ, there is a $\gamma \in E - \bigcup_{\alpha \in \Phi} P_\alpha^+$ such that $\Delta = \Delta(\gamma)$.*

Proof Let Δ be a basis of the root system Φ. From the above proposition, there is $\gamma \in E$ such that $< \gamma, \alpha >$ is greater than 0 for all $\alpha \in \Delta$. Since every member of Φ is a nonnegative linear combination or a nonpositive linear combination of members of Δ, $\gamma \in E - \bigcup_{\alpha \in \Phi} P_\alpha$, $\Phi^+ \subseteq \Phi^+(\gamma)$, and $\Phi^- \subseteq -\Phi^+(\gamma)$. Since Φ is a disjoint union of Φ^+ and Φ^-, it follows that $\Phi^+ = \Phi^+(\gamma)$ and $\Phi^- = -\Phi^+(\gamma)$. Consequently, $\Delta \subseteq \Delta(\gamma)$. Since $\Delta(\gamma)$ is a basis, $\Delta = \Delta(\gamma)$. ♯

Our next aim is to study the Weyl group of a root system Φ and its action on the set of all bases of Φ. The following few propositions are crucial for the purpose.

Proposition 2.2.16 *Let Φ be a root system and Δ be a basis of Φ. Let $\alpha \in \Phi^+$ be a positive root with respect to the basis Δ. Then there is a finite sequence $\alpha_1, \alpha_2, \ldots, \alpha_r$ of elements of Δ such that $\alpha = \alpha_1 + \alpha_2 + \cdots + \alpha_r$ and for all $i \leq r$, $\alpha_1 + \alpha_2 + \cdots + \alpha_i$ is a root.*

Proof The proof is by the induction on the height $ht\alpha$ of α. If $ht\alpha = 1$, then $\alpha \in \Delta$ and there is nothing to do. Assume that the assertion holds for all positive roots of height n, $n \geq 1$. Let $\alpha \in \Phi^+$ with $ht\alpha = n + 1$. Evidently, $\alpha \notin \Delta$. We show that $< \alpha, \beta >$ is greater than 0 for some $\beta \in \Delta$. Suppose not. Then $< \alpha, \gamma > \leq 0$ for all $\gamma \in \Delta$. Suppose that $\alpha = \sum_{\gamma \in \Delta} a_\gamma \gamma$, where $a_\gamma \geq 0$ for all $\gamma \in \Delta$. Since $ht\alpha \geq 2$, at least one $a_\gamma > 0$. But then $< \alpha, \alpha >= \sum_{\gamma \in \Delta} a_\gamma < \gamma, \alpha >$ is less than 0. This is a contradiction. Thus, there is a $\beta \in \Delta$ such that $< \alpha, \beta >$ is greater than 0. Consequently, $\alpha - \beta$ belongs to Φ. Since $\alpha \in \Phi^+$, $\beta \in \Delta$, and $\alpha = \alpha - \beta + \beta$, it follows that $\alpha - \beta \in \Phi^+$. Thus, $\alpha = \beta + (\alpha - \beta)$. Clearly, $ht(\alpha - \beta) = n$. By the induction hypothesis, there is a finite sequence $\alpha_1, \alpha_2, \ldots, \alpha_r$ such that $\alpha - \beta = \alpha_1 + \alpha_2 + \cdots + \alpha_r$ and $\alpha_1 + \alpha_2 + \cdots + \alpha_i$ is a root for all i. Clearly, $\alpha = \alpha_1 + \alpha_2 + \cdots + \alpha_r + \beta$ has the required property. ♯

Proposition 2.2.17 *Let Φ be a root system and Δ be a basis of Φ. Let $\alpha \in \Delta$. Then σ_α permutes the elements of $\Phi^+ - \{\alpha\}$.*

Proof Let $\beta \in \Phi^+ - \{\alpha\}$. Then $\beta = \sum_{\gamma \in \Delta} a_\gamma \gamma$, where $a_\gamma \geq 0$ for all γ. Since $\beta \neq \pm\alpha$, there is a $\gamma \in \Delta, \gamma \neq \alpha$ such that $a_\gamma > 0$. Now $\sigma_\alpha(\beta) = \beta - \prec \beta, \alpha \succ \alpha = \sum_{\gamma \in \Delta - \{\alpha\}} a_\gamma \gamma - (a_\alpha - \prec \beta, \alpha \succ)\alpha$. Since one of the coefficients is positive, $\sigma_\alpha(\beta) \in \Phi^+ - \{\alpha\}$. ♯

Corollary 2.2.18 *Let Φ be a root system and Δ be a basis of Φ. Let δ denote the element $\frac{1}{2}\sum_{\gamma \in \Phi^+} \gamma$ of E. Then $\sigma_\alpha(\delta) = \delta - \alpha$ for all $\alpha \in \Delta$.*

Proof From the above proposition, σ_α permutes $\Phi^+ - \{\alpha\}$. Consequently, since $\sigma_\alpha(\alpha) = -\alpha$, $\sigma_\alpha(\delta) = \delta - \alpha$ for all $\alpha \in \Delta$. ♯

Proposition 2.2.19 *Let Δ be a basis of a root system Φ and let $\alpha_1, \alpha_2, \ldots, \alpha_r$ be a finite-ordered sequence of elements of Δ such that $(\sigma_{\alpha_1}\sigma_{\alpha_2}\cdots\sigma_{\alpha_{r-1}})(\alpha_r)$ is a negative root. Then there is an integer s, $1 \leq s < r$, such that $\sigma_{\alpha_1}\sigma_{\alpha_2}\cdots\sigma_{\alpha_r} = \sigma_{\alpha_1}\sigma_{\alpha_2}\cdots\sigma_{\alpha_{s-1}}\sigma_{\alpha_{s+1}}\sigma_{\alpha_{s+2}}\cdots\sigma_{\alpha_{r-1}}$.*

Proof It follows by the induction on $r - 1$ and the fact that σ_α permutes $\Phi^+ - \{\alpha\}$ for all $\alpha \in \Delta$. ♯

Corollary 2.2.20 *Let Φ be a root system with Δ as a basis. Let σ be a member of $W(\Phi)$. Let r be the smallest natural number such that σ can be expressed as the product $\sigma_{\alpha_1}\sigma_{\alpha_2}\cdots\sigma_{\alpha_r}$ of simple reflections ($\alpha_i \in \Delta$ for all i). Then $\sigma(\alpha_r)$ is a negative root.*

Proof Suppose that $\sigma(\alpha_r)$ is a positive root. Since $\sigma_{\alpha_r}(\alpha_r) = -\alpha_r$, it follows that $(\sigma_{\alpha_1}\sigma_{\alpha_2}\cdots\sigma_{\alpha_{r-1}})(\alpha_r)$ is a negative root. From the above proposition, $\sigma_{\alpha_1}\sigma_{\alpha_2}\cdots\sigma_{\alpha_r} = \sigma_{\alpha_1}\sigma_{\alpha_2}\cdots\sigma_{\alpha_{s-1}}\sigma_{\alpha_{s+1}}\sigma_{\alpha_{s+2}}\cdots\sigma_{\alpha_{r-1}}$ for some $s < r$. This is a contradiction to the choice of r. ♯

Let Φ be a root system in E. Consider the open subspace $E - \bigcup_{\alpha\in\Phi} P_\alpha$ of E. The connected components of $E - \bigcup_{\alpha\in\Phi} P_\alpha$ are called the **Weyl Chambers** of Φ. They are all open connected subsets of E and partition $E - \bigcup_{\alpha\in\Phi} P_\alpha$ into mutually disjoint classes. Evidently, each $\gamma \in E - \bigcup_{\alpha\in\Phi} P_\alpha$ belongs to a unique Weyl Chamber which we denote by $C(\gamma)$. $\Phi^+(\gamma) = \{\alpha \in \Phi \mid <\gamma, \alpha> \textit{ is greater than zero}\}$. Clearly, $C(\gamma) = C(\gamma')$ if and only if γ and γ' both lie on the same side of P_α for all $\alpha \in \Phi$. Note that $-\Phi^+(\gamma) = \Phi^+(-\gamma)$. It is also clear that $\Delta(\gamma) \mapsto C(\gamma)$ defines a bijective map from the set of all bases of Φ to the set of all Weyl chambers of Φ. If Δ is a basis of Φ, then there is a $\gamma \in E - \bigcup_{\alpha\in\Phi} P_\alpha$ such that $\Delta = \Delta(\gamma)$. The corresponding $C(\gamma)$ is called the **Fundamental Weyl Chamber associated with** Δ. Since each member σ of the Weyl group is an orthogonal transformation, $\sigma(C(\gamma)) = C(\sigma(\gamma))$. Thus, the Weyl group $W(\Phi)$ acts on the set of all Weyl chambers, and, in turn, on the set of all bases of Φ. We shall discuss the properties of this action.

Theorem 2.2.21 *Let Φ be a root system in E with Δ as a basis. Then the following hold:*

(i) The subgroup $W^0(\Phi)$ of the Weyl group $W(\Phi)$ generated by the set $\{\sigma_\alpha \mid \alpha \in \Delta\}$ of simple reflections acts simply transitively on the set of all Weyl chambers/bases of Φ.
(ii) If $\alpha \in \Phi$, then there is an element σ of $W^0(\Phi)$ such that $\sigma(\alpha) \in \Delta$.
(iii) $W(\Phi)$ is generated by the set $\{\sigma_\alpha \mid \alpha \in \Delta\}$ of simple reflections (members of Δ are called the simple roots).
(iv) $W(\Phi)$ acts simply transitively on the set of all Weyl chambers/bases of Φ.

Proof (i) We first show that $W^0(\Phi)$ acts transitively on the set of Weyl chambers/bases. Let $C(\gamma)$ be a Weyl Chamber of Φ, where $\gamma \in E - \bigcup_{\alpha\in\Phi} P_\alpha$. We need to show the existence of a $\sigma \in W^0(\Phi)$ such that $\Delta(\sigma(\gamma)) = \Delta$ (note that $\sigma(\gamma) \in E - \bigcup_{\alpha\in\Phi} P_\alpha$ for all $\sigma \in W(\Phi)$). Equivalently, we need to show the existence of a $\sigma \in W^0(\Phi)$ such that $< \sigma(\gamma), \alpha >$ is greater than 0 for all $\alpha \in \Delta$. Let σ be a member of $W^0(\Phi)$ for which $< \sigma(\gamma), \delta >$ is largest, where $\delta = \frac{1}{2}\sum_{\beta\in\Phi^+} \beta$. For each $\alpha \in \Delta$, $\sigma_\alpha\sigma \in W^0(\Phi)$. Hence

$$< \sigma(\gamma), \delta > \geq < \sigma_\alpha\sigma(\gamma), \delta >=, < \sigma(\gamma), \sigma_\alpha(\delta) >=, < \sigma(\gamma), \delta > - < \sigma(\gamma), \alpha >$$

for each $\alpha \in \Delta$. Thus, $< \sigma(\gamma), \alpha > \geq 0$ for all $\alpha \in \Delta$. Further, $< \sigma(\gamma), \alpha > \neq 0$ for otherwise $\sigma(\gamma)$ will belong to P_α. This means that $< \sigma(\gamma), \alpha > > 0$ for all $\alpha \in \Delta$.

Finally, we show that the action is simply transitive. Suppose not. Then there exists a $\sigma \in W^0(\Phi)$, $\sigma \neq I$, and a basis Δ' such that $\sigma(\Delta') = \Delta'$. From

Corollary 2.2.20, expressing σ as the minimal product $\sigma = \sigma_{\alpha_1}\sigma_{\alpha_2}\cdots\sigma_{\alpha_r}$ of simple reflection, $\sigma(\alpha_r) < 0$. This is a contradiction, since $\sigma(\Delta') = \Delta'$.

(ii) Since $W^0(\Phi)$ acts transitively on the set of Weyl chambers/bases, it is sufficient to show that each root belongs to a basis of Φ. Again, since $P_\alpha \neq \bigcup_{\beta\neq\pm\alpha}(P_\beta \bigcap P_\alpha)$, there is a $\gamma \in P_\alpha$ such that $\gamma \notin P_\beta$ for all $\beta \neq \pm\alpha$. Take γ' sufficiently near γ with $< \gamma', \alpha > > 0$ and $|< \gamma', \beta >| > < \gamma', \alpha >$ for all $\beta \neq \pm\alpha$. Then $\alpha \in \Phi^+(\gamma')$ and it cannot be expressed as two distinct members of $\Phi^+(\gamma')$. Thus, $\alpha \in \Delta(\gamma')$.

(iii) We need to show that $\sigma_\alpha \in W^0(\Phi)$ for all $\alpha \in \Phi$. Consider σ_α, $\alpha \in \Phi$. From (ii), there is a $\sigma \in W^0(\Phi)$ such that $\sigma(\alpha) \in \Delta$. Clearly, $\sigma_\alpha = \sigma^{-1}\sigma_{\sigma(\alpha)}\sigma \in W^0(\Phi)$.

(iv) Follows from (i) and (iii).

♯

Definition 2.2.22 Let Φ be a root system in E with a basis Δ. Then each member $\sigma \in W(\Phi)$ is expressible as a product $\sigma = \sigma_{\alpha_1}\sigma_{\alpha_2}\cdots\sigma_{\alpha_r}$ of simple reflections. The smallest r such that $\sigma = \sigma_{\alpha_1}\sigma_{\alpha_2}\cdots\sigma_{\alpha_r}$ is called the **length** of σ and it is denoted by $l(\sigma)$. Evidently, $l(I) = 0$.

Proposition 2.2.23 $l(\sigma) =| \{\alpha \in \Phi^+ \mid \sigma(\alpha) \in \Phi^-\} |$.

Proof The proof is by the induction on $l(\sigma)$. If $l(\sigma) = 0$, then $\sigma = I$ and there is no $\alpha \in \Phi^+$ such that $I(\alpha) \in \Phi^-$. Assume that the result is true for all $\tau \in W(\Phi)$ for which $l(\tau) < l(\sigma)$. We prove it for σ. Write σ in the reduced form as $\sigma = \sigma_{\alpha_1}\sigma_{\alpha_2}\cdots\sigma_{\alpha_{l(\sigma)}}$. Then by Corollary 2.2.20, $\sigma(\alpha_{l(\sigma)}) \in \Phi^-$. Clearly, $l(\sigma\sigma_{\alpha_{l(\alpha)}}) = l(\sigma) - 1$ and it follows from Proposition 2.2.17 that $| \{\beta\in\Phi^+ \mid \sigma\sigma_{\alpha_{l(\sigma)}}(\beta) \in \Phi^-\} |=| \{\beta \in \Phi^+ \mid \sigma(\beta) \in \Phi^-\} - \{\alpha_{l(\sigma)}\} |=| \{\beta \in \Phi^+ \mid \sigma(\beta) \in \Phi^-\} | -1$. The result follows from the induction hypothesis. ♯

Proposition 2.2.24 *Let Δ be a basis of a root system Φ in E. Then the following hold:*

(i) $\sigma(\gamma) \leq \gamma$ for all $\gamma \in \overline{C(\Delta)}$, where $C(\Delta)$ is the fundamental chamber associated with the basis Δ.

(ii) $C(\Delta)$ is a fundamental domain of the action of the Weyl group $W(\Phi)$ on $E - \bigcup_{\alpha\in\Phi} P_\alpha$ in the sense that $E - \bigcup_{\alpha\in\Phi} P_\alpha = \bigcup_{\gamma\in C(\Delta)} W(\Phi)(\gamma)$, where $W(\Phi)(\gamma) \bigcap W(\Phi)(\eta) = \emptyset$ for all $\gamma, \eta \in C(\Delta)$, $\gamma \neq \eta$, and whenever $\sigma(\gamma) = \tau(\gamma)$, $\sigma = \tau$.

Proof (i) We prove it by the induction on $l(\sigma)$. Suppose that $l(\sigma) = 1$. Then $\sigma = \sigma_\alpha$, where $\alpha \in \Delta$. Given $\gamma \in \overline{C(\Delta)}$, $< \gamma, \alpha > \geq 0$. Consequently,

$$\sigma(\gamma) = \sigma_\alpha(\gamma) = \gamma - \frac{2 < \gamma, \alpha >}{< \alpha, \alpha >}\alpha \leq \gamma.$$

Assume that the result holds for all those $\tau\in W(\Phi)$ for which $l(\tau) < l(\sigma)$, $l(\sigma) \geq 2$. We prove it for σ. Express σ in reduced form as a product $\sigma_{\alpha_1}\sigma_{\alpha_{i_2}}\cdots\sigma_{\alpha_{i_r}}$ of simple reflections, where $l(\sigma)=r$. Then σ^{-1} has the representation $\sigma_{\alpha_{i_r}}\sigma_{\alpha_{i_{r-1}}}$

$\cdots\sigma_{\alpha_{i_1}}$ as a reduced product of simple reflections. Evidently, $\sigma_{\alpha_{i_r}}\sigma_{\alpha_{i_{r-1}}}\cdots\sigma_{\alpha_{i_2}}$ is a reduced product. By the induction hypothesis, $\tau(\gamma) \leq \gamma$ for all $\gamma \in \overline{C(\Delta)}$, where $\tau = \sigma_{\alpha_{i_r}}\sigma_{\alpha_{i_{r-1}}}\cdots\sigma_{\alpha_{i_2}}$. In particular, $\tau(\sigma_{\alpha_1}(\gamma)) \leq \sigma_{\alpha_1}(\gamma)$. But already $\sigma_{\alpha_1}(\gamma) \leq \gamma$.

(ii) Since $W(\Phi)$ acts simply transitively on the set of Weyl chambers, it is sufficient to show that $\sigma(\gamma) = \gamma$ implies that $\sigma = I$. Suppose $\sigma \neq I$ but $\sigma(\gamma) = \gamma$. As above, if σ is expressed in reduced form $\sigma = \sigma_{\alpha_{i_1}}\sigma_{\alpha_{i_2}}\cdots\sigma_{\alpha_{i_r}}$, then $\sigma(\gamma) = \gamma$ implies that $\sigma_{\alpha_{i_r}}(\gamma) = \gamma$. This means that $<\gamma, \alpha_{i_r}>= 0$. This is a contradiction, since $\gamma \in C(\Delta)$. ♯

Definition 2.2.25 A root system Φ in E is said to be **irreducible** if it cannot be expressed as a disjoint union of two proper subsets Φ_1 and Φ_2 such that $<\alpha, \beta>= 0$ for all $\alpha \in \Phi_1$ and $\beta \in \Phi_2$. We say that it is reducible, otherwise.

Thus, A_1, A_2, B_2, and G_2 are all irreducible whereas $A_1 \times A_1$ is reducible.

Proposition 2.2.26 *If Φ is irreducible and Δ is a basis, then Δ is also irreducible in the same sense. Conversely, if a basis of Φ is irreducible, then Φ is also irreducible.*

Proof Suppose that Δ is irreducible, $\Phi = \Phi_1 \bigcup \Phi_2$, $\Phi_1 \bigcap \Phi_2 = \emptyset$, and $<\alpha, \beta>= 0$ for all $\alpha \in \Phi_1$ and $\beta \in \Phi_2$. Then $\Delta = \Delta_1 \bigcup \Delta_2$, where $\Delta_1 = \Delta \bigcap \Phi_1$ and $\Delta_2 = \Delta \bigcap \Phi_2$. Since Δ is irreducible, $\Delta_1 = \Delta$ or $\Delta_2 = \Delta$. Since $\Phi_1 \perp \Phi_2$, $\Delta = \Delta_1 \subseteq \Phi_1$ implies that no member of Φ_2 can be expressed as a linear combination of members of Δ. Since Δ is a basis, $\Phi_1 = \Phi$ and $\Phi_2 = \emptyset$. Similarly, if $\Delta_2 = \Delta$, then $\Phi_2 = \Phi$ and $\Phi_1 = \emptyset$.

Conversely, suppose that Φ is irreducible and Δ is a basis of Φ. Suppose further that $\Delta = \Delta_1 \bigcup \Delta_2$, where $\Delta_1 \bigcap \Delta_2 = \emptyset$ and $<\alpha, \beta>= 0$ for all $\alpha \in \Delta_1$ and $\beta \in \Delta_2$. Let $\Phi_1 = \{\alpha \in \Phi \mid \sigma(\alpha) \in \Delta_1 \; for\ some\ \sigma \in W(\Phi)\}$ and $\Phi_2 = \{\beta \in \Phi \mid \sigma(\beta) \in \Delta_2 \; for\ some\ \sigma \in W(\Phi)\}$. Evidently, $\Phi = \Phi_1 \bigcup \Phi_2$. Suppose that $\gamma \in \Phi_1$. Then there exists an $\alpha \in \Delta_1$ and $\sigma \in W(\Phi)$ such that $\sigma(\alpha) = \gamma$. If $\alpha \in \Delta_1$ and $\beta \in \Delta_2$, then $<\alpha, \beta>= 0$. Consequently, $\sigma_\alpha\sigma_\beta = \sigma_\beta\sigma_\alpha$ and any σ in $W(\Phi)$ can be expressed as $\sigma = \sigma_{\beta_1}\sigma_{\beta_2}\cdots\sigma_{\beta_r}\sigma_{\alpha_1}\sigma_{\alpha_2}\cdots\sigma_{\alpha_s}$, where $\beta_i \in \Delta_2$ and $\alpha_j \in \Delta_1$. If $\gamma = \sigma(\alpha) \in \Phi_1, \alpha \in \Delta_1$, then $\gamma = \sigma(\alpha) = \alpha - \sum_{j=1}^{s} a_j\alpha_j$ belongs to the subspace generated by Δ_1. Thus, $\Phi_1 \subseteq <\Delta_1>$. It follows that $\Phi_1 =, <\Delta_1>$ and $\Phi_2 =, <\Delta_2>$. Since $\Delta_1 \perp \Delta_2$, $\Phi_1 \perp \Phi_2$. Since Φ is irreducible, $<\Delta_1>= \Phi$ or $<\Delta_2>= \Phi$. Consequently, $\Delta_1 = \Delta$ or $\Delta_2 = \Delta$. ♯

Proposition 2.2.27 *Let Φ be an irreducible root system in E with a basis Δ. Then there is a largest element β in Φ with respect to the partial order $\leq$. Further, if $\beta = \sum_{\alpha\in\Delta} a_\alpha\alpha$, then $a_\alpha > 0$ for all α. Also $<\beta, \alpha> \geq 0$ for all $\alpha \in \Delta$.*

Proof Obviously, a maximal root exists. Let $\beta = \sum_{\alpha\in\Delta} a_\alpha\alpha$ be a maximal root. Clearly, $a_\alpha \geq 0$ for all α and at least one $a_\alpha > 0$. Put $\Delta_1 = \{\alpha \in \Delta \mid a_\alpha > 0\}$ and $\Delta_2 = \{\alpha \in \Delta \mid a_\alpha = 0\}$. Clearly $\{\Delta_1, \Delta_2\}$ is a partition of Δ. We need to show that $\Delta_2 = \emptyset$. Suppose not. Since Δ is a basis $<\mu, \nu>\leq 0$ for all $\mu, \nu \in \Delta$. In particular, $<\beta, \alpha>\leq 0$ for all $\alpha \in \Delta_2$. Since Φ is irreducible, there is a pair α, α' with $\alpha \in \Delta_1$, $\alpha' \in \Delta_2$ such that $<\alpha, \alpha'>\neq 0$. Consequently, $<\alpha, \alpha'>< 0$ and so $<\beta, \alpha'><0$. This implies that $\beta+\alpha'$ is a root, a contradiction to the maximality of β. Next, we show the uniqueness. Let β' be another maximal element. Then as in the case of β, $\beta'=\sum_{\alpha\in\Delta}a'_{\alpha}\alpha$ with $a'_{\alpha}>0$. Evidently, $<\beta', \beta>>0$. This means that $\beta=\beta'$ for otherwise $\beta-\beta'$ is a root and $\beta'<\beta$ or $\beta<\beta'$. ♯

Every root system Φ in E can be expressed as a pairwise disjoint union $\Phi_1\bigcup\Phi_2\bigcup\cdots\bigcup\Phi_r$ of irreducible root systems with $\Phi_i\perp\Phi_j$ for $i\neq j$. If $<\Phi_i>=E_i$, then $E=E_1\oplus E_2\oplus\cdots\oplus E_r$, where the sum is the orthogonal sum.

Proposition 2.2.28 *Let Φ be an irreducible root system in E. Then the representation of $W(\Phi)$ on E induced by the action of $W(\Phi)$ on E is an irreducible representation. More explicitly, no nontrivial proper subspace of E is invariant under $W(\Phi)$.*

Proof Let V be a nontrivial subspace of E which is invariant under $W(\Phi)$. Since the members of $W(\Phi)$ are orthogonal transformations, the orthogonal complement $V^{\perp}$ is also invariant under $W(\Phi)$ and $E=V\oplus V^{\perp}$. Suppose that $\alpha\in\Phi-V$. Let $v\in V$. Then $\sigma_{\alpha}(v)=v-\prec v,\alpha\succ\alpha$. Since $\sigma_{\alpha}(v)\in V$ and $\alpha\notin V$, $<v,\alpha>=0$. This means that $V\subseteq P_{\alpha}$. Consequently, $\Phi-V\subseteq V^{\perp}$. Thus, $\Phi=(\Phi\bigcap V)\bigcup(\Phi\bigcap V^{\perp})$. Since Φ is irreducible, $\Phi\bigcap V=\Phi$. Again, since $<\Phi>=E$, $V=E$. ♯

Proposition 2.2.29 *Let Φ be an irreducible system in E. Then at the most two root lengths appear in Φ. More explicitly, $1\leq|\{||\alpha||\mid\alpha\in\Phi\}|\leq 2$. Further, roots of the same lengths are conjugates in the sense that if $||\alpha||=||\beta||$, then there is a $\sigma\in W(\Phi)$ such that $\sigma(\beta)=\alpha$.*

Proof Since members of $W(\Phi)$ are orthogonal transformations, they preserve root lengths. Since $\{\sigma(\alpha)\mid\sigma\in W(\Phi)\}$ generate E (Proposition 2.2.28), for any pair $\alpha,\beta\in\Phi$, there is a $\sigma\in W(\Phi)$ such that $<\sigma(\alpha),\beta>\neq 0$. For any pair $\alpha,\beta\in\Phi$, the only possible values of $\frac{||\alpha||^2}{||\beta||^2}$ are $1, 2, 3, \frac{1}{2}, \frac{1}{3}$ (see the discussions prior to Proposition 2.2.7). If there are three different root lengths, then obviously, $\frac{2}{3}$ will appear as a root length ratio. Thus, there are at the most two root lengths which appear in Φ.

Next, let $\alpha,\beta\in\Phi$ be of equal lengths. We may assume that $<\alpha,\beta>\neq 0$. Again, it follows from the discussions prior to Proposition 2.2.7 that $\prec\alpha,\beta\succ=\pm 1$. If $\prec\alpha,\beta\succ=-1$, then replacing β by $\sigma_{\beta}(\beta)=-\beta$, we may assume that $\prec\alpha,\beta\succ=1$. But then $\sigma_{\alpha}\sigma_{\beta}\sigma_{\alpha}(\beta)=\alpha$. This proves the assertion. ♯

If an irreducible root system has two root lengths, then the larger roots are called long roots while others are called short roots.

Proposition 2.2.30 *If Φ is an irreducible root system with a basis Δ, then the maximal root β is long.*

Proof Let β be the largest root of Φ and $\alpha\in\Phi$. We need to show that $<\alpha,\alpha>\leq<\beta,\beta>$. Replacing α by a $W(\Phi)$-conjugate, we may assume that $\alpha\in\overline{C(\Delta)}$. Since $\beta-\alpha>0$, $<\gamma,\beta-\alpha>\geq 0$ for all $\gamma\in\overline{C(\Delta)}$. In particular, $<\beta,\beta-\alpha>\geq 0$ and also $<\alpha,\beta-\alpha>\geq 0$. Thus, $<\beta,\beta>\geq<\beta,\alpha>\geq<\alpha,\alpha>$. ♯

Cartan Matrices

Let Φ be a root system with an ordered basis $\Delta = \{\alpha_1, \alpha_2, \ldots, \alpha_r\}$. Then the matrix $A = [a_{ij}]$ is called the **Cartan matrix** of Φ relative to the ordered basis Δ, where $a_{ij} = \prec \alpha_i, \alpha_j \succ = \frac{2<\alpha_i,\alpha_j>}{<\alpha_j,\alpha_j>}$. Evidently, the Cartan matrix is an integral matrix and its entries are called **Cartan integers**. Thus, the Cartan matrix of $A_1 \times A_1$, A_2, B_2, and G_2 are

$$A = \begin{bmatrix} 2 & 0 \\ 0 & 2 \end{bmatrix}, \begin{bmatrix} 2 & -1 \\ -1 & 2 \end{bmatrix}, \begin{bmatrix} 2 & -2 \\ -1 & 2 \end{bmatrix}, \; and \; \begin{bmatrix} 2 & -1 \\ -3 & 2 \end{bmatrix},$$

respectively.

Not all matrices are Cartan matrices. Indeed, the diagonal entries of all Cartan matrices are 2, while the possibilities for off-diagonal entries are 0, −1, −2, or −3.

Since $W(\Phi)$ acts transitively on the set of bases of Φ, the Cartan matrix is unique up to the conjugation by the permutation matrices. Again, since Δ is a basis, the matrix $[< \alpha_i, \alpha_j >]$ is non-singular. Consequently, the Cartan matrix $[\prec \alpha_i, \alpha_j \succ] = \frac{2<\alpha_i,\alpha_j>}{<\alpha_j,\alpha_j>}$ is a non-singular integral matrix.

Proposition 2.2.31 *If (E, Φ) and (E', Φ') are isomorphic root systems, then they have the same Cartan matrices. Conversely, if the Cartan matrix of (E, Φ) with respect to an ordered basis $\Delta = \{\alpha_1, \alpha_2, \ldots, \alpha_r\}$ is the same as the Cartan matrix of Φ' with respect to an ordered basis $\Delta' = \{\alpha'_1, \alpha'_2, \ldots, \alpha'_r\}$ of Φ', then (E, Φ) and (E', ϕ') are isomorphic.*

Proof Suppose that (E, Φ) and (E', Φ') are isomorphic root systems. Then there is a vector space isomorphism η from E to E' such that $\prec \eta(\alpha), \eta(\beta) \succ = \prec \alpha, \beta \succ$ for all $\alpha, \beta \in \Phi$ and $\eta(\Phi) = \Phi'$. If Δ is a base of Φ, then $\eta(\Phi)$ is a base of Φ'. Clearly, the Cartan matrix of (E, Φ) is the same as that of (E', Φ'). Conversely, suppose that the Cartan matrix of (E, Φ) with respect to an ordered basis $\Delta = \{\alpha_1, \alpha_2, \ldots, \alpha_r\}$ is the same as the Cartan matrix of Φ' with respect to an ordered basis $\Delta' = \{\alpha'_1, \alpha'_2, \ldots, \alpha'_r\}$ of Φ'. Then $\prec \alpha_i, \alpha_j \succ = \prec \alpha'_i, \alpha'_j \succ$ for all i, j. Thus, we have a vector space isomorphism η from E to E' given by $\eta(\alpha_i) = \alpha'_i$. All that we need to show is that $\eta(\Phi) = \Phi'$ and $\prec \eta(\alpha), \eta(\beta) \succ = \prec \alpha, \beta \succ$ for all $\alpha, \beta \in \Phi$. Clearly, $\sigma_{\eta(\alpha)}(\eta(\beta)) = \eta(\beta) - \prec \eta(\beta), \eta(\alpha) \succ \eta(\alpha) = \eta(\beta) - \prec \beta, \alpha \succ \eta(\alpha) = \eta(\sigma_\alpha(\beta))$ for all $\alpha, \beta \in \Delta$. Since $W(\Phi)$ and $W(\Phi')$ are generated by simple reflections, η induces an isomorphism $\overline{\eta}$ from $W(\Phi)$ to $W(\Phi')$ defined by $\overline{\eta}(\sigma) = \sigma o \eta o \sigma^{-1}$. Evidently, $\overline{\eta}(\sigma_\alpha) = \sigma_{\eta(\alpha)}$ for all $\alpha \in \Delta$. Given a $\beta \in \Phi$, there is a $\sigma \in W(\Phi)$ and a $\alpha \in \Delta$ such that $\beta = \sigma(\alpha)$. Consequently, $\eta(\beta) = (\eta o \sigma o \eta^{-1})(\eta(\alpha))$ belongs to Φ'. This shows that η is an isomorphism from (E, Φ) to (E', Φ'). ♯

It follows from the above proposition that a root system is uniquely determined (up to isomorphism) by its Cartan matrix. One can easily construct a root system associated with a Cartan matrix.

Proposition 2.2.32 *Let Φ be a root system in E with a basis Δ. Then the following hold:*

(i) There exists a unique element $\hat{\sigma} \in W(\Phi)$ such that $\hat{\sigma}(\Phi^+) = \Phi^-$.
(ii) $\hat{\sigma}^2 = I$ and $l(\hat{\sigma}) = \mid \Delta \mid$.

Proof (i) Let β_0 be the unique maximal member of Φ. It follows from Proposition 2.2.27 that $< \beta_0, \alpha > \geq 0$ for all $\alpha \in \Delta$. Clearly, $\sigma_{\beta_0}(\alpha) = \alpha - \prec \beta_0, \alpha \succ \alpha \in \Phi^-$ for all $\alpha \in \Delta$. Since σ_{β_0} is linear, $\sigma_{\beta_0}(\Phi^+) = \Phi^-$. If η is another such map, then $\sigma_{\beta_0}\eta^{-1}(\Phi^+) = \Phi^+$. Consequently, $\sigma_{\beta_0}\eta^{-1}(\Delta) = \Delta$. Since $W(\Phi)$ acts simply transitively on the set of bases, it follows that $\eta = \sigma_{\beta_0}$.

(ii) Since σ_{β_0} is a reflection, $(\sigma_{\beta_0})^2 = I$. Further, it follows from Proposition 2.2.23 that $l(\sigma_{\beta_0}) = \mid \Delta \mid$. ♯

Exercises

2.2.1. Let Φ be an irreducible root system with Δ as a basis. Suppose that $\sigma \in W(\Phi)$ is a product of t simple reflections. Show that $l(\sigma) \equiv t(mod 2)$.
2.2.2. If Δ is a basis of Φ, then show that Δ^ν is a basis of Φ^ν.
2.2.3. Show that the map $\sigma \mapsto (-1)^{l(\sigma)}$ is a surjective homomorphism from $W(\Phi)$ to $\{1, -1\}$.
2.2.4. Suppose that $\sigma \in W(\Phi)$ is a reflection. Show that $\sigma = \sigma_\alpha$ for some $\alpha \in \Phi$.
2.2.5. Let Φ be a root system. Let c be a positive real number. Show that $\Phi_c = \{\alpha \in \Phi \mid < \alpha, \alpha >= c\}$ is a root system provided that it is nonempty.

2.3 Dynkin Diagram and the Classification of Root Systems

A group G having a presentation of the type $< x_1, x_2, \ldots, x_r;\ (x_ix_j)^{m_{ij}} = 1,\ m_{ii} = 1, m_{ij} \in \mathbb{Z} >$ is called a **Coxeter group**. For example, the symmetric group S_n is a Coxeter group. Indeed, S_n has a presentation

$$< x_1, x_2, \ldots, x_{n-1};\ x_i^2, (x_jx_{j+1})^3, (x_kx_l)^2 \mid 1 \leq i \leq n-1, 1 \leq j \leq n-2, \mid k - l \mid \geq 2, 1 \leq k, l \leq n-1 >$$

(see Algebra 2, Chap. 10). It will be shown that the Weyl group $W(\Phi)$ is also a Coxeter group. We shall classify Weyl groups, and also the root systems, with the help of the Coxeter graphs and the Dynkin diagrams.

Let Φ be a root system with $\Delta = \{\alpha_1, \alpha_2, \ldots, \alpha_l\}$ as a basis. Recall that the possible values of $\prec \alpha_i, \alpha_j \succ$ for $i \neq j$ are 0, 1, 2, or 3. The Coxeter graph of Φ relative to the base Δ is a graph with l vertexes $v_1, v_2, \ldots, v_l$ corresponding to the simple roots $\alpha_1, \alpha_2, \ldots, \alpha_l$, where the ith vertex v_i is joined to the jth vertex v_j by $\prec \alpha_i, \alpha_j \succ \cdot \prec \alpha_j, \alpha_i \succ$ edges. Thus, the Coxeter graph of $A_1 \times A_1$ is

$$v_1 \qquad v_2,$$

the Coxeter graph for A_2 is

the Coxeter graph of B_2 is

and that of G_2 is

In case only one root length occurs, $\prec \alpha_i, \alpha_j \succ = \prec \alpha_j, \alpha_i \succ$ and consequently, $\prec \alpha_i, \alpha_j \succ \prec \alpha_j, \alpha_i \succ$ completely determines $\prec \alpha_i, \alpha_j \succ$ for all i, j, for example, in the case of A_2. Thus, in this case, the Coxeter graph completely determines the Cartan matrix, and in turn, the root system. However, in case there are two root lengths, we can still determine the Cartan matrix provided that we know which root is long and which root is short. For example in B_2, $\prec \alpha, \beta \succ = 1$ and $\prec \beta, \alpha \succ = 2$, where α is short and β is long. This prompts us to think of a more special type of Coxeter graphs termed Dynkin diagrams. In a Coxeter graph, whenever a double or triple edge occurs, let us put an arrow pointing toward the shorter root. Such a graph is termed as a **Dynkin diagram**. Evidently, the Dynkin diagram completely determines the Cartan matrix, and in turn the root system. Thus, for example, the Dynkin diagram of B_2 is 

and of G_2 is 

We have another Dynkin diagram

Clearly, the Cartan matrix associated with this Dynkin diagram is

$$\begin{bmatrix} 2 & -1 & 0 & 0 \\ -1 & 2 & -2 & 0 \\ 0 & -1 & 2 & -1 \\ 0 & 0 & -1 & 2 \end{bmatrix}.$$

The corresponding root system is denoted by F_4.

Clearly, the Dynkin diagram associated with the irreducible root system is a connected Dynkin diagram.

Theorem 2.3.1 *Let Φ be an irreducible root system of rank l. Then the Dynkin diagram of Φ is one and only one of the following:*

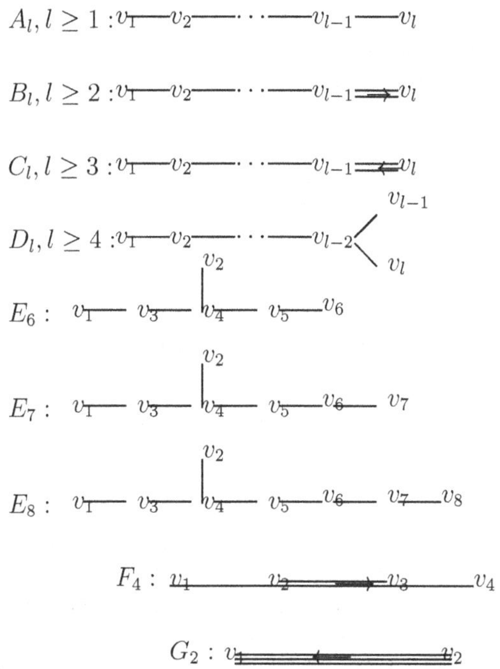
$A_l, l \geq 1$: v_1 — v_2 — · · · — v_{l-1} — v_l
$B_l, l \geq 2$: v_1 — v_2 — · · · — v_{l-1} ⇒ v_l
$C_l, l \geq 3$: v_1 — v_2 — · · · — v_{l-1} ⇐ v_l
$D_l, l \geq 4$: v_1 — v_2 — · · · — v_{l-2} v_{l-1} v_l
E_6 : v_1 v_3 v_4 v_5 v_6 v_2
E_7 : v_1 v_3 v_4 v_5 v_6 v_7 v_2
E_8 : v_1 v_3 v_4 v_5 v_6 v_7 v_8 v_2
F_4 : v_1 v_2 v_3 v_4
G_2 : v_1 v_2

Proof With our earlier remarks, it is sufficient to classify Coxeter graphs of root systems. For this purpose, without loss of generality, we may consider a basis $\{\alpha_1, \alpha_2, \ldots, \alpha_l\}$ of an Euclidean space E consisting of unit vectors such that $< \alpha_i, \alpha_j > \leq 0$ and $4 < \alpha_i, \alpha_j >^2 = 0, 1, 2,$ *or* 3 for $i \neq j$ (basis of a root system with each vector made a unit vector by dividing by its length). The connected Coxeter graph of such a basis is the graph with vertexes $\{v_1, v_2, \ldots, v_l\}$ corresponding to $\{\alpha_1, \alpha_2, \ldots, \alpha_l\}$ together with $4 < \alpha_i, \alpha_j >^2$ edges joining v_i with $v_j, i \neq j$. We classify them.

Let us call a linearly independent subset S of E to be an admissible set if $< \alpha, \beta > \leq 0$ and $4 < \alpha, \beta >^2 = 0, 1, 2,$ *or* 3 for all distinct $\alpha, \beta \in S$. Evidently, the subset of an admissible set is an admissible set. As usual, we can talk of the Coxeter graph of S and denote it by Γ_S. The following are some properties of Coxeter graphs of admissible sets:

(i) The number of pairs of distinct vertexes in Γ_S which are connected by edges is strictly less than the number $\mid S \mid$ of elements of S: Suppose that $S = \{\alpha_1, \alpha_2, \ldots, \alpha_m\}$, $m \leq l$. Consider $\beta = \sum_{i=1}^{m} \alpha_i$. Since S is linearly independent, $\beta \neq 0$. Hence $0 << \beta, \beta >= \sum_{i=1}^{m} < \alpha_i, \alpha_i > +2\sum_{1 \leq k < l \leq m} < \alpha_k, \alpha_l >= m + 2\sum_{1 \leq k < l \leq m} < \alpha_k, \alpha_l >$. Since $< \alpha_k, \alpha_l > \leq 0$ and for nonzero terms $< \alpha_k, \alpha_l >$, $4 < \alpha_k, \alpha_l >^2 = 1, 2,$ *or* 3, it follows that the nonzero terms under summation in RHS are ≤ -1. Consequently, there are at the most $m - 1$ nonzero terms.

(ii) Γ_S contains no cycle: If Γ_S contains a cycle, then S contains an admissible subset S' such that $\Gamma_{S'}$ is a cycle. This contradicts (i).

(iii) Not more than three edges can originate from a vertex of Γ_S: Let $\alpha \in S$ and $\beta_1, \beta_2, \ldots, \beta_k$ be vectors in S which are connected to α by 1, 2, or 3 edges. Then $< \alpha, \beta_i > < 0$ for all i. Clearly, no β_i is connected to β_j for otherwise Γ_S will have a cycle. This means that $< \beta_i, \beta_j >= 0$ for all $i \neq j$. Let β_0 be a unit vector orthogonal to the subspace generated by $\{\beta_1, \beta_2, \ldots, \beta_k\}$. Clearly, $< \alpha, \beta_0 > \neq 0$, $\alpha = \sum_{i=0}^{k} < \alpha, \beta_i > \beta_i$, and $1 =, < \alpha, \alpha >= \sum_{i=0}^{k} < \alpha, \beta_i >^2$. Thus, $\sum_{i=1}^{k} < \alpha, \beta_i >^2 < 1$. Consequently, $\sum_{i=1}^{k} 4 < \alpha, \beta_i >^2 < 4$. This shows that $k \leq 3$.

(iv) From (iii), it follows that the only connected graph Γ_S which contains a triple edge is

(v) Let S be an admissible set and $S' = \{\beta_1, \beta_2, \ldots, \beta_k\}$ be an admissible subset of S such that the Coxeter graph $\Gamma_{S'}$ is a simple chain. Then $T = (S - \{\beta_1, \beta_2, \ldots, \beta_k\}) \bigcup \{\beta\}$ is an admissible subset, where $\beta = \sum_{i=1}^{k} \beta_i$: Evidently, T is a linearly independent set. We need to show that β is a unit vector and $4 < \gamma, \beta >^2 = 0, 1, 2, \textit{or}\ 3$ for all γ in $S - \{\beta_1, \beta_2, \ldots, \beta_k\}$. Since $\Gamma_{S'}$ is a simple chain, $2 < \beta_i, \beta_{i+1} >= -1$ for all i, $1 \leq i \leq k-1$ and $< \beta_i, \beta_j >= 0$ whenever $| i - j | \geq 2$. Hence $< \beta, \beta >= k + \sum_{1 \leq i < j \leq k} 2 < \beta_i, \beta_j >= k - (k-1) = 1$. This shows that β is a unit vector. Let γ be a member of $S - \{\beta_1, \beta_2, \ldots, \beta_k\}$. Since Γ_S has no cycle, γ can be connected to at the most one β_i. This means that $< \gamma, \beta >= 0$ or $< \gamma, \beta >=, < \gamma, \beta_i >$ for some i. Consequently, $4 < \gamma, \beta >^2 = 0, 1, 2, \textit{or}\ 3$. Thus, the graph Γ_T of T is just the graph obtained by shrinking the subgraph $\Gamma_{S'}$ of Γ_S to a point.

(vi) There is no subgraph of Γ_S of the form

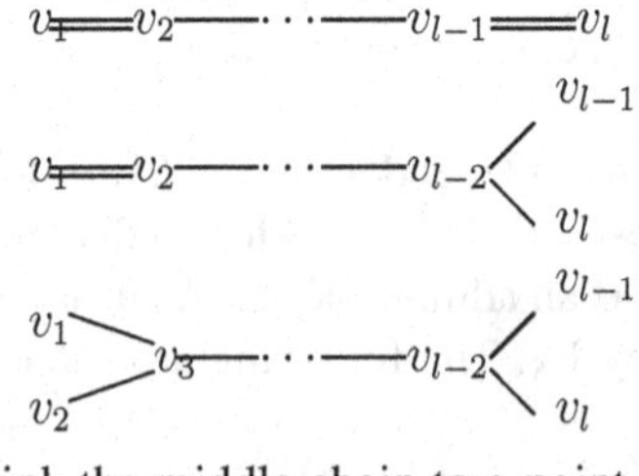

This is because if we shrink the middle chain to a point, then we get a subgraph in which there is a vertex with four incident edges. This, however, is a contradiction to (iii).

(vii) The possible connected Coxeter graphs of a root system are of the following forms:

$$v_1 \text{---} v_2 \text{---} \cdots \text{---} v_{l-1} \text{---} v_l,$$

$$v_1 \text{---} v_2 \cdots v_{k-1} \text{---} v_k =\!\!=\!\!= w_l \text{---} w_{l-1} \cdots w_2 \text{---} w_1$$

$$v_1 \equiv\!\!\equiv\!\!\equiv v_2$$

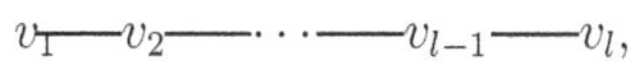

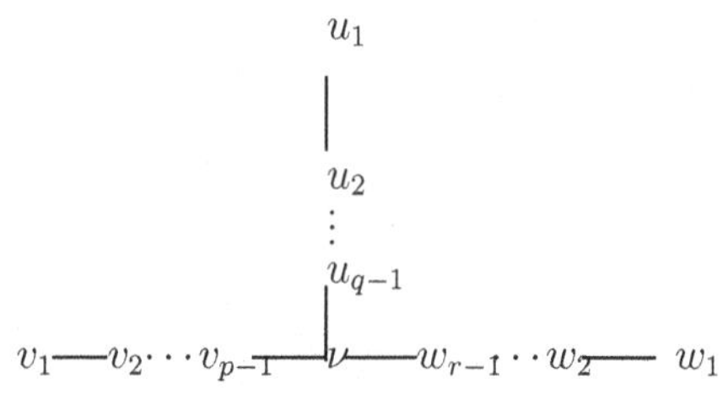

This is because a Coxeter graph Γ_S containing two double edges will contain a subgraph of the form

$$v_1 =\!\!=\!\!= v_2 \text{---} \cdots \text{---} v_{l-1} =\!\!=\!\!= v_l,$$

a Coxeter graph containing a double edge and a node will have a subgraph of the form

$$v_1 =\!\!=\!\!= v_2 \text{---} \cdots \text{---} v_{l-2} {\Big\langle}{}^{\textstyle v_{l-1}}_{\textstyle v_l,}$$

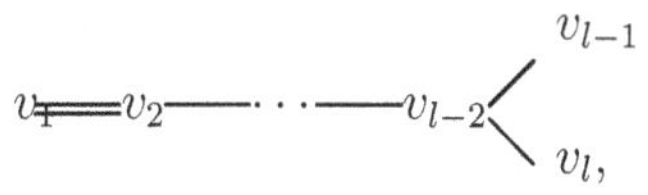

and a Coxeter graph having two nodes will have a subgraph of the form

$${}^{\textstyle v_1}_{\textstyle v_2}{\Big\rangle} v_3 \text{---} \cdots \text{---} v_{l-2} {\Big\langle}{}^{\textstyle v_{l-1}}_{\textstyle v_l}$$

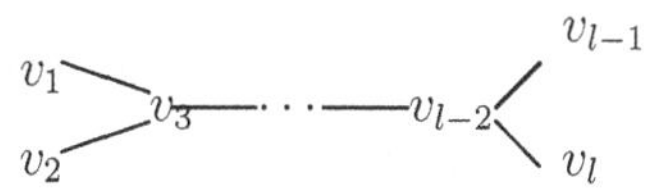

This is impossible because of (vi). Further, the only connected Coxeter graph with a triple edge is G_2.

(viii) The Coxeter graph of a root system containing a double edge is the Coxeter graph

$$v_1 \text{---} v_2 =\!\!=\!\!= v_3 \text{---} v_4$$

of F_4 or the Coxeter graph

v_1 v_2 v_{l-1} v_l

of B_l (same as that of C_l): From (vii) the only possible form of a Coxeter graph of a root system containing a double edge is

v_1 v_2 v_{k-1} v_k w_l w_{l-1} w_2 w_1

Let α_i be the root corresponding to the vertex v_i, and β_j be the root corresponding to the vertex w_j. Let α denote the vector $\sum_{i=1}^{k} i\alpha_i$ and β denote the vector $\sum_{j=1}^{l} j\beta_j$. Under the assumption,

(a) $2 < \alpha_i, \alpha_{i+1} >= -1 = 2 < \beta_j, \beta_{j+1} >$,
(b) the rest of the pairs of roots are orthogonal to each other, and
(c) $4 < \alpha_k, \beta_l >^2 = 2$.
Consequently, $< \alpha, \alpha >= \sum_{i=1}^{k} i^2 - \sum_{i=1}^{k-1} i(i+1) = \frac{k(k+1)}{2}$ and $< \beta, \beta >= \sum_{j=1}^{l} j^2 - \sum_{j=1}^{l-1} j(j+1) = \frac{l(l+1)}{2}$. Also using (c), we get $< \alpha, \beta >^2 = \frac{k^2l^2}{2}$. Since $\{\alpha, \beta\}$ is linearly independent, $< \alpha, \beta >^2$ is less than $< \alpha, \alpha >^2 < \beta, \beta >^2$. Thus, $\frac{k^2l^2}{2} < \frac{k(k+1)}{2}\frac{l(l+1)}{2}$. This means that $(k-1)(l-1) < 2$. It follows that $k = 2 = l$, or k is anything and $l = 1$. The assertion follows.

(ix) The Coxeter graph having a node is one of the following forms:

$D_l, l \geq 4$:
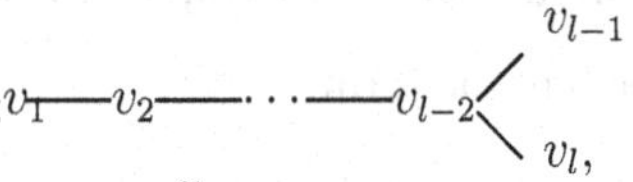
v_1 v_2 v_{l-2} v_{l-1} v_l,

E_6 :
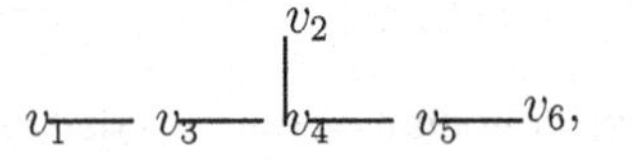
v_2
v_1 v_3 v_4 v_5 v_6,

E_7 :
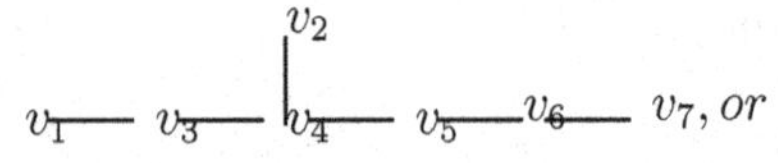
v_2
v_1 v_3 v_4 v_5 v_6 v_7, or

E_8 :
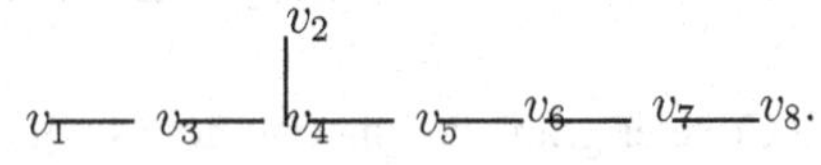
v_2
v_1 v_3 v_4 v_5 v_6 v_7 v_8.

We establish this:

From (vii), a Coxeter graph having a node is of the form

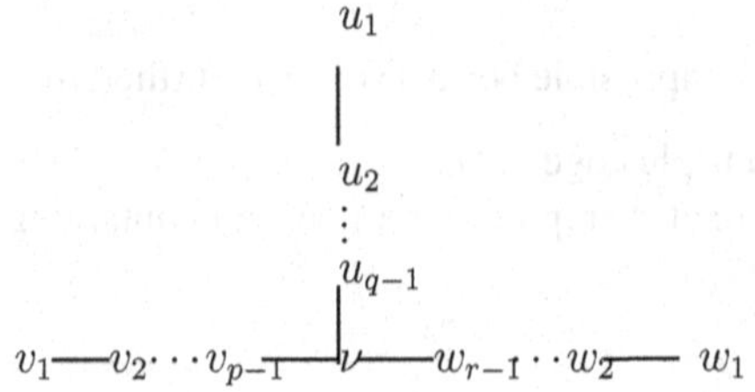
u_1
u_2
u_{q-1}
v_1 v_2 v_{p-1} v w_{r-1} w_2 w_1

Let v_i correspond to the root α_i, ν correspond to the root δ, w_j correspond to β_j, and u_k correspond to γ_k. Put $\alpha = \sum_{i=1}^{p-1} i\alpha_i$, $\beta = \sum_{j=1}^{r-1} j\beta_j$, and $\gamma = \sum_{k=1}^{q-1} k\gamma_k$. Clearly, $\{\alpha, \beta, \gamma\}$ is a pairwise orthogonal nonzero set of vectors, and δ is outside the subspace generated by $\{\alpha, \beta, \gamma\}$. Let $\hat{\alpha}$ denote the unit vector along α. Since δ lies outside the subspace generated by $\{\alpha, \beta, \gamma\}$, it follows that $1 \;\; > < \delta, \hat{\alpha} >^2 + < \delta, \hat{\beta} >^2 + < \delta, \hat{\gamma} >^2$. Now, $< \delta, \hat{\alpha} >^2 = \frac{<\delta,\alpha>^2}{<\alpha,\alpha>}$. Further, $< \delta, \alpha >^2 = (p-1)^2 < \delta, \alpha_{p-1} >^2 = \frac{(p-1)^2}{4}$ and as in (viii), $< \alpha, \alpha > = \frac{p(p-1)}{2}$. Thus, $< \delta, \hat{\alpha} >^2 = \frac{2(p-1)^2}{4p(p-1)} = \frac{1}{2}(1 - \frac{1}{p})$. Similarly, $< \delta, \hat{\beta} >^2 = \frac{2(q-1)^2}{4q(q-1)} = \frac{1}{2}(1 - \frac{1}{q})$, and $< \delta, \hat{\gamma} >^2 = \frac{2(r-1)^2}{4r(r-1)} = \frac{1}{2}(1 - \frac{1}{r})$. Consequently, we get that $\frac{1}{p} + \frac{1}{q} + \frac{1}{r} \;\; > \;\; 1$. Without any loss, we may assume that $p \geq \;\; q \geq r$. If $r = 1$, then for an arbitrary $p, q \geq 1$, $\frac{1}{p} + \frac{1}{q} + \frac{1}{r} \;\; > \;\; 1$. This is the case of D_l. Suppose that $r \geq 2$. Then necessarily $r = 2$, and possible solutions are $(p, 2, 2)$ (corresponding to D_p), $(3, 3, 2)$ (corresponding to E_6), $(4, 3, 2)$ (corresponding to E_7), and $(5, 3, 2)$ (corresponding to E_8). Evidently, in each case the Coxeter graph uniquely determines the Dynkin diagram except B_l and C_l which can be distinguished with the help of arrows. ♯

The above theorem determines the possible Dynkin diagrams of an irreducible root system. It, however, does not guarantee the existence of an irreducible root system corresponding to each Dynkin diagram.

Theorem 2.3.2 *There is a root system associated with each Dynkin diagram A_l – G_2 as given in the above theorem.*

Proof We show their existence by constructing them.

A_l. Consider the standard Euclidean space $\mathbb{R}^{l+1}$ of dimension $l + 1$ together with the standard basis $\{e_1, e_2, \ldots, e_{l+1}\}$. Let $L = \mathbb{Z}e_1 + \mathbb{Z}e_2 + \cdots + \mathbb{Z}e_{l+1}$ denote the Lattice generated by $\{e_1, e_2, \ldots, e_{l+1}\}$. Let E denote the l-dimensional subspace of $\mathbb{R}^{l+1}$ which is the orthogonal complement of $\mathbb{R}e$, where $e = e_1 + e_2 + \cdots + e_{l+1}$. Let $\Phi = \{\alpha \in L \bigcap E \mid < \alpha, \alpha > = 2\}$. A nonzero vector $\alpha = \sum_{i=1}^{l+1} a_i e_i$ belongs to $L \bigcap E$ if and only if $\sum_{i=1}^{l+1} a_i = 0$ and $\sum_{i=1}^{l+1} a_i^2 = 2$, where each a_i is an integer. Thus, $\Phi = \{e_i - e_j \mid i \neq j\}$. Take $\Delta = \{e_i - e_{i+1} \mid 1 \leq l \leq l\}$. Then Δ is linearly independent and generates E. Also for $i < j$, $e_i - e_j = (e_i - e_{i+1}) + (e_{i+1} - e_{i+2}) + \cdots + (e_{j-1} - e_j)$. Thus, every element of Φ is an integral linear combination of members of Δ. It is easily observed that the Cartan matrix of Φ relative to Δ is that of A_l. Further, $\sigma_{e_i - e_{i+1}}$ is an orthogonal transformation and it is uniquely determined by its effect on e_i, $i \leq l + 1$. Clearly, $\sigma_{e_i - e_{i+1}}(e_i) = e_{i+1}$, $\sigma_{e_i - e_{i+1}}(e_{i+1}) = e_i$, and $\sigma_{e_i - e_{i+1}}(e_j) = e_j$ for all $j \neq i, i + 1$. Thus, $\sigma_{e_i - e_{i+1}}$ may be identified with the transposition (e_i, e_{i+1}). It follows that the Weyl group of A_l is the symmetric group S_{l+1}.

$B_l, l \geq 2$. Take E to be $\mathbb{R}^l$ and $\Phi = \{\alpha \in L \mid < \alpha, \alpha > = 1 \; or \; 2\}$, where L is a Lattice in $\mathbb{R}^l$ generated by the standard basis $\{e_1, e_2, \ldots, e_l\}$. Clearly, $\Phi = \{\pm e_i \mid 1 \leq i \leq l\} \bigcup \{(\pm(e_i \pm e_j) \mid i \neq j\}$. Clearly, the subset $\Delta = \{e_i - e_{i+1} \mid 1 \leq$

$i \leq l-1\}\bigcup\{e_l\}$ is a basis of E and every element of Φ is an integral linear combination of members of Δ. It can be easily seen that Φ is a root system with Δ as a basis such that the Dynkin diagram is that of B_l. e_i, $1 \leq i \leq l$ are short roots, while $e_i \pm e_j, i \neq j$ are long roots. We find the Weyl group. As in case of A_l, the subgroup of $W(\Phi)$ generated by $\{\sigma_{e_i - e_{i+1}} \mid 1 \leq i \leq l-1\}$ is S_l. Further, $\sigma_{e_i}(\pm e_i) = \mp e_i$ and $\sigma_{e_i}(e_j) = e_j$ for $j \neq i$. Thus, the subgroup of $W(\Phi)$ generated by $\{\sigma_{e_i} \mid 1 \leq i \leq l\}$ is isomorphic to $\mathbb{Z}_2^l$ on which S_l acts. It follows that $W(\Phi) = W(B_l)$ is the semi-direct product $\mathbb{Z}_2^l \succ S_l$.

$C_l, l \geq 3$. The root system C_l is the dual to B_l, and long and short roots are interchanged. Thus, Φ can be taken to be $\{\pm(e_i \pm e_j \mid i \neq j\}\bigcup\{2e_i \mid 1 \leq i \leq l\}$. Evidently, $W(C_l) \approx W(B_l)$.

$D_l, l \geq 4$. Take $E = \mathbb{R}^l$, $\Phi = \{\alpha \in L \mid < \alpha, \alpha >= 2\} = \{\pm(e_i \pm e_j) \mid i \neq j\}$. $\Delta = \{e_i - e_{i+1} \mid 1 \leq i \leq l-1\}\bigcup\{e_{l-1} + e_{l+1}\}$ is a basis. It can be seen that Φ is a root system with Δ as a basis for which the Cartan matrix is the Cartan matrix of D_l. Observe that $\sigma_{e_{l-1}+e_l}(e_{l-1}) = -e_l$, $\sigma_{e_{l-1}+e_l}(e_l) = -e_{l-1}$ and $\sigma_{e_{l-1}+e_l}(e_j) = e_j$ for $j \notin \{l-1, l\}$. Thus, $W(D_l) \approx \mathbb{Z}_2^{l-1} \succ S_l$.

$E_6, E_7,$ *and* E_8. Since E_6 and E_7 are admissible Coxeter subgraphs of E_8, it is sufficient to construct a root system associated with E_8. Take $E = \mathbb{R}^8$. Put $L' = L + \mathbb{Z}\frac{e}{2}$, where $e = \sum_{i=1}^{8} e_i$. Consider the subgroup L'' of L' consisting of elements of the type $\sum_{i=1}^{8} a_i e_i + \frac{a}{2}e$, where $a + \sum_{i=1}^{8} a_i$ is an even integer. Finally, let us take Φ to be the subset $\{\alpha \in L'' \mid < \alpha, \alpha >= 2\}$. It can be shown that $\Phi = \{\pm(e_i \pm e_j) \mid 1 \leq i, j \leq 8, i \neq j\}\bigcup\{\frac{1}{2}\sum_{i=1}^{8} \epsilon_i e_i \mid \epsilon_i = \pm 1 \ with\ even\ number\ of\ \epsilon_i\ being\ -1\}$. It is straightforward to verify that Φ is a root system with $\Delta = \{\frac{1}{2}(e_1 + e_8 - e_2 + e_3 + \cdots + e_7), e_1 + e_2, e_2 - e_1, e_3 - e_2, \ldots, e_7 - e_6\}$ as a basis such that the corresponding Cartan matrix is that of E_8. Looking at the action of the Weyl group on the standard basis, it can be shown that the order of the Weyl group $W(E_8)$ is $2^{14}3^5 5^2 7$.

F_4. For F_4, take $E = \mathbb{R}^4$. Consider $L' = L + \mathbb{Z}\frac{e}{2}$, where, as usual, $e = e_1 + e_2 + e_3 + e_4$. It is straightforward to verify that $\Phi = \{\alpha \in L' \mid < \alpha, \alpha >= 1\ or\ 2\} = \{\pm e_i \mid 1 \leq i \leq 4\}\bigcup\{\pm(e_i - e_j) \mid i \neq j\}\bigcup\{\pm\frac{1}{2}(e_1 \pm e_2 \pm e_3 \pm e_4), \mid \pm\ are\ chosen\ arbitrarily\}$ is a root system with $\Delta = \{e_2 - e_3, e_3 - e_4, e_4, \frac{1}{2}(e_1 - e_2 - e_3 - e_4)\}$ as a basis such that the Cartan matrix is that of F_4. Looking at the action of members of $W(\Phi)$ on the standard basis, one can show that the order of $W(F_4)$ is $2^7 3^2$.

G_2. We have already constructed G_2 in the beginning of Sect. 2.1. However, we give another construction. Take E as the orthogonal complement of $\mathbb{R}e$ in $\mathbb{R}^3$, where $e = e_1 + e_2 + e_3$. Then $\Phi = \{\alpha \in L \bigcap E \mid < \alpha, \alpha >= 2\ or\ 6\} = \{\pm(e_1 - e_2), \pm(e_2 - e_3), \pm(e_1 - e_3), \pm(2e_1 - e_2 - e_3), \pm(2e_2 - e_1 - e_3), \pm(2e_3 - e_1 - e_2)\}$ is a root system with $\{e_1 - e_2, -2e_1 + e_2 + e_3\}$ as a basis such that the corresponding Cartan matrix is that of G_2. Looking at the action of $W(G_2)$ on the standard basis of $\mathbb{R}^3$, we can show that $W(G_2) \approx D_6$. ♯

Now, we describe the group $Aut(\Phi)$ of automorphisms of a root system Φ. If η is an automorphism of Φ, then $\prec \eta(\alpha), \eta(\beta) \succ = \prec \alpha, \beta \succ$. Hence, η induces automorphisms of the corresponding Dynkin diagram. We have already seen that the Weyl group $W(\Phi)$ is a normal subgroup of $Aut(\Phi)$. Clearly, $Aut(\Phi)$ acts on the set of all bases of Φ. Fix a basis Δ of Φ. Consider the subgroup $\Gamma = \{\eta \in Aut(\Phi) \mid \eta(\Delta) = \Delta\}$. If η is a member of $Aut(\Phi)$, then since $W(\Phi)$ acts transitively on the set of all bases, there is a member $\sigma \in W(\Phi)$ such that $\eta(\Delta) = \sigma(\Delta)$. Consequently, $\sigma^{-1}\eta \in \Gamma$. This means that $Aut(\Phi) = W(\Phi)\Gamma$. If $\eta \in W(\Phi)\bigcap\Gamma$, then since $W(\Phi)$ acts simply transitively on the set of bases, $W(\Phi)\bigcap\Gamma = \{I\}$. Hence $Aut(\Phi) = W(\Phi) \succ \Gamma$. The automorphisms fixing a base is precisely the automorphisms of Dynkin diagrams. They are called the diagram automorphisms or also the graph automorphisms of the root system Φ. If the root system has two root lengths, then there is no graph automorphism. Thus in this case $\Gamma = \{I\}$ and $Aut(\Phi) = W(\Phi)$. Looking at the diagrams, we observe that the group $\Gamma(A_l)$ of graph automorphisms of A_l is $\mathbb{Z}_2$, $l \geq 2$. $\Gamma(D_4) \approx S_3$ and $\Gamma(D_l) \approx \mathbb{Z}_2$ for $l > 4$. Further $\Gamma(E_6) \approx \mathbb{Z}_2$.

Exercises

2.3.1. Determine Cartan matrices for each Dynkin diagram.
2.3.2. Determine the root system B_3 and C_3.
2.3.3. Determine the root system F_4.
2.3.4. Show that the dual of B_l is C_l provided that $l \neq 2$. If Φ is an irreducible system different from B_l and C_l, then show that Φ^{ν} is isomorphic to Φ.
2.3.5. Show that $\alpha \mapsto -\alpha$ is an automorphism of Φ. When does it belong to $W(\Phi)$?

2.4 Conjugacy Theorem, Existence and Uniqueness Theorems

In this section, we introduce the Cartan and Borel subalgebras, and prove the Conjugacy theorem. We also establish the important theorem of Serre about the existence of a semi-simple Lie algebra associated with a root system. A semi-simple Lie algebra is described in terms of generators and relations. All fields are assumed to be algebraically closed fields.

Cartan Subalgebras

Let L be a semi-simple Lie algebra over an algebraically closed field of characteristic 0. Recall that a maximal toral subalgebra H of L determines (uniquely) a root system Φ. However, if we take another maximal toral subalgebra H' of L, then the corresponding root system ϕ' may differ apparently from Φ. Our aim is to show that Φ is unique up to an isomorphism between the root systems. For this purpose, we need to show that H and H' are conjugate in L in the sense that there is an automorphism η of L such that $\eta(H) = H'$. More generally, we introduce the concept of Cartan subalgebras of a Lie algebra (not necessarily a semi-simple Lie algebra) which agree with that of maximal toral subalgebras in the case of semi-simple Lie algebras, and then show that any two Cartan subalgebras are conjugate to each other.

Let L be a finite-dimensional Lie algebra over an algebraically closed field (not necessarily of characteristic 0). Let $x \in L$. Consider the endomorphism $ad(x)$ of L.

For $\lambda \in F$, let m_λ denote the multiplicity of λ as a root of the characteristic polynomial $\phi^\lambda(t)$ of $ad(x)$ (note that if λ is not a root, then $m_\lambda = 0$). Let $L_\lambda(ad(x))$ denote the kernel of $(ad(x) - \lambda I)^{m_\lambda}$. In case λ is not a root, $L_\lambda(ad(x)) = \{0\}$. Clearly, $ad(x)$ restricted to $L_\lambda(ad(x))$ is the sum of λI and a nilpotent endomorphism. It is a well-known fact from Linear algebra that

$$L = \oplus \sum_{\lambda \in F} L_\lambda(ad(x)) = L_0(ad(x)) \oplus \left(\oplus \sum_{\lambda \in F^\star} L_\lambda(ad(x)) \right).$$

More generally, if A is a subalgebra of L such that $[x, A] \subseteq A$, then

$$A = A_0(ad(x)) \oplus \oplus \sum_{\lambda \in F^\star} A_\lambda(ad(x)),$$

where $A_\lambda(ad(x)) = A \bigcap L_\lambda(ad(x))$.

Proposition 2.4.1 $[L_\lambda(ad(x)), L_\mu(ad(x))] \subseteq L_{\lambda+\mu}(ad(x))$ *for all* $\lambda, \mu \in F$.

Proof The following identity can be easily established by the induction on m:

$$(ad(x) - \lambda I - \mu I)^m([y, z]) = \sum_{i=0}^{m} {}^mC_i[(ad(x)) - \lambda I)^i(y), (ad(x)) - \mu I)^{m-i}(z)]$$

for all $x, y,$ *and* $z \in L$. Let $u \in L_\lambda(ad(x))$ and $v \in L_\mu(ad(x))$. Then $(ad(x) - \lambda I)^{m_\lambda}(u) = 0 = (ad(x) - \mu I)^{m_\mu}(v)$. It follows from the above identity that $(ad(x) - (\lambda + \mu)I)^{m_\lambda + m_\mu}([u, v]) = 0$. Thus, $[u, v] \in L_{\lambda+\mu}(ad(x))$. ♯

The following corollary is immediate.

Corollary 2.4.2 *$L_0(ad(x))$ is a subalgebra of L.* ♯

Corollary 2.4.3 *For each $\lambda \in F^\star$, each element of $L_\lambda(ad(x))$ is ad-nilpotent.*

Proof Since L is finite dimensional, $L_{n\lambda}(ad(x)) = \{0\}$ for some n. The result follows from Proposition 2.4.1. ♯

Definition 2.4.4 A subalgebra of L of the form $L_0(ad(x))$ is called an **Engel subalgebra** of L. More concretely, a subalgebra A is said to be an Engel subalgebra if there is an element $x \in L$ such that $u \in A$ if and only if $ad(x)^r(u) = 0$ for some r.

Lemma 2.4.5 *Let A be a subalgebra of a Lie algebra L. Let u be an element of A such that $x \in A$ and $L_0(ad(x)) \subseteq L_0(ad(u)))$ implies that $L_0(ad(x)) = L_0(ad(u))$. Suppose further that $A \subseteq L_0(ad(u))$. Then $L_0(ad(u)) \subseteq L_0(ad(x))$ for all $x \in A$.*

Proof For simplicity, put $A_0 = L_0(ad(u))$. For $x \in A$, consider the family $\{ad(u + \lambda x) \mid \lambda \in F\}$. Since $A \subseteq A_0$, for each $\lambda \in F$, $ad(u + \lambda x)$ induces an endomorphism on A_0 and also on L/A_0. Thus, the characteristic polynomial $\phi^\lambda(t)$ of $ad(u + \lambda x)$

is the product $\psi^\lambda(t)\eta^\lambda(t)$, where $\psi^\lambda(t)$ is the characteristic polynomial of the restriction $ad(u + \lambda x)|_{A_0}$ to A_0 and $\eta^\lambda(t)$ is that of the induced endomorphism on L/A_0 (here, t is indeterminate). Suppose that the dimension of A_0 is r. Then $DimL/A_0 = l - r$, where $l = Dim\ L$. Let $\{x_1, x_2, \ldots, x_r, x_{r+1}, \ldots, x_l\}$ be a basis of L with $\{x_1, x_2, \ldots, x_r\}$ a basis of A_0. Evidently, in the matrix of $ad(u + \lambda x)$ with respect to this basis, the entries in the first r row are multiples of λ and also the rest of the entries are linear polynomials in λ. As such,

$$\psi^\lambda(t) = t^r + a_1(\lambda)t^{r-1} + \cdots + a_r(\lambda),$$

and

$$\eta^\lambda(t) = t^{l-r} + b_1(\lambda)t^{l-r-1} + \cdots + b_{l-r}(\lambda),$$

where $a_i(t)$ and $b_j(t)$ are polynomials in t of degrees at the most i and at the most j, respectively. Since all the eigenvectors of $ad(u)$ corresponding to the eigenvalue 0 belong to A_0, the polynomial $b_{l-r}(t)$ is nonzero. Since there are at the most $l - r$ roots of $b_{l-r}(t)$, we have distinct members $\lambda_1, \lambda_2, \ldots, \lambda_{r+1}$ of F such that $b_{l-r}(\lambda_j) \neq 0$ for all j, $1 \leq j \leq r + 1$. Consequently, 0 is not an eigenvalue of $ad(u + \lambda_j x)$ on L/A_0 for each j. This means that $L_0(ad(u + \lambda_j x)) \subseteq A_0 = L_0(ad(u))$. Hence, by the hypothesis $L_0(ad(u)) = L_0(ad(u + \lambda_j x))$ for all j. Thus, $ad(u + \lambda_j x)$ has 0 as the only eigenvalue on A_0. In other words, $\psi^{\lambda_j}(t) = t^r$ for all j. Hence $\lambda_1, \lambda_2, \ldots, \lambda_{r+1}$ are zeros of $a_i(t)$ for all i, $1 \leq i \leq r$. Since the degree of $a_i(t)$ is at the most $i \leq r$, it follows that $a_i(t) = 0$ for all i, $1 \leq i \leq r$. Consequently, for each $\lambda \in F$, the characteristic polynomial of $ad(u + \lambda x)$ is t^r. Thus, $A_0 \subseteq L_0(ad(u + \lambda x))$ for all $\lambda \in F$. Since x is arbitrary, replacing x by $x - u$ and putting $\lambda = 1$, we obtain that $L_0(ad(u)) \subseteq L_0(ad(x))$ for all $x \in A$. ♯

Lemma 2.4.6 *Let A be a subalgebra of a Lie algebra L which contains an Engel subalgebra. Then A is self-normalizing in the sense that $N_L(A) = A$. In particular, all Engel subalgebras are self-normalizing.*

Proof Suppose that $L_0(ad(x)) \subseteq A$. Then $x \in A$. If $N_L(A) \neq A$, then the endomorphism on $N_L(A)/A$ induced by $ad(x)$ has a nonzero eigenvalue. This is a contradiction, since $[x, N_L(A)] \subseteq A$. ♯

Definition 2.4.7 A self-normalizing nilpotent subalgebra of a Lie algebra L is called a **Cartan subalgebra** (**CSA**) of L.

Example 2.4.8 All maximal toral subalgebras (being self-normalizing abelian subalgebras) of a semi-simple Lie algebra over an algebraically closed field are Cartan subalgebras.

Remark 2.4.9 We shall show that a Cartan subalgebra of Lie algebra over an algebraically closed field exists. Indeed, it also exists for Lie algebras over any infinite field. However, the existence problem for Lie algebras over finite fields is still an open problem.

Proposition 2.4.10 *Cartan subalgebras of a finite-dimensional Lie algebra over an algebraically closed field are precisely the minimal Engel subalgebras. In particular, Cartan subalgebra of a finite-dimensional Lie algebra over an algebraically closed field exists.*

Proof Let $H = L_0(ad(u))$ be a minimal Engel subalgebra of L. By Lemma 2.4.6, H is self-normalizing. Further, H satisfies the hypothesis of Lemma 2.4.5 and hence $H \subseteq L_0(ad(x))$ for all $x \in H$. This means that $ad(x)|_H$ is nilpotent for all $x \in H$. By the theorem of Engel, H is nilpotent. Consequently, H is a Cartan subalgebra.

Conversely, let H be a Cartan subalgebra of L. Then H is nilpotent. Hence $ad(x)|_H$ is nilpotent for all $x \in H$. This means that $H \subseteq L_0(ad(x))$ for all $x \in H$. We need to show that $H = L_0(ad(x))$ for some $x \in H$. Let $u \in L$ be such that $L_0(ad(u))$ is minimal among $\{L_0(ad(x)) \mid x \in H\}$. We show that $H = L_0(ad(u))$. Suppose the contrary. Then $L_0(ad(u))/H$ is a nonzero space. From Lemma 2.4.5, $L_0(ad(u)) \subseteq L_0(ad(x))$ for all $x \in H$. Thus, each $h \in H$ induces the endomorphism $\overline{ad(h)}$ of $L_0(ad(u))/H$ which is given by $\overline{ad(h)}(v + H) = ad(h)(v) + H$. Clearly $\overline{ad(h)}$ is a nilpotent endomorphism for each $h \in H$. By Theorem 1.3.14, there is a nonzero element $v + H \in L_0(ad(u))/H$ such that $\overline{ad(h)}(v + H) = H$. This means that $v \notin H$ and $[v, H] \subseteq H$. This is a contradiction, since H is self-normalizing. ♯

Corollary 2.4.11 *Cartan subalgebras of a semi-simple Lie algebra over an algebraically closed field of characteristic 0 are precisely the maximal toral subalgebras.*

Proof Since every maximal toral subalgebra of a semi-simple Lie algebra over an algebraically closed field is an abelian and self-normalizing subalgebra, it is a Cartan subalgebra. Conversely, let H be a Cartan subalgebra of a semi-simple algebra L over an algebraically closed field. We need to show that H is a maximal toral subalgebra. From Proposition 2.4.10, $H = L_0(ad(u))$ is a minimal Engel subalgebra. Let $u = u_s + u_n$ be the Jordan decomposition of u. Then since $ad(u_s)$ and $ad(u_n)$ commute and $ad(u_n)$ is nilpotent, $L_0(ad(u_s)) \subseteq L_0(ad(u))$. Consequently, $H = L_0(ad(u_s)) = C_L(u_s)$. Since u_s is semi-simple, $C_L(u_s)$ contains a maximal toral subalgebra which is a Cartan subalgebra. In turn, this maximal toral subalgebra is minimal Engel subalgebra. It follows that $H = C_L(u_s)$ is a maximal toral subalgebra. ♯

Corollary 2.4.12 *Every maximal toral subalgebra of a semi-simple Lie algebra over an algebraically closed field of characteristic 0 is of the form $C_L(s)$, where s is a semi-simple element.* ♯

A semi-simple element s of a semi-simple Lie algebra L over an algebraically closed field of characteristic 0 is termed as a **regular semi-simple** element if $C_L(s)$ is a maximal toral subalgebra. For example, $Diag(\lambda_1, \lambda_2, \ldots, \lambda_n) \in sl(n, F)$ is a regular semi-simple element of $sl(n, F)$ if and only if all λ_i are distinct. Indeed, a matrix A in $sl(n, F)$ is regular semi-simple if and only if all eigenvalues of A are distinct (cf Exercise 2.4.1).

Proposition 2.4.13 *Image of a Cartan subalgebra under an epimorphism is a Cartan subalgebra.*

Proof Let η be an epimorphism from L to L'. Let H be a Cartan subalgebra of L. Then H is nilpotent. Since the image of a nilpotent subalgebra under a homomorphism is nilpotent, $\eta(H)$ is nilpotent. Let $v \in L'$ such that $[v, \eta(H)] \subseteq \eta(H)$. Since η is an epimorphism, there is an element $u \in L$ such that $\eta(u) = v$. Since $H \subseteq \eta^{-1}(\eta(H))$ and H is a Cartan subalgebra, it follows that $\eta^{-1}(\eta(H))$ contains a minimal Engel subalgebra. By Lemma 2.4.6, $N_L(\eta^{-1}(\eta(H))) = \eta^{-1}(\eta(H))$. Since $[\eta(u), \eta(H)] \subseteq \eta(H)$, $[u, \eta^{-1}(\eta(H))] \subseteq \eta^{-1}(\eta(H))$. Thus, $u \in N_L(\eta^{-1}(\eta(H))) = \eta^{-1}(\eta(H))$. Consequently, $v=\eta(u) \in \eta(H)$. This shows that $\eta(H)$ is self-normalizing also. ♯

Proposition 2.4.14 *Let η be an epimorphism from L to L'. Let H' be a Cartan subalgebra of L'. Then any Cartan subalgebra H of $\eta^{-1}(H')$ is also a Cartan subalgebra of L.*

Proof Let H be a Cartan subalgebra of $\eta^{-1}(H')$. Then H is already nilpotent. From the previous proposition, $\eta(H)$ is a Cartan subalgebra of $\eta(L) \subseteq L'$. Let $x \in L$ such that $[x, H] \subseteq H$. Then $[\eta(x), \eta(H)] \subseteq \eta(H)$. This means that $\eta(x) \in \eta(H)$. Hence $x \in H + Ker\eta \subseteq \eta^{-1}(H')$. Since H is Cartan subalgebra of $\eta^{-1}(H')$, $x \in H$. ♯

Let δ be a nilpotent derivation of A. Then $exp(\delta)$ is an automorphism of A. In particular, if $ad(x)$ is nilpotent, then $exp(ad(x))$ is an automorphism of L. This automorphism is called an inner automorphism of L. The subgroup of $Aut(L)$ that generated all inner automorphisms is called an inner automorphism group of L, and it is denoted by $Inn(L)$. Since $\sigma ad(x)\sigma^{-1} = ad(\sigma(x))$, it follows that $Inn(L)$ is a normal subgroup of $Aut(L)$. The group $Aut(L)/Inn(L)$ is called the outer automorphism group and it is denoted by $Out(L)$. We shall show that any two Cartan subalgebras H and H' are conjugate in the sense that there is a member η of $Inn(L)$ such that $\eta(H) = H'$. In particular, any two maximal toral subalgebras of a semi-simple Lie algebra over an algebraically closed field of characteristic 0 are conjugate to each other. This, therefore, implies that the root system associated with a semi-simple Lie algebra over an algebraically closed field is independent (up to an isomorphism) of the choice of maximal toral subalgebra.

Let L be a Lie algebra. An element $x \in L$ is termed as a **strongly ad-nilpotent** element of L if $x \in L_0(ad(y))$ for some $y \in L$. From Corollary 2.4.3, it follows that strongly ad-nilpotent elements are ad-nilpotent elements. However, an ad-nilpotent element need not be strongly ad-nilpotent. Let $Inn_0(L)$ denote the subgroup of $Aut(L)$ generated by the set $\{exp(ad(x)) \mid x \ is \ strongly \ ad - nilpotent\}$. Evidently, $Inn_0(L) \subseteq Inn(L)$. If x is strongly ad-nilpotent, then for each $\sigma \in Aut(L)$, $\sigma(x)$ is also strongly ad-nilpotent. This is because $\sigma(L_0(ad(y)) = L_0(ad(\sigma(y))$. Thus, $Inn_0(L)$ is also a normal subgroup of $Aut(L)$. For semi-simple Lie algebras, $Inn_0(L) = Inn(L)$ (for a proof, see the Lie algebra by Jacobson).

Proposition 2.4.15 *Let η be an epimorphism from L to L'. Let $\sigma' \in Inn_0(L')$. Then there is a $\sigma \in Inn_0(L)$ such that the diagram*

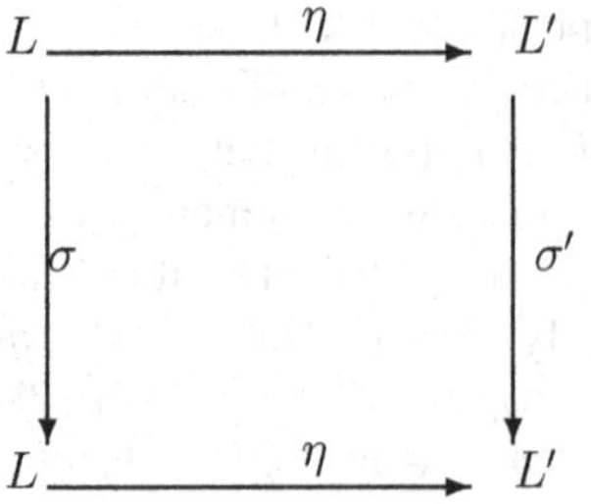

is commutative.

Proof Let $x' \in L'_0(ad(y'))$ be a strongly ad-nilpotent element of L'. Since η is an epimorphism, there is an element y in L such that $\eta(y) = y'$ and also $\eta(L_\lambda(ad(y))) = L'_\lambda(ad(\eta(y)))$ for all $\lambda \in F$. In particular, $\eta(L_0(ad(y))) = L'_0(ad(\eta(y)))$. Consequently, there is a strongly ad-nilpotent element $x \in L$ such that $\eta(x) = x'$. Further, it is easy to observe that $\eta o exp(ad(x)) = exp(ad(\eta(x))) o\eta$. Since $Inn_0(L)$ is generated by $\{exp(ad(x)) \mid x\ is\ strongly\ nilpotent\}$, the result follows. ♯

Our aim is to show that if H and H' are Cartan subalgebras of L, then there is an element $\sigma \in Inn_0(L)$ such that $\sigma(H) = H'$. We proceed as follows. We first prove it for solvable Lie algebras. Next, we shall show that any two maximal solvable subalgebras (termed as Borel subalgebras) of a Lie algebra are conjugate under $Inn_0(L)$. Consequently, the assertion will follow.

Theorem 2.4.16 *Let L be a finite-dimensional solvable Lie algebra. Let H and H' be Cartan subalgebras of L. Then there exists a $\sigma \in Inn_0(L)$ such that $\sigma(H) = H'$.*

Proof The proof is by the induction on the $DimL$. If $DimL = 1$, then there is nothing to do. Assume that the result is true for all Lie algebras whose dimensions are less than $DimL$, where $DimL > 1$. We need to prove it for L. If L is nilpotent, then L is the only Cartan subalgebra of L and there is nothing to do. Assume that L is not nilpotent. Let H_1 and H_2 be Cartan subalgebras of L. Since L is solvable (and non-nilpotent), it has a proper nontrivial abelian ideal. Let A be a nonzero abelian ideal of the smallest dimension. From Proposition 2.4.13, $H'_1 = (H_1 + A)/A$ and $H'_2 = (H_2 + A)/A$ are Cartan subalgebras of L/A. By the induction hypothesis, there is a $\hat{\sigma} \in Inn_0(L/A)$ such that $\hat{\sigma}(H'_1) = H'_2$. By Proposition 2.4.15, there is a $\sigma \in Inn_0(L)$ such that $\nu o\sigma = \hat{\sigma} o\nu$, where ν is the quotient map from L to L/A. This means that $\sigma(H_1 + A) = H_2 + A$.

Suppose that $H_2 + A \neq L$. Since $\sigma(H_1)$ and H_2 are Cartan subalgebras of $H_2 + A$, by the induction hypothesis, there is an element $\tau \in Inn_0(H_2 + A)$ such that $\tau(\sigma(H_1)) = H_2$. It can be seen from the definition that every member of $Inn_0(H_2 + A)$ is a restriction of a member of $Inn_0(L)$. Thus, there is a member $\hat{\tau}$ of $Inn_0(L)$ such that $\hat{\tau}|_{H_2+A} = \tau$. This means that $\hat{\tau}(\sigma(H_1)) = H_2$, and we are done.

Next, suppose that $H_2 + A = L = \sigma(H_1 + A)$. Since H_2 is a Cartan subalgebra of L, $H_2 = L_0(ad(x))$ is a minimal Engel subalgebra of L, $x \in L$. Since A is an ideal, $A = A_0(ad(x)) \oplus A_\star(ad(x))$, where $A_\star(ad(x)) = \oplus \sum_{\lambda \in F^\star} A_\lambda(ad(x))$. Clearly,

$A_0(ad(x))$ and $A_\star(ad(x))$ are $ad(L) = ad(H_2 + A)$-stable. Since A is a minimal abelian ideal, $A = A_0(ad(x))$ or $A = A_\star(ad(x))$. If $A = A_0(ad(x))$, then $A \subseteq H_2$ and so $H_2 = L$. Since H_2 is nilpotent (being a Cartan subalgebra), L will turn out to be nilpotent. This is a contradiction to our assumption. Hence $A = A_\star(ad(x))$. Further, since σ is an automorphism, $L = H_1 + A$. Thus, $x = y + z$ for some $y \in H_1$ and $z \in A = L_\star(ad(x))$. Now, $[x, z] = az$ for some $a \neq 0$. Hence $[x, a^{-1}z] = z,\ a^{-1}z \in A$. Since A is abelian, $(ad(a^{-1}z))^2 = 0$. Hence $exp(ad(a^{-1}z)) = I + ad(a^{-1}z)$. Now, $exp(ad(a^{-1}z))(x) = x - [x, a^{-1}z] = x - z = y$. Thus, $H = L_0(ad(y))$ is also a Cartan subalgebra. Since $y \in H_1$, $H_1 \subseteq H$. Consequently, $H = H_1$. This means that $exp(ad(a^{-1}z))(H_1) = H_2$. Since $a^{-1}z \in A = L_\star(ad(x))$, $a^{-1}z$ is a finite sum $\sum_i z_i$ of an element of A. Clearly, all z_i are strongly nilpotent elements of L. Since A is abelian, $exp(ad(a^{-1}z)) = \prod_i exp(ad(z_i))$ belongs to $Inn_0(L)$. ♯

A maximal solvable subalgebra of a Lie algebra L is called a **Borel subalgebra** of L. Thus, every solvable subalgebra of a finite-dimensional Lie algebra is contained in a Borel subalgebra. In particular, every Cartan subalgebra is contained in a Borel subalgebra. The subalgebra $t(n, F)$ of upper triangular matrices is a Borel subalgebra of $gl(n, F)$.

Proposition 2.4.17 *Any Borel subalgebra is self-normalizing.*

Proof Let B be a Borel subalgebra of L which is not self-normalizing. Then there is an element $x \in L - B$ such that $[x, B] \subseteq B$. Consider the subspace $B + Fx$ of L. Clearly, $B + Fx$ is a subalgebra of L such that $[B + Fx, B + Fx] \subseteq B$. Consequently, $B + Fx$ is a solvable subalgebra containing B properly. This is a contradiction. ♯

Let L be a Lie algebra. Since $RadL$ is a maximal solvable ideal of L, for any Borel subalgebra B of L, $B + Rad(L)$ is also a solvable subalgebra of L. Since B is a maximal solvable subalgebra, $B + Rad(L) = B$. This means that all Borel subalgebras contain $Rad(L)$. Thus, every Borel subalgebra B of L determines and is determined uniquely by the Borel subalgebra $B/Rad(L)$ of $L/Rad(L)$. The correspondence $B \mapsto B/Rad(L)$ is a bijective correspondence between the set of Borel subalgebras of L to the set of Borel subalgebras of $L/Rad(L)$.

Proposition 2.4.18 *Let L be a semi-simple Lie algebra over an algebraically closed field of characteristic 0. Let H be a Cartan subalgebra of L with corresponding root system Φ. Let Δ be a basis of Φ. Then*
(i) $N(\Delta) = \oplus\sum_{\alpha>0} L_\alpha$ is a nilpotent subalgebra of L and
(ii) $B(\Delta) = H \oplus (\oplus\sum_{\alpha>0} L_\alpha)$ is a Borel subalgebra of L such that $[B(\Delta), B(\Delta)] = N(\Delta)$.

Proof (i) Clearly, $[N(\Delta), N(\Delta)] \subseteq \oplus\sum_{ht(\alpha)=2} L_\alpha$, $[N(\Delta), [N(\Delta), N(\Delta)]] \subseteq \oplus\sum_{ht(\alpha)=3} L_\alpha$, and so on. Evidently, $N(\Delta)$ is nilpotent of the index at the most m, where m is the largest among the heights of the roots of ϕ with respect to Δ.

(ii) Since $[H, L_\alpha] = L_\alpha$ and $[H, H] = \{0\}$, $[B(\Delta), B(\Delta)] = N(\Delta)$. Since $N(\Delta)$ is nilpotent, $B(\Delta)$ is a solvable subalgebra. Let A be a subalgebra containing $B(\Delta)$

properly. Then there is $\alpha < 0$ such that $L_\alpha \subseteq A$. But then the subalgebra S_α generated by $L_\alpha \bigcup L_{-\alpha}$ is contained in A. Consequently, A contains a subalgebra isomorphic to $sl(2, F)$. Since $sl(2, F)$ is simple, A cannot be solvable. This shows that $B(\Delta)$ is a Borel subalgebra. ♯

A Borel subalgebra $B(\Delta)$ is called a **Standard Borel subalgebra** relative to H, where Δ is a basis of the root system Φ associated with H.

Proposition 2.4.19 *Any two standard Borel subalgebras $B(\Delta_1)$ and $B(\Delta_2)$ relative to the Cartan subalgebra H are conjugate under $Inn_0(L)$.*

Proof For $\alpha \in \Phi$, consider the automorphism τ_α of L given by $\tau_\alpha = exp(ad(x_\alpha))exp(ad(-y_\alpha))exp(ad(x_\alpha))$. Since $x_\alpha \in L_{\frac{1}{2}}(ad(h_\alpha)$ and $y_\alpha \in L_{-\frac{1}{2}}(ad(h_\alpha)$, x_α and y_α are strongly ad-nilpotent elements of L. As such $\tau_\alpha \in Inn_0(L)$. Further, it is easily observed that $\tau_\alpha|_H = \sigma_\alpha$ on H. Clearly, $\tau_\alpha(B(\Delta)) = B(\sigma_\alpha(\Delta))$. Since $W(\Phi)$ acts transitively on the set of all bases, the result follows. ♯

We have the following Conjugacy theorems:

Theorem 2.4.20 *Any two Borel subalgebras of a Lie algebra L are conjugate under $Inn_0(L)$.*

Before proving this theorem, we use it to establish the following corollary:

Corollary 2.4.21 *Any two Cartan subalgebras of a Lie algebra L are conjugate under $Inn_0(L)$.*

Proof Let H_1 and H_2 be Cartan subalgebras of L. Then they are solvable (being nilpotent). Hence there are Borel algebras B_1 and B_2 such that $H_1 \subseteq B_1$ and $H_2 \subseteq B_2$. From the above theorem, there is a member $\sigma \in Inn_0(L)$ such that $\sigma(B_1) = B_2$. Since $\sigma(H_1)$ and H_2 are Cartan subalgebras of a solvable Lie algebra B_2, by Theorem 2.4.16, there is a member $\eta \in Inn_0(B_2)$ such that $\eta(\sigma(H_1)) = H_2$. We already know that η is a restriction of $\hat{\eta} \in Inn_0(L)$ to B_2. Hence $\hat{\eta} o \sigma \in Inn_0(L)$ such that $\hat{\eta} o \sigma(H_1) = H_2$. ♯

Proof of the Theorem 2.4.20. The proof is by the induction on $DimL$. If $dimL = 1$, then L is the only Borel subalgebra of L and there is nothing to do. Assume that the result is true for all those Lie algebras whose dimensions are less than the $DimL$, where $DimL > 1$. If $Rad(L) \neq \{0\}$, then $DimL/Rad(L) < DimL$, and hence by the induction assumption, the result is true for $L/Rad(L)$. Using the discussions which follow the proof of the Proposition 2.4.17 and the proof of the Proposition 2.4.15, the result follows for L. We can, therefore, assume that L is semi-simple.

Take a standard Borel subalgebra $B(\Delta)$ of L. It is sufficient to show that any Borel subalgebra B of L is conjugate to $B(\Delta)$ under $Inn_0(L)$. Further, if $B(\Delta) \bigcap B = B(\Delta)$ or it is B, then $B(\Delta) = B$ and there is nothing to do. We assume that $B(\Delta) \bigcap B$ is properly contained in $B(\Delta)$ and it is also properly contained in B. We, again, use a second downward induction on $DimB(\Delta) \bigcap B$, where $B(\Delta) \bigcap B$ is properly contained in $B(\Delta)$.

Let us first assume that $B(\Delta)\bigcap B \neq \{0\}$. Consider the set N of all ad-nilpotent elements of $B(\Delta)\bigcap B$. We observe that N is an ideal of $B(\Delta)\bigcap B$: Clearly, N is a subspace. Let $x \in B(\Delta)\bigcap B$ and $y \in N$. Then $[x, y] \in [B(\Delta)\bigcap B, B(\Delta)\bigcap B]$. Since $B(\Delta)\bigcap B$ is solvable, by Corollary 1.3.18, $[x, y]$ is an ad-nilpotent element of $B(\Delta)\bigcap B$. Hence $[x, y] \in N$. This shows that N is an ideal of $B(\Delta)\bigcap B$. However, if $N \neq \{0\}$, N is not an ideal of L for L, being semi-simple, has no nontrivial solvable ideal. Now, there are two cases: (i) $N \neq \{0\}$ and (ii) $N = \{0\}$.

Consider the case (i). Since $N \neq \{0\}$, $N_L(N)$ is a proper nontrivial subalgebra of L. Further, we show that $B(\Delta)\bigcap B$ is properly contained in $B(\Delta)\bigcap N_L(N)$ and it is also properly contained in $B\bigcap N_L(N)$: For each $x \in N$, we have an endomorphism $\overline{ad(x)}$ of $\frac{B(\Delta)}{B(\Delta)\bigcap B}$ which is given by $\overline{ad(x)}(u + B(\Delta)\bigcap B) = ad(x)(u) + B(\Delta)\bigcap B$. Since the elements of N are ad-nilpotent, $\overline{ad(x)}$ is nilpotent for each $x \in N$. By Theorem 1.3.14, there is a nonzero element $u + B(\Delta)\bigcap B$ of $\frac{B(\Delta)}{B(\Delta)\bigcap B}$ such that $\overline{ad(x)}(u + B(\Delta)\bigcap B) = B(\Delta)\bigcap B$ for all $x \in N$. This means that $[x, u] \in N$ for all $x \in N$, whereas $u \notin B(\Delta)\bigcap B$. But $[x, u]$ belongs to $[B(\Delta), B(\Delta)]$ and so it is ad-nilpotent. Thus, $[x, u] \in N$ for all $x \in N$. Hence, u belongs to $B(\Delta)\bigcap N_L(N)$ and consequently $B(\Delta)\bigcap B$ is properly contained in $B(\Delta)\bigcap N_L(N)$. Similarly, $B(\Delta)\bigcap B$ is properly contained in $B\bigcap N_L(N)$.

Now $B(\Delta)\bigcap N_L(N)$ and $B\bigcap N_L(N)$ are solvable subalgebras of $N_L(N)$. Let C and D be Borel subalgebras of $N_L(N)$ containing $B(\Delta)\bigcap N_L(N)$ and $B\bigcap N_L(N)$, respectively. Since $DimN_L(N) < DimL$, by the induction hypothesis, there is a member $\sigma \in Inn_0(N_L(N))$ such that $\sigma(D) = C$. Let $\hat{\sigma} \in Inn_0(L)$ such that $\hat{\sigma}|_{N_L(N)} = \sigma$. Note that $B(\Delta)\bigcap B$ is a proper subalgebra of $B(\Delta)$ and also of B. Since C is solvable, there is a Borel subalgebra $\hat{B}$ of L such that $C \subseteq \hat{B}$. Evidently, $B(\Delta)\bigcap B$ is properly contained in $B(\Delta)\bigcap \hat{B}$. Hence by the second induction assumption (which is downward), there is a member $\tau \in Inn_0(L)$ such that $\tau(\hat{B}) = B(\Delta)$. In turn, $\tau\hat{\sigma}(D) \subseteq B(\Delta)$. Thus,

$$\tau\sigma(B(\Delta)\bigcap B) \subset \tau\sigma(D\bigcap N_L(N)) \subseteq (\tau\hat{\sigma}(D))\bigcap(\tau\hat{\sigma}(B)) \subseteq B(\Delta)\bigcap(\tau\hat{\sigma}(B))$$

(the notation $\subset$ stands for proper containment). Again we apply the second induction assumption to conclude that $B(\Delta)$ and $\tau\hat{\sigma}(B)$ are conjugate under $Inn_0(L)$. This means that $B(\Delta)$ and B are conjugate under $Inn_0(L)$.

Now, consider the case (ii). $B(\Delta)\bigcap B$ has no nonzero ad-nilpotent elements. Since a Borel subalgebra is self-normalizing, it contains semi-simple and nilpotent parts of each of its elements. Thus, $B(\Delta)\bigcap B$ contains only semi-simple elements and so it is a toral subalgebra. Denote $B(\Delta)\bigcap B$ by T. Since $N(\Delta) = [B(\Delta), B(\Delta)]$ is nilpotent, $T\bigcap N(\Delta) = \{0\}$. Note that $B(\Delta) = H \oplus N(\Delta)$. We show the existence of a Borel subalgebra $\hat{B}$ of L which is conjugate to B under $Inn_0(L)$ and which is such that $B(\Delta)\bigcap \hat{B} = T \subseteq H$. Suppose that $x \in N_{B(\Delta)}(T)$. Then $[x, t] \in T$ for all $t \in T$. Since $[x, t] \in N(\Delta)$, it is nilpotent as well as semi-simple. Hence $[x, t] = 0$ for all $t \in T$. This means that $x \in C_{B(\Delta)}(T)$. Thus, $N_{B(\Delta)}(T) = C_{B(\Delta)}(T)$. Let C be a Cartan subalgebra of $C_{B(\Delta)}(T)$ containing T. Then C is nilpotent and $T \subseteq N_{C_{B(\Delta)}(T)}(C) = C$. Using the Jacobi identity, it follows that $N_{B(\Delta)}(T) = N_{C_{B(\Delta)}(T)}(C) = C$. Since C

is a self-normalizing nilpotent subalgebra of $B(\Delta)$, it is a Cartan subalgebra of $B(\Delta)$ which contains T. By Theorem 2.4.16, it follows that C is a maximal toral subalgebra of $B(\Delta)$ which is conjugate to H under $Inn_0(B(\Delta))$ and so also under $Inn_0(L)$. Consequently, we get a Borel subalgebra $\hat{B}$ of L which is conjugate to B, and $B(\Delta) \bigcap \hat{B}$ is a toral subalgebra contained in H, where $B(\Delta) = H \oplus (\oplus \sum_{\alpha \in \Phi^+} L_\alpha)$. Thus, we can assume, without any loss, that $B(\Delta) \bigcap B = T \subseteq H$.

Suppose that $T = H$. Then, if $B \neq B(\Delta)$, then there is an element $\alpha < 0$ such that $H \oplus L_\alpha \subseteq B$. Evidently, $B(\sigma_\alpha(\Delta)) \bigcap B$ contains $H \oplus L_\alpha \subseteq B$. By the second induction hypothesis, $B(\sigma_\alpha(\Delta))$ is conjugate to B under $Inn_0(L)$. By Proposition 2.4.19, $B(\sigma_\alpha(\Delta))$ is conjugate to $B(\Delta)$ and we are done.

Now, suppose that $T \subset H$. There are two cases: (i) $B \subseteq C_L(T)$ and (ii) $B \nsubseteq C_L(T)$. Consider the case (i). Since $T \neq \{0\}$ and L is semi-simple, $C_L(T)$ is a proper subalgebra of L. Also B is a Borel subalgebra of $C_L(T)$. Since $H \subseteq C_L(T)$, we can embed H into a Borel subalgebra $\hat{B}$ of $C_L(T)$. By the induction hypothesis, there is $\sigma \in Inn_0(C_L(T))$ and so also a $\hat{\sigma} \in Inn_0(L)$ such that $\hat{\sigma}(\hat{B}) = B$. It follows that $\hat{B}$ is a Borel subalgebra of L. Evidently, $B(\Delta) \bigcap \hat{B} \supseteq H$. By the second induction hypothesis, $B(\Delta)$ and $\hat{B}$ are conjugate under $Inn_0(L)$. Consequently, $B(\Delta)$ and B are conjugate. Thus, in this case we are done.

Now, assume that $B \nsubseteq C_L(T)$. Then there is a common eigenvector $x \in B - C_L(T)$ of the set $\{ad(u)|_B \mid u \in T\}$ and also an element $t \in T$ such that $[t, x] = \lambda x \neq 0$. Replacing t by $\lambda^{-1}t$, we may assume that $[t, x] = x$. Let $\hat{\Phi}$ denote the set $\{\alpha \in \Phi^+ \mid \alpha(t) \in \mathbb{Q}^\star\}$. Clearly, $\alpha, \beta \in \hat{\Phi}$ implies that $\alpha + \beta \in \hat{\Phi}$. Consequently $A = H \oplus (\oplus \sum_{\alpha \in \hat{\Phi}} L_\alpha)$ is a subalgebra of L. Suppose that $x = h_x + \sum_{\alpha \in \Phi} x_\alpha$, where $h_x \in H$ and $x_\alpha \in L_\alpha$. Further, $[t, x] = x$, $[t, h_x] = 0$, and $[t, x_\alpha] = \alpha(t)x_\alpha$. Thus, $x \in A$. Clearly, A is solvable and so it can be embedded in a Borel subalgebra $\hat{B}$ of L. Clearly, $\hat{B} \bigcap B$ contains $T + Fx$. Hence $B(\Delta) \bigcap B = T$ is properly contained in $\hat{B} \bigcap B$. By the second induction hypothesis, $\hat{B}$ is conjugate to B as well as to $B(\Delta)$ under $Inn_0(L)$. The conjugacy of $B(\Delta)$ and B follows.

Lastly, let us assume that $B(\Delta) \bigcap B = \{0\}$. Since $B(\Delta) = H \oplus \oplus \sum_{\alpha > 0} L_\alpha$, $DimB(\Delta) > \frac{1}{2}DimL$. Again, since $B(\Delta) \bigcap B = \{0\}$, $DimB(\Delta) + DimB = Dim(B(\Delta) + B) \leq DimL$. Thus, $DimB < \frac{1}{2}DimL$. Let T be a maximal toral subalgebra of B. Suppose that $T = \{0\}$. Then B contains no semi-simple element. In other words, all the elements of B are ad-nilpotent. By the Engel theorem, B is nilpotent. Already, B is self-normalizing. Consequently, B turns out to be a Cartan subalgebra of L. This is a contradiction (Corollary 2.4.11). Thus, $T \neq \{0\}$. Take a Cartan subalgebra (maximal toral subalgebra, by Corollary 2.4.11) $\hat{H}$ of L containing T, and the standard Borel subalgebra $\hat{B}$ with respect to $\hat{H}$. Evidently, $\hat{B} \bigcap B \neq \{0\}$. From the previous case, $\hat{B}$ and B are conjugate. In particular, $DimB > \frac{1}{2}DimL$. This is a contradiction. The proof of the conjugacy of Borel subalgebras is complete. ♯

Corollary 2.4.22 *Let L be a semi-simple Lie algebra over an algebraically closed field of characteristic 0. Let H be a Cartan subalgebra of L with the corresponding root system Φ. Then any Borel subalgebra containing H is a standard Borel subalgebra associated with Φ.*

Proof Let Δ be a basis of Φ and $B(\Delta)$ the corresponding standard Borel subalgebra of L. Let B be a Borel subalgebra of L containing H. Let $\sigma \in Inn_0(L)$ such that $\sigma(B(\Delta)) = B$. Since $\sigma(L_\alpha) = L_{\sigma(\alpha)}$, it follows that $\sigma(\Delta)$ is a basis of Φ and $\sigma(B(\Delta)) = B(\sigma(\Delta))$. ♯

Definition 2.4.23 The dimension of a Cartan subalgebra of a finite-dimensional Lie algebra L over an algebraically closed field of characteristic 0 is termed as the **rank** of L.

Thus, the rank of a semi-simple Lie algebra is the rank of the associated root system Φ.

Definition 2.4.24 A subalgebra P of L containing a Borel subalgebra is called a **Parabolic subalgebra**.

Evidently, every parabolic subalgebra is self-normalizing. Let L be a semi-simple Lie algebra with H as a Cartan subalgebra and Φ the corresponding root system. Let Δ be a basis of Φ and Δ' a subset of Δ. Then the subalgebra $P(\Delta')$ of L generated by $\bigcup_{\alpha \in \Delta \bigcup -\Delta'} L_\alpha$ is a parabolic subalgebra of L containing $B(\Delta)$. This is called a **Standard Parabolic** subalgebra relative to Δ

Most of the parts of the following proposition have already been established.

Proposition 2.4.25 *Let L be a semi-simple Lie algebra with H as a maximal toral subalgebra (Cartan subalgebra), Φ the corresponding root system, $\Delta = \{\alpha_1, \alpha_2, \ldots, \alpha_l\}$ a basis of Φ, $\{t_\alpha \mid \alpha \in \Phi\}$ the subset of H in bijective correspondence with Φ defined by the property $\alpha(h) = \kappa(t_\alpha, h)$ for all $h \in H$, $h_\alpha = \frac{2t_\alpha}{\kappa(t_\alpha, t_\alpha)} \in H$, $x_\alpha \in L_\alpha$, $y_\alpha \in L_{-\alpha}$ such that $[x_\alpha, y_\alpha] = h_\alpha$, $[h_\alpha, x_\beta] = \alpha(h_\beta)x_\beta = \prec \beta, \alpha \succ x_\beta$, and $[h_\alpha, y_\beta] = \alpha(h_{-\beta})y_\beta = - \prec \beta, \alpha \succ y_\beta$. Then L is generated by the set $\{x_{\alpha_i}, y_{\alpha_i}, h_{\alpha_i} \mid 1 \leq i \leq l\}$ and it satisfies the following relations:*

R_1. $[h_{\alpha_i}, h_{\alpha_j}] = 0, 1 \leq i, j \leq l.$
R_2. $[x_{\alpha_i}, y_{\alpha_i}] = h_{\alpha_i}$ *for all* i.
R_3. $[x_{\alpha_i}, y_{\alpha_j}] = 0$ *for all* $i \neq j$.
R_4. $[h_{\alpha_i}, x_{\alpha_j}] = \prec \alpha_j, \alpha_i \succ x_{\alpha_j}$.
R_5. $[h_{\alpha_i}, y_{\alpha_j}] = - \prec \alpha_j, \alpha_i \succ y_{\alpha_j}$, *for all* i, j.
R_6. $(ad(x_{\alpha_i}))^{-\prec \alpha_j, \alpha_i \succ +1}(x_{\alpha_j}) = 0$ *for all* $i \neq j$.
R_7. $(ad(y_{\alpha_i}))^{-\prec \alpha_j, \alpha_i \succ +1}(y_{\alpha_j}) = 0$ *for all* $i \neq j$.

Proof Everything is already established except the fact that L is generated by the set $\{x_{\alpha_i}, y_{\alpha_i}, h_{\alpha_i} \mid 1 \leq i \leq l\}$ and it satisfies the relations R_6 and R_7. Since Δ is a basis of Φ, $\{h_{\alpha_i} \mid 1 \leq i \leq l\}$ is a basis of H. Further, $L = H \oplus (\oplus \sum_{\alpha \in \Phi} L_\alpha)$, $L_\alpha = , < x_\alpha >$, and $L_{-\alpha} =, < y_\alpha >$. Thus, it is sufficient to show that for each $\alpha \in \Phi$, x_α and y_α belong to the subalgebra generated by $\{x_{\alpha_i}, y_{\alpha_i}, h_{\alpha_i} \mid 1 \leq i \leq l\}$ (observe that $x_{-\alpha} = y_\alpha$ and $y_{-\alpha} = x_\alpha$). Let α be a positive root. From Proposition 2.2.16, $\alpha = \alpha_{i_1} + \alpha_{i_2} + \cdots + \alpha_{i_r}$, $1 \leq i_j \leq l$ such that each partial sum is a root. By induction on r, we show that x_α belongs to the subalgebra generated by $\{x_{\alpha_i} \mid 1 \leq i \leq l\}$. If $r = 1$, then there is nothing to do. Assume that the result holds for r. Let $\alpha = \alpha_{i_1} + \alpha_{i_2} + \cdots + \alpha_{i_r} + \alpha_{i_{r+1}}$, where each partial sum is a root. Put $\beta = \alpha_{i_1} + \alpha_{i_2} +$

$\cdots + \alpha_{i_r}$. Then by the induction hypothesis, x_β belongs to the subalgebra generated by $\{x_{\alpha_i}, y_{\alpha_i}, h_{\alpha_i} \mid 1 \leq i \leq l\}$. Hence $[x_\beta, x_{\alpha_{r+1}}]$ belongs to the subalgebra generated by $\{x_{\alpha_i}, y_{\alpha_i}, h_{\alpha_i} \mid 1 \leq i \leq l\}$. By Proposition 2.1.9, L_α is contained in the subalgebra generated by $\{x_{\alpha_i}, y_{\alpha_i}, h_{\alpha_i} \mid 1 \leq i \leq l\}$. Similarly, if α is a negative root, then $y_{-\alpha}$ lies in the subalgebra generated by $\{x_{\alpha_i}, y_{\alpha_i}, h_{\alpha_i} \mid 1 \leq i \leq l\}$.

Since Δ is basis, $\prec \alpha_i, \alpha_j \succ$ is less than 0 for all i, j. Hence $\alpha_j - \alpha_i$ is not a root for all $j \neq i$. It follows that the α_i-string through α_j is $\alpha_j, \alpha_j + \alpha_i, \cdots \alpha_j + q\alpha$, where $-q = \prec \alpha_j, \alpha_i \succ$ (Proposition 2.2.10). Now, $ad(x_{\alpha_i})(x_{\alpha_j}) \in L_{\alpha_j+\alpha_i}$, $(ad(x_{\alpha_i}))^2(x_{\alpha_j}) \in L_{\alpha_j+2\alpha_i}$ and so on. The relation R_6 is immediate. Similarly, the relation R_7 holds. ♯

Consider the set $\Omega = \{\hat{x}_i, \hat{y}_i, \hat{h}_i \mid 1 \leq i \leq l\}$ of symbols which is in bijective correspondence with the set $\{x_{\alpha_i}, y_{\alpha_i}, h_{\alpha_i} \mid 1 \leq i \leq l\}$ of generators of L (Proposition 2.4.25). Let $L(\Omega)$ denote the free Lie algebra on Ω. Let Γ denote the ideal of $L(\Omega)$ generated by the relators of the types $[\hat{h}_i, \hat{h}_j]$, $[\hat{x}_i, \hat{y}_i] - \hat{h}_i$, $[\hat{x}_i, \hat{y}_j], i \neq j$, $[\hat{h}_i, \hat{x}_j] - c_{ji}\hat{x}_j$, *and* $[\hat{h}_i, \hat{y}_j] + c_{ji}\hat{y}_j$, where c_{ji} are the Cartan integers $\prec \alpha_j, \alpha_i \succ$. Denote $L(\Omega)/\Gamma$ by $\overline{L}$ and the image $\nu(u) = u + \Gamma$ of u under the quotient map ν by $\overline{u}$. Evidently, the association $\overline{x_i} \mapsto x_{\alpha_i}$, $\overline{y_i} \mapsto y_{\alpha_i}$, *and* $\overline{h_i} \mapsto h_i$ extends to a surjective Lie algebra homomorphism from $\overline{L} = L(\Omega)/\Gamma$ to L. Let V denote the vector space with $\{v_1, v_2, \ldots, v_l\}$ as a basis. Let $T(V)$ denote the tensor algebra on V (essentially the polynomial algebra in non-commuting variables $\{v_1, v_2, \ldots, v_l\}$) and $gl(T(V))$ the general linear Lie algebra on $T(V)$. Clearly, $B = \{1\} \bigcup \{v_{i_1} \otimes v_{i_2} \otimes \cdots \otimes v_{i_t} \mid 1 \leq i_j \leq l, t \in \mathbb{N}\}$ is a basis of $T(V)$. A linear transformation on $T(V)$ is uniquely determined by its effect on B. For each $j, 1 \leq j \leq l$, consider the linear transformations $\rho(\hat{h}_j)$, $\rho(\hat{y}_j)$, $\rho(\hat{x}_j)$ on $T(V)$ defined as follows:

(i) Put $\rho(\hat{h}_j)(1) = 0$ and

$$\rho(\hat{h}_j)(v_{i_1} \otimes v_{i_2} \otimes \cdots \otimes v_{i_t}) = -(c_{i_1 j} + c_{i_2 j} + \cdots + c_{i_t j})v_{i_1} \otimes v_{i_2} \otimes \cdots \otimes v_{i_t}.$$

(ii) Put $\rho(\hat{y}_j)(1) = v_j$ and

$$\rho(\hat{y}_j)(v_{i_1} \otimes v_{i_2} \otimes \cdots \otimes v_{i_t}) = v_j \otimes v_{i_1} \otimes v_{i_2} \otimes \cdots \otimes v_{i_t}.$$

(iii) Define $\rho(\hat{x}_j)(v_{i_1} \otimes v_{i_2} \otimes \cdots \otimes v_{i_t})$ inductively by putting $\rho(\hat{x}_j)(1) = 0 = \rho(\hat{x}_j)(v_i)$ and

$$\rho(\hat{x}_j)(v_{i_1} \otimes v_{i_2} \otimes \cdots \otimes v_{i_t}) =$$
$$v_{i_1} \otimes \rho(\hat{x}_j)(v_{i_2} \otimes v_{i_3} \otimes \cdots \otimes v_{i_t}) \;-\; \delta_{i_1 j}(c_{i_2 j} + \cdots + c_{i_t j})(v_{i_2} \otimes v_{i_3} \otimes \cdots \otimes v_{i_t}).$$

From the universal property of a free Lie algebra, ρ can be extended uniquely to a Lie algebra homomorphism from $L(\Omega)$ to $gl(T(V))$ which we again denote by ρ.

Proposition 2.4.26 $\Gamma \subseteq Ker\,\rho$ *and* ρ *induces a unique Lie algebra homomorphism* $\overline{\rho}$ *from* $\overline{L}$ *to* $gl(T(V))$ *given by* $\overline{\rho}(\overline{h_j}) = \rho(\hat{h}_j)$, $\overline{\rho}(\overline{x_j}) = \rho(\hat{x}_j)$, *and* $\overline{\rho}(\overline{y_j}) = \rho(\hat{y}_j)$.

Proof By the definition, each $\rho(\hat{h}_j)$ is a diagonal linear transformation. As such, $\rho([\hat{h}_i, \hat{h}_j]) = \rho(\hat{h}_i)\rho(\hat{h}_j) - \rho(\hat{h}_j)\rho(\hat{h}_i) = 0$. This shows that $[\hat{h}_i, \hat{h}_j] \in Ker\ \rho$.

Now, consider $[\rho(\hat{x}_i), \rho(\hat{y}_j)]$. Clearly,

$$[\rho(\hat{x}_i), \rho(\hat{y}_j)](1) = (\rho(\hat{x}_i)\rho(\hat{y}_j))(1) - (\rho(\hat{y}_j)\rho(\hat{x}_i))(1) = 0 = \rho(\delta_{ji}\hat{h}_i)(1).$$

Further, by the definition, we have
$\rho(\hat{x}_i)\rho(\hat{y}_j)(v_{i_1} \otimes v_{i_2} \otimes \cdots \otimes v_{i_t})$
$= \rho(\hat{x}_i)(v_j \otimes v_{i_1} \otimes v_{i_2} \otimes \cdots \otimes v_{i_t})$
$= v_j \otimes \rho(\hat{x}_i)(v_{i_1} \otimes v_{i_2} \otimes \cdots \otimes v_{i_t}) - \delta_{ji}(c_{i_1 i} + c_{i_2 i} + \cdots + c_{i_t i})v_{i_1} \otimes v_{i_2} \otimes \cdots \otimes v_{i_t}$
$= v_j \otimes v_{i_1} \otimes \rho(\hat{x}_i)(v_{i_2} \otimes \cdots \otimes v_{i_t}) - \delta_{i_1 i}(c_{i_2 i} + \cdots + c_{i_t i})v_j \otimes v_{i_2} \otimes \cdots \otimes v_{i_t} -$
$\delta_{ji}(c_{i_1 i} + c_{i_2 i} + \cdots + c_{i_t i})v_{i_1} \otimes v_{i_2} \otimes \cdots \otimes v_{i_t}$,
and
$\rho(\hat{y}_j)\rho(\hat{x}_i)(v_{i_1} \otimes v_{i_2} \otimes \cdots \otimes v_{i_t})$
$= \rho(\hat{y}_j)(v_{i_1} \otimes \rho(\hat{x}_i)(v_{i_2} \otimes \cdots \otimes v_{i_t}) - \delta_{i_1 i}(c_{i_2 i} + \cdots + c_{i_t i})v_{i_2} \otimes \cdots \otimes v_{i_t})$
$= v_j \otimes v_{i_1} \otimes \rho(\hat{x}_i)(v_{i_2} \otimes \cdots \otimes v_{i_t}) - \delta_{i_1 i}(c_{i_2 i} + \cdots + c_{i_t i})v_j \otimes v_{i_2} \otimes \cdots \otimes v_{i_t}$.
Hence,

$$[\rho(\hat{x}_i), \rho(\hat{y}_j)](v_{i_1} \otimes v_{i_2} \otimes \cdots \otimes v_{i_t}) = \delta_{ji}\rho(\hat{h}_i)(v_{i_1} \otimes v_{i_2} \otimes \cdots \otimes v_{i_t}).$$

This shows that $[\hat{x}_i, \hat{y}_i] - \hat{h}_i \in Ker\ \rho$ and also $[\hat{x}_i, \hat{y}_j] \in Ker\ \rho$ for all $i \neq j$.

Next, we consider $[\rho(\hat{h}_i), \rho(\hat{x}_j)]$. Using induction on t, we first show that

$$\rho(\hat{h}_i)\rho(\hat{x}_j)(v_{i_1} \otimes v_{i_2} \otimes \cdots \otimes v_{i_t}) =$$
$$(c_{ji} - c_{i_1 i} - c_{i_2 i} - \cdots - c_{i_t i})\rho(\hat{x}_j)(v_{i_1} \otimes v_{i_2} \otimes \cdots \otimes v_{i_t}) \quad (\star)$$

for all i, j. Equivalently, we show that $\rho(\hat{x}_j)(v_{i_1} \otimes v_{i_2} \otimes \cdots \otimes v_{i_t})$ is an eigenvector of $\rho(\hat{h}_i)$ corresponding to the eigenvalue $(c_{ji} - c_{i_1 i} - c_{i_2 i} - \cdots - c_{i_t i})$. Clearly, $\rho(\hat{h}_i)\rho(\hat{x}_j)(v_{i_1}) = 0 = (c_{ji} - c_{i_1 i})\rho(\hat{x}_j)(v_{i_1})$. Thus, the assertion is true for $t = 1$. Assume the result for t. Then

$$\rho(\hat{h}_i)\rho(\hat{x}_j)(v_{i_2} \otimes v_{i_3} \otimes \cdots \otimes v_{i_{t+1}}) =$$
$$(c_{ji} - c_{i_2 i} - c_{i_3 i} - \cdots - c_{i_{t+1} i})\rho(\hat{x}_j)(v_{i_2} \otimes v_{i_3} \otimes \cdots \otimes v_{i_{t+1}}).$$

This means that $\rho(\hat{x}_j)(v_{i_2} \otimes v_{i_3} \otimes \cdots \otimes v_{i_{t+1}})$ is an eigenvector of $\rho(\hat{h}_i)$ corresponding to eigenvalue $c_{ji} - c_{i_2 i} - c_{i_3 i} - \cdots - c_{i_{t+1} i}$. It follows from the definition of $\rho(\hat{h}_i)$ that $\rho(\hat{x}_j)(v_{i_2} \otimes v_{i_3} \otimes \cdots \otimes v_{i_{t+1}}) = a(v_{j_1} \otimes v_{j_2} \otimes \cdots \otimes v_{j_p})$ for some $v_{j_1}, v_{j_2}, \ldots, v_{j_p}$ and $a \in F$ such that

$$-(c_{j_1 i} + c_{j_2 i} + \cdots + c_{j_p i}) = c_{ji} - c_{i_2 i} - c_{i_3 i} - \cdots - c_{i_{t+1} i}.$$

Hence,

$$v_{i_1} \otimes \rho(\hat{x_j})(v_{i_2} \otimes v_{i_3} \otimes \cdots \otimes v_{i_{t+1}}) = a(v_{i_1} \otimes v_{j_1} \otimes v_{j_2} \otimes \cdots \otimes v_{j_p}).$$

Thus, $v_{i_1} \otimes \rho(\hat{x_j})(v_{i_2} \otimes v_{i_3} \otimes \cdots \otimes v_{i_{t+1}})$ is an eigenvector of $\rho(\hat{h_i})$ corresponding to the eigenvalue $-(c_{i_1 i} + c_{j_1 i} + c_{j_2 i} + \cdots + c_{j_p i}) = (c_{ji} - c_{i_1 i} - c_{i_2 i} - \cdots - c_{i_{t+1} i})$. Now, by the definition

$$\rho(\hat{x_j})(v_{i_1} \otimes v_{i_2} \otimes \cdots \otimes v_{i_t}) =$$
$$v_{i_1} \otimes \rho(\hat{x_j})(v_{i_2} \otimes v_{i_3} \otimes \cdots \otimes v_{i_t}) \; - \; \delta_{i_1 j}(c_{i_2 j} + \cdots + c_{i_t j})(v_{i_2} \otimes v_{i_3} \otimes \cdots \otimes v_{i_t}).$$

If $i_1 \neq j$, then the second term in the RHS is 0 and consequently, $\rho(\hat{x_j})(v_{i_1} \otimes v_{i_2} \otimes \cdots \otimes v_{i_t})$ is an eigenvector of $\rho(\hat{h_i})$ corresponding to the eigenvalue $(c_{ji} - c_{i_1 i} - c_{i_2 i} - \cdots - c_{i_{t+1} i})$. Suppose that $i_1 = j$. Then the second term in the RHS is also an eigenvector of $\rho(\hat{h_i})$ corresponding to the eigenvalue $-(c_{i_2 i} - c_{i_3 i} - \cdots - c_{i_{t+1} i}) = (c_{ji} - c_{i_1 i} - c_{i_2 i} - \cdots - c_{i_{t+1} i})$. This proves $\star$. Again, by the definition

$$\rho(\hat{x_i})\rho(\hat{h_i})(v_{i_1} \otimes v_{i_2} \otimes \cdots \otimes v_{i_t}) = -(c_{i_1 i} + c_{i_2 i} + \cdots + c_{i_t i})\rho(\hat{x_i})(v_{i_1} \otimes v_{i_2} \otimes \cdots \otimes v_{i_t}).$$

Using the $\star$ and the above identity, we obtain

$$[\rho(\hat{h_i}), \rho(\hat{x_j})] = c_{ji}\rho(\hat{x_j}).$$

This shows that $[\hat{h_i}, \hat{x_j}] - c_{ji}\hat{x_j}$ belongs to $Ker\ \rho$. That $[\hat{h_i}, \hat{y_j}] + c_{ji}\hat{y_j}$ belongs to $Ker\ \rho$ is still a straightforward and easy verification. The rest follows from the fundamental theorem of homomorphism. ♯

Proposition 2.4.27 *Let Φ be a root system with $\Delta = \{\alpha_1, \alpha_2, \ldots, \alpha_l\}$ as a basis. Then $\{\overline{h_i} \mid 1 \leq i \leq l\}$ forms a basis for an abelian subalgebra $\overline{H}$ of $\overline{L}$. Let $\overline{X}$ denote the subalgebra of $\overline{L}$ generated by the set $\{\overline{x_i} \mid 1 \leq i \leq l\}$, and $\overline{Y}$ denote the subalgebra of $\overline{L}$ generated by the set $\{\overline{y_i} \mid 1 \leq i \leq l\}$. Then $\overline{L}$ is vector space direct sum $\overline{H} \oplus \overline{X} \oplus \overline{Y}$.*

Proof Suppose that $\sum_{i=1}^{l} a_i\overline{h_i} = 0$ in $\overline{L}$. Then $\rho(\sum_{i=1}^{l} a_i\hat{h_i}) = \overline{\rho}(\sum_{i=1}^{l} a_i\overline{h_i}) = 0$ in $gl(T(V))$. This means that $\sum_{i=1}^{l} a_i\rho(\hat{h_i}) = 0$. Consequently, $-\sum_{i=1}^{l} a_i c_{ji} v_j = \sum_{i=1}^{l} a_i\rho(\hat{h_i})(v_j) = 0$ for all j. Since the Cartan matrix $[c_{ji}]$ is non-singular, it follows that $a_i = 0$ for all i. It follows that $\overline{H}$ is an l-dimensional subspace of $\overline{L}$ with $\{\overline{h_i} \mid 1 \leq i \leq l\}$ as a basis. Again, since $[\hat{h_i}, \hat{h_j}] \in \Gamma$, it follows that $\overline{H}$ is an abelian Lie subalgebra. It also follows that the quotient map ν induces a vector space isomorphism from $\sum_{i=1}^{l} F\hat{h_i}$ to $\overline{H}$.

For each i, $[\hat{x_i}, \hat{y_i}] - \hat{h_i}$, $[\hat{h_i}, \hat{x_i}] - 2\hat{x_i}$, and $[\hat{h_i}, \hat{y_i}] + 2\hat{y_i}$ all belong to Γ. Hence, we get a surjective Lie algebra homomorphism χ_i from $sl(2, F)$ to the subalgebra $F\overline{x_i} + F\overline{y_i} + F\overline{h_i}$ which is given by $\chi_i(e_{12}) = \overline{x_i}$, $\chi_i(e_{21}) = \overline{y_i}$, and $\chi_i(e_{11} - e_{22}) = \overline{h_i}$. Since $\overline{h_i} \neq 0$ ($\{\overline{h_i} \mid 1 \leq i \leq l\}$ is linearly independent) and $sl(2, F)$ is simple, χ_i is an isomorphism. Further, using the relations defining Γ and applying eigenvalue and eigenvector arguments for $ad(\overline{h_i})$, we can easily see that $\{\overline{x_i}, \overline{y_i}, \overline{h_i} \mid 1 \leq i \leq l\}$

is linearly independent. Again, using induction on t, we can easily establish

$$\text{(i)}\ [\overline{h_j}, [\overline{x_{i_1}}, [\overline{x_{i_2}}, \ldots, [\overline{x_{i_{t-1}}}, \overline{x_{i_t}}] \cdots]]] = \\ (c_{i_1 i}+c_{i_2 i} + \cdots + c_{i_t i})[\overline{x_{i_1}}, [\overline{x_{i_2}}, \ldots, [\overline{x_{i_{t-1}}}, \overline{x_{i_t}}] \cdots]],$$

$$\text{(ii)}\ [\overline{h_j}, [\overline{y_{i_1}}, [\overline{y_{i_2}}, \ldots, [\overline{y_{i_{t-1}}}, \overline{y_{i_t}}] \cdots]]] = \\ -(c_{i_1 i}+c_{i_2 i} + \cdots + c_{i_t i})[\overline{y_{i_1}}, [\overline{y_{i_2}}, \ldots, [\overline{y_{i_{t-1}}}, \overline{y_{i_t}}] \cdots]],$$

and also the fact that $[\overline{y_j}, [\overline{x_{i_1}}, [\overline{x_{i_2}}, \ldots, [\overline{x_{i_{t-1}}}, \overline{x_{i_t}}] \cdots]]]$ belongs to $\overline{X}$ whereas $[\overline{x_j}, [\overline{y_{i_1}}, [\overline{y_{i_2}}, \ldots, [\overline{y_{i_{t-1}}}, \overline{y_{i_t}}] \cdots]]]$ belongs to $\overline{Y}$. Thus, $\overline{H} + \overline{X} + \overline{Y}$ is a subalgebra of $\overline{L}$. Consequently, $\overline{L} = \overline{H} + \overline{X} + \overline{Y}$. That the sum is a direct sum of subspaces is a consequence of our earlier discussions. ♯

Remark 2.4.28 Recall that if V is an L-module, then for $\lambda \in H^\star$, V_λ denotes the subspace $\{v \in V \mid h \cdot v = \lambda(h)v \ for\ all\ h \in H\}$. If $V_\lambda \neq \{0\}$, then λ is called a weight, and V_λ is called a weight space associated with the weight λ. Every Lie algebra L is an L-module by putting $x \cdot y = [x, y]$. A weight λ is called a positive weight with respect to a basis Δ of the root system Φ if it is a positive linear combination of members of Δ. Indeed, it follows from the proof of the above proposition that $\overline{H} = \overline{L}_0$, $\overline{X} = \oplus \sum_{\lambda>0} \overline{L}_\lambda$, and $\overline{Y} = \oplus \sum_{\lambda<0} \overline{L}_\lambda$. Consequently, $\overline{L} = \overline{L}_0 \oplus (\oplus \sum_{\lambda>0} \overline{L}_\lambda) \oplus (\oplus \sum_{\lambda<0} \overline{L}_\lambda)$.

The following lemma is a crucial lemma needed to establish the theorem of Serre.

Lemma 2.4.29 *Put* $\overline{x_{ij}} = ad(\overline{x_i})^{-c_{ji}+1}(\overline{x_j})$ *and* $\overline{y_{ij}} = ad(\overline{y_i})^{-c_{ji}+1}(\overline{y_j})$, $i \neq j$. *Then* $ad(\overline{x_k})(\overline{y_{ij}}) = 0$ *for each* k *and* $ad(\overline{y_k})(\overline{x_{ij}}) = 0$ *for each* k.

Proof Suppose that $k \neq i$. Then $[\overline{x_k}, \overline{y_i}] = 0$. This means that $ad(\overline{x_k})$ and $ad(\overline{y_i})$ commute with each other. Consequently, $ad(\overline{x_k})(\overline{y_{ij}}) = (ad(\overline{y_i})^{-c_{ji}+1}ad(\overline{x_k})(\overline{y_j})$. If $k \neq j$, then it is 0, and we are done. Suppose that $k = j$. Then it results in $(ad(\overline{y_i})^{-c_{ji}+1}(\overline{h_j})$. Now, $ad(\overline{y_i})(\overline{h_j}) = c_{ij}\overline{y_i}$. If $c_{ij} = 0$, then $(ad(\overline{y_i})^{-c_{ji}+1}(\overline{h_j}) = 0$, and we are done. If $c_{ij} \neq 0$, then $c_{ji} \neq 0$ and it is negative. Hence $-c_{ji} + 1 \geq 2$. Since $ad(\overline{y_i})^2(\overline{h_j}) = 0$, we are done.

Next, suppose that $k = i$. Clearly, $\overline{L}$ is an S-module under the adjoint action, where S is the subalgebra $F\overline{x_i} + F\overline{y_i} + F\overline{h_i}$ which is isomorphic to $sl(2, F)$ (cf. the above proposition). Using the arguments of Sect. 2.4 (although $\overline{L}$ may be infinite dimensional), we see that $[\overline{x_i}, \overline{y_j}] = 0$ for $j \neq i$ and $[\overline{h_i}, \overline{y_j}] = -c_{ji}\overline{y_j}$. Using the induction on r, we can show that $ad(\overline{x_i})(ad(\overline{y_i})^r(\overline{y_j}) = r(-c_{ji} - r + 1)ad(\overline{y_i})^{r-1}(\overline{y_j})$. Putting $r = -c_{ji} + 1$, we get the desired result. ♯

Recall that an endomorphism ρ of a vector space is called a **locally nilpotent** endomorphism if for each $x \in V$, there is a natural number n_x such that $\rho^{n_x}(x) = 0$.

Evidently, ρ is locally nilpotent if and only if the restriction of ρ to each finite-dimensional subspace is nilpotent. Consequently, given a locally nilpotent endomorphism ρ on V, we can talk of the endomorphism $exp(\rho) = 1 + \rho + \frac{\rho^2}{2!} + \frac{\rho^3}{3!} + \cdots$. Since $exp(\rho)exp(-\rho) = I$, it follows that $exp(\rho)$ is an automorphism of V.

Theorem 2.4.30 (Serre) *Let Φ be a root system with a basis $\Delta = \{\alpha_1, \alpha_2, \ldots, \alpha_l\}$. Let L be a Lie algebra having a presentation with the set $\{\hat{x}_i, \hat{y}_i, \hat{h}_i \mid 1 \leq i \leq l\}$ of symbols as the set of generators and the set consisting of relations (i) $[\hat{h}_i, \hat{h}_j] = 0$, $1 \leq i, j \leq l$, (ii) $[\hat{x}_i, \hat{y}_i] = \hat{h}_i$ for all i, (iii) $[\hat{x}_i, \hat{y}_j] = 0$ for all $i \neq j$, (iv) $[\hat{h}_i, \hat{x}_j] = \prec \alpha_j, \alpha_i \succ \hat{x}_j$, (v) $[\hat{h}_i, \hat{y}_j] = - \prec \alpha_j, \alpha_i \succ \hat{y}_j$, for all i, j, (vi) $(ad(\hat{x}_i))^{-\prec\alpha_j,\alpha_i\succ+1}(\hat{x}_j) = 0$ for all $i \neq j$, and (vii) $(ad(\hat{y}_i))^{-\prec\alpha_j,\alpha_i\succ+1}(\hat{y}_j) = 0$ for all $i \neq j$. Then L is a $3l$-dimensional semi-simple Lie algebra generated by the set $\{x_i, y_i, h_i \mid 1 \leq i, j \leq l\}$ of representatives of $\{\hat{x}_i, \hat{y}_i, \hat{h}_i \mid 1 \leq i, j \leq l\}$ in L. Further, the subalgebra H of L generated by the set $\{h_i \mid 1 \leq i \leq l\}$ is a maximal toral subalgebra of L such that the corresponding root system is Φ.*

Proof Without any loss, we can take L to be $\overline{L}/\overline{I}$, $x_i = \overline{x_i} + \overline{I}$, $y_i = \overline{y_i} + \overline{I}$, and $h_i = \overline{h_i} + \overline{I}$, where $\overline{I}$ is the ideal of $\overline{L}$ generated by the set $\{\overline{x_{ij}}, \overline{y_{ij}} \mid 1 \leq i, j \leq l\}$. Thus, we need to show that $\overline{L}/\overline{I}$ is a $3l$-dimensional semi-simple Lie algebra with $\{x_i, y_i, h_i \mid 1 \leq i \leq l\}$ as a basis and with the subalgebra $H = < \{h_i \mid 1 \leq i \leq l\} >$ as a maximal toral subalgebra such that the associated root system is Φ. We prove it in several steps:

Step 1. Let $\overline{J}$ be the ideal of $\overline{X}$ generated by the set $\{\overline{x_{ij}} \mid 1 \leq i, j \leq l\}$ (note that $\overline{x_{ij}} \in \overline{X}$ for all i, j) and $\overline{K}$ be the ideal of $\overline{Y}$ generated by the set $\{\overline{y_{ij}} \mid 1 \leq i, j \leq l\}$ (note that $\overline{y_{ij}} \in \overline{Y}$ for all i, j). Then $\overline{J}$ and $\overline{K}$ are ideals of $\overline{L}$ such that $\overline{I} = \overline{J} + \overline{K}$: We prove it. Consider the ideal $\overline{J}$ of $\overline{X}$. Using the identity (i) in the proof of Proposition 2.4.27, we observe that $\overline{x_{ij}}$ is an eigenvector of $ad(\overline{h_k})$ corresponding to the eigenvalue $-c_{jk} - (1 - c_{ji})c_{ik}$. Again, from the proof of Proposition 2.4.27, it follows that $ad(\overline{h_k})(\overline{X}) \subseteq \overline{X}$. Further, using induction and the Jacobi identity, we see that $ad(\overline{h_k})(\overline{J}) \subseteq \overline{J}$. Also, from the above lemma, $ad(\overline{y_k})(\overline{x_{ij}}) = 0$. Again, using Proposition 2.4.27, we observe that $ad(\overline{y_k})(\overline{X}) = \overline{H} + \overline{X}$. The Jacobi identity together with the fact that $ad(\overline{h_k})(\overline{J}) \subseteq \overline{J}$ implies that $ad(\overline{y_k})(\overline{J}) \subseteq \overline{J}$. Since $\{\overline{x_k}, \overline{y_k} \mid 1 \leq k \leq l\}$ generates $\overline{L}$, $ad(\overline{u})(\overline{J}) \subseteq \overline{J}$ for each $\overline{u} \in \overline{L}$. This shows that $\overline{J}$ is an ideal of L. Similarly, $\overline{K}$ is an ideal of L. Clearly, $\overline{J} + \overline{K} \subseteq \overline{I}$ and since $\overline{I}$ is the smallest ideal of $\overline{L}$ containing $\overline{x_{ij}}$ and $\overline{y_{ij}}$, $\overline{I} = \overline{J} + \overline{K}$.

Step 2. $L = H \oplus X \oplus Y$ (direct sum of vector spaces), where $X = \frac{\overline{X}}{\overline{J}}$, $Y = \frac{\overline{Y}}{\overline{K}}$. Further, the quotient map ν from $\overline{L}$ to L and maps $\overline{H}$ isomorphically on to H: This follows from the fact that $\overline{L} = \overline{H} \oplus \overline{X} \oplus \overline{Y}$ (Proposition 2.4.27) and the fact that $\overline{H} \bigcap \overline{I} = \{0\}$.

Step 3. The quotient map ν maps $\sum_{i=1}^{l} F\overline{h_i} + \sum_{i=1}^{l} F\overline{x_i} + \sum_{i=1}^{l} F\overline{y_i}$ isomorphically in to L: This follows again from Proposition 2.4.27.

Step 4. For each $\lambda \in H^\star$, $L_\lambda = \{x \in L \mid [h, x] = \lambda(h)x\}$ is finite dimensional. Further, $H = L_0$, $X = \sum_{\lambda > 0} L_\lambda$ and $Y = \sum_{\lambda < 0} L_\lambda$: It follows from the identities (i), (ii) of Proposition 2.4.27, and steps 1, 2, and 3 that L_λ is finite dimensional. The rest of the statement in this step follows from the remark following Proposition 2.4.27.

Step 5. For each i, $1 \leq i \leq l$, $ad(x_i)$ and $ad(y_i)$ are locally nilpotent endomorphisms. Consider $ad(x_i)$. Let V denote the subset $\{x \in L \mid (ad(x_i))^r(x) = 0 \; for \; some \; r\}$. Evidently, $0 \in V$. Let $x, y \in V$. Then there exists $r, s \in \mathbb{N}$ such that $(ad(x_i))^r(x) = 0 = (ad(x_i))^s(y)$. Clearly, $(ad(x_i))^{max(r,s)}(x + y) = 0$ and $(ad(x_i))^r(ax) = 0$ for all $a \in F$. Further, since $(ad(x_i))^m([y, z]) = \sum_{i+j=m} {}^mC_i(ad(x_i))^i(y)(ad(x_i))^{m-i}(z)$, it follows that $(ad(x_i))^{r+s+1}([x, y]) = 0$. This shows that V is a Lie subalgebra of L. Since $\overline{x_{ij}} \in \overline{I}$, $(ad(x_i))^{-c_{ji}+1}(x_j) = 0$. Further, it is also clear that $ad(x_i)(y_j) = 0$ for all $j \neq i$ and $(ad(x_i))^{-c_{ji}+3}(y_i) = 0$. Thus, $x_i \in V$ for each i. Similarly, $y_i \in V$ for each i. Since L is the Lie algebra generated by $\{x_i, y_i \mid 1 \leq i \leq l\}$, $V = L$. This shows that $ad(x_i)$ is a locally nilpotent endomorphism on L. Similarly, $ad(y_i)$ is also a locally nilpotent endomorphism on L.

Step 6. Let $\sigma \in W(\Phi)$. Then $DimL_\lambda = DimL_{\sigma\lambda}$: Since $W(\Phi)$ is generated by the simple reflections, it is sufficient to show that $DimL_{\sigma_{\alpha_i}\lambda} = DimL_\lambda$. For each i, let τ_i denote the automorphism $exp(ad(x_i))exp(-(ad(y_i)))exp(ad(x_i))$ of L. It can be easily observed that τ_i maps the members of L_λ to $L_{\sigma_{\alpha_i}\lambda}$, and in turn, the members of $L_{\sigma_{\alpha_i}\lambda}$ to L_λ. The assertion follows, since τ_i is an automorphism.

Step 7. For each $\alpha \in \Phi$, $Dim\ L_\alpha = 1$ and $L_{a\alpha} \neq \{0\}$ if and only if $a \in \{0, 1, -1\}$: For each $\alpha \in \Phi$, there is a member $\sigma \in W(\Phi)$ such that $\sigma(\alpha) \in \Delta$. Thus, it is sufficient to assume that $\alpha = \alpha_i \in \Delta$. However, it follows from Proposition 2.4.27 that $Dim\ \overline{L}_{\alpha_i} = 1$ and $\overline{L}_{a\alpha_i} \neq \{0\}$ if and only if $a \in \{0, 1, -1\}$. The result follows from step 3.

Step 8. Suppose that $L_\lambda \neq \{0\}$. Then $\lambda = 0$ or else $\lambda \in \Phi$: If $\lambda \neq 0$ and $L_\lambda \neq \{0\}$, then it follows from Remark 2.4.28 and step 2 that $\lambda = \pm\sum_{i=1}^l a_i\alpha_i$ with all $a_i \geq 0$ and at least one $a_i \neq 0$. It is sufficient to observe that λ is an integral multiple of an element of Φ (step 7). Suppose not. Then there is a vector $\gamma \in P_\lambda - \bigcup_{\alpha\in\Phi} P_\alpha$. Thus, we can find a $\sigma \in W(\Phi)$ such that $< \sigma(\gamma), \alpha_i >$ is greater than 0 for all i. Since $< \lambda, \gamma >= 0$, $0 =, < \sigma(\lambda), \sigma(\gamma) >= \sum_{i=1}^l a_i' < \alpha_i, \sigma(\gamma) >$, where $\sigma(\lambda) = \sum_{i=1}^l a_i'\alpha_i$. This implies that $a_i' > 0$ for some i and $a_j' < 0$ for some j. Hence $L_{\sigma(\lambda)} = \{0\}$ (this is because $\sigma(\lambda)$ is neither positive nor negative). This is a contradiction to step 7, since $DimL_\lambda \neq 0$.

Step 9. L is of finite dimension, and indeed, $DimL = l + \mid \Phi \mid$. Also $L = H \oplus (\oplus\sum_{\alpha\in\Phi} L_\alpha)$: It follows from steps 4, 7, and 8.

Step 10. L is a semi-simple Lie algebra with H as a maximal toral subalgebra such that the corresponding root system is Φ: We need to show that L contains no nontrivial abelian ideal. Let A be an abelian ideal of L. Since $L = H \oplus (\oplus\sum_{\alpha\in\Phi} L_\alpha)$ and A is an ideal, $A = (H \bigcap A) \oplus \oplus\sum_{\alpha\in\Phi}(L_\alpha \bigcap A)$. If $(L_\alpha \bigcap A) \neq \{0\}$, then $L_\alpha \subseteq A$. Since A is an ideal, $[L_\alpha, L_{-\alpha}] \subseteq A$. Consequently, $A \bigcap S_\alpha$ is a nontrivial ideal of $S_\alpha \approx sl(2, F)$. This is a contradiction. Hence $L_\alpha \bigcap A = \{0\}$ for all $\alpha \in \Phi$. Therefore, $A = A \bigcap H \subseteq H$. Since $[L_\alpha, A] \subseteq A \subseteq H$, $A \subseteq Ker\alpha$ for each $\alpha \in \Phi$. This shows that $A \subseteq \bigcap_{\alpha\in\Phi} Ker\alpha = \{0\}$.

Step 11. H is a maximal toral subalgebra of L such the Φ is the corresponding root system: We have already seen that H is abelian (so nilpotent subalgebra) and the elements of H are semi-simple. Since $L = H \oplus (\oplus\sum_{\alpha\in\Phi} L_\alpha)$, $C_L(H) = H$. This means that H is maximal toral subalgebra. The rest is evident. ♯

Corollary 2.4.31 (Existence Theorem) *Given a root system Φ, there is a pair (L, H), where L is a semi-simple Lie algebra, H a maximal toral subalgebra of L such that the associated root system is Φ.* ♯

Let Φ be a root system in E with a basis $\Delta = \{\alpha_1, \alpha_2, \ldots, \alpha_l\}$ and Φ' be another root system in E'. Let η be an isomorphism from Φ to Φ'. Then $\eta(\Delta) = \Delta' = \{\alpha'_1, \alpha'_2, \ldots, \alpha'_l\}$ is a basis of Φ', where $\eta(\alpha_i) = \alpha'_i$ for all i. Further, η induces a vector space isomorphism $\hat{\eta}$ from E to E', such that $\prec \hat{\eta}(\alpha_i), \hat{\eta}(\alpha_j) \succ = \prec \alpha_i, \alpha_j \succ$ for all i, j. Observe that $\alpha \mapsto -\alpha$ from Φ to itself is an automorphism of Φ (sufficient to observe for the rank 2 case).

Theorem 2.4.32 (Isomorphism Theorem) *Let (L, H) and (L', H') be pairs, where L and L' are semi-simple Lie algebras, H and H' being maximal toral subalgebras of L and L', respectively. Let Φ be the root system associated with (L, H) and Φ' be that associated with (L', H'). Let η be an isomorphism from Φ to Φ' which sends a basis Δ to a basis Δ'. Then η induces an isomorphism π from H to H' given by $\pi(h_\alpha) = h'_{\alpha'}$, where $h_\alpha = \frac{2t_\alpha}{\kappa(t_\alpha, t_\alpha)}$ and $h'_{\alpha'} = \frac{2t_{\alpha'}}{\kappa'(t_{\alpha'}, t_{\alpha'})}$. For each $\alpha \in \Delta$, choose a nonzero element $x_\alpha \in L_\alpha$, and a nonzero element $x'_{\alpha'} \in L_{\alpha'}$, where $\alpha' = \eta(\alpha)$. Then there is a unique extension $\overline{\pi}$ of π which takes x_α to $x'_{\alpha'}$ for each $\alpha \in \Delta$. In particular, let L, H, Φ, and Δ be as usual. Then given $x_\alpha \in L_\alpha - \{0\}$, and $y_\alpha \in L_{-\alpha} - \{0\}$ for each $\alpha \in \Delta$, there is a unique automorphism σ of L of order 2 such that $\sigma(x_\alpha) = y_\alpha$ for each $\alpha \in \Delta$.*

Proof For x_α and $x'_{\alpha'}$, we have unique y_α and $y'_{\alpha'}$ such that $[x_\alpha, y_\alpha] = h_\alpha$ and $[x'_{\alpha'}, y'_{\alpha'}] = h'_{\alpha'}$ for all $\alpha \in \Delta$ and $\alpha' \in \Delta'$. Since $\{x'_{\alpha'}, y'_{\alpha'}, h'_{\alpha'}\}$ satisfy the same relations which are satisfied by $\{x_\alpha, y_\alpha, h_\alpha\}$, there is a unique Lie algebra homomorphism $\overline{\pi}$ from L to L' given by $\overline{\pi}(x_\alpha) = x'_{\alpha'}$, $\overline{\pi}(y_\alpha) = y'_{\alpha'}$ and $\overline{\pi}(h_\alpha) = h'_{\alpha'}$. Similarly, we have a unique homomorphism $\overline{\overline{\pi}}$ from L' to L such that $\overline{\pi} o \overline{\overline{\pi}}$ and $\overline{\overline{\pi}} o \overline{\pi}$ are respective identity maps. The rest of the assertion also follows, since $\alpha \mapsto -\alpha$ is an automorphism of Φ. ♯

Theory of Weights in Euclidean Spaces

As observed, the root system and roots in a finite-dimensional real Euclidean spaces have faithful intrinsic relationship with the structure theory (adjoint representations)of semi-simple Lie algebras. On the same lines (see Chap. 3), the weights and the weight lattice associated with a root system in a Euclidean space (to be described below) have a deep intrinsic relationship with the representation theory of semi-simple Lie algebras. We develop the abstract theory of weights in Euclidean spaces for its application in the next chapter on the representation theory.

Let Φ be a root system in a Euclidean space E. Let $\Delta = \{\alpha_1, \alpha_2, \ldots, \alpha_l\}$ be a basis of Φ. An element $\lambda \in E$ is called an **abstract weight** associated with Φ if $\prec \lambda, \alpha \succ = \frac{2<\lambda, \alpha>}{<\alpha, \alpha>}$ belong to $\mathbb{Z}$ for all $\alpha \in \Phi$. In particular, members of Φ are also weights. The set Λ of weights associated with Φ forms a subgroup of E. The subgroup Λ_r of Λ generated by Φ is a lattice in E called a **root lattice**. A weight $\lambda \in \Lambda$ is called a **dominant weight** if $\prec \lambda, \alpha \succ \geq 0$ for all $\alpha \in \Delta$. It is said to be **strongly dominant** if $\prec \lambda, \alpha \succ > 0$ for all $\alpha \in \Delta$. The set of dominant weights will be denoted by Λ^+. Evidently, $\Lambda^+ \subseteq \overline{C(\Delta)}$ and $\Lambda \bigcap C(\Delta)$ are precisely the set of strongly dominant weights.

Consider the basis $\Delta^{\nu} = \{\hat{\alpha_i} = \frac{2\alpha_i}{<\alpha_i,\alpha_i>} \mid 1 \leq i \leq l\}$ of E. Consider the basis $\{\lambda_1, \lambda_2, \ldots, \lambda_l\}$ of E which is dual to Δ^{ν}. Thus,

$$< \lambda_i, \hat{\alpha_j} >=\prec \lambda_i, \alpha_j \succ= \frac{2 < \lambda_i, \alpha_j >}{< \alpha_j, \alpha_j >} = \delta_{ij}.$$

Evidently, λ_i is a dominant weight for each i. We term λ_i as a **fundamental dominant weight**. Clearly, $< \lambda - \sum_{i=1}^{l} \prec \lambda, \alpha_i \succ \lambda_i, \alpha_j >= 0$ for each j. Hence

$$\lambda = \sum_{i=1}^{l} \prec \lambda, \alpha_i \succ \lambda_i.$$

Thus, Λ is a lattice with basis $\{\lambda_i \mid 1 \leq i \leq l\}$ and $\lambda \in \Lambda^+$ if and only if $\prec \lambda, \alpha_i \succ \geq 0$ (or equivalently, $< \lambda, \alpha_i > \geq 0$) for all i. Since Λ_r is a sub-lattice of Λ of the same rank, Λ/Λ_r is a finite group called the **fundamental group** of Φ. In this case, we can determine the order of Λ/Λ_r as follows: Each α_i is an integral linear combination of $\{\lambda_j \mid 1 \leq j \leq l\}$. Suppose that $\alpha_i = \sum_{k=1}^{l} m_{ik}\lambda_k, m_{ik} \in \mathbb{Z}$. Then

$$\prec \alpha_i, \alpha_j \succ= \sum_{k=1}^{l} m_{ik} \prec \lambda_k, \alpha_j \succ= m_{ij}.$$

Thus, the matrix of transformation from the basis Δ to $\{\lambda_1, \lambda_2, \ldots, \lambda_l\}$ is the Cartan matrix A of L relative to Δ. It follows that $Det A \lambda_i \in \Lambda_r$ for each i. Evidently, $Det A$ is a nonnegative integer. For example, in the case of A_2, the determinant of the Cartan matrix is 3 and $3\lambda_1 = 2\alpha_1 + \alpha_2$ and $3\lambda_2 = \alpha_1 + 2\alpha_2$. It follows that the order of Λ/Λ_r in this case is 3.

The following lemmas will be used in the next chapter:

Lemma 2.4.33 *The Weyl group $W(\Phi)$ acts on Λ. Further, each orbit of the action of $W(\Phi)$ contains a unique dominant weight. The isotopy group $W(\Phi)_{\lambda}$ of a strongly dominant weight is trivial.*

Proof Since the members of $W(\Phi)$ are orthogonal transformations, $W(\Phi)(\Lambda) = \Lambda$ and hence $W(\Phi)$ acts on Λ. If λ is dominant, then $\sigma(\lambda) < \lambda$ for all $\sigma \in W(\Phi)$. Consequently, in each orbit there is a unique dominant weight. Further, if λ is strongly dominant and $\sigma \neq I$, then since $\sigma(\lambda) < \lambda$, $\sigma(\lambda)$ cannot be strongly dominant. ♯

Lemma 2.4.34 *Let $\lambda \in \Lambda^+$. Then the set $\{\mu \in \Lambda^+ \mid \mu < \lambda\}$ is finite.*

Proof Let $\lambda = \sum_{i=1}^{l} a_i\alpha_i$ and $\mu = \sum_{i=1}^{l} b_i\alpha_i$ be dominant weights with $\mu < \lambda$. Since a_i and b_i are positive rationals with $\mu < \lambda$, and $a_i - b_i$ belong to $\mathbb{N}\bigcup\{0\}$, there are only finitely many sets $\{b_i, 1 \leq i \leq l\}$ such that $\sum_{i=1}^{l} b_i\alpha_i < \lambda$. ♯

Lemma 2.4.35 $\delta = \frac{1}{2}\sum_{\alpha\in\Phi^+}\alpha$ *is a strongly dominant weight, and indeed,* $\delta = \sum_{i=1}^{l}\lambda_i$.

Proof Since $\sigma_{\alpha_i}(\delta) = \delta - \alpha_i$,

$$< \delta - \alpha_i, \alpha_i >=, < \sigma_{\alpha_i}(\delta), \alpha_i >=, < (\sigma_{\alpha_i})^2(\delta), \sigma_{\alpha_i}(\alpha_i) >=, < \delta, -\alpha_i > .$$

This means that $< 2\delta - \alpha_i, \alpha_i >= 0$ or equivalently, $2 < \delta, \alpha_i >=, < \alpha_i, \alpha_i >$ for each i. Thus, $\prec \delta, \alpha_i \succ = 1$ for each i. Consequently, $\delta = \sum_{i=1}^{l} \prec \delta, \alpha_i \succ \lambda_i = \sum_{i=1}^{l}\lambda_i$. ♯

Lemma 2.4.36 *Let* $\lambda \in \Lambda^+$ *and* $\lambda = \sigma(\mu), \sigma \in W(\Phi)$. *Then* $< \mu+\delta, \mu+\delta > \leq < \lambda+\delta, \lambda+\delta >$. *Equality holds if and only if* $\lambda = \mu$ *or* $\lambda = 0 = \mu$.

Proof Since σ is an orthogonal transformation,

$$< \mu+\delta, \mu+\delta > = < \sigma(\mu+\delta), \sigma(\mu+\delta) >=< \lambda+\sigma(\delta), \lambda+\sigma(\delta) >=, < \lambda+\delta, \lambda+\delta > - 2 < \lambda, \delta - \sigma(\delta) > .$$

Again, since δ is dominant, it follows that $\sigma(\delta) < \delta$. Hence $\delta - \sigma(\delta)$ is the sum of positive roots. Further, since λ is dominant, $< \lambda, \delta - \sigma(\delta) > \geq 0$. It follows that $< \mu+\delta, \mu+\delta > \leq < \lambda+\delta, \lambda+\delta >$ and equality holds if and only if $\lambda = 0$ or $\delta = \sigma(\delta)$. Observe that $\delta = \sigma(\delta)$ if and only if $\sigma = I$. ♯

A subset Π of a weight lattice Λ is called a **saturated subset** of Λ if $\lambda \in \Pi$ and $\alpha \in \Phi$ imply that $\lambda - i\alpha$ belongs to Π for all $i, 0 \leq i \leq \prec \lambda, \alpha \succ$. If $\lambda \in \Pi$ and $\alpha \in \Phi$, then $\sigma_\alpha(\lambda) = \lambda - \prec \lambda, \alpha \succ \alpha$ belongs to Π. Since $W(\Phi)$ is generated by simple reflections, $W(\Phi)(\Pi) = \Pi$. A member λ of Π is called the **highest weight** of Π if $\mu \in \Pi$ implies that $\mu \leq \lambda$. Thus, given a root system Φ, $\Phi \bigcup\{0\}$ is a saturated subset of Λ. Further, if Φ is irreducible, then it has unique largest root which is the highest weight of $\Phi \bigcup\{0\}$. It follows from Lemma 2.4.34 that a saturated subset of Λ with the highest weight is finite.

Lemma 2.4.37 *Let* Π *be a saturated subset of* Λ *with the highest weight* λ. *If* $\mu \in \Lambda^+$ *such that* $\mu \leq \lambda$, *then* $\mu \in \Pi$.

Proof Suppose that $\lambda = \sum_{\alpha\in\Delta} a_\alpha\alpha$. Then $\mu = \lambda - \sum_{\alpha\in\Delta} b_\alpha\alpha$, where $b_\alpha \leq a_\alpha$ for each α. The proof is by the induction on $\sum_{\alpha\in\Delta} b_\alpha$. If $\sum_{\alpha\in\Delta} b_\alpha = 0$, then $\mu = \lambda$ and there is nothing to do. Assume that $\nu \in \Pi$ whenever $\nu = \lambda - \sum_{\alpha\in\Delta} c_\alpha\alpha$, where $c_\alpha \leq a_\alpha$ for each α and $\sum_{\alpha\in\Delta} c_\alpha = m \geq 0$. Let $\mu = \lambda - \sum_{\alpha\in\Delta} b_\alpha\alpha$, where $b_\alpha \leq a_\alpha$ for each α and $\sum_{\alpha\in\Delta} b_\alpha = m+1$. We need to show that $\mu \in \Pi$. Evidently, $b_\alpha > 0$ for some $\alpha \in \Delta$. Hence $< \sum_{\alpha\in\Delta} b_\alpha\alpha, \sum_{\alpha\in\Delta} b_\alpha\alpha > > 0$. Consequently, for some $\beta \in \Delta, b_\beta > 0$ and $< \sum_{\alpha\in\Delta} b_\alpha\alpha, \beta > > 0$. In turn, $\prec \sum_{\alpha\in\Delta} b_\alpha\alpha, \beta \succ > 0$. Thus, $\mu - \beta \in \Lambda^+$ and by the induction hypothesis, $\mu - \beta$ belongs to Π. Since Π is saturated, $\mu \in \Pi$. ♯

Remark 2.4.38 Using the arguments in the proof of the above lemma, we can easily see that if $\mu \in \Pi$ and $\mu \neq \lambda$, then for some $\beta \in \Delta$, $\mu + \beta$ belongs to Π.

Remark 2.4.39 It is clear from the previous lemma that a saturated set Π with the highest weight λ is precisely $\{\sigma(\mu) \mid \mu \in \Lambda^+, \sigma \in W(\Phi)$ *and* $\mu \leq \lambda\}$. Thus, for a given $\lambda \in \Lambda^+$, there is a unique saturated set with the highest weight λ as described above. This will be denoted by $\Pi(\lambda)$.

The following lemma will be used in the next chapter on the representation theory.

Lemma 2.4.40 *Let Π be a saturated subset of Λ with the highest weight λ ($\Pi = \Pi(\lambda)$) and $\mu \in \Pi$. Then $< \mu + \delta, \mu + \delta > \leq < \lambda + \delta, \lambda + \delta >$ and equality holds if and only if $\mu = \lambda$.*

Proof From Lemma 2.4.36, $< \sigma(\lambda) + \delta, \sigma(\lambda) + \delta > \leq < \lambda + \delta, \lambda + \delta >$ for all $\sigma \in W(\Phi)$. As such, it is sufficient to establish the result when μ is dominant and $\mu \neq \lambda$. From the remarks above, $\mu + \alpha$ belongs to Π for some $\alpha \in \Delta$. Now,

$$< \mu + \alpha + \delta, \mu + \alpha + \delta > =, < \mu + \delta, \mu + \delta > + 2 < \mu + \delta, \alpha > + < \alpha, \alpha > .$$

Since $\mu + \delta$ is dominant, $< \mu + \delta, \mu + \delta > < < \mu + \alpha + \delta, \mu + \alpha + \delta >$. In turn,

$$< \mu + \delta, \mu + \delta > < < \mu + \alpha + \delta, \mu + \alpha + \delta > < < \mu + \alpha + \beta + \delta, \mu + \alpha + \beta + \delta > .$$

Proceeding inductively, we arrive at the desired result, since Π is finite. ♯

Exercises

2.4.1. Describe Cartan subalgebras of $sl(n, F)$.
2.4.2. Let L be a semi-simple Lie algebra over an algebraically closed field of characteristic 0. Show that a semi-simple element $x \in L$ is regular if and only if it belongs to a unique Cartan subalgebra of L.
2.4.3. Let L be a semi-simple Lie algebra over an algebraically closed field of characteristic 0. Show that a solvable Engel subalgebra of L is a Cartan subalgebra of L.
2.4.4. Show by means of an example that the image of a Cartan subalgebra under a homomorphism need not be a Cartan subalgebra.
2.4.5. Show by means of an example that the inverse image of a Cartan subalgebra under an epimorphism need not be a Cartan subalgebra.
2.4.6. Show that a Cartan subalgebra is a maximal nilpotent subalgebra. Show by means of an example that the converse is not true.
2.4.7. Give an example of an ad-nilpotent element which is not strongly ad-nilpotent. What happens if L is a semi-simple Lie algebra over an algebraically closed field of characteristic 0?
2.4.8. Suppose that $Inn_0(L) = \{I_L\}$. Show that L is nilpotent.

2.4.9. Let L, H, Φ, and Δ be as usual. Show that any maximal subalgebra $\hat{N}$ of L subject to the condition that all elements of $\hat{N}$ are nilpotent is conjugate under $Inn_0(L)$ to $N(\Delta) = [B(\Delta), B(\Delta)]$.

2.4.10. Show that any Borel subalgebra contains a Cartan subalgebra. Show also that the intersection of any two Borel subalgebras contains a Cartan subalgebra.

2.4.11. Show that Cartan subalgebra of a semi-simple Lie algebra is abelian.

2.4.12. Show that any parabolic subalgebra of L containing $B(\Delta)$ is of the form $P(\Delta')$ for some $\Delta' \subseteq \Delta$.

2.4.13. Show that any parabolic subalgebra of a semi-simple Lie algebra over an algebraically closed field of characteristic 0 is conjugate to a standard parabolic subalgebra under $Inn_0(L)$.

2.4.14. Determine all parabolic subalgebras of $sl(2, \mathbb{C})$ and $sl(3, \mathbb{C})$.

2.4.15. Show that for each $r \leq l$, the subalgebra of L generated by $\{x_i, y_i, h_i \mid 1 \leq i \leq r\}$ is a semi-simple Lie subalgebra of L. Deduce that E_6 is a subalgebra of E_7 and E_7 is a subalgebra of E_8.

2.4.16. Show that the subalgebra $\overline{X}$ and the subalgebra $\overline{Y}$ of $\overline{L}$ in Proposition 2.4.27 are free Lie algebras.

2.4.17. Suppose that the rank of Φ is 1. Show that L is isomorphic to $sl(2, F)$. What is the Lie algebra $\overline{L}$ in this case?

2.4.18. Find the fundamental group of A_l, B_4, C_4, E_6, E_7, E_8, G_2, and F_4.

2.4.19. Show that any subgroup of Λ containing Λ_r is $W(\Phi)$ invariant.

2.4.20. Show that $\lambda_1 = \frac{1}{3}(2\alpha_1 + 4\alpha_2)$ and $\lambda_2 = \frac{1}{3}(\alpha_1 + 2\alpha_2$ are the fundamental dominant weights for A_2.

2.2.21. Show that $\lambda_1 = 2\alpha_1 + \alpha_2$ and $\lambda_2 = 3\alpha_1 + 2\alpha_2$ are the fundamental dominant weights for G_2.

2.4.22. More generally, for each of the Lie algebras A_l- G_2, using the Dynkin diagram and the Cartan matrix associated with them, compute their fundamental dominant weights in terms of their standard root systems (see Theorem 2.3.2).

Chapter 3
Representation Theory of Lie Algebras

Our main concern in this chapter will be to study and develop the representation theory and the character theory of semi-simple Lie algebras over an algebraically closed field of characteristic 0 including the theorem of Harish-Chandra and the Weyl, Kostant, and Steinberg formulas. However, we shall start with the first basic result on the representation theory of Lie algebras, viz., the theorems of Ado and of Iwasawa which assert that any finite-dimensional Lie algebra over an arbitrary field has a finite-dimensional faithful representation. More concretely, they assert that any finite-dimensional Lie algebra over a field F is isomorphic to a sub-Lie algebra of $gl(n, F)$ for some n.

3.1 Theorems of Ado and Iwasawa

Our aim in this section is to establish the results of Ado and Iwasava which assert that any finite-dimensional Lie algebra has a faithful finite-dimensional representation. For this purpose, we establish some preliminary results concerning subalgebras of an associative algebra A generated by Lie subalgebras of A_L.

Proposition 3.1.1 *Every finite-dimensional associative algebra with identity (all associative algebras are assumed to be with identities) over a field F has a faithful finite-dimensional representation.*

Proof Let A be a finite-dimensional associative algebra with identity. Let $End(A)$ denote the algebra of endomorphisms of A considered as a vector space. Clearly, $DimEnd(A) = (Dim(A))^2$ and so it is finite. Let $x \in A$. Let l_x denote the left multiplication on A by x. Then $l_x \in End(A)$. Also $l_{\alpha x+\beta y} = \alpha l_x + \beta l_y$ and $l_x o l_y = l_{xy}$. Thus, the map ρ from A to $End(A)$ defined by $\rho(x) = l_x$ is a homomorphism. Suppose that $\rho(x) = \rho(y)$. Then $l_x = l_y$. In particular, $x = x \cdot 1 = l_x(1) = l_y(1) = y \cdot 1 = y$. This shows that ρ is a finite-dimensional faithful representation of A on $End(A)$. ♯

R. Lal, *Algebra 4*, Infosys Science Foundation Series,
https://doi.org/10.1007/978-981-16-0475-1_3

Let V be a vector space over a field F. Then the vector space $gl(V) = End(V)$ of endomorphisms on V is an associative algebra with respect to the composition of endomorphisms, and it is also a Lie algebra with respect to the Lie product $[,]$ given by $[\mu, \nu] = \mu o \nu - \nu o \mu$. Recall that a representation ρ of a Lie algebra L on a vector space V is a Lie algebra homomorphism from L to $gl(V)$. It is said to be a finite-dimensional representation if V is of finite dimension ($DimV$ is called the degree of the representation ρ). The representation ρ is said to be a faithful representation if it is injective.

Let A be an associative algebra with identity over a field F. For a subset S of A, $< S >$ denotes the associative subalgebra generated by S. Note that $< S >$ need not contain the identity of A. The subalgebra $< \{a\} >$ is also denoted by $< a >$. Evidently, $< a > = \{f(a) \mid f(x) \in F[x]\; with\; f(0) = 0\}$. More generally, $< S > = \{f(a_1, a_2, \ldots, a_m) \mid f(x_1, x_2, \ldots, x_m) is\; a\; polynomial\; without\; nonzero\; constant\; term\; and\; a_i \in S\}$. We can also view $< S >$ as the set of linear combinations of nonzero monomials in S. For $a \in S$, it is clear that $<< a > \bigcap S > = < a >$.

Proposition 3.1.2 *(i) Let A be an associative algebra. Let $T \subseteq S \subseteq A$ be such that $[x, y] \in S$ for all $x, y \in S$ and also $[x, y] \in T$ for all $x, y \in T$. Let $u \in S$ be in the normalizer of T in S in the sense that $[T, u] \subseteq < T >$. Then*

$$< T > u \subseteq u < T > + < T > .$$

(ii) Suppose that T and S are as in (i). Suppose further that $< T >$ is nilpotent and $< T > \neq < S >$. Then there is an element $u \in S - < T >$ such that $[t, u] \in < T >$ for all $t \in T$.

Proof (i) Any element of $< T >$ is a linear combination of monomials in T. Thus, it is sufficient to show that $t_1 t_2 \cdots t_r u = uv + w$, where $v, w \in < T >$. We prove it by the induction on r. Now, $t_1 u = ut_1 + [t_1, u]$. Since $[t_1, u] \in < T >$, the result is true for $r = 1$. Assume the result for r. Consider $t_1 t_2 \cdots t_{r+1} u$. By the induction hypothesis,

$$t_2 t_3 \cdots t_{r+1} u = uw + v,$$

where $w, v \in < T >$. Thus,

$$t_1 t_2 \cdots t_{r+1} u = t_1 uw + t_1 v = ut_1 w + [t_1, u]w + t_1 v.$$

Clearly, $t_1 w$, $[t_1, u]$, and $t_1 v$ belong to $< T >$. The result follows for $r + 1$.

(ii) Since $< T > \neq < S >$, there is an element w_0 in $S - < T >$. If $[t, w_0] \in < T >$ for all $t \in T$, then we can take u to be w_0. If not, then there is $t_0 \in T$ such that $w_1 = [t_0, w_0]$ belongs to $S - < T >$. If $[t, w_1]$ belongs to $< T >$ for all t in T, then we can take u to be w_1. If not, then there is a t_1 in T such that $w_2 = [t_1, w_1]$ belongs to $S - < T >$. If not, proceed. We claim that this process terminates after finitely many steps giving us an element u with the required property. Suppose that the process does not terminate. Then we arrive at a sequence $t_0, t_2, \ldots, t_r, t_{r+1}, \cdots$ in T and a sequence $w_0, w_1, \ldots, w_r, w_{r+1} \cdots$ of elements in $S - < T >$, where

$w_r = [t_{r-1}, w_{r-1}]$ for all r. Clearly, for each r, w_r is a linear combination of monomials $T \bigcup \{w_0\}$. Since $< T >$ is nilpotent, there is a natural number m such that $< T >^m = \{0\}$. Now, as observed, w_{2m} is a linear combination of monomials of the form $u_1 u_2 \cdots u_p w_0 v_1 v_2 \cdots v_q$ such that $p + q = 2m - 1$. It turns out that $p \geq m$ or $q \geq m$. Thus, $w_{2m} = 0$. This means that $w_{2m} \in < T >$, a contradiction to the supposition that $w_r \in S - < T >$ for all r. ♯

Let B and C be subalgebras of an associative algebra A over a field F. Let BC denote the subspace generated by the set $\{bc \mid b \in B \; and \; c \in C\}$. A subalgebra B of A is called a **nilpotent subalgebra** if there is a natural number n such that $B^n = \{0\}$ or equivalently, $b_1 b_2 \cdots b_n = 0$ for all n-tuple $(b_1, b_2, \cdots, b_n)$ of elements of B. If B and C are nilpotent ideals of A, then $B + C$ is also a nilpotent ideal. As such, every finite-dimensional associative algebra has the largest nilpotent ideal. The largest nilpotent ideal is called the **radical** of A and it is denoted by $R(A)$.

Proposition 3.1.3 *Let V be a finite-dimensional vector space over a field F. Let S be a subset of $gl(V) = End(V)$ such that $[\rho, \eta] = \rho\eta - \eta\rho$ belongs to S for all $\rho, \eta \in S$. Suppose that each ρ in S is nilpotent in the sense that $\rho^n = 0$ for some n. Then the associative subalgebra $< S >$ of $gl(V)$ generated by S is nilpotent. In particular, if S is a Lie subalgebra of $gl(V)$ consisting of nilpotent endomorphisms of V, then $< S >$ is nilpotent.*

Proof The proof is by the induction on $DimV$. If $DimV = 0$, then $V = \{0\}$ and there is nothing to do. Assume that the result is true for all vector spaces W with $DimW < DimV$, $DimV > 0$. If $S = \{0\}$, then there is nothing to do. Suppose that $S \neq \{0\}$. Let $\Sigma = \{T \subseteq S \mid [\rho, \eta] \in T \forall \; \rho, \eta \in T \; and \; < T > \; is \; nilpotent \; subalgebra \; of gl(V)\}$. Clearly, $\Sigma \neq \emptyset$. Take a $T_0 \in \Sigma$ such that $Dim < T_0 >$ is maximum. We shall show that $< T_0 > = < S >$ and thereby establish that $< S >$ is nilpotent. Let W denote the subspace of V generated by the set $< T_0 > (V) = \{\rho(x) \mid \rho \in < T_0 >, \; x \in V\}$. Let $\rho \in S - \{0\}$. Take $T = < \rho > \bigcap S$. Then $T \neq \{0\}$ and $< T > = < \rho > \neq \{0\}$. Since $\rho \in S$, ρ is nilpotent. Suppose that $\rho^m = 0$. Then $f(\rho)^m = 0$ whenever $f(x)$ is a polynomial without a nonzero constant term. This means that $< T >$ is nilpotent. Hence $< T > \in \Sigma$. Thus, $T_0 \neq \{0\}$, $Dim < T_0 > \geq 1$, and $W \neq \{0\}$. Suppose that $W = V$. Then every member v of V is a linear combination of members of $< T_0 > (V)$. Expressing each v_i again as a linear combination of members of $< T_0 > (V)$, we see that each member of V is a linear combination of members of $< T_0 >^2 (V)$. Iterating the process, we see that each member of V is a linear combination of members of $< T_0 >^m (V)$ for each m. Since $< T_0 >$ is nilpotent, we arrive at the contradiction that $V = \{0\}$. Thus, $\{0\} \neq W \neq V$.

Let U denote the subset $\{\rho \in S \mid \rho(W) \subseteq W\}$ of S. Clearly, $T_0 \subseteq U$ and $[\rho, \eta] \in U$ for all $\rho, \eta \in U$. Since the members of S are nilpotent, the members of U induce nilpotent endomorphisms on W as well as on V/W. Let $\hat{U}$ denote $\{\rho|_W \in gl(W) \mid \rho \in U\}$ and $\overline{U}$ denote the set $\{\overline{\rho} \in gl(V/W), \; where \; \overline{\rho}(v + W) = \rho(v) + W, \; \rho \in U\}$. Then $\hat{U}$ and $\overline{U}$ consist of nilpotent endomorphisms in $gl(W)$ and $gl(V/W)$, respectively, and which are closed under $[,]$. Since $DimW < DimV$ and also

$Dim(V/W) < DimV$, by the induction hypothesis, $< \hat{U} >$ and $< \overline{U} >$ are nilpotent subalgebras of $gl(W)$ and $gl(V/W)$, respectively. Thus, there is a natural number p such that $\rho_1\rho_2 \cdots \rho_p(x) \in W$ for all $\rho_i \in U$ and all $x \in V$. Also, there is a natural number q such that $\eta_1\eta_2 \cdots \eta_q(y) = 0$ for all $\eta_j \in U$ and $y \in W$. This implies that $\rho_1\rho_2 \cdots \rho_{p+q} = 0$ on V for all $\rho_i \in U$. Thus, U is a nilpotent subalgebra of $gl(V)$ and so $U \in \Sigma$.

Now, we show that $< T_0 > = < S >$. Suppose not. Then by Proposition 3.1.2 (ii), there is a $\rho \in S- < T_0 >$ such that $[\rho, \eta] \in < T_0 >$ for all $\eta \in T_0$. Again by Proposition 3.1.2(i), $x \in V$ and $\rho \in < T_0 >$ implies that $\rho\eta(x) = \eta\rho(x) + \chi(x)$, where $\chi(x) \in < T_0 >$. Consequently, $\rho(W) \subseteq W$ and so $\rho \in U$. Already, $U \in \Sigma$. Since $\rho \notin < T_0 >$, $DimU > Dim < T_0 >$. This is a contradiction. Hence $< T_0 > = < S >$ and we are done. ♯

Proposition 3.1.4 *Let V be a finite-dimensional vector space. Let $T \subseteq S \subseteq gl(V)$ such that $[\rho, \eta] \in S$ for all $\rho, \eta \in S$ and $[\rho, S] \subseteq T$ for all $\rho \in T$. Suppose that every element of T is nilpotent. Then $< T >$ is contained in the radical $R(< S >)$ of the subalgebra $< S >$ of $gl(V)$.*

Proof Clearly, $(< T >< S > + < T >) < S > \subseteq (< T >< S > + < T >)$. Also from our earlier propositions $< S > (< T >< S > + < T >) \subseteq (< T >< S > + < T >)$. Thus, $(< T >< S > + < T >)$ is an ideal of $< S >$. By Proposition 3.1.3, $< T >$ is nilpotent. Hence, there is a natural number m such that $(< T >< S > + < T >)^m \subseteq < T >< S >$. Further, $(< T >< S >)^r \subseteq < T >^r < S >$. Thus, we get a natural number l such that $(< T >< S > + < T >)^l = 0$. This implies that $(< T >< S > + < T >)$ is a nilpotent ideal of $< S >$. Consequently, $(< T >< S > + < T >)$ is contained in the radical $R(< S >)$ of $< S >$. In particular, $< T > \subseteq R(< S >)$. ♯

The Engel theorem follows as an immediate corollary to the above proposition: Let L be a finite-dimensional Lie algebra which is ad-nilpotent in the sense that there is a natural number n such that $ad(x)^n = 0$ for all $x \in L$. Consider the subset $ad(L) = \{ad(x) \mid x \in L\}$ of $gl(L)$. Since $[ad(x), ad(y)] = ad([x, y])$ belongs to $ad(L)$ for all $x, y \in L$ and also $ad(x)$ is nilpotent for all $x \in L$, it follows from the above proposition that the associative subalgebra $< ad(L) >$ of $gl(L)$ generated by $ad(L)$ is nilpotent. Hence, there is a natural number n such that $ad(x_1)ad(x_2) \cdots ad(x_n) = 0$ for all $x_1, x_2, \ldots, x_n \in L$. This means that L is a nilpotent Lie algebra. ♯

More generally, we have the following corollary:

Corollary 3.1.5 *Let L be a finite-dimensional Lie algebra. Then an ideal N of L is a nil radical of L if and only if for each $x \in N$, $ad(x)$ is a nilpotent linear transformation on L and if A is any ideal of L with this property, then $A \subseteq N$. Indeed, $N = ad^{-1}(R(< ad(L) >) \bigcap ad(L)) = \{x \in L \mid ad(x) \in R(< ad(L) >)\}$.*

Proof Suppose that N is the nil radical of L. Let $x \in N$. Then for each $y \in L$, $[x, y] \in N$. Since N is nilpotent, $ad(x)^m([x, y]) = 0$ for some $m \in \mathbb{N}$. This means that $ad(x)^{m+1}(y) = 0$ for all $y \in L$. Thus, for each $x \in N$, $ad(x)$ is a nilpotent linear transformation on L. Further, if A is an ideal of L with the same property, then $ad(y)|_A$ is nilpotent for all $y \in A$. By the theorem of Engel, A is a nilpotent ideal of L. Since N is nil radical of L, $A \subseteq N$.

Conversely, let N be an ideal of L satisfying the given conditions. For each $x \in N$, $ad(x)$ is nilpotent on L. In particular, $ad(x)|_N$ is nilpotent for each $x \in N$. Hence N is a nilpotent ideal of L. Further, if A is a nilpotent ideal of L, then as above, $ad(y)$ is nilpotent for each $y \in A$. From the given condition, $A \subseteq N$. This shows that N is the nil radical of L.

Finally, since $R(< ad(L) >) \bigcap ad(L)$ is an ideal of the Lie subalgebra $ad(L)$ of $gl(L)$, and $ad^{-1}(R(< ad(L) >) \bigcap ad(L))$ is an ideal of L. Since $ad(x)$ is nilpotent for all $x \in ad^{-1}(R(< ad(L) >) \bigcap ad(L))$, $ad^{-1}(R(< ad(L) >) \bigcap ad(L)) \subseteq N$. Already, $ad(N) \subseteq R(< ad(L) >) \bigcap ad(L)$. Hence $N \subseteq ad^{-1}(R(< ad(L) >) \bigcap ad(L))$. ♯

Next, we discuss the structure of a Lie subalgebra L of $gl(V)$ for which the associative subalgebra $< L >$ of $gl(V)$ generated by L is a semi-simple associative subalgebra. Here, V is a finite-dimensional vector space over a field of characteristic 0. The following basic fact from linear algebra will be used.

Proposition 3.1.6 *A linear transformation ρ on a finite-dimensional vector space V of dimension n over a field F of characteristic 0 is nilpotent if and only if $Tr(\rho^k) = 0$ for all $k \in \mathbb{N}$.*

Proof Suppose that ρ is nilpotent. Then all the eigenvalues of ρ and also of ρ^k, $k \in \mathbb{N}$ are 0. Consequently $Tr(\rho^k) = 0$ for all $k \in \mathbb{N}$. Conversely, suppose that $Tr(\rho^k) = 0$ for all $\rho \in \mathbb{N}$. Let $\lambda_1, \lambda_2, \cdots, \lambda_n$ be n eigenvalues of ρ in the algebraic closure $\overline{F}$ of F. Then $0 = Tr(\rho^k) = \sum_{i=1}^{n} \lambda_i^k$ for all $k \in \mathbb{N}$. Using Newton's identity relating $\sum_{i=1}^{n} \lambda_i^k$ and the elementary symmetric polynomials in $\lambda_1, \lambda_2, \cdots, \lambda_n$, we obtain that all the coefficients of the characteristic polynomial of ρ are 0. Consequently, the characteristic polynomial of ρ is x^n. By the Cayley Hamilton theorem, $\rho^n = 0$. ♯

Corollary 3.1.7 *Let $\rho \in gl(V)$, where V is as above. Suppose that $\rho = \sum_{i=1}^{r}[\mu_i, \nu_i]$, $\mu_i, \nu_i \in gl(V)$ such that $[\rho, \mu_i] = 0$ for all i. Then ρ is nilpotent.*

Proof Since $[\rho, \mu_i] = 0$, $\rho\mu_i = \mu_i\rho$ and $\rho^m\mu_i = \mu_i\rho^m$ for each $m \in \mathbb{N}$. Now, for each $m \in \mathbb{N}$,

$$\begin{aligned} \rho^m = \rho^{m-1}\rho &= \sum_{i=1}^{r} \rho^{m-1}[\mu_i, \nu_i] = \sum_{i=1}^{r} \rho^{m-1}(\mu_i\nu_i - \nu_i\mu_i) \\ &= \sum_{i=1}^{r} \mu_i\rho^{m-1}\nu_i - \rho^{m-1}\nu_i\mu_i = \sum_{i=1}^{r}[\mu_i, \rho^{m-1}\nu_i]. \end{aligned}$$

Consequently, $Tr(\rho^m) = 0$ for each $m \in \mathbb{N}$. It follows from the above proposition that ρ is nilpotent. ♯

Theorem 3.1.8 *Let L be a Lie subalgebra of $gl(V)$, where V is a finite-dimensional vector space over a field F of characteristic 0. Suppose that the associative subalgebra $< L >$ of $gl(V)$ generated by L is semi-simple associative subalgebra of $gl(V)$. Then there is a semi-simple Lie ideal L_1 of the Lie algebra L such that $L = L_1 \oplus Z(L)$.*

Proof We first show that the radical $R(L)$ of the Lie algebra L is the center $Z(L)$ of L. Suppose not. Then $R_1(L) = [L, R(L)]$ is a nonzero solvable ideal of L. Hence, there is a natural number $m > 1$ such that $R_1(L)^{m-1} \neq \{0\}$ and $R_1(L)^m = \{0\}$, where $R_1(L)^r$ denotes the rth term of the derived series of $R_1(L)$. Consider $[L, R_1(L)^{m-1}]$. Let $\rho \in [L, R_1(L)^{m-1}]$. Then $\rho = \sum_{i=1}^{r}[\mu_i, \nu_i]$, where $\mu_i \in R_1(L)^{m-1}$ and $\nu_i \in L$. Clearly $[\rho, \mu_i] = 0$ for all i. By the previous corollary, ρ is a nilpotent linear transformation. This shows that every element of $[L, R_1(L)^{m-1}]$ is a nilpotent transformation. By Proposition 3.1.4, $[L, R_1(L)^{m-1}]$ is contained in the radical of the associative subalgebra $< L >$ of $gl(V)$. Since $< L >$ is assumed to be semi-simple associative subalgebra, $[L, R_1(L)^{m-1}] = \{0\}$. Consequently $R_1(L)^{m-1} \subseteq Z(L)$. Further, since $R_1(L)^{m-1} \subseteq [L, R(L)]$, every element ρ of $R_1(L)^{m-1}$ is expressible as $\rho = \sum_{i=1}^{r}[\mu_i, \nu_i]$, where $[\rho, \mu_i] = 0$ ($R_1(L)^{m-1} \subseteq Z(L)$). Again using the above argument, we obtain that $R_1(L)^{m-1} = \{0\}$. This is a contradiction. Hence $R_1(L) = [L, R(L)] = 0$. Consequently $R(L) \subseteq Z(L)$. Already by the definition of $R(L)$, $Z(L) \subseteq R(L)$. This shows that $R(L) = Z(L)$. A similar argument implies that $R(L) \bigcap [L, L] = \{0\}$. Thus, we can find a vector subspace L_1 of L such that L is the vector space direct sum $R(L) \oplus L_1$ and $L_1 \supseteq [L, L]$. Evidently, since $L_1 \supseteq [L, L]$), L_1 is an ideal. Finally, $L_1 \approx L/R(L)$ is semi-simple (the fact $L = L_1 \oplus Z(L)$ follows from Theorem 1.5.16 also). ♯

Corollary 3.1.9 *Let L be a center-less Lie subalgebra of $gl(V)$, V is a finite-dimensional vector space over a field of characteristic 0. Suppose that $< L >$ is a semi-simple associative subalgebra. Then L is semi-simple.* ♯

Corollary 3.1.10 *Let L be a solvable Lie subalgebra of $gl(V)$, where V is a finite-dimensional vector space over a field F of characteristic 0. Suppose that $< L >$ is semi-simple associative algebra. Then L is abelian Lie algebra. More generally, if $R(< L >)$ is the radical of the associative algebra $< L >$, then $< L > /R(< L >)$ is a commutative associative algebra.*

Proof Applying the above theorem, we observe that L_1 in this case is solvable semi-simple Lie algebra. Consequently, $L_1 = \{0\}$. This means that $L = Z(L)$ is abelian Lie algebra. Evidently, $< L >$ is commutative associative algebra. More generally, consider $R(< L >)$ the radical of the associative algebra $< L >$. Treating $R(< L >)$ as a Lie subalgebra of $< L >$, we see that $R(< L >)$ is an ideal of the Lie algebra $< L >$. It follows that Lie algebra $(L + R(< L >))/R(< L >)$, being homomorphic image of L, is solvable. Evidently, the associative subalgebra $< L + R(< L >) > /R(< L >)$ is semi-simple. Consequently, as a Lie algebra $(L + R(< L >))/R(< L >)$ is abelian. Hence, as an associative algebra $< L > /R(< L >) \approx < (L + R(L))/R(L) >$ is commutative associative algebra. ♯

Corollary 3.1.11 *Let L be a Lie subalgebra of $gl(V)$, where V is a finite-dimensional vector space over a field of characteristic 0. Then $L \bigcap R(< L >) = R(L)_0$, where $R(L)_0 = \{\rho \in R(L) \mid \rho \text{ is nilpotent}\}$. Further, $[R(L), L] \subseteq R(< L >)$.*

Proof Since $R(< L >)$ is nilpotent associative subalgebra, it is nilpotent considered as Lie algebra (check it). This implies that $L \bigcap R(< L >) \subseteq R(L)$. Also, since the elements of $R(< L >)$ are nilpotent, $L \bigcap R(< L >)$ is contained in the set $R(L)_0$ of nilpotent elements of $R(L)$. Consider $< R(L) >$. By the above corollary, $< R(L) > /R(< R(L) >)$ is a commutative associative algebra. Now, any nilpotent element in a commutative associative algebra generates an ideal which is nilpotent. Consequently, every nilpotent element of a commutative associative algebra is contained in the radical of the algebra. Since $< R(L) > /R(< R(L) >)$ is semi-simple associative algebra, it has no nonzero nilpotent element. Thus, $R(< R(L) >)$ consists of nilpotent elements of $< R(L) >$ and hence $R(< R(L) >) \bigcap R(L)$ is a subspace of L. Now, consider $(L + R(< L >))/R(< L >)$. Clearly, the associative algebra generated by $(L + R(< L >))/R(< L >)$ is $< L > /R(< L >)$ which is semi-simple. From Theorem 3.1.8, it follows that the radical of the Lie algebra $(L + R(< L >))/R(< L >)$ is contained in its center. Further, since $(R(L) + R(< L >))/R(< L >)$ is a solvable ideal of $(L + R(< L >))/R(< L >)$, it follows that $[(R(L) + R(< L >))/R(< L >), L + R(< L >))/R(< L >)] = \{0\}$. Consequently, $[R(L), L] \subseteq R(< L >)$. In turn, $[R(L), L] \subseteq L \bigcap R(< L >) \subseteq R(L)_0$. Thus, $[R(L)_0, L] \subseteq R(L)_0$ and so $R(L)_0$ is an ideal. Since the elements of $R(L)_0$ are nilpotent, it follows from Proposition 3.1.4 that $R(L)_0 \subseteq R(< L >)$. This shows that $R(L)_0 \subseteq L \bigcap R(< L >)$. Thus, $R(L)_0 = L \bigcap R(< L >)$. The rest is also clear. ♯

Corollary 3.1.12 *Let L be a Lie subalgebra of $gl(V)$, where V is a finite-dimensional vector space over a field of characteristic 0. Suppose that V is completely reducible in the sense that for every L-invariant subspace U of V, there is an L-invariant subspace W of V such that $V = U \oplus W$. Then $< L >$ and also $< L \bigcup\{1\} >$ are semi-simple associative algebras. Further, $L = L_1 \oplus Z(L)$, where L_1 is a semi-simple ideal of L and the elements of $Z(L)$ are semi-simple elements.*

Proof It is evident that a subspace of V is L-invariant if and only if it is $< L >$ $(< L \bigcup\{1\} >)$-invariant. Consequently, V is a completely reducible $< L >$ $(< L \bigcup\{1\} >)$-module. Since V is finite dimensional over a field of characteristic 0, $V = V_1 \oplus V_2 \oplus \cdots \oplus V_r$, where each V_i is an irreducible $< L >$ $(< L \bigcup\{1\} >)$-submodule. Consider the radical $R(< L >)$ $(R(< L \bigcup\{1\} >))$ of $< L >$ $(< L \bigcup\{1\} >)$ and the subspaces $R(< L >)V_i$ $(R(< L \bigcup\{1\} >)V_i)$ of V. Evidently, these are $< L >$ $(< L \bigcup\{1\} >)$-subspaces contained in V_i. Since $R(< L >)$ $(R(< L \bigcup\{1\} >))$ is nilpotent, there is a natural number m such that $R(< L >)^m = \{0\}$ $(R(< L \bigcup\{1\} >)^m = \{0\})$. Consequently, $R(< L >)V_i$ $(R(< L \bigcup\{1\} >)V_i)$ is a proper $< L >$ $(< L \bigcup\{1\} >)$-invariant subspace of V_i. Since V_i is irreducible, $R(< L >)V_i$ $(R(< L \bigcup\{1\} >)V_i)$ is zero for each i. Since $V = \oplus\sum_{i=1}^{r} V_i$, $R(< L >)V$ $(R(< L \bigcup\{1\} >)V)$ is zero. This means that $R(< L >)$ $(R < L \bigcup\{1\} >)$ is zero and hence $< L >$ $(< L \bigcup\{1\} >)$ is semi-simple. From Theorem 3.1.8, $L = L_1 \oplus Z(L)$, where L_1 is a semi-simple Lie algebra. Now, we show that the elements of $Z(L)$ are semi-simple. Suppose not. Then there is an element $\rho \in Z(L)$ which is not semi-simple. Consider the subalgebra $< \{\rho, 1\} > = F[\rho]$ of $gl(V)$. Since ρ is not semi-simple, there is a nonzero nilpotent element $\eta \in F[\rho]$. This means that η belongs to the center of the subalgebra $< L \bigcup \{1\} >$. Consequently, $\eta < L \bigcup \{1\} >$ is an ideal of $< L \bigcup \{1\} >$. Since η is nilpotent, $\eta < L \bigcup \{1\} >$ is a nilpotent ideal of $< L \bigcup \{1\} >$ containing η. Thus, $\eta < L \bigcup \{1\} >$ is a nonzero nilpotent ideal of $< L \bigcup \{1\} >$. This is a contradiction, since $< L \bigcup \{1\} >$ is semi-simple. ♯

We use the above discussions to give another proof of Lie's theorem.

Proposition 3.1.13 *Let L be an abelian Lie subalgebra of $gl(V)$, where V is a finite-dimensional vector space over an algebraically closed field F. Suppose that V is L-irreducible. Then V is one dimensional.*

Proof Let ρ be a member of L. Let λ be an eigenvalue of ρ and V_λ the λ-eigensubspace of V. Let $\eta \in L$. Since $\rho\eta = \eta\rho$, V_λ is invariant under η. Thus, V_λ is a nontrivial L-submodule of V. Since V is L-irreducible, $V_\lambda = V$. Indeed, every subspace of V is invariant under L. Thus, V is without nonzero proper subspaces. Hence V is one dimensional. ♯

Proof of Lies's Theorem Let L be a solvable Lie subalgebra of $gl(V)$, where V is a finite-dimensional vector space over an algebraically closed field of characteristic 0. Let

$$V = V_0 \supseteq V_1 \supseteq V_2 \supseteq \cdots \supseteq V_m = \{0\}$$

be an L-composition series of V. For each $\rho \in L$ and for each i, we have the induced linear transformation $\overline{\rho_i}$ on V_i/V_{i+1} given by $\overline{\rho_i}(x + V_{i+1}) = \rho(x) + V_{i+1}$. Let $\hat{L}_i$ denote $\{\overline{\rho_i} \mid \rho \in L\}$. Clearly, $\rho \mapsto \overline{\rho_i}$ is a surjective homomorphism from L to the Lie algebra $\hat{L}_i$. Consequently, $\hat{L}_i$ is solvable. Clearly, V_i/V_{i+1} is an irreducible $\hat{L}_i$-subspace. It follows from Corollary 3.1.12 that $\hat{L}_i$ is abelian. From the above proposition, $Dim V_i/V_{i+1} = 1$. Thus, there is a basis $\{x_0, x_1, x_2, \ldots, x_{m-1}\}$ of V such that $\{x_i, x_{i+1}, x_{i+2}, \ldots, x_{m-1}\}$ is a basis of V_i for each i. Clearly, the matrix of each member of L is triangular with respect to this basis. ♯

Proposition 3.1.14 *Let L be a finite-dimensional Lie algebra over a field of characteristic 0. Then $[L, R(L)] \subseteq N(L)$, where $N(L)$ denotes the nil radical of L.*

Proof Applying Corollary 3.1.11 to $ad(L) \subseteq gl(L)$, we obtain that $[ad(L), ad(R(L))] \subseteq R(< ad(L) >)$. Thus, there is a natural number m such that $\prod_{i=1}^{m} [ad(x_i), ad(y_i)] = 0$ for all $x_1, x_2, \ldots, x_m$ in L and $y_1, y_2, \ldots, y_m$ in $R(L)$. Consequently, $\prod_{i=1}^{m} ad([x_i, y_i]) = 0$ for all $x_1, x_2, \ldots, x_m$ in L and $y_1, y_2, \ldots, y_m$ in $R(L)$. It follows that for all $x \in L$, $\prod_{i=1}^{m} ad([x_i, y_i])(x) = 0$. This means that $[L, R(L)]$ is nilpotent. Hence $[L, R(L)] \subseteq N(L)$. ♯

In particular, we get another proof of the already proved fact that $[L, L]$ is nilpotent for all finite-dimensional solvable Lie algebra L over a field of characteristic 0.

Corollary 3.1.15 *Let L be a finite-dimensional solvable Lie algebra over a field of characteristic 0. Then $d(L) \subseteq N(L)$ for all derivation d on L, where $N(L)$ denotes the nilpotent radical of L.*

Proof Consider the vector space $\hat{L} = L \oplus Fd$, where Fd is the one-dimensional space generated by d. Define a product [,] on $\hat{L}$ by

$$[(x, \alpha d), (y, \beta d)] = ([x, y] + \beta d(x) - \alpha d(y), 0).$$

It can be easily checked that $\hat{L}$ is a Lie algebra with respect to this Lie product. Further,

$$[(x, 0), (y, \alpha d)] = ([x, y] + \alpha d(x), 0) \in L \oplus \{0\},$$

for all $x, y \in L$ and $\alpha \in F$. This means that $L \oplus \{0\}$ is an ideal of $\hat{L}$. Since $\hat{L}/(L \oplus \{0\})$ is one dimensional, it is solvable. Hence $\hat{L}$ is solvable. It follows that $[\hat{L}, \hat{L}] = [L, L] \oplus \{0\} + d(L) \oplus \{0\} = ([L, L] + d(L)) \oplus \{0\}$ is nilpotent. This means that $[L, L] + d(L)$ is nilpotent. Hence, $d(L) \subseteq [L, L] + d(L) \subseteq N(L)$. ♯

Let ρ be a representation of a Lie algebra L on V. Consider the universal enveloping algebra $(U(L), j_L)$ of L. Note that j_L is injective linear map which is a Lie homomorphism from L to $U(L)_L$ (Corollary 1.2.6). From the universal property of universal enveloping algebra, we have a unique algebra homomorphism $\overline{\rho}$ from $U(L)$ to $gl(V)$ such that $\overline{\rho} o j_L = \rho$. Let A denote the kernel of $\overline{\rho}$. Then $\nu o j_L$ is a Lie algebra homomorphism from L to $\overline{U(L)}_L = (U(L)/A)_L$. From Proposition 3.1.1, we have an injective algebra homomorphism i from $\overline{U(L)}$ to $gl(\overline{U(L)})$ Consequently, $i o \nu o j_L$ is a representation of L on $U(L)/A$. Evidently, $i o \nu o j_L$ is a faithful representation if and only if $A \bigcap j_L(L) = \{0\}$. Since every associative algebra has a faithful representation, we obtain the following propositions.

Proposition 3.1.16 *A Lie algebra L has a finite-dimensional faithful representation if and only if there is an ideal A of $U(L)$ of finite co-dimension in $U(L)$ such that $A \bigcap j_L(L) = \{0\}$.* ♯

Proposition 3.1.17 *Let A be an associative algebra over a field F. Let $a \in A$. Then the following conditions are equivalent:*

(i) The element a is algebraic in the sense that there is a polynomial $f(x) \in F[x]$ such that $f(a) = 0$.

(ii) The subalgebra $< a > = F[a]$ of A generated by $\{a\}$ is finite dimensional.

Proof $((i) \Rightarrow (ii))$ Suppose that a is an algebraic element of A. Let $f(x)$ be the smallest degree monic polynomial such that $f(a) = 0$. Suppose that $degf(x) = m$. The subalgebra of A generated by $\{a\}$ is $F[a] = \{\phi(a) \mid \phi(x) \in F[x]\}$. Using the division algorithm, we see that $F[a] = \{\phi(a) \mid deg\phi(x) < m\}$. Clearly, $\{1, a, a^2, \cdots, a^{m-1}\}$ is a basis of $F[a]$.

$((ii) \Rightarrow (i))$ Assume that the dimension of $F[a]$ is m. Then $\{1, a, a^2, \cdots, a^m\}$ is linearly dependent. Hence, there is a nonzero polynomial $f(x)$ of degree at the most m such that $f(a) = 0$. This means that a is algebraic. ♯

Proposition 3.1.18 *Let L be a finite-dimensional Lie algebra with a basis $\{x_1, x_2, \ldots, x_m\}$. Let A be an ideal of $U(L)$. Then $\overline{U(L)} = U(L)/A$ is finite dimensional if and only if $\overline{j_L(x_i)} = j_L(x_i) + A$ is algebraic in $\overline{U(L)}$ for each i.*

Proof We know that the set $\{(j_L(x_1))^{\alpha_1}(j_L(x_2))^{\alpha_2}\cdots(j_L(x_m))^{\alpha_m} \mid \alpha_i \geq 0\}$ of standard monomials in $\{j_L(x_1), j_L(x_2), \ldots, j_L(x_m)\}$ is a basis of $U(L)$ (Corollary 1.2.7). Thus, the set $\{(\overline{j_L(x_1)})^{\alpha_1}(\overline{j_L(x_2)})^{\alpha_2}\cdots(\overline{j_L(x_m)})^{\alpha_m} \mid \alpha_i \geq 0\}$ of standard monomials in $\{\overline{j_L(x_1)}, \overline{j_L(x_2)}, \ldots, \overline{j_L(x_m)}\}$ generates the algebra $\overline{U(L)}$. If $\overline{U(L)}$ is of finite dimension, then all elements of $\overline{U(L)}$ are algebraic. In particular, $\overline{j_L(x_i)}$ are algebraic for each i. Conversely, suppose that $\overline{j_L(x_i)}$ is algebraic for each i. Suppose that t_i is the smallest natural number such that $\overline{j_L(x_i)}$ is a root of a monic polynomial of degree t_i. Evidently, $\{(\overline{j_L(x_1)})^{\beta_1}(\overline{j_L(x_2)})^{\beta_2}\cdots(\overline{j_L(x_m)})^{\beta_m} \mid \beta_i \leq t_i - 1\}$ is a set of generators of $\overline{U(L)}$ as a vector space. Consequently, $\overline{U(L)}$ is finite dimensional. ♯

Proposition 3.1.19 *Let L be a finite-dimensional Lie algebra with a basis $\{x_1, x_2, \ldots, x_m\}$. Let A and B be ideals of $U(L)$ such that $U(L)/A$ and $U(L)/B$ are finite dimensional. Then $U(L)/AB$ is finite dimensional.*

Proof If $j_L(x_i) + A$ is a root of a polynomial $f_i(x)$, then $f_i(j_L(x_i)) \in A$. Similarly, if $j_L(x_i) + B$ is a root of a polynomial $g_i(x)$, then $g_i(j_L(x_i)) \in B$. Consequently, $f_i(j_L(x_i))g_i(j_L(x_i)) \in AB$ for all i. This shows that $f_i(j_L(x_i))g_i(j_L(x_i)) + AB$ is algebraic in $U(L)/AB$. The result follows from the earlier proposition. ♯

Proposition 3.1.20 *Let A be an associative algebra. Let S be a set of generators of A. Let d be a derivation of A. Suppose that for each $x \in S$, there is a natural number n_x such that $d^{n_x}(x) = 0$. Then for all $a \in A$, there is a natural number n_a such that $d^{n_a}(a) = 0$. If S is finite, then d is nilpotent.*

Proof Consider the subset $B = \{x \in A \mid d^{n_x} = 0 \ for\ some\ n_x \in \mathbb{N}\}$. By our hypothesis, $S \subseteq B$. If $x, y \in B$, then there are natural numbers n_x and n_y such that $d^{n_x}(x) = 0 = d^{n_y}(y)$. Take $n = max(n_x, n_y)$. Then $d^n(x + y) = 0$. Also $d^{n_x}(\alpha x) = 0$. This shows that B is a subspace of A. Further, by Leibniz theorem

$$d^n(xy) = \sum_{r=0}^{n} {}^nC_r d^r(x)d^{n-r}(y).$$

Thus, $d^n(xy) = 0$, where $n = n_x + n_y - 1$. This shows that B is a subalgebra of A containing S. Consequently $B = A$.

It further follows from the Leibniz theorem that if $d^{n_x}(x) = 0$, then $d^{n_x}(x^t) = 0$ for all $t \in \mathbb{N}$. Suppose that $S = \{x_1, x_2, \cdots, x_r\}$ is finite. Let $N = \sum_{i=1}^{r} n_{x_i} - 1$. Then from the above observation and representations of elements of A in terms of elements of S, it follows that $d^N(x) = 0$ for all $x \in A$. This means that d is nilpotent. ♯

Proposition 3.1.21 *Let L be a finite-dimensional Lie algebra over a field of characteristic 0. Let A be an ideal of the $U(L)$ such that $U(L)/A$ is finite dimensional and* *every member of $j_L(N(L))$ is nilpotent modulo A, where $(U(L), j_L)$ is the universal enveloping algebra of L and $N(L)$ is the nil radical of L. Then there is an ideal B of $U(L)$ such that the following hold:*

(i) $B \subseteq A$ and $U(L)/B$ is finite dimensional.
(ii) $\hat{d}(B) \subseteq B$ for all derivations d of L (refer to Theorem 5.2.3 (3.6): For each derivation d of L, we have a unique derivation $\hat{d}$ of $U(L)$ subject to $\hat{d}oj_L = j_Lod$).
(iii) Every element of $j_L(N(L))$ is nilpotent modulo B.

Proof Consider the ideal $C = < A \bigcup j_L(N(L)) >$ of $U(L)$ (considered as an associative algebra) generated by $A \bigcup j_L(N(L))$. Clearly, C/A is the ideal of the associative algebra $U(L)/A$ which is generated by the Lie ideal $(j_L(N(L)) + A)/A$ of $(U(L)/A)_L$. The elements of $(j_L(N(L)) + A)/A$ are nilpotent in $U(L)/A$. From Proposition 3.1.4, it follows that $(j_L(N(L)) + A)/A$ is contained in the radical $R(U(L)/A)$ of the associative algebra $U(L)/A$. Hence $C/A \subseteq R(U(L)/A)$. Thus, there is a natural number m such that $B = C^m \subseteq A$. Since $U(L)/C$ is finite dimensional, it follows from Proposition 3.1.19 that $U(L)/B$ is finite dimensional. This proves (i). From Corollary 3.1.15, it follows that $\hat{d}(U(L)) \subseteq j_L(N(L))$ for all derivations d of L. This proves (ii). Let $x \in N(L)$. Consider $j_L(x)$. Clearly, $j_L(x)^r$ belongs to A for some r. In turn, $j_L(x)^r$ belongs to C and so $(j_L(x))^{mr}$ belongs to B. This proves (iii). ♯

Theorem 3.1.22 *Let L be a finite-dimensional Lie algebra over a field of characteristic 0. Let P be a solvable ideal of L and Q be a Lie subalgebra of L such that L is the semi-direct product $P \succ Q$. Suppose that there is a finite-dimensional representation η of P on V such that $\eta(x)$ is nilpotent for all x in the nil radical $N(P)$ of P. Then there is a finite-dimensional representation ρ of L such that the following hold: (i) If $x \in P$, then $\rho(x) = 0$ only when $\eta(x) = 0$. (ii) $\rho(y)$ is nilpotent whenever $y = z + u$, where $z \in N(P)$ and $u \in Q$ such that $ad(u)|_P$ is nilpotent.*

Proof η induces an associative algebra homomorphism $\hat{\eta}$ from the universal enveloping algebra $U(P)$ of P to $gl(V)$ such that $\hat{\eta}oj_P = \eta$. Since V is finite dimensional, the kernel A of $\hat{\eta}$ is of finite co-dimension in $U(P)$. Further, if $x \in N(P)$, then $\eta(x)^n = 0$ for some n. Thus, $j_P(x)^n \in A$ for each $x \in N(P)$ and the hypothesis of the previous proposition is satisfied with P in place of L. Hence there is an ideal B of $U(P)$ such that $B \subseteq A$, $U(P)/B$ is finite dimensional, $\hat{d}(B) \subseteq B$ for all derivation d of P, and every element of $j_P(N(P))$ is nilpotent modulo B. We construct a representation ρ of L on $U(P)/B$ which satisfies the required properties stated in the theorem. We first construct a representation $\hat{\rho}$ of L on $U(P)$. For $x \in P$, we put $\hat{\rho}(x) = r_{j_P(x)}$, where $r_{j_P(x)}$ denotes the right multiplication by $j_P(x)$. For $y \in Q$, put $\hat{\rho}(y) = \widehat{ad}(y)$, where $\widehat{ad}(y)$ is the unique derivation on $U(P)$ subject to the condition $\widehat{ad}(y)oj_P = ad(y)|_P$ on P. It can be easily checked that $\hat{\rho}$ defines a representation of L on $U(P)$. Since $\hat{d}(B) \subseteq B$ for each derivation d of P, it follows that B is a subspace of $U(P)$ which is invariant under $\hat{\rho}$. In turn, $\hat{\rho}$ induces a representation ρ of L on $U(P)/B$. Let $x \in P$ such that $\rho(x) = 0$. Then $r_{j_P}(U(P)) \subseteq B$. This means that $j_P(x) \in B \subseteq A$. Thus, $\hat{\eta}(j_P(x)) = 0$. Consequently, $\eta(x) = 0$. Let $z \in N(P)$, then from the above proposition, $j_P(z)$ is nilpotent modulo B. Hence $\rho(z)$ is nilpotent. Since $N(P)$ is an ideal of L, it follows that $\hat{\rho}(j_P(z))$ is in the radical $R(<\rho(L)>)$. Suppose that $y = z + u$, where $z \in N(P)$ and $u \in Q$ such that $ad(u)|_P$ is nilpotent. We need to show that $\rho(y)$ is nilpotent. It is sufficient to show that $\rho(u)$ is nilpotent, since $\rho(z) \in R(<\rho(L)>)$. Already, by the definition, $\hat{\rho}(u)$ is a derivation on $U(P)$ which is $\hat{ad(u)}$ on $U(P)$ and $ad(u)|_P$ is nilpotent. Since $<j_L(P)> = U(P)$, from Proposition 3.1.20, it follows that for each $a \in U(P)$, there is a natural number n_a such that $\rho(u)^{n_a}(\nu(a)) = 0$. Thus, for each $\overline{a} \in U(P)/B$, we have a natural number n_a such that $(\rho(u)(\overline{a}))^{n_a} = 0$. Since $U(P)/B$ is finite dimensional, $\rho(u)$ is nilpotent. Thus, ρ satisfies the required condition. ♯

Theorem 3.1.23 (Ado) *Every finite-dimensional Lie algebra over a field of characteristic 0 has a faithful finite-dimensional representation.*

Proof Suppose that we are able to construct a finite-dimensional representation ρ from L to $gl(V)$ such that the restriction $\rho|_{Z(L)}$ of ρ on $Z(L)$ is faithful. Then the representation $\hat{\rho} = \rho \oplus ad$ from L to $gl(V \oplus L)$ is a finite-dimensional representation of L such that

$$Ker\ \hat{\rho} = \{x \in L \mid 0 = \hat{\rho}(x) = \rho(x) \oplus ad(x)\} = \{x \in L \mid x \in Z(L)\ and\ \rho(x) = 0\} = \{0\}.$$

This means that $\hat{\rho}$ is a faithful finite-dimensional representation of L.

Thus, it is sufficient to construct a finite-dimensional representation ρ of L which is faithful on $Z(L)$. Since $Z(L) \subseteq N(L)$, we have an ascending chain

$$Z(L) = N_1(L) \subset N_2(L) \subset \cdots \subset N_m(L) = N(L)$$

such that $N_i(L)$ is an ideal of $N_{i+1}(L)$ and $DimN_{i+1}(L)/N_i(L) = 1$ for each i. Suppose that the $Dim\ Z(L) = r$. Let $W_1 = <\{x_1, x_2, \ldots, x_r, x_{r+1}\}>$ be an $r+1$-dimensional vector space over F. Consider the linear transformation η_1 on W_1 given by $\eta_1(x_i) = x_{i+1}, i \leq r$ and $\eta_1(x_{r+1}) = 0$. Evidently, $\eta_1^r \neq 0$ but $\eta_1^{r+1} = 0$. Clearly, $\{\eta_1, \eta_1^2, \ldots \eta_1^r\}$ is a linearly independent subset of $End W_1$. Let U_1 denote the subspace of $End W_1$ with basis $\{\eta_1, \eta_1^2, \ldots \eta_1^r\}$. Evidently, U_1 is an abelian Lie subalgebra of $gl(W_1)$ which is isomorphic to the Lie subalgebra $Z(L)$ of L. Thus, we get a faithful finite-dimensional representation ρ_1 of $Z(L)$ on W_1. Since $N_2(L)$ is nilpotent and $N_2(L) = Z(L) \succ Fu_2$ for some $u_2 \in N_2(L)$, from the previous theorem, we obtain a finite-dimensional representation ρ_2 of $N_2(L)$ such that ρ_2 is faithful on $Z(L)$. Proceeding inductively, we obtain a finite-dimensional representation ρ_r of $N(L)$ such that $\rho|_{Z(L)}$ is faithful. Again, we have an ascending chain

$$N(L) = R_1(L) \subseteq R_2(L) \subseteq \cdots \subseteq R_s(L) = R(L),$$

where $R_j(L)$ is an ideal of $R_{j+1}(L)$ and $Dim\ R_{j+1}(L)/R_j(L) = 1$ for each j. Indeed, $N(L) = N(R_j(L))$ for each j. Starting with ρ_r and using the previous theorem again and again, by the induction, we get a finite-dimensional representation ρ of $R(L)$ which agrees with ρ_r on $N(L)$ and so it is faithful when restricted to $Z(L)$. Further, by the Radical splitting theorem due to Levi (Theorem 1.5.16), $L = R(L) \succ L_1$, where L_1 is a semi-simple Lie algebra. Using the previous theorem, we obtain a faithful finite-dimensional representation of L. ♯

Remark 3.1.24 It follows from the proof of the above theorem that we can find a finite-dimensional faithful representation ρ of L such that $\rho(x)$ is nilpotent for all $x \in N(L)$. Consequently, $j_L(x) \in R(< L >)$ for all $x \in N(L)$.

Now, we shall establish the result for Lie algebras over fields of the characteristic p.

Theorem 3.1.25 (Iwasawa) *Every finite-dimensional Lie algebra L over a field of characteristic $p \neq 0$ has a faithful finite-dimensional representation.*

Before proving the theorem, we establish a few lemmas needed for the purpose.

Lemma 3.1.26 *Let A be an associative algebra over a field of characteristic $p \neq 0$. Then $ad(a)^{p^r} = ad(a^{p^r})$ for each $r \in \mathbb{N}$, where $ad(a)$ is the linear transformation on A given by $ad(a)(x) = ax - xa$.*

Proof The following identity can be easily established by using induction on m:

$$(ad(a))^m(x) = \sum_{i=0}^{m}(-1)^{i+1}\, {}^mC_i x^i a x^{m-i}.$$

Putting $m = p^r$ and observing that ${}^{p^r}C_i$ is divisible by p for all $i \neq 0$, we obtain the required result. ♯

Lemma 3.1.27 *Let L be a finite-dimensional Lie algebra over a field of characteristic $p \neq 0$. Then for each $a \in L$, there is a p-polynomial $m_a(x) \in F[x]$ (a polynomial of the form*

$$\alpha_0 x + \alpha_1 x^p + \alpha_2 x^{p^2} + \cdots + \alpha_{m-1} x^{p^{m-1}} + x^{p^m}$$

is termed as a p-polynomial) such that $m_a(j_L(a))$ belongs to the center $Z(U(L))$ of the universal enveloping associative algebra $(U(L), j_L)$ of L.

Proof Let a be a member of L. Let $\chi(x)$ denote the characteristic polynomial of $ad(a) \in gl(L)$. Then $\chi(ad(a)) = 0$. Suppose that $deg\chi(x) = Dim\ L = m$. For each i, $0 \leq i \leq m$, there are polynomials $q_i(x)$ and $r_i(x)$ such that

$$x^{p^i} = \chi(x)q_i(x) + r_i(x),$$

where $degr_i(x) < m$. Since the space of polynomials of degrees less than m is a vector space of dimension m, $\{r_0(x), r_1(x), \ldots, r_m(x)\}$ is linearly dependent. Consequently, there exist scalars $\alpha_0, \alpha_1, \ldots, \alpha_m$ not all zero such that $\sum_{i=0}^{m} \alpha_i r_i(x) = 0$. Let t be the largest such that $\alpha_t \neq 0$. We may assume that $\alpha_t = 1$. Take

$$m_a(x) = \alpha_0 x + \alpha_1 x^p + \alpha_2 x^{p^2} + \cdots + \alpha_{t-1} x^{p^{t-1}} + x^{p^t}.$$

Then $m_a(x) = \chi(x)\sum_{i=0}^{t} \alpha_i q_i(x)$. Thus, $m_a(x)$ is a p-polynomial such that $m_a(ad(a)) = 0$ or equivalently,

$$(\alpha_0 ad(a) + \alpha_1 (ad(a))^p + \alpha_2 (ad(a))^{p^2} + \cdots + \alpha_{t-1}(ad(a))^{p^{t-1}} + (ad(a))^{p^t})(x) = 0$$

for all $x \in L$. Using the previous lemma, we obtain that

$$ad(\alpha_0 a + \alpha_1 a^p + \alpha_2 a^{p^2} + \cdots + \alpha_{t-1} a^{p^{t-1}} + a^{p^t}) = 0$$

on L. By the PBW theorem, it follows that

$$[\alpha_0 j_L(a) + \alpha_1 j_L(a)^p + \alpha_2 j_L(a)^{p^2} + \cdots + \alpha_{t-1} j_L(a)^{p^{t-1}} + j_L(a)^{p^t}, j_L(x)] = 0$$

for all $x \in L$. Since $j_L(L)$ generates $U(L)$, it follows that $m_a(j_L(a)) \in Z(U(L))$. ♯

Let L be a Lie algebra with an ordered basis $B = \{x_i \mid i \in I\}$. Then by the PBW theorem, the set $\{j_L(x_{i_1})^{m_1} j_L(x_{i_2})^{m_2} \cdots j_L(x_{i_r})^{m_r} \mid i_1 < i_2 < \cdots < i_r, m_j \geq 0\}$ of standard monomials forms a basis of the universal enveloping associative algebra $(U(L), j_L)$ of L. For each $m \geq 0$, let $U(L)^{(m)}$ denote the subspace $F + j_L(L) + j_L(L)^2 + \cdots + j_L(L)^m$ of $U(L)$, where $j_L(L)^k$ is the subspace of $U(L)$ generated by the set of standard monomials $j_L(x_{i_1})^{m_1} j_L(x_{i_2})^{m_2} \cdots j_L(x_{i_r})^{m_r}$ with $m_1 + m_2 + \cdots + m_r = k$. We have the filtration

$$U(L)^{(0)} \subseteq U(L)^{(1)} \subseteq \cdots \subseteq U(L)^{(m)} \subseteq \cdots$$

of $U(L)$. The following proposition follows from the PBW theorem and a clever use of induction.

Proposition 3.1.28 *Suppose that for each x_i in an ordered basis $B = \{x_i \mid i \in I\}$ of L, there is a natural number n_i such that $j_L(x_i)^{n_i} = j_L(y_i) + j_L(z_i)$, where $j_L(y_i) \in U(L)^{(n_i-1)}$ and $j_L(z_i) \in Z(U(L))$. Then the set $\{j_L(x_{i_1})^{\alpha_1} j_L(x_{i_2})^{\alpha_2} \cdots j_L(x_{i_r})^{\alpha_r} j_L(z_{i_1})^{\beta_1} j_L(z_{i_2})^{\beta_2} \cdots j_L(z_{i_r})^{\beta_r} \mid i_1 < i_2 < \cdots < i_r, m_j \geq 0, 0 \leq \alpha_j < n_{i_j}\}$ also forms a basis of $U(L)$.* ♯

Proof of the Iwasawa theorem. Let $\{x_1, x_2, \ldots, x_n\}$ be an ordered basis of L. For each x_k, let $m_k(x)$ be the polynomial described in Lemma 3.1.27 such that $m_k(j_L(x_k))$ belongs to the center $Z(U(L))$ of $U(L)$. Suppose that the degree of $m_k(x)$ is p^{t_k}. Then $j_L(x_k)^{p^{t_k}} = j_L(y_k) + j_L(z_k)$, where $j_L(y_k) \in U(L)^{p^{m_k-1}} \subseteq U(L)^{(p^{m_k}-1)}$. From the previous proposition, the set $\{j_L(x_1)^{\alpha_1} j_L(x_2)^{\alpha_2} \cdots j_L(x_n)^{\alpha_n} j_L(z_1)^{\beta_1} j_L(z_2)^{\beta_2} \cdots j_L(z_n)^{\beta_n} \mid 0 \leq \alpha_k < p^{m_k}\}$ forms a basis of $U(L)$. Let C be the ideal of $U(L)$ generated by the set $\{j_L(z_k) \mid 1 \leq k \leq n\}$. Observe that $\{\overline{j_L(x_1)}^{\alpha_1} \overline{j_L(x_2)}^{\alpha_2} \cdots \overline{j_L(x_n)}^{\alpha_n} \mid 0 \leq \alpha_j < p^{m_j}\}$ forms a basis $U(L)/C$. It also follows that the map χ from L to the Lie algebra $(U(L)/C)_L$ given by $\chi(a) = j_L(a) + C$ is an injective Lie algebra homomorphism. As such, it is a faithful finite-dimensional representation of L. ♯We obtain the following corollary.

Corollary 3.1.29 *Every finite-dimensional Lie algebra over a field F is isomorphic to a Lie subalgebra of $gl(n, F)$ for some n.* ♯

Remark 3.1.30 There is no connection between the structure of a Lie algebra over a field of characteristic $p \neq 0$ and the complete reducibility of its representations. Indeed, every finite-dimensional Lie algebra over a field of characteristic $p \neq 0$ has a faithful finite-dimensional completely reducible representation and also it has a faithful finite-dimensional representation which is not completely reducible.

Exercises

3.1.1. Show that every finite-dimensional Lie algebra over a field F is isomorphic to a Lie subalgebra of $sl(n, F)$ for some n.

3.1.2. Let ρ be a faithful finite-dimensional representation on V of a finite-dimensional Lie algebra L over a field of characteristic 0. Consider the tensor power representation $\otimes^r \rho$ of L on $\otimes^r V$. Let A_r denote the kernel of the induced associative algebra homomorphism $U(\otimes^r \rho)$ from $U(L)$ to $gl(\otimes^r V)$. Show that $\bigcap_{r=0}^{\infty} A_r = \{0\}$.

3.1.3. Show that any finite-dimensional Lie algebra over a field of characteristic $p \neq 0$ has an indecomposable finite-dimensional representation of arbitrarily high degree.

3.1.4. Let L be a Lie algebra over a field of characteristic $p \neq 0$ with an ordered basis $\{x_1, x_2, \ldots, x_m\}$. Let $j_L(z_k) = m_k(j_L(x_k))$ be a member of $Z(U(L))$ as described in the proof of the Iwasawa theorem. Let A denote the ideal of $U(L)$ generated by the set $\{j_L(z_1)^2, j_L(z_2), \ldots, j_L(z_k), \ldots, j_L(z_m)\}$. Show that $U(L)/A$ is not a semi-simple L-module. Deduce that every finite-dimensional Lie algebra over a field of characteristic $p \neq 0$ has a faithful finite-dimensional representation which is not completely reducible.

3.2 Cyclic Modules and Weights

This is a small section in which we shall be concerned with cyclic modules over semi-simple Lie algebras. All Lie algebras are assumed to be over the complex field $\mathbb{C}$ (or over any algebraically closed field of characteristic 0).

Let L be a semi-simple Lie algebra. Let V (not necessarily finite dimensional) be an L-module. Let H be a Cartan subalgebra of L. Let λ be a member of the dual space $H^\star$ of H. Let $V_\lambda = \{v \in V \mid h \cdot v = \lambda(h)v \ \forall\, h \in H\}$. V_λ may be $\{0\}$. If $V_\lambda \neq \{0\}$, then it is called a **weight space** of the L-module V and λ is called the corresponding **weight**. If we consider L as an L-module through the adjoint action, then the weight spaces are precisely the root spaces L_α and weights are precisely roots $\alpha \in \Phi$. If V is a finite-dimensional vector space, then it is the direct sum $\oplus \sum_{\lambda \in H^\star} V_\lambda$ of its weight spaces. However, if V is infinite dimensional, then it need not be sum of its weight spaces (Example 3.2.4).

Proposition 3.2.1 *Let V be a simple L-module with a weight space. Then V is the direct sum of its weight spaces.*

Proof Let W be the sum of its weight spaces. Under the hypothesis, $W \neq \{0\}$. Clearly, $h(W) \subseteq W$ for all $h \in H$. Let $x \in L_\alpha$, where $\alpha \in \Phi$. Let V_λ be a weight space. Let $v \in V_\lambda$. Then $h \cdot (x \cdot v) = x \cdot (h \cdot v) + [h, x] \cdot v = \lambda(h) x \cdot v + \alpha(h) x \cdot v = (\lambda + \alpha)(h) x \cdot v$. This shows that $x \cdot v$ belongs to $V_{\lambda+\alpha} \subseteq W$. Since W is the sum of weight spaces and $L = H \oplus (\oplus \sum_{\alpha \in \Phi} L_\alpha)$, it follows that W is a nonzero L-submodule. Since V is simple, $W = V$. Further, since $V_\lambda \bigcap \sum_{\mu \in H^\star - \{\lambda\}} V_\mu = \{0\}$ for all $\lambda \in H^\star$, it follows that V is the direct sum of its weight spaces. ♯

Proposition 3.2.2 *Let V be a simple L-module and H a Cartan subalgebra of L. Then the following conditions are equivalent:*

(i) V has a weight space (nonzero).
(ii) $U(H)v$ is finite dimensional for all $v \in V$.
(iii) $< \{1, j_H(h)\} > v$ is finite dimensional for all $v \in V$ and $h \in H$, where $< \{1, j_H(h)\} >$ is the subalgebra of $U(H)$ generated by $\{1, j_H(h)\}$.

Proof $((i) \Rightarrow (ii))$. Assume (i). It follows from the previous proposition that V is the direct sum of its weight spaces. As such, it is sufficient to show that $U(H)v$ is finite dimensional whenever $v \in V_\lambda$ for some λ. Evidently, $U(H)v \subseteq < \{\lambda(h)(v) \mid h \in H\} >$. Since H is finite dimensional, $U(H)v$ is finite dimensional.

$((ii) \Rightarrow (iii))$ is evident.

$((iii) \Rightarrow (i))$. Assume (iii). Then the subspace $\sum_{i=1}^{l} < \{1, j_H(h_i)\} > v$ is a nonzero finite-dimensional H-module, where $\{h_1, h_2, \cdots, h_l\}$ is a basis of H. Hence it has a nonzero weight space. ♯

Let H be a Cartan subalgebra of L and Φ be the root system associated with H. Let Δ be a basis of Φ. Let V be an L-module. A nonzero vector $v^+ \in V_\lambda$ is called a **maximal vector** (with weight λ) of the L-module V relative to Δ if for each $\alpha \in \Delta$, $L_\alpha \cdot v^+ = \{0\}$ and $h \cdot v^+ = \lambda(h)v^+$. Equivalently, v^+ is a maximal vector of V of weight λ if $j_L(L_\alpha)v^+ = 0$ and $j_L(h)v^+ = \lambda(h)v^+$ for all $\alpha \in \Delta$ and $h \in H$. Thus, if L is a simple Lie algebra and α^+ is a maximal root of Φ relative to Δ, then any nonzero member of the root space L_{α^+} is a maximal vector of the L-module L. Evidently, in this case all maximal vectors are of this type.

Proposition 3.2.3 *Every finite-dimensional L-module has a maximal vector.*

Proof Let V be a finite-dimensional L-module. Then V is also a finite-dimensional $B(\Delta)$-module, where $B(\Delta) = H \oplus (\oplus \sum_{\alpha > 0} L_\alpha)$ is the standard Borel subalgebra related to the basis Δ of Φ. Since $B(\Delta)$ is solvable, by the theorem of Lie, there is a common eigenvector $v^+ \in V$ of members of $B(\Delta)$. We show that v^+ is a maximal vector. We have a $\lambda \in H^\star$ such that $hv^+ = \lambda(h)v^+$ for all $h \in H$. Suppose that $x_\alpha \in L_\alpha, \alpha > 0$. Then there is a $\mu(\alpha) \in F$ such that $x_\alpha v^+ = \mu(\alpha)v^+$. Now,

$$\begin{aligned}\mu(\alpha)\lambda(h)v^+ &= h\mu(\alpha)v^+ = h(x_\alpha v^+) = [h, x_\alpha]v^+ + x_\alpha(hv^+)\\ &= \alpha(h)x_\alpha v^+ + \lambda(h)x_\alpha v^+ = (\lambda(h) + \alpha(h))\mu(\alpha)v^+.\end{aligned}$$

Since $v^+ \neq 0$, $(\lambda(h) + \alpha(h))\mu(\alpha) = \lambda(h)\mu(\alpha)$ for all $\alpha > 0$ and $h \in H$. Again, since $\alpha > 0$, $\mu(\alpha) = 0$ for all $\alpha > 0$. This shows that $x_\alpha v^+ = 0$ for all $\alpha > 0$. ♯

However, the following example asserts that if V is an infinite-dimensional L-module, then it need not have any nonzero weight space and so it has no maximal vector.

Example 3.2.4 Take L to be $sl(2, \mathbb{C})$ with standard basis $\{x, y, h\}$, where $x = e_{12}$, $y = e_{21}$ and $h = e_{11} - e_{22}$. Since $j_L(x)$ is non-nilpotent and $U(L)$ is without zero divisors, $1 - j_L(x)$ is non-invertible in $U(L)$. Consequently, by Krull's theorem (note that $U(L)$ is with identity), there is a maximal left ideal I of $U(L)$ which contains $1 - j_L(x)$. Consider the simple $U(L)$ (and so also L)-module $V = U(L)/I$. Using induction on r and s, we can easily establish that

$$(1 - j_L(x))^r j_L(h)^s \equiv 0(mod I)$$

whenever $s < r$ and

$$(1 - j_L(x))^r j_L(h)^r \equiv r!2^r(mod I).$$

Consider the set $X = \{j_L(h)^m \mid m \geq 0\}$. Suppose that $\sum_{i=0}^{t} a_i(j_L(h)^{m_i} + I) = I$, $m_1 < m_2 < \cdots < m_t$. It follows from the above identities that $(1 - j_L(x))^{m_t}$ $(\sum_{i<t} a_i j_L(h)^{m_i}) \in I$ whereas $(1 - j_L(x))^{m_t} j_L(h)^{m_t} \notin I$. Thus, $a_t = 0$. Proceeding inductively, we see that $a_i = 0$ for all i. This shows that X is linearly independent and hence V is infinite dimensional. Further, since V is simple (for I is maximal left ideal) and $U(L)h$ is infinite dimensional, it follows from Proposition 3.2.2 that V has no nonzero weight space and so it has no maximal vector also.

In what follows, L is a semi-simple Lie algebra, H is a Cartan subalgebra of L, Φ is the root system associated with H, and Δ is a given basis of Φ. Let v^+ be a maximal vector with weight λ in an L-module V. Since L-submodules of V are precisely $U(L)$-submodules of V $(x \cdot v = j_L(x) \cdot v)$, the L-submodules of V generated by v^+ are precisely $U(L)v^+$.

Definition 3.2.5 An L-module V is said to be a **Standard Cyclic Module** of weight λ if $V = U(L)v^+$, where v^+ is a maximal vector with weight λ.

Thus, $sl(2, \mathbb{C}) = U(sl(2, \mathbb{C}))e_{12}$ is a standard cyclic $sl(2, \mathbb{C})$-module with e_{12} as a maximal vector with weight λ, where $\lambda(h) = 2$.

Theorem 3.2.6 *Let $V = U(L)v^+$ be a standard cyclic L-module with maximal vector v^+ and weight λ. Let Φ^+ denote the set $\{\beta_1, \beta_2, \cdots, \beta_r\}$ of positive roots of Φ relative to the base $\Delta = \{\alpha_1, \alpha_2, \cdots, \alpha_l\}$. Then the following hold:*

(i) V is generated as a vector space by the set

$$\{(j_L(y_{\beta_1}))^{m_1}(j_L(y_{\beta_2}))^{m_2}\cdots(j_L(y_{\beta_r}))^{m_r}v^+ \mid m_i \in \mathbb{N}\bigcup\{0\}\}.$$

Further, $V = \oplus\sum_{\mu\leq\lambda} V_\mu$ is the direct sum of its weight spaces. The weights of V are nonzero members of the type $\lambda - \sum_{i=1}^{l} a_i\alpha_i, a_i \geq 0$. In particular, λ is the highest weight of $V = U(L)v^+$.

(ii) For each $\mu \in H^\star$, V_μ is finite dimensional and $Dim V_\lambda = 1$.
(iii) V is indecomposable with a unique maximal submodule.
(iv) All nontrivial homomorphic images and nontrivial quotients of V are also standard cyclic modules with weight λ.

Proof (i). Using the Cartan decomposition, $L = N^- \oplus B(\Delta)$, where $N^- = \oplus_{\alpha<0}L_\alpha$. From the PBW theorem and its consequence, it follows that $U(L) = U(N^-)U(B(\Delta))$. Since v^+ is a common eigenvector of $B(\Delta)$, $U(B(\Delta))v^+ = Fv^+$. Thus, $U(L)v^+ = U(N^-)Fv^+$. Again, using the PBW theorem, we see that $\{(j_L(y_{\beta_1}))^{m_1}(j_L(y_{\beta_2}))^{m_2}\cdots(j_L(y_{\beta_r}))^{m_r} \mid m_i \in \mathbb{N}\bigcup\{0\}\}$ is a basis of $U(N^-)$. This shows that V is generated as a vector space by the set

$$\{(j_L(y_{\beta_1}))^{m_1}(j_L(y_{\beta_2}))^{m_2}\cdots(j_L(y_{\beta_r}))^{m_r}v^+ \mid m_i \in \mathbb{N}\bigcup\{0\}\}.$$

The rest of the assertions in (i) follows if we observe that $y_{\beta_1}^{m_1}y_{\beta_2}^{m_2}\cdots y_{\beta_r}^{m_r}\cdot v^+$ has weight $\lambda - \sum_{k=1}^{r} m_k\beta_k$, where each β_k is a nonnegative linear combination of members of Δ.

(ii) Evidently, for each $\mu = \lambda - \sum_{i=1}^{l} a_i\alpha_i$, there are only finitely many vectors of the type $y_{\beta_1}^{m_1}y_{\beta_2}^{m_2}\cdots y_{\beta_r}^{m_r}\cdot v^+$ such that $\lambda - \sum_{k=1}^{r} m_k\beta_k = \mu$. This means that V_μ is finite dimensional and $V_\lambda = Fv^+$.

(iii) Suppose that $V = U \oplus W$ is the direct sum of L-submodules. We need to show that $v^+ \in U$ or $v^+ \in W$. Suppose not. Then $v^+ = u + w$, where $u \in U$, $w \in W$, and $\{u, w\}$ is linearly independent. Since v^+ is a common eigenvector of $B(\Delta)$, for each $x \in B(\Delta) - \{0\}$, there is a scalar $a_x \neq 0$ such that $xu + xw = xv^+ = a_xv^+ = a_xu + a_xw$. Since $\{u, w\}$ is linearly independent, $xu = a_xu$ and $xw = a_xw$. Consequently, u and w are both maximal vectors. This is a contradiction, since $Dim\ V_\lambda = 1$. This shows that V is indecomposable. Finally, the sum of weight spaces different from V_λ is the unique maximal proper submodule.

(iv) It is immediate. ♯

Corollary 3.2.7 *Let V be a standard cyclic simple L-module. If v^+ and w^+ are maximal vectors with weights λ and λ', respectively, then there is a nonzero scalar a such that $w^+ = av^+$ and $\lambda = \lambda'$.*

Proof Since V is simple, $U(L)v^+ = V = U(L)w^+$. From the part (ii) of the above theorem, $\lambda' \leq \lambda$ and $\lambda \leq \lambda'$. This means that $\lambda = \lambda'$. Since $Dim V_\lambda = 1$, there is a nonzero scalar a such that $w^+ = av^+$. ♯

Now, for each $\lambda \in H^\star$, we construct and describe the standard cyclic L-modules with the highest weight λ. We also describe the standard cyclic simple L-module with the highest weight λ which is unique up to isomorphism.

Observe that any standard cyclic L-module $U(L)v^+$ of maximal weight λ is also a $B(\Delta)$-module of which Fv^+ is a one-dimensional $B(\Delta)$-submodule. This prompts us to construct standard cyclic L-modules as follows.

Let λ be a given member of $H^\star - \{0\}$. Treat F as a vector space over F. Define the external multiplication $\cdot$ from $B(\Delta) \times F$ to F by

$$\left(h + \sum_{\alpha>0} x_\alpha\right) \cdot a = \lambda(h)a,$$

$a \in F$. It follows easily that F is a one-dimensional $B(\Delta)$-module. In turn, we have a $U(B(\Delta))$-module structure on F given by

$$j_{B(\Delta)}\left(h + \sum_{\alpha>0} x_\alpha\right) \cdot a = \lambda(h)a,$$

$a \in F$. Now, $U(L)$ is a $(U(L), U(B_\Delta))$ bi-module, since $U(B(\Delta))$ is a subalgebra of $U(L)$. This gives us the induced $U(L)$-module

$$U(L) \otimes_{U(B(\Delta))} F = U(L)(1 \otimes 1) = U(L)(1 \otimes a), a \neq 0.$$

Note that $1 \otimes 1 \neq 0$, since $U(L)$ is a free $U(B(\Delta))$-module. Further, for $\alpha > 0$,

$$j_L(x_\alpha)(1 \otimes 1) = j_L(x_\alpha) \otimes 1 = 1 \otimes j_{B(\Delta)}(x_\alpha) \cdot 1 = 1 \otimes \lambda(0)1 = 1 \otimes 0 = 0.$$

Also,

$$j_L(h)(1 \otimes 1) = j_L(h) \otimes 1 = 1 \otimes \lambda(h)1 = \lambda(h)(1 \otimes 1).$$

This shows that $1 \otimes 1$ (so also $1 \otimes a, a \neq 0$) is a maximal vector of the $U(L)$-module $U(L)(1 \otimes 1)$ with the highest weight λ. Thus, $U(L)(1 \otimes 1)$ is a standard cyclic module with the highest weight λ. We denote it by $Z(\lambda)$. The map χ from $U(L)$ to $Z(\lambda)$ defined by $\chi(u) = u \otimes 1$ is easily seen to be a surjective $U(L)$-module (so also L-module) homomorphism. The $Ker\chi$ is a proper left ideal of $U(L)$ generated by the set $\{j_L(x_\alpha) \mid \alpha > 0\} \bigcup \{j_L(h) - \lambda(h)1 \mid h \in H\}$. We denote the $Ker\chi$ by $I(\lambda)$. Thus, the module $Z(\lambda)$ is isomorphic to $U(L)/I(\lambda)$.

Proposition 3.2.8 *The module $Z(\lambda)$ is universal in the sense that for any standard cyclic L-module $V = U(L)v_0$ with maximal vector v_0 of the highest weght λ, there is a unique surjective $U(L)$-homomorphism $\overline{\psi}$ from $Z(\lambda)$ to V such that $\overline{\psi}(1 \otimes 1) = v_0$.*

Proof Consider the map ψ from $U(L) \times F$ to V given by $\psi(u, a) = auv_0$, $u \in U(L), a \in F$. Clearly, ψ is a balanced map and also $\psi(uw, a) = u \cdot \psi(w, a)$. Hence ψ induces a $U(L)$-homomorphism $\overline{\psi}$ from $Z(\lambda) = U(L) \otimes_{B(\Delta)} F$ to V which is given by $\overline{\psi}(u \otimes a) = auv_0$. Evidently, $\overline{\psi}(1 \otimes 1) = v_0$. Further, such a $\overline{\psi}$ is unique, since $Z(\lambda)$ is generated by $1 \otimes 1$ and V is generated by v_0. ♯

Corollary 3.2.9 *An L-module $V \neq \{0\}$ is a standard cyclic L-module of weight λ if and only if it is a homomorphic image of $Z(\lambda)$.*

Proof If $V \neq \{0\}$ and ϕ is an epimorphism from $Z(\lambda)$ to V, then it is easily observed that $\phi(1 \otimes 1)$ is a maximal vector of weight λ which generates V. ♯

Theorem 3.2.10 *For each $\lambda \in H^\star - \{0\}$, there is a unique (up to isomorphism) standard cyclic simple module of the highest weight λ.*

Proof $Z(\lambda) \approx U(L)/I(\lambda)$, where $I(\lambda)$ is a proper left ideal of $U(L)$ as described before. By Krull's theorem, $I(\lambda)$ is contained in a maximal left ideal $M(\lambda)$ of $U(L)$. Clearly, $V(\lambda) = U(L)/M(\lambda)$ is a simple $U(L)$-module which is a homomorphic image of $Z(\lambda)$. From the previous corollary, $V(\lambda)$ is a simple standard cyclic module of the highest weight λ.

Next, we show that any two standard cyclic simple modules of the highest weight λ are isomorphic. Let V and W be two standard cyclic simple modules of the highest weight λ. Consider the L-module $P = V \oplus W$. Let v^+ be a maximal vector of V and w^+ be that of W. The vector $x^+ = (v^+, w^+)$ of $V \oplus W$ is also a maximal vector with weight λ. Let Q be the L-submodule of P generated by x^+. The projection maps p_1 and p_2 restricted to Q are surjective homomorphisms from Q to V and from Q to W, respectively. Now, $Kerp_2 = V \times \{0\} \bigcap Q$ is an L-submodule of $V \times \{0\}$. Since V is simple, $V \times \{0\} \bigcap Q$ is $V \times \{0\}$ or it is $\{(0, 0)\}$. Suppose that $V \times \{0\} \bigcap Q$ is $V \times \{0\}$. Then $(v^+, 0) = a(v^+, w^+)$. This leads to a contradiction. Thus, p_2 is injective. Similarly, p_1 is injective. This means that $W \approx Q \approx V$. This shows that V and W are isomorphic. ♯

A natural question, now, is to look at those $\lambda \in H^\star - \{0\}$ for which the simple standard cyclic module $V(\lambda)$ is finite dimensional and then to find the weight spaces with their multiplicities.

Let V be a finite-dimensional simple L-module. Then V has a maximal vector and any two maximal vectors have the same weight λ (say). Indeed, then, V is the standard cyclic simple L-module $V(\lambda)$ as described above. Let $\alpha \in \Delta$ be a simple root and S_α the Lie subalgebra of L generated by $\{x_\alpha, y_\alpha\}$. Then S_α is a simple Lie algebra isomorphic to $sl(2, F)$. Evidently, V is also a S_α-module and has a maximal vector with maximal weight λ. Consequently (cf. Theorem 1.4.28), we have the following proposition:

Proposition 3.2.11 *Let $V = V(\lambda)$ be a simple L-module. Let μ be a weight of V and $\alpha \in \Delta$ be a root. Then $\mu(h_\alpha) = \prec \mu, \alpha \succ$ is an integer.* ♯

Thus, the weights appearing in a finite-dimensional simple L-module can also be treated as abstract weights discussed in Sect. 2.4, and the results in that section can be used in this situation also. The set of weights of an L-module V is denoted by $\Gamma(V)$, and the lattice generated by $\Gamma(V)$ is denoted by $\Lambda(V)$. Weights in this situation are also termed as integrals. A member $\lambda \in H^\star$ such that $\lambda(h_\alpha) = \prec \lambda, \alpha \succ \in \mathbb{Z}$ for each $\alpha \in \Phi$ is called an integral. Thus, weights are integrals. Further, the set Λ_{univ} of all integrals forms a lattice generated by the set $\{\lambda_1, \lambda_2, \ldots, \lambda_l\}$ (dual to the basis $\{\alpha_1^\nu = \frac{2\alpha_1}{<\alpha_1,\alpha_1>}, \alpha_2^\nu = \frac{2\alpha_2}{<\alpha_2,\alpha_2>}, \cdots, \alpha_l^\nu = \frac{2\alpha_l}{<\alpha_l,\alpha_l>}\}$ of $H^\star$) of fundamental weights (integrals). A dominant weight is also called a dominant integral. The set of dominant integrals in V is denoted by $\Lambda^+(V)$. $\Lambda(V(\lambda))$ is also denoted by $\Lambda(\lambda)$ and $\Lambda^+(V(\lambda))$ by $\Lambda^+(\lambda)$. We have already seen (see Sect. 2.4, theory of weights in Euclidean spaces) that

$$\Lambda_{univ} \supseteq \Lambda_r,$$

where Λ_r is the lattice generated by the root system Φ. The fundamental group Λ_{univ}/Λ_r is a finite group whose order is the determinant of the Cartan matrix. Further, let V be an L-module, then since $\prec \mu, \alpha \succ$ is an integer for each $\mu \in \Gamma(V)$, it follows that $\Lambda(V) \subseteq \Lambda_{univ}$. If in addition V is a faithful L-module, then for each $\alpha \in \Phi$, $L_\alpha V \neq \{0\}$. Hence, there is a $\mu \in \Gamma(V)$ such that $L_\alpha V_\mu \neq \{0\}$. This means that $\mu + \alpha \in \Gamma(V)$. In turn, $\alpha \in \Lambda(V)$. Thus, if V is a faithful L-module, then

$$\Lambda_r \subseteq \Lambda(V).$$

For example, if $L = sl(2, F)$ with $\Phi = \{\alpha, -\alpha\}$, then $\Lambda_r = \mathbb{Z}\alpha$, whereas $\Lambda_{univ} = \mathbb{Z}\lambda = \mathbb{Z}\frac{\alpha}{2}$.

Lemma 3.2.12 *Let L be a semi-simple Lie algebra, H a Cartan subalgebra of L, Φ the corresponding root system, and $\Delta = \{\alpha_1, \alpha_2, \ldots, \alpha_l\}$ a basis of Φ. Then in the universal enveloping algebra $U(L)$, the following identities are satisfied for all i, j; $1 \leq i, j \leq l$ and $k \geq 0$:*

(i) $[j_L(x_{\alpha_i}), j_L(y_{\alpha_j})^{k+1}] = 0,\ i \neq j,$
(ii) $[j_L(h_{\alpha_i}), j_L(y_{\alpha_j})^{k+1}] = -(k+1)\alpha_j(h_{\alpha_i})j_L(y_{\alpha_j})^{k+1}$, *and*
(iii) $[j_L(x_{\alpha_i}), j_L(y_{\alpha_i})^{k+1}] = -(k+1)j_L(y_{\alpha_i})^k(k \cdot 1 - j_L(h_{\alpha_i}))$, *where* 1 *represents the identity of* $U(L)$ *and* $[u, v] = uv - vu,\ u, v \in U(L)$.

Proof Let us first recall that j_L is injective Lie algebra homomorphism from L to $U(L)_L$. The proof in each case is by the induction on k.

(i) For $k = 0$, $[j_L(x_{\alpha_i}), j_L(y_{\alpha_j})] = j_L([x_{\alpha_i}, y_{\alpha_j}]) = 0$ for $i \neq j$. Assume the identity for k. Then $[j_L(x_{\alpha_i}), j_L(y_{\alpha_j})^{k+1}] = 0$. Now,

$$[j_L(x_{\alpha_i}), j_L(y_{\alpha_j})^{k+2}] =$$
$$[j_L(x_{\alpha_i}), j_L(y_{\alpha_j})^{k+1}]j_L(y_{\alpha_j}) + j_L(y_{\alpha_j})^{k+1}[j_L(x_{\alpha_i}), j_L(y_{\alpha_j})] = 0.$$

This proves (i).

(ii) For $k = 0$,

$$[j_L(h_{\alpha_i}), j_L(y_{\alpha_j})] = j_L([h_{\alpha_i}, y_{\alpha_j}]) = j_L(-\alpha_j(h_{\alpha_i})y_{\alpha_j}) = -\alpha_j(h_{\alpha_i})j_L(y_{\alpha_j}).$$

Assume the identity for k. Then

$$[j_L(h_{\alpha_i}), j_L(y_{\alpha_j})^{k+1}] = -(k+1)\alpha_j(h_{\alpha_i})j_L(y_{\alpha_j})^{k+1}.$$

Now, as in (i),

$$[j_L(h_{\alpha_i}), j_L(y_{\alpha_j})^{k+2}] =$$
$$[j_L(h_{\alpha_i}), j_L(y_{\alpha_j})^{k+1}]j_L(y_{\alpha_j}) + j_L(y_{\alpha_j})^{k+1}[j_L(h_{\alpha_i}), j_L(y_{\alpha_j})] =$$
$$-(k+1)\alpha_j(h_{\alpha_i})j_L(y_{\alpha_j})^{k+2} - \alpha_j(h_{\alpha_i})j_L(y_{\alpha_j})^{k+2} = -(k+2)\alpha_j(h_{\alpha_i})j_L(y_{\alpha_j})^{k+2}.$$

This proves (ii).

(iii) For $k = 0$,

$$[j_L(x_{\alpha_i}), j_L(y_{\alpha_i})] = j_L([x_{\alpha_i}, y_{\alpha_i}]) = j_L(h_{\alpha_i}) = -(0+1)j_L(y_{\alpha_i})^0(0 \cdot 1 - j_L(h_{\alpha_i})).$$

Thus, the identity holds for $k = 0$. Assume the result for k. Then

$$[j_L(x_{\alpha_i}), j_L(y_{\alpha_i})^{k+1}] = -(k+1)j_L(y_{\alpha_i})^k(k \cdot 1 - j_L(h_{\alpha_i})).$$

As before,

$$[j_L(x_{\alpha_i}), j_L(y_{\alpha_i})^{k+2}] =$$
$$[j_L(x_{\alpha_i}), j_L(y_{\alpha_i})^{k+1}]j_L(y_{\alpha_i}) + j_L(y_{\alpha_i})^{k+1}[j_L(x_{\alpha_i}), j_L(y_{\alpha_i})] =$$
$$-(k+1)j_L(y_{\alpha_i})^k(k \cdot 1 - j_L(h_{\alpha_i}))j_L(y_{\alpha_i}) + j_L(y_{\alpha_i})^{k+1}j_L(h_{\alpha_i}) =$$
$$-(k+2)j_L(y_{\alpha_i})^{k+1}((k+1) \cdot 1 - j_L(h_{\alpha_i})).$$

This proves (iii). ♯

The set of weights of an L-module V is denoted by $\Gamma(V)$. For a dominant integral $\lambda \in H^+$, we have L-module $V(\lambda)$. $\Gamma(V(\lambda))$ is also denoted by $\Gamma(\lambda)$.

Theorem 3.2.13 *Let L be a finite-dimensional semi-simple Lie algebra and H a Cartan subalgebra of L. Let Φ be the associated root system and $\Delta = \{\alpha_1, \alpha_2, \ldots, \alpha_l\}$ a basis of Φ. Let $\lambda \in H^\star$ be a dominant integral. Then the L-module $V(\lambda)$ is finite dimensional and the Weyl group $W(\Phi)$ permutes the set $\Gamma(\lambda)$ of weights of $V(\lambda)$. Further, $Dim\ V_\mu = Dim\ V_{\sigma(\mu)}$ for all $\sigma \in W(\Phi)$ and $\mu \in \Gamma(\lambda)$.*

Proof $V(\lambda)$ has a nonzero weight space $V(\lambda)_\lambda$ which is of dimension 1. By Proposition 3.2.1, $V(\lambda)$ is the direct sum of its weight spaces. Since all weight spaces are finite dimensional, to show that $V(\lambda)$ is finite dimensional, it is sufficient to show that the set $\Gamma(\lambda)$ of weights of $V(\lambda)$ is finite. Further, it follows from the discussions in Sect. 2.4 of Chap. 2 that for any dominant integral μ, the set $\{\nu \mid \nu \ is\ dominant\ integral\ with\ \nu < \mu\}$ is finite. Again, since $\sigma(\mu) < \mu$ for all $\sigma \in W(\Phi)$, it is sufficient to show that $W(\Phi)$ permutes $\Gamma(\lambda)$ and also $DimV(\lambda)_{\sigma(\mu)} = DimV(\lambda)_\mu$. Now, we proceed to prove this in the following four steps. Let v^+ be a maximal vector. Then for each i, $a_i = \lambda(h_{\alpha_i}) \in \mathbb{Z}$. Since λ is a dominant integral, $a_i \geq 0$ for all i.

Step 1: $j_L(y_{\alpha_j})^{a_j+1} \cdot v^+ = 0$ *for each* j: Put $w = j_L(y_{\alpha_j})^{a_j+1} \cdot v^+$. Then by Lemma 3.2.12 (i), for $i \neq j$,

$$j_L(x_{\alpha_i}) \cdot w = j_L(x_{\alpha_i}) j_L(y_{\alpha_j})^{a_j+1} \cdot v^+ = [j_L(x_{\alpha_i}), j_L(y_{\alpha_j})^{a_j+1}](v^+) = 0,$$

since $j_L(x_{\alpha_i})v^+ = 0$. Further, by Lemma 3.2.12 (ii, iii),

$$j_L(x_{\alpha_j}) j_L(y_{\alpha_j})^{a_j+1} \cdot v^+ = [j_L(x_{\alpha_j}), j_L(y_{\alpha_j})^{a_j+1}](v^+) =$$
$$-(a_j+1) j_L(y_{\alpha_j})^{a_j}(a_j v^+ - \lambda(h_{\alpha_j})v^+) = 0.$$

This implies that $j_L(x_{\alpha_j}) \cdot w = 0$. Thus, $w = 0$, for if not, it is a maximal vector in $V(\lambda)$ with weight $\lambda - (a_j+1)\alpha_j \neq \lambda$.

Step 2: *For each j, $V(\lambda)$ is the direct sum of finite dimensional S_{α_j}-modules*: Let V_0 denote the sum of all finite-dimensional S_{α_j}-modules. Let W be the subspace of V generated by the set $\{v^+, j_L(y_{\alpha_j}) \cdot v^+, \ldots, j_L(y_{\alpha_j})^{a_j} \cdot v^+\}$. It follows from Lemma 3.2.12 that W is a nonzero S_{α_j}-module. Thus, V_0 is a nonzero S_{α_j}-module. If W' is any finite-dimensional S_{α_j}-module, then $\oplus \sum_{\alpha \in \Phi} L_\alpha W'$ is also a finite-dimensional S_{α_j}-module. Thus, V_0 is an L-submodule of $V(\lambda)$. Since $V(\lambda)$ is simple, it follows that that $V_0 = V(\lambda)$.

Step 3: *For each j, $\rho(x_{\alpha_j})$ and $\rho(y_{\alpha_j})$ are locally nilpotent, where ρ is the representation afforded by the L-module structure on $V(\lambda)$*: It follows from Step 2 that every element $v \in V(\lambda)$ belongs to a finite-dimensional S_{α_j}-module W. Evidently, $\rho(x_{\alpha_j})$ and $\rho(y_{\alpha_j})$ are nilpotent on W.

Final Step: It follows from Step 3 that for each j,
$\chi_j = exp(\rho(x_{\alpha_j}))exp(\rho(-y_{\alpha_j})exp(\rho(x_{\alpha_j}))$ makes sense and it is an automorphism of $V(\lambda)$. If $\mu \in \Gamma(\lambda)$, then $V(\lambda)_\mu$ is finite dimensional, and it is contained in a finite-dimensional S_{α_j}-module. Consequently, $\chi_j(V(\lambda)_\mu) = V(\lambda)_{\sigma(\alpha_j)(\mu)}$. Since $W(\Phi)$ is generated by the simple reflections, it follows that $W(\Phi)$ permutes $\Gamma(\lambda)$ and $DimV(\lambda)_{\sigma(\mu)} = DimV(\lambda)_\mu$. ♯

Remark 3.2.14 Consider $V(\lambda)$, where $\lambda \in \Lambda^+$. Let $\mu \in \Gamma(\lambda)$ and $\alpha \in \Phi$. Then the subspace $W = \sum_{i \in \mathbb{Z}} V(\lambda)_{\mu+i\alpha}$ of $V(\lambda)$ is a S_α-module. It follows from the representation theory of $sl(2, F)$ that we have a connected α-string $\{\mu + i\alpha, -r \leq i \leq q\}$ through μ, where $r - q = \prec \mu, \alpha \succ$. Thus, $\mu \in \Gamma(\lambda)$ if and only if $\mu \leq \lambda$. In other words, $\Gamma(\lambda)$ is saturated.

Exercises

3.2.1. Describe the $sl(2, \mathbb{C})$-module $Z(\lambda)$ and the simple $sl(2, \mathbb{C})$-module $V(\lambda)$, where $\lambda(h) = 4$.

3.2.2. Describe maximal vectors and maximal weights of the natural representations of the Lie algebras $sl(l+1, \mathbb{C})$, $sp(2l, \mathbb{C})$, $o(2l+1, \mathbb{C})$, and also $o(2l, \mathbb{C})$.

3.2.3. For a fixed $m \in \mathbb{N}$, show that the set $\{\lambda \in H^\star \mid DimV(\lambda) \leq m\}$ is finite. Deduce that the set of isomorphism classes of L-modules of dimensions at the most m is finite.

3.2.4. Suppose that $Z(\lambda)$ has a maximal vector of weight μ. Show that $Z(\mu)$ is isomorphic to a submodule of $Z(\lambda)$.

3.2.5. Let L, H, Φ, and Δ be as usual. Let λ, μ be members of $H^\star$. Describe a basis of $Z(\lambda)_\mu$. Conclude that the dimension of $Z(\lambda)_\mu$ is the number of distinct sets $\{a_\alpha \geq 0 \mid \alpha > 0\}$ of nonnegative integers for which $\lambda - \mu = \sum_{\alpha>0} a_\alpha \alpha$.

3.3 Characters and Harish-Chandra's Theorem

In group representations, characters play a very important and crucial role. Indeed, two group representations over algebraically closed fields of characteristic 0 are equivalent if and only if their characters are the same (cf. Algebra 2, Corollary 9.3.9). Here, we introduce the notion of characters of representations of semi-simple Lie algebras, develop its theory, and see as to how far it classifies representations. Our aim in this section is to establish the theorem of Harish-Chandra.

Let $\lambda \in \Lambda^+$ be a dominant integral and $\mu \in H^\star$. Consider the L-module $V(\lambda)$. The $Dim V(\lambda)_\mu$ is termed as the **multiplicity** $m(\mu)$ of μ in $V(\lambda)$. If $V(\lambda)_\mu = \{0\}$, then the multiplicity of μ is 0. As usual $\mathbb{Z}(\Lambda)$ denotes the integral group ring on the group Λ. Thus, $\mathbb{Z}(\Lambda)$ consists of the set of all functions r from Λ to $\mathbb{Z}$ which are 0 at all but finitely many members of Λ. The addition is coordinate-wise and the product $r \cdot s$ is given by $(r \cdot s)(\mu) = \sum_{\rho+\eta=\mu} r(\rho)s(\eta)$. Clearly, $\mathbb{Z}(\Lambda)$ is a commutative ring with identity. The group Λ can be identified with a subgroup of the group of units of $\mathbb{Z}(\Lambda)$ through the embedding e from Λ to $\mathbb{Z}(\Lambda)$ given by $e(\lambda)(\mu) = 0$ if $\mu \neq \lambda$ and $e(\lambda)(\lambda) = 1$. $e(0)$ is the identity of $\mathbb{Z}(\Lambda)$ which is denoted by 1. Note that the additive group $(\mathbb{Z}(\Lambda), +)$ is simply the free abelian group on $e(\Lambda)$. Further, any nonzero element r of $\mathbb{Z}(\Lambda)$ is uniquely determined by a finite subset J of Λ together with the set $\{a_\lambda \in \mathbb{N} \mid \lambda \in J\}$ in the sense that $r = \sum_{\lambda\in J} a_\lambda e(\lambda)$. In general, we write an element in $\mathbb{Z}(\Lambda)$ as $\sum_{\lambda\in\Lambda} a_\lambda e(\lambda)$, where it is understood that it is a finite sum. Evidently, $W(\Phi)$ acts as automorphisms of $\mathbb{Z}(\Lambda)$. Indeed, if $r = \sum_{\lambda\in\Lambda} a_\lambda e(\lambda)$, then $\sigma(r) = \sum_{\lambda\in\Lambda} a_\lambda e(\sigma(\lambda))$.

For each dominant integral λ, we have a unique member $Ch_{V(\lambda)}$ of $\mathbb{Z}(\Lambda)$ given by

$$Ch_{V(\lambda)} = \sum_{\mu\in\Gamma(\lambda)} m_\lambda(\mu)e(\mu).$$

This member $Ch_{V(\lambda)}$ is called the **formal character** of $V(\lambda)$. In short, we denote $Ch_{V(\lambda)}$ by Ch_λ. Thus, if $L = sl(2, F)$ and $\Delta = \{\alpha\}$, where $\alpha(h) = 2$, then for any $\lambda \in \Lambda^+$,

$$Ch_\lambda = e(\lambda) + e(\lambda - \alpha) + \cdots + e(\lambda - m\alpha),$$

where $m = \prec \lambda, \alpha \succ$.

Next, by Weyl's theorem, any finite-dimensional L-module V has a unique decomposition

$$V = V(\lambda_1) \oplus V(\lambda_2) \oplus \cdots \oplus V(\lambda_r),\ \lambda_i \in \Lambda^+$$

as simple modules. The **formal character** Ch_V of V is defined by

$$Ch_V = Ch_{\lambda_1} + Ch_{\lambda_2} + \cdots + Ch_{\lambda_r}.$$

Since $W(\Phi)$ acts on the weight spaces of simple modules, Ch_λ is invariant under $W(\Phi)$ for each $\lambda \in \Lambda^+$.

Proposition 3.3.1 *Let $r = \sum_{\lambda\in\Lambda} r(\lambda)e(\lambda)$ be a member of $\mathbb{Z}(\Lambda)$ such that $\sigma(r) = r$ for each $\sigma \in W(\Phi)$. Then there is a unique finite subset J of Λ^+ and set $\{a(\lambda) \in \mathbb{N} \mid \lambda \in J\}$ such that $r = \sum_{\lambda\in J} a(\lambda)Ch_\lambda$.*

Proof It follows from the hypothesis that

$$r = \sum_{\lambda\in\Lambda} r(\lambda)e(\lambda) = \sum_{\lambda\in\Lambda^+} r(\lambda) \sum_{\sigma\in W(\Phi)} e(\sigma(\lambda)).$$

Let A_r denote the set $\{\lambda \in \Lambda^+ \mid r(\lambda) \neq 0\}$. Consider the set $M(r) = \bigcup_{\lambda\in A_r}\{\mu \in \Lambda^+ \mid \mu < \lambda\}$. Evidently, $M(r)$ is finite (Lemma 2.4.34). The proof is by the induction on the cardinality $\mid M(r) \mid$ of $M(r)$. Suppose that $\mid M(r) \mid = 1$. Then there is only one dominant weight λ appearing in the expression for r. Consequently,

$$r = r(\lambda) \sum_{\sigma\in W(\Phi)} e(\sigma(\lambda)) = r(\lambda)Ch_\lambda.$$

Thus, the result holds when $\mid M(r) \mid = 1$. Assume the result for all $s \in \mathbb{Z}(\Lambda)$ satisfying the hypothesis for which $\mid M(s) \mid < \mid M(r) \mid$, where $r = \sum_{\lambda\in\Lambda} r(\lambda)e(\lambda)$ which satisfies the hypothesis. Let $\lambda \in \Lambda^+$ be the largest such that $r(\lambda) \neq 0$. Then $r' = r - r(\lambda)Ch_\lambda$ is such that $\sigma(r') = r'$ for all $\sigma \in W(\Phi)$ and $M(r') < M(r)$. By the induction hypothesis, there is a unique finite subset J' of Λ^+ and the set $\{a(\mu) \in \mathbb{N} \mid \mu \in J'\}$ such that $r' = \sum_{\mu\in J'} a(\mu)Ch_\mu$. Clearly,

$$r = \sum_{\mu\in J'} a(\mu)Ch_\mu + r(\lambda)Ch_\lambda.$$

The rest of the assertion is easily observed by using the fact that $\mathbb{Z}(\Lambda)$ is a free abelian group on Λ. ♯

Corollary 3.3.2 *$Ch_V = Ch_W$ if and only if $V \approx W$.* ♯

Corollary 3.3.3 *$Ch_{V\oplus W} = Ch_V + Ch_W$ and $Ch_{V\otimes W} = Ch_V \star Ch_W$, where $\star$ is the convolution product.*

Proof The first part is evident. For the second, observe that the weights of $V \otimes W$ are of the form $\lambda + \mu$, where λ is a weight of V and μ is a weight of W. Clearly, the multiplicity $m(\lambda + \mu)$ of $\lambda + \mu$ in $V \otimes W$ is $m(\lambda)m(\mu)$. ♯

The tensor algebra

$$T(V) = F \oplus V \oplus \otimes^2 V \oplus \cdots \oplus \otimes^n V \oplus \cdots$$

of a vector space V, where $\otimes^n V = \underbrace{V \otimes V \otimes \cdots \otimes V}_{n}$. The product in $T(V)$ is obtained by extending linearly the partial product given by

$$(x_{i_1} \otimes x_{i_2} \otimes \cdots \otimes x_{i_r}) \cdot (y_{j_1} \otimes y_{j_2} \otimes \cdots \otimes y_{j_s}) =$$
$$x_{i_1} \otimes x_{i_2} \otimes \cdots \otimes x_{i_r} \otimes y_{j_1} \otimes y_{j_2} \otimes \cdots \otimes y_{j_s}.$$

Indeed, if V is a finite-dimensional vector space with $\{e_1, e_2, \ldots, e_n\}$ as a basis, then $T(V)$ is isomorphic to the polynomial algebra over F in non-commuting set $\{x_1, x_2, \ldots, x_n\}$ of variables which is in bijective correspondence with a basis of V. Further, the symmetric algebra $S(V)$ of V is $T(V)/A$, where A is the ideal of $T(V)$ generated by $\{x \otimes y - y \otimes x \mid x, y \in V\}$. Consequently, the symmetric algebra is isomorphic to the polynomial algebra $F[x_1, x_2, \ldots, x_n]$ in commuting $\{x_1, x_2, \ldots, x_n\}$ variables. The following lemma will be used in our discussion.

Lemma 3.3.4 *Any polynomial in $F[x_1, x_2, \ldots, x_n]$ is expressible as a linear combination of powers of linear polynomials. Indeed, if $f(x_1, x_2, \ldots, x_n)$ is of degree d, then it can be expressed as a linear combination of the set $\{(a_1x_1 + a_2x_2 + \cdots + a_nx_n)^t \mid a_i \in \mathbb{Z}, t \leq d\}$.*

Proof It is sufficient to show that each monomial $x_1^{i_1} x_2^{i_2} \cdots x_n^{i_n}$ is a linear combination of powers of linear polynomials. The proof is by the induction on n. If $n = 1$, then there is nothing to do. Assume that the result is true for n. Consider $x_1^{i_1} x_2^{i_2} \cdots x_{n+1}^{i_{n+1}}$. If $i_1 = 0$, then the result follows by the induction hypothesis. Suppose that $i_1 \neq 0$. By the induction hypothesis, $x_2^{i_2} x_3^{i_3} \cdots x_{n+1}^{i_{n+1}}$ is a linear combination of powers of integral linear polynomials in $x_2, x_3, \ldots, x_{n+1}$. It is sufficient, therefore, to show that $x_1^{i_1} g(x_2, x_3, \cdots x_{n+1})^{j_1}$ is a linear combination of powers of integral linear combination of x_1 and $g(x_2, x_3, \ldots, x_{n+1})$. Put $y_1 = g(x_2, x_3, \ldots, x_{n+1})$. Suppose that $i_1 + j_1 = m$. Let W be the space of polynomials in x_1 and y_1 of the degree at the most m. Then the dimension of W is $m + 1$. Let $a_1, a_2, \ldots, a_m$ be distinct integers in F (characteristic F is 0). Consider the $(m + 1) \times (m + 1)$ Vandermonde matrix A whose first row entries are 1, and the (r+1)th row is $(x_1 + a_r y_1)^r$. Evidently, A is non-singular and $\{Ae_1, Ae_2, \ldots, Ae_{m+1}\}$ generates W. The result follows, since Ae_i are powers of integral linear polynomials. ♯

Let L be a semi-simple Lie algebra, H a Cartan subalgebra, and Φ the associated root system with $\Delta = \{\alpha_1, \alpha_2, \ldots, \alpha_l\}$ a basis of Φ. It follows from the above lemma that $S(H^\star)$ is the linear span of the set $\{\lambda^k \mid \lambda \in \Lambda, k \in \mathbb{N} \bigcup \{0\}\}$, where Λ is the group of lattices in $H^\star$ generated by Δ. Thus, the action of $W(\Phi)$ on $H^\star$ can be extended to $S(H^\star)$ in a natural manner. The subalgebra $S(H^\star)^{W(\Phi)} = \{f \in S(H^\star) \mid \sigma(f) = f \; \forall \, \sigma \in W(\Phi)\}$ of $S(H^\star)$ is called the the algebra of $W(\Phi)$-invariant polynomials on H.

Example 3.3.5 Consider $L = sl(2, F)$. Then $H = < h >, H^{\star} = < \alpha >, S(H^{\star}) = F[\lambda]$, where $\lambda = \frac{\alpha}{2}$ is the fundamental dominant weight. Since $\sigma_{\alpha}(\lambda) = -\lambda$, it follows that $S(H^{\star})^{W(\Phi)} = F[\lambda^2]$. More generally, since all $\lambda \in \Lambda$ are conjugate to a member of Λ^+, it follows that $\bigcup_{\lambda \in \Lambda^+}(\bigcup_{k \geq 0} W(\Phi)(\lambda^k))$ generates $S(H^{\star})^{W(\Phi)}$ as a vector space.

Next, consider L. If $\rho \in GL(L)$ and $f \in S(L)$, then the map $\rho \cdot f$ given by $(\rho \cdot f)(x) = f(\rho^{-1}(x))$ is a member of $S(L)$. Thus, $GL(L)$ acts on $S(L)$. In particular, the subgroup G of $GL(L)$ generated by the set $\{exp(ad(x)) \mid ad(x)\ is\ nilpotent\}$ also acts on $S(L)$. We denote by $S(L)^G$ the subalgebra $\{f \in S(L) \mid \rho \cdot f = f \ \forall \rho \in G\}$ of $S(L)$ of G-invariant polynomial functions on L. To have examples of G-invariant polynomial functions on L, start with a simple L-module $V(\lambda)$ affording a finite-dimensional irreducible representation ρ of L on $V(\lambda)$ with highest weight $\lambda \in \Lambda^+$. Let z be a member of $N^+ = \oplus \sum_{\alpha \in \Phi^+} L_{\alpha}$. Then $ad(z)$ is nilpotent. Put $\eta = exp(ad(z))$. We can easily check that the map ρ^{η} from L to $gl(V(\lambda))$ given by

$$\rho^{\eta}(x) = \rho\left(\eta(x)\right) = \rho\left((1 + ad(z) + \frac{ad(z)^2}{2!} + \cdots\right)(x)\right)$$

is a representation which is irreducible. Let v^+ be a maximal vector for the representation ρ and $\beta \in \Phi^+$. Since $\eta(x_{\beta})$ is still a member of N^+, $\rho^{\eta}(x_{\beta})(v^+) = 0$. For each $h \in H$, $[z, h] \in N^+$ and so $\rho([z, h])(v^+) = 0$. Thus, if $h \in H$, then $\rho^{\eta}(h)(v^+) = \rho(h + [z, h])(v^+) = \lambda(h)(v^+)$. This shows that v^+ is also a maximal vector for the representation ρ^{η} with weight λ. From Theorem 3.2.10, the representations ρ and ρ^{η} are equivalent. Hence there exists a vector space automorphism χ of V such that $\chi(\rho(x)(v)) = \rho^{\eta}(x)(\chi(v))$ for all $x \in L$ and $v \in V$. It follows that the maps $x \mapsto Tr(\rho(x))$ and, in turn, $x \mapsto Tr(\rho(x)^r)$ are polynomial functions which are invariant under $exp(ad(z))$ for all r and for all $z \in N^+$. Since any subalgebra of L consisting of nilpotent elements is conjugate to a subalgebra of $N(\Delta)$ under $Inn_0(L)$ (cf. Exercise 2.4.9), it follows that the polynomial function $x \mapsto Tr(\rho(x)^r)$ is G-invariant. Such polynomial functions are called **Trace Polynomial functions** associated with the representation ρ.

Our next aim is to discuss the relationship between $S(L)^G$ and $S(H^{\star})^{W(\Phi)}$. Indeed, they are isomorphic.

Consider the Killing form κ on L which is a nondegenerate symmetric form; its restriction on the Cartan subalgebra H is also nondegenerate. Let $\{h_1, h_2, \cdots, h_l\}$ be a basis of H. For each $\alpha \in \Phi$, select a nonzero element x_{α} of the root space L_{α}. Then $\{h_i, 1 \leq i \leq l\} \bigcup \{x_{\alpha} \mid \alpha \in \Phi\}$ is a basis of L. Let $\{k_1, k_2, \ldots, k_l\}$ be the basis of H which is dual to $\{h_1, h_2, \cdots, h_l\}$ with respect to $\kappa|_H$. Further, for each $\alpha \in \Phi$, we have a unique $z_{\alpha} \in L_{-\alpha}$ such that $\kappa(x_{\alpha}, z_{\alpha}) = 1$. Evidently, $\kappa(h, z_{\alpha}) = 0$ for all $h \in H$ and $\alpha \in \Phi$. Consequently, $B = \{k_i \mid 1 \leq i \leq l\} \bigcup \{z_{\alpha} \mid \alpha \in \Phi\}$ is a basis of L which is dual to the basis $\{h_i, 1 \leq i \leq l\} \bigcup \{x_{\alpha} \mid \alpha \in \Phi\}$ of L. A member $f \in S(L)$ is a polynomial function of B. Further, B can be treated as a basis of $L^{\star}$. As such $f|_H$ can also be viewed as a member of $S(H^{\star})$. Thus, $f \mapsto f|_H$ defines a map from $S(L)$ to $S(H^{\star})$. Further, if $f \in S(L)^G$, then $exp(ad(x_{\alpha}))(f) = f$ for all $\alpha \in \Phi$. Since

$\sigma_\alpha = exp(ad(x_\alpha)exp(-ad(x_{-\alpha}))exp(ad(x_\alpha))$, it follows that $\sigma_\alpha(f|_H) = f|_H$. This means that $f|_H \in S(H^\star)^{W(\Phi)}$. In turn, we obtain an algebra homomorphism χ from $S(L)^G$ to $S(H^\star)^{W(\Phi)}$ given by $\chi(f) = f|_H$.

Proposition 3.3.6 (Chevalley) *χ is an epimorphism.*

Proof For each $\gamma \in S(H^\star)$, let $Sym(\gamma)$ denote the sum $\sum_{\sigma \in W(\Phi)} \sigma(\gamma)$. Evidently, $Sym(\gamma) \in S(H^\star)^{W(\Phi)}$ and $\{Sym(\lambda^k) \mid \lambda \in \Lambda^+, k \geq 0\}$ is a generating set of $S(H^\star)^{W(\Phi)}$ (Lemma 3.3.4). It is sufficient, therefore, to show that $Sym(\lambda^k) \in Image\chi$ for each $\lambda \in \Lambda^+$ and $k \geq 0$. Note that for each $\lambda \in \Lambda^+$, the set $\{\mu \in \Lambda^+ \mid \mu \leq \lambda\}$ is finite. We prove the result by the induction on $\mid \{\mu \in \Lambda^+ \mid \mu \leq \lambda\} \mid$. Suppose that $\mid \{\mu \in \Lambda^+ \mid \mu \leq \lambda\} \mid = 1$. Then (Theorem 3.2.13), each weight of $V(\lambda)$ is $W(\Phi)$-conjugate to λ and is of multiplicity 1. Thus, the map ϕ defined by $\phi(x) = Tr(\rho(x)^k)$ is a G-invariant polynomial function on L such that $\chi(\phi) = Sym(\lambda^k)$ (here ρ is the representation afforded by the L-module $V(\lambda)$). Assume that $Sym(\mu)^k \in Image\chi$ whenever $\mu \in \Lambda^+$ such that $\mid \{\nu \in \Lambda^+ \mid \nu \leq \mu\} \mid < \mid \{\eta \in \Lambda^+ \mid \eta \leq \lambda\} \mid$. Let ρ be the representation associated with the L-module $V(\lambda)$ and $k \geq 0$. Let f be a member of $S(L)^G$ given by $f(x) = Tr(\rho(x)^k)$. Clearly,

$$f|_H = Sym(\lambda^k) + \sum_{\mu \in \Lambda^+, \mu < \lambda} m_\lambda Sym(\mu^k).$$

By the induction hypothesis, $Sym(\mu^k) \in Image\chi$. Also $f|_H \in S(H^\star)^{W(\Phi)}$. Hence $Sym(\lambda^k) \in Image\chi$. ♯

Using some basic algebraic geometry, we establish the following theorem:

Theorem 3.3.7 *The map χ is an isomorphism from $S(L)^G$ to $S(H^\star)^{W(\Phi)}$.*

Proof We have already seen that χ is an epimorphism. We need to show that it is a monomorphism. Identify L with the affine space $A_n(F) = F^n$ with the help of the basis $\{k_1, k_2, \ldots, k_l\} \bigcup \{z_\alpha \mid \alpha \in \Phi\}$ of L as described before, where $n = DimL = l + \mid \Phi \mid$. We have polynomial maps $\{p_1, p_2, \ldots, p_n\}$ in $S(L)$ such that for each $x \in L$, the polynomial

$$p_x(X) = \sum_{i=1}^{n} p_i(x) X^i$$

in $F[X]$ is the characteristic polynomial of $\rho(x)$. The smallest m such that the polynomial function $p_m \neq 0$ is called the ρ-**rank** of L. An element $x \in L$ is called ρ-regular if $p_n(x) \neq 0$. Since $\rho(x)$ and $\rho(x_s)$ have the same characteristic polynomial, x is ρ-regular if and only if x_s is ρ-regular. Let R denote the set of ρ-regular elements in $L \approx A_n(F)$. Then $R \neq \emptyset$ and it, being a complement of a zero set of a set of polynomials, is an open subset of $A_n(F)$ with respect to the Zariski topology. Since every nonempty open subset in Zariski topology is dense, R is dense. Since any semi-simple element is contained in a Cartan subalgebra and any two Cartan subalgebras are conjugate under G, any semi-simple element is G-conjugate to an element of H. If $h \in H$, then $C_L(h) \supseteq H$ and so $DimC_L(h) \geq l = rankL$. By Corollary 2.4.12, there is a semi-simple element h in H such that $C_L(h) = H$. Such an element is called a regular semi-simple element of L. As observed earlier, ρ-regular semi-simple elements exist and they are precisely regular semi-simple elements with $m = rankL = l$. Since only a nilpotent element which commutes with a regular semi-simple element is 0, it follows that R is precisely the set of regular semi-simple elements. Consequently, $R \subseteq H$. Now, if $f \in S(L)^G$ and $\chi(f) = f|_H = 0$, then $f(R) = \{0\}$. Since R is dense, $f = 0$. This proves that χ is injective. ♯

Corollary 3.3.8 *$S(L)^G$ is generated by the set of trace polynomials.*

Proof Indeed, the proof of Proposition 3.3.6 asserts that χ restricted to the subalgebra of $S(L)^G$ generated by trace polynomials is surjective. The result is immediate. ♯

Universal Casimir Element

Recall (Definition 1.4.15) the concept of the Casimir element c_ρ of a faithful representation ρ of a semi-simple Lie algebra L on a vector space V relative to a basis of L. Consider the adjoint representation ad of L. Since L is semi-simple, ad is faithful. The Casimir element c_{ad} of ad with respect to the basis $\{h_i \mid 1 \leq i \leq l\} \bigcup \{x_\alpha \mid \alpha \in \Phi\}$ is given by

$$c_{ad} = \sum_{i=1}^{l} ad(h_i)ad(k_i) + \sum_{\alpha \in \Phi} ad(x_\alpha)ad(z_\alpha),$$

where $\{k_i \mid 1 \leq i \leq l\} \bigcup \{z_\alpha \mid \alpha \in \Phi\}$ is the dual basis of $\{h_i \mid 1 \leq i \leq l\} \bigcup \{x_\alpha \mid \alpha \in \Phi\}$ as described earlier. From the universal property of universal enveloping algebra, we have a unique associative algebra homomorphism $\overline{ad}$ from $U(L)$ to $gl(L)$ such that $\overline{ad} o j_L = ad$. The element

$$c_{univ} = \sum_{i=1}^{l} j_L(h_i) j_L(k_i) + \sum_{\alpha \in \Phi} j_L(x_\alpha) j_L(z_\alpha)$$

of $U(L)$ will be termed as the **Universal Casimir** element of L. Evidently, $\overline{ad}(c_{univ}) = c_{ad}$.

The following lemma, which is essentially Exercise 1.4.7, will be used to relate $\overline{\rho}(c_{univ})$ with c_ρ, where ρ is a faithful representation of L on V, and $\overline{\rho}$ is the unique homomorphism from $U(L)$ to $gl(V)$ such that $\overline{\rho} o j_L = \rho$.

Lemma 3.3.9 *Let f and g be two nondegenerate symmetric and associative bi-linear forms on a simple Lie algebra L. Then there is a nonzero scalar a such that $f = ag$. Consequently, every nondegenerate symmetric and associative bi-linear form on a simple Lie algebra is $a\kappa$ for some $a \neq 0$.*

Proof We have two vector space isomorphisms χ_f and χ_g from L to $L^\star$ given by $\chi_f(y)(x) = f(x, y)$ and $\chi_g(y)(x) = g(x, y)$. Recall that $L^\star$ is also an L-module with respect to the product $\cdot$ given by $(x \cdot \eta)(y) = -\eta([x, y])$, $\eta \in L^\star$. Using the associativity of f and g, we can easily verify that χ_f and χ_g are module isomorphisms. Consequently, $\chi_f \chi_g^{-1}$ is an automorphism of L. Since L is simple, $\chi_f \chi_g^{-1}$ is the multiplication by a nonzero scalar a. Hence $\chi_f = a\chi_g$ for some nonzero scalar a. The rest is evident. ♯

Corollary 3.3.10 *Let L be a simple Lie algebra and ρ a nontrivial representation of L on V. Then ρ is faithful and* $\overline{\rho}(c_{univ}) = ac_\rho$ *for some nonzero scalar a.*

Proof Since L is simple, ρ is an injective Lie homomorphism from L to $gl(V)$. Thus, ρ is faithful. It follows from the above lemma that the Killing form $\kappa = a\beta_\rho$ for some nonzero scalar a. Consequently, $\overline{\rho}(c_{univ}) = ac_\rho$. ♯

More generally, if L is semi-simple Lie algebra, then we have distinct simple ideals $L_1, L_2, \cdots, L_r$ of L such that $L = L_1 \oplus L_2 \oplus \cdots \oplus L_r$ and $\kappa(L_i, L_j) = \{0\}$ for all $i \neq j$. Further, $\kappa|_{L_i} = \kappa_i$ is the Killing form of L_i. For each i, let B_i and C_i be bases of L_i which are dual to each other with respect to κ_i. Then $B = \bigcup_{i=1}^r B_i$ and $C = \bigcup_{i=1}^r C_i$ are the bases of L which are dual to each other with respect to κ. Consequently, with respect to these choices of bases,

$$c_{univ} = c_{univ}^1 + c_{univ}^2 + \cdots + c_{univ}^r,$$

where c_{univ}^i denotes the universal Casimir element of L_i with respect to the basis B_i. We have scalars a_i such that $\overline{\rho|_{L_i}}(c_{univ}^i) = a_i c_{\rho|_{L_i}}^i$ and $\overline{\rho}(c_{univ}) = \sum_{i=1}^r a_i c_{\rho|_{L_i}}^i$. Evidently, $\overline{\rho}(c_{univ})$ commutes with each element of $\rho(L)$.

Corollary 3.3.11 *Let L be a simple Lie algebra. Then* $\overline{ad}(c_{univ}) = I_L$.

Proof If L is a simple Lie algebra, then ad is an irreducible representation. Consequently, c_{ad} is the multiplication by $DimL/DimL = 1$ (Corollary 1.4.19). The result follows from the universal property of $U(L)$. ♯

Theorem 3.3.12 *A universal Casimir element* c_{univ} *of a semi-simple Lie algebra L is a member of the center* $Z(U(L))$ *of the universal enveloping algebra* $U(L)$ *of L.*

Proof Let c_{univ} be the universal Casimir element of L with respect to a basis $\{x_1, x_2, \ldots, x_n\}$ of L. Then by the definition

$$c_{univ} = \sum_{i=1}^{n} j_L(x_i) j_L(y_i),$$

where $\{y_1, y_2, \ldots, y_n\}$ is the basis of L which is dual to $\{x_1, x_2, \ldots, x_n\}$ with respect to $\beta_{ad} = \kappa$. Since $U(L)$ is generated by $j_L(L)$, it is sufficient to show that $c_{univ} j_L(x) = j_L(x) c_{univ}$ for all $x \in L$. We imitate the proof of Proposition 1.4.17. Suppose that $[x, x_i] = \sum_{k=1}^n a_{ik} x_k$ and $[x, y_i] = \sum_{k=1}^n b_{ik} x_k$. Then

$[j_L(x), j_L(x_i)] = \sum_{k=1}^{n} a_{ik} j_L(x_k)$ and $[j_L(x), j_L(y_i)] = \sum_{k=1}^{n} b_{ik} j_L(y_k)$. Using the associativity of $\beta_{ad} = \kappa$,

$$a_{il} = \sum_{k=1}^{n} a_{ik}\kappa(x_k, y_l) = \kappa([x, x_i], y_l) =$$

$$\kappa(-[x_i, x], y_l) = \kappa(x_i, -[x, y_l]) = -\sum_{p=1}^{n} b_{lp}\kappa(x_i, y_p) = -b_{li}.$$

Now, using the identity $[u, vw] = [u, v]w + v[u, w]$ in $U(L)$ (indeed, in any associative algebra) and the fact that $a_{il} = -b_{li}$ for all i, l, we see that

$$[j_L(x), c_{univ}] = [j_L(x), \sum_{i=1}^{n} j_L(x_i) j_L(y_i)] = \sum_{i=1}^{n}[j_L(x), j_L(x_i) j_L(y_i)] =$$

$$\sum_{i=1}^{n}[j_L(x), j_L(x_i)] j_L(y_i) + \sum_{i=1}^{n} j_L(x_i)[j_L(x), j_L(y_i)] = 0.$$

This shows that c_{univ} belongs to $Z(U(L))$. ♯

We have already observed that a simple standard cyclic module $V(\lambda)$ with maximal weight λ is uniquely determined by its formal character Ch_λ. To determine the formal character Ch_λ, we need to determine the multiplicities $m_\lambda(\mu)$ for each $\mu \in \Gamma(\lambda)$ (the set of weights of $V(\lambda)$). For this purpose, we shall describe the formulas of Weyl, Kostant, and Steinberg in the next section. Here in this section, we establish the theorem of Harish-Chandra about the characters (to be introduced) of infinite-dimensional standard cyclic modules $Z(\lambda)$.

Let L be a finite-dimensional semi-simple Lie algebra. Consider the subgroup $G = Inn(L)$ of the group $Aut(L)$ of automorphisms of L generated by the set $\{exp(ad(x) \mid x \textit{ is ad nilpotent element of } L\}$. From the universal property of the universal enveloping algebra $U(L)$, for each ad-nilpotent element x of L, $exp(ad(x))$ determines a unique automorphism of $U(L)$ denoted again by $exp(ad(x))$ such that

$$exp(ad(x)) o j_L = j_L o exp(ad(x))$$

for each ad-nilpotent $x \in L$. More explicitly,

$$exp(ad(x))(j_L(y)) = j_L(exp(ad(x))(y)) =$$

$$j_L(y + ad(x)(y) + \frac{1}{2!}(ad(x))^2(y) + \cdots\cdots) = exp(ad(j_L(x))(j_L(y)))$$

for each ad-nilpotent element $x \in L$ and for each element $y \in L$. Since $\{j_L(y) \mid y \in L\}$ generates $U(L)$,

$$exp(ad(x))(u) = exp(ad(j_L(x)))(u)$$

for all $u \in U(L)$. Note that $ad(w)(u) = wu - uw$ for all $u, w \in U(L)$ and

$$exp(ad(j_L(x)))(u) = u + ad(j_L(x))(u) + \frac{1}{2!}(ad(j_L(x)))^2(u) + \cdots.$$

Thus, G can also be viewed as a group of automorphisms of $U(L)$ as described above and as such it acts on $U(L)$ as a group of automorphisms. Since $Z(U(L))$ is a characteristic subalgebra of $U(L)$, $Z(U(L))$ is invariant under the action of G. Indeed, we have the following lemma:

Lemma 3.3.13 *The center $Z(U(L))$ is precisely the subalgebra of G-invariants of $U(L)$. More precisely, $Z(U(L)) = \{u \in U(L) \mid \sigma(u) = u \ \forall \sigma \in G\}$.*

Proof Let $u \in Z(U(L))$. Then u commutes with $j_L(x)$ for all $x \in L$. In particular, $ad(j_L(x))(u) = 0$ for all $x \in L$. Hence, $exp(ad(j_L(x)))(u) = u$ for all ad-nilpotent elements x of L. Since G is generated by the set $\{exp(ad(x)) \mid x \ is \ ad \ nilpotent \ element \ of \ L\}$, it follows that $\sigma(u) = u$ for all $\sigma \in G$.

Conversely, let u be a nonzero member of $U(L)$ such that $\sigma(u) = u$ for all $\sigma \in G$. Then $exp(adj_L(x))(u) = u$ for all ad-nilpotent elements x of L. Since u is a finite linear combination of elements of the types $j_L(x_1)j_L(x_2)\cdots j_L(x_r)$ and $ad(j_L(x))(j_L(x_1)j_L(x_2)\cdots j_L(x_r)) = \sum_{i=1}^{r} j_L(x_1)j_L(x_2)\cdots ad(j_L(x))(j_L(x_i))\cdots j_L(x_r)$, it follows that u belongs to a finite-dimensional L-submodule W of $U(L)$. For $\alpha \in \Phi$, let x_α be a nonzero member of L_α. Then $ad(x_\alpha)$ is nilpotent and so $z = ad(j_L(x_\alpha))$ is nilpotent on W. Let m be the largest integer such that $z^m \neq 0$ but $z^{m+1} = 0$. Let us take $m + 1$ distinct members $a_1, a_2, \ldots a_{m+1}$ of F (F is infinite). By our hypothesis,

$$(exp(a_i z))(u) = \left(1 + a_i z + \frac{1}{2!}(a_i z)^2 + \cdots + \frac{1}{m!}(a_i z)^m\right)(u) = u$$

for all i. Consider the $(m + 1) \times (m + 1)$ matrix A given by

$$A = \begin{bmatrix} 1 & 1 & \cdots\cdot & 1 \\ a_1 & a_2 & \cdots\cdot & a_{m+1} \\ \frac{a_1^2}{2!} & \frac{a_2^2}{2!} & \cdots\cdot & \frac{a_{m+1}^2}{2!} \\ \cdot\cdot & \cdot & \cdots\cdot & \\ \cdot & \cdot & \cdots\cdot & \cdot \\ \cdot & \cdot & \cdots\cdot & \cdot \\ \frac{a_1^m}{m!} & \frac{a_2^m}{m!} & \cdots\cdot & \frac{a_{m+1}^m}{m!} \end{bmatrix}.$$

Evidently,

$$Det A = \frac{\prod_{i<j}(a_i - a_j)}{2!3!\cdots m!} \neq 0.$$

Hence A is non-singular. Thus, the rows and columns of A are linearly independent. It follows that z belongs to the subspace of $gl(U(L))$ generated by the set $\{exp(ad(a_i z)) \mid i \leq m\}$. We have scalars b_i, $1 \leq i \leq m$ such that

$$z = \sum_{i=}^{m} b_i exp(a_i z).$$

Consequently, $z(u)=ad(j_L(x_\alpha))(u)=\sum_{i=1}^{m} b_i exp(ada_i j_L(x_\alpha))(u) = (\sum_{i=1}^{m} b_i)u$. Again, since $z = ad(j_L(x_\alpha))$ is nilpotent, $\sum_{i=1}^{m} b_i = 0$. Hence $[j_L(x_\alpha), u] = 0$ for each α. Since $\{j_L(x_\alpha) \mid \alpha \in \Phi\}$ generates $U(L)$, it follows that $u \in Z(U(L))$. ♯

Consider the infinite-dimensional standard cyclic $U(L)$ (also L)-module $Z(\lambda) = U(L)v^+$ with maximal vector v^+ of weight $\lambda \in H^\star$. The center $Z(U(L))$ also acts on $Z(\lambda)$. Let $z \in Z(U(L))$. Then for each $h \in H$, $j_L(h)zv^+ = zj_L(h)v^+ = \lambda(h)zv^+$ and $j_L(x_\alpha)zv^+ = zj_L(x_\alpha)v^+ = 0$ for all $x_\alpha \in L_\alpha$, where $\alpha > 0$. This means that zv^+ is another maximal vector of $Z(\lambda)$ of weight λ. Hence, there is a unique scalar $\chi_\lambda(z)$ such that $zv^+ = \chi_\lambda(z)v^+$. Let A_z denote the subset $\{v \in Z(\lambda) \mid zv = \chi_\lambda(z)v\}$. Since $z \in Z(U(L))$, A_z is a $U(L)$-submodule of $Z(\lambda)$ which contains v^+. This means that $A_z = Z(\lambda)$ for all $z \in Z(U(L))$. Thus, $Z(U(L))$ acts on $Z(\lambda)$ by the multiplication by scalars. This defines a map χ_λ from $Z(U(L))$ to F given by $zv = \chi_\lambda(z)v$, $v \in Z(\lambda)$. It can be easily observed that χ_λ is algebra homomorphism. The homomorphism χ_λ is called the **character** of the infinite-dimensional representation of L afforded by the standard cyclic module $Z(\lambda)$.

Unlike in the representation theory of groups, χ_λ may be the same as χ_μ even if $\lambda \neq \mu$.

Proposition 3.3.14 *Let $\lambda \in H^\star - \{0\}$. Let v_0 be a maximal vector of $Z(\lambda)$ of weight μ. Then $\chi_\lambda = \chi_\mu$.*

Proof Consider the submodule $U(L)v_0$ of $Z(\lambda)$. Clearly, $U(L)v_0$ is a standard cyclic $U(L)$-module of the highest weight μ. By Corollary 3.2.9 (see also Exercise 3.2.4), we have a surjective $U(L)$-homomorphism ϕ from $Z(\mu)$ to $U(L)v_0$ such that $\phi(v^+) = v_0$, where v^+ is a maximal vector of $Z(\mu)$ of the highest weight μ. Let $z \in Z(U(L))$. Then $zv^+ = \chi_\mu(z)v^+$. Hence

$$\chi_\mu(z)\phi(v^+) = \phi(\chi_\mu(z)v^+) = \phi(zv^+) = z\phi(v^+) = zv_0 = \chi_\lambda(z)v_0 = \chi_\lambda(z)\phi(v^+).$$

This shows that $\chi_\mu(z) = \chi_\lambda(z)$. Since z is an arbitrary member of $Z(U(L))$, $\chi_\lambda = \chi_\mu$. ♯

Corollary 3.3.15 *Let λ be a member of $\Lambda(H^\star)$. Let α be a member of Δ such that $m = \prec \lambda, \alpha \succ$ is a nonnegative integer. Then $\chi_\lambda = \chi_\mu$, where $\mu = \lambda - (m+1)\alpha$.*

Proof We first show that the member $(j_L(y_\alpha))^{m+1} \otimes 1$ of $Z(\lambda)$ is a maximal vector in $Z(\lambda)$ of weight $\lambda - (m+1)\alpha$. For $\beta > 0$,

$$\begin{aligned} j_L(x_\beta)(j_L(y_\alpha)^{m+1} \otimes 1) &= j_L(x_\beta) j_L(y_\alpha)^{m+1} \otimes 1 = \\ [j_L(x_\beta), j_L(y_\alpha)^{m+1}] \otimes 1 &+ j_L(y_\alpha)^{m+1} j_L(x_\beta) \otimes 1 = \\ [j_L(x_\beta), j_L(y_\alpha)^{m+1}] \otimes 1 + j_L(y_\alpha)^{m+1} &\otimes j_{B(\Delta)}(x_\beta) \cdot 1 = [j_L(x_\beta), j_L(y_\alpha)^{m+1}] \otimes 1. \end{aligned}$$

If $\beta \neq \alpha$, then using the above identity and Lemma 3.2.12(i), we obtain that

$$j_L(x_\beta)(j_L(y_\alpha)^{m+1} \otimes 1) = 0.$$

Again, by Lemma 3.2.12(iii),we obtain that

$$j_L(x_\alpha)(j_L(y_\alpha)^{m+1} \otimes 1) = -(m+1) j_L(y_\alpha)^{m+1}(m1 - j_L(h_\alpha)) \otimes 1.$$

Now,

$$\begin{aligned} (m1 - j_L(h_\alpha)) \otimes 1 = m1 \otimes 1 - j_L(h_\alpha) \otimes 1 &= m1 \otimes 1 - 1 \otimes j_{B(\Delta)}(h_\alpha) \cdot 1 = \\ m1 \otimes 1 - 1 \otimes \lambda(h_\alpha) = (m1 \otimes 1) - (1 \otimes m) &= 0. \end{aligned}$$

Thus,

$$j_L(x_\beta)(j_L(y_\alpha)^{m+1} \otimes 1) = 0$$

for all $\beta > 0$. Further,

$$\begin{aligned} j_L(h)(j_L(y_\alpha)^{m+1} \otimes 1) &= j_L(h) j_L(y_\alpha)^{m+1} \otimes 1 = \\ [j_L(h), j_L(y_\alpha)^{m+1}] \otimes 1 + j_L(y_\alpha)^{m+1} j_L(h) \otimes 1 &= -(m+1)\alpha(h) j_L(y_\alpha)^{m+1} \otimes \\ 1 + j_L(y_\alpha)^{m+1} \otimes j_{B(\Delta)}(h) \cdot 1 &= (\lambda - (m+1)\alpha)(h)(j_L(y_\alpha)^{m+1} \otimes 1). \end{aligned}$$

This shows that $j_L(y_\alpha)^{m+1} \otimes 1$ is a maximal vector of $Z(\lambda)$ of weight $\lambda - (m+1)\alpha$, and $U(L)(j_L(y_\alpha)^{m+1} \otimes 1)$ is a standard cyclic submodule of $Z(\lambda)$ of weight $\mu = \lambda - (m+1)\alpha$. It follows from Proposition 3.3.14 that $\chi_\lambda = \chi_\mu$. ♯

Corollary 3.3.16 *Let λ and μ be members of $\Lambda(H^\star)$ (note that $\Lambda(H^\star)$ is the lattice generated by the set of fundamental weights). Suppose that $\lambda + \delta$ and $\mu + \delta$ are $W(\Phi)$-conjugate. Then $\chi_\lambda = \chi_\mu$.*

Proof Since $W(\Phi)$ is generated by the set $\{\sigma_\alpha \mid \alpha \in \Delta\}$ of simple reflections, it is sufficient to show that $\chi_\lambda = \chi_\mu$ whenever $\sigma_\alpha(\lambda + \delta) = \mu + \delta$, $\alpha \in \Delta$. Suppose that $\sigma_\alpha(\lambda + \delta) = \mu + \delta$. Then

$$\begin{aligned} \mu = \sigma_\alpha(\lambda + \delta) - \delta = \sigma_\alpha(\lambda) + \sigma_\alpha(\delta) - \delta &= \sigma_\alpha(\lambda) - \alpha \\ &= \lambda - (\prec \lambda, \alpha \succ + 1)\alpha. \end{aligned}$$

Since $\lambda \in \Lambda(H^\star)$, $\prec \lambda, \alpha \succ$ is an integer. If $\prec \lambda, \alpha \succ$ is nonnegative, then the result follows from the above corollary. If $\prec \lambda, \alpha \succ = -1$, then $\lambda = \mu$ and we are done.

Suppose that $\prec \lambda, \alpha \succ$ is negative and different from -1. Then $\prec \mu, \alpha \succ = -\prec \lambda, \alpha \succ - 2$ is a nonnegative integer and so $\chi_\mu = \chi_\lambda$. ♯

We shall see soon that the conclusion of the corollary holds for all pairs $\lambda, \mu \in H^\star$.

Our next aim will be to establish the following theorem of Harish-Chandra of which the sufficiency part is already established in the form of the above corollary.

Theorem 3.3.17 (Harish-Chandra) *Let λ and μ be members of $\Lambda(H^\star)$. Then $\chi_\lambda = \chi_\mu$ if and only if $\lambda + \delta$ is conjugate to $\mu + \delta$ under the Weyl group $W(\Phi)$.*

Let L, H, Φ, and $\Delta = \{\alpha_1, \alpha_2, \ldots, \alpha_l\}$ be as usual. Then

$$S = \{y_\alpha \mid \alpha > 0\} \bigcup \{h_{\alpha_1}, h_{\alpha_2}, \ldots, h_{\alpha_l}\} \bigcup \{x_\alpha \mid \alpha > 0\}$$

is a basis of L. Well order S subject to the condition $y_\alpha \leq h_{\alpha_i} \leq x_\beta$ for all $\alpha, \beta > 0$, $\alpha_i \in \Delta$ and $h_{\alpha_i} \leq h_{\alpha_j}$ for $i \leq j$. Consider the corresponding PBW basis $\tilde{S}$ of $U(L)$ consisting of the monomials of the form

$$\Pi_{\alpha>0} j_L(y_\alpha)^{r_\alpha} \Pi_{i=1}^l j_L(h_{\alpha_i})^{t_i} \Pi_{\alpha>0} j_L(x_\alpha)^{s_\alpha}, \; r_\alpha, t_i, s_\alpha \geq 0.$$

We have a unique linear map ξ from $U(L)$ to $U(H)$ which maps the basis members $\Pi_{i=1}^l j_L(h_{\alpha_i})^{t_i} = \Pi_{i=1}^l j_H(h_{\alpha_i})^{t_i}$ in $U(H)$ to itself and all other basis members to 0.

Let ϕ be a linear map from H to F. Then ϕ is a Lie algebra homomorphism from H to F, since H is abelian. Thus, ϕ induces an associative algebra homomorphism $\overline{\phi}$ from $U(H)$ to F given by $\overline{\phi}(j_H(h)) = \phi(h)$. If ϕ is nonzero linear map, then $\overline{\phi}(1) = 1$. Indeed, $\overline{\phi}(\Pi_{i=1}^l j_L(h_{\alpha_i})^{t_i}) = \Pi_{i=1}^l \phi(h_{\alpha_i})^{t_i}$. Note that $\overline{\phi + \psi} \neq \overline{\phi} + \overline{\psi}$, since $\overline{(\phi + \psi)}(1) = 1 \neq 1 + 1 = \overline{\phi}(1) + \overline{\psi}(1)$. Let $\lambda \in H^\star$. Then λ is a linear map from H to F. Consider $Z(\lambda) = U(L)(1 \otimes 1)$. Note that $\Pi_{\alpha>0} j_L(y_\alpha)^{r_\alpha} \Pi_{i=1}^l j_L(h_{\alpha_i})^{t_i} \Pi_{\alpha>0} j_L(x_\alpha)^{s_\alpha}(1 \otimes 1)$ is 0 if $s_\alpha \neq 0$ for some $\alpha > 0$, it is a vector of weight lower than λ if $s_\alpha = 0$ for all $\alpha > 0$ and $r_\alpha \neq 0$ for some $\alpha > 0$, and it is $\overline{\lambda}(\Pi_{i=1}^l j_L(h_{\alpha_i})^{t_i})(1 \otimes 1) = \Pi_{i=1}^l \lambda(h_{\alpha_i})^{t_i}(1 \otimes 1)$ if $r_\alpha = 0 = s_\alpha$ for all α. Thus, for $u \in U(L)$,

$$u \cdot (1 \otimes 1) = \overline{\lambda}(\xi(u))(1 \otimes 1) + \textit{linear combination of elements of weights less than } \lambda.$$

Since, for $z \in Z(U(L))$, $z(1 \otimes 1) = \chi_\lambda(z)(1 \otimes 1)$, it follows that

$$\chi_\lambda(z)(1 \otimes 1) = z \cdot (1 \otimes 1) = \overline{\lambda}(\xi(z))(1 \otimes 1).$$

Consequently,

$$\chi_\lambda(z) = \overline{\lambda}(\xi(z)) \quad \cdots \; (1)$$

for all $z \in Z(U(L))$. In turn, it follows that $\xi|_{Z(U(L))}$ from $Z(U(L))$ to $U(H)$ is an algebra homomorphism.

Next, observe that $U(H)$ is a commutative algebra isomorphic to the polynomial algebra $F[x_1, x_2, \cdots, x_l]$ under the map induced by $j_H(h_{\alpha_i}) \mapsto x_i$. The correspondence $h_{\alpha_i} \mapsto j_H(h_{\alpha_i}) - 1$ from H to $U(H)$ induces a linear map which is a Lie algebra homomorphism, since H is abelian and $U(H)$ is commutative algebra. In turn, we obtain a unique homomorphism η from $U(H)$ to itself subject to $\eta(j_H(h_{\alpha_i})) = j_H(h_{\alpha_i}) - 1$. Evidently, η is an automorphism of $U(H)$. Let ψ denote the homomorphism $\eta o \xi|_{Z(U(L))}$ from $Z(U(L))$ to $U(H)$. Now,

$$\overline{\lambda+\delta}(j_H(h_{\alpha_i}) - 1) = \overline{\lambda+\delta}(j_H(h_{\alpha_i})) - \overline{\lambda+\delta}(1) = (\lambda+\delta)(h_{\alpha_i}) - 1 =$$

$$\lambda(h_{\alpha_i}) + \delta(h_{\alpha_i}) - 1 = \lambda(h_{\alpha_i}) + (\sum_{j=1}^{l} \lambda_j)(h_{\alpha_i}) - 1 =$$

$$\lambda(h_{\alpha_i}) + \lambda_i(h_{\alpha_i}) - 1 = \lambda(h_{\alpha_i}).$$

Hence, using (3.2), we obtain that

$$\overline{\lambda+\delta}(\psi(z)) = \overline{\lambda}(\xi(z)) = \chi_\lambda(z) \cdots (2) \tag{3.1}$$

for all $\lambda \in H^\star$ and $z \in Z(U(L))$.

We have an isomorphism $\lambda \mapsto t_\lambda$ from $H^\star$ to H given by $\lambda(h) = \kappa(t_\lambda, h)$, since the Killing form κ is nondegenerate on H. An element $\sigma \in W(\Phi)$ induces an automorphism $\hat{\sigma}$ of H which is defined by $\hat{\sigma}(t_\lambda) = t_{\sigma(\lambda)}$. Thus, $W(\phi)$ acts on H also as a group of automorphisms and the action is given by $\sigma(t_\lambda) = t_{\sigma(\lambda)}$, $\sigma \in W(\Phi)$. In turn, $W(\Phi)$ acts on $U(H)$ and the action is given by $\sigma(j_H(h)) = j_H(\sigma(h))$, $\sigma \in W(\Phi)$.

Proposition 3.3.18 *Let $\sigma \in W(\Phi)$ and $\lambda \in H^\star$. Then $\overline{(\sigma(\lambda))} = \overline{\lambda} o \sigma$ on $U(H)$.*

Proof Since $W(\Phi)$ is generated by $\{\sigma_\alpha \mid \alpha \in \Phi\}$, it is sufficient to show that $\overline{\sigma_\alpha(\lambda)} = \overline{\lambda} o \sigma_\alpha$ for all $\alpha \in \Phi$. First note that

$$\prec \lambda, \alpha \succ = \frac{2 < \lambda, \alpha >}{< \alpha, \alpha >} = \frac{2\kappa(t_\lambda, t_\alpha)}{\kappa(t_\alpha, t_\alpha)}$$

for all $\lambda \in H^\star$ and $\alpha \in \Phi$. Now,

$$\overline{\sigma_\alpha(\lambda)}(j_H(t_\mu)) = \overline{\lambda - \prec \lambda, \alpha \succ \alpha}(j_H(t_\mu)) = (\lambda - \prec \lambda, \alpha \succ \alpha)(t_\mu) =$$

$$\lambda(t_\mu) - \prec \lambda, \alpha \succ \alpha(t_\mu) = \lambda(t_\mu) - \frac{2\kappa(t_\lambda, t_\alpha)}{\kappa(t_\alpha, t_\alpha)}\kappa(t_\alpha, t_\mu)$$

for all $\lambda, \mu \in H^\star$ and $\alpha \in \Phi$. On the other hand,

$$(\overline{\lambda} o \sigma_\alpha)(j_H(t_\mu)) = \overline{\lambda}(j_H(t_{\sigma_\alpha(\mu)})) = \lambda(t_{\sigma_\alpha(\mu)}) = \lambda(t_{\mu - \prec \mu, \alpha \succ \alpha}) =$$

$$\lambda(t_\mu - \prec \mu, \alpha \succ t_\alpha) = \lambda(t_\mu) - \frac{2\kappa(t_\mu, t_\alpha)}{\kappa(t_\alpha, t_\alpha)}\kappa(t_\lambda, t_\alpha)$$

for all $\lambda, \mu \in H^\star$ and $\alpha \in \Phi$. Since κ is symmetric and $U(H)$ is generated by the set $\{j_H(t_\mu) \mid \mu \in H^\star\}$, $\overline{\sigma_\alpha(\lambda)} = \overline{\lambda} o \sigma_\alpha$. Since $W(\Phi)$ is generated by $\{\sigma_\alpha \mid \alpha \in \Phi\}$, $\overline{\sigma(\lambda)} = \overline{\lambda} o \sigma$ for all $\sigma \in W(\Phi)$. ♯

Proposition 3.3.19 *Let* λ *be a member of* Λ. *Then*

(i) $\overline{\sigma(\lambda+\delta)}(\psi(z)) = \overline{\tau(\lambda+\delta)}(\psi(z))$ *for all* $\sigma, \tau \in W(\Phi)$,
(ii) $(\overline{(\lambda+\delta)} o \sigma)(\psi(z)) = (\overline{(\lambda+\delta)} o \tau)(\psi(z))$ *for all* $\sigma, \tau \in W(\Phi)$,
(iii) $\sigma(\psi(z)) = \psi(z)$ *for all* $\sigma \in W(\Phi)$ *and* $z \in Z(U(L))$, *and*
(iv) $\chi_\lambda = \chi_\mu$ *for all* $\lambda, \mu \in H^\star$ *when ever* $\lambda+\delta$ *is* $W(\Phi)$ *conjugate to* $\mu+\delta$.

Proof (i) From the Eq. (3.1), $\chi_\lambda(z) = \overline{\lambda+\delta}(\psi(z))$ for all $\lambda \in H^\star$ and $z \in Z(U(L))$. Suppose that λ is integral. Since δ is integral, $\mu = \sigma(\lambda+\delta) - \delta$ is integral. Since $\sigma(\lambda+\delta) = \mu+\delta$, it follows from Corollary 3.3.16 that

$$\overline{\lambda+\delta}(\psi(z)) = \chi_\lambda(z) = \chi_\mu(z) = \overline{\mu+\delta}(\psi(z)) = \overline{\sigma(\lambda+\delta)}(\psi(z))$$

for all $\lambda \in \Lambda$, $\sigma \in W(\Phi)$, and $z \in Z(U(L))$. This proves (i).

(ii) It follows from (i) and Proposition 3.3.18.

(iii) From (ii), it follows that $\overline{\lambda+\delta}(\sigma(\psi(z))) = \overline{\lambda+\delta}(\psi(z))$ for all $\lambda \in \Lambda$ (take $\tau = I$). Since δ is integral, any $\mu \in \Lambda$ is of the form $\lambda+\delta$ for some $\lambda \in \Lambda$. Thus, $\overline{\mu}(\sigma(\psi(z))) = \overline{\mu}(\psi(z))$ for all $\mu \in \Lambda$. This implies that $\sigma(\psi(z)) = \psi(z)$ for all $\sigma \in W(\Phi)$ and for all $z \in Z(U(L))$.

(iv) We have already seen (Eq. 3.1) that $\chi_\lambda(\psi(z)) = \overline{\lambda+\delta}(\psi(z))$ for all $\lambda \in H^\star$ and $z \in Z(U(L))$. Since $\psi(z)$ is $W(\Phi)$-invariant, it follows (Proposition 3.3.18) that $\overline{\lambda+\delta}(\psi(z)) = \overline{\sigma(\lambda+\delta)}(\psi(z))$. Consequently, $\chi_\lambda(z) = \chi_\mu(z)$, whenever $\mu+\delta = \sigma(\lambda+\delta)$ for some $\sigma \in W(\Phi)$. ♯

Since H is abelian, we have a natural isomorphism η from $U(H)$ to the symmetric algebra $S(H)$ of H which respects the $W(\Phi)$ actions. Indeed, we can identify $S(H)$ with the polynomial algebra $F[x_1, x_2, \ldots, x_l]$ and the isomorphism η is uniquely determined by the requirement $\eta(j_H(t_{\alpha_i})) = x_i$, where $\Delta = \{\alpha_1, \alpha_2, \ldots, \alpha_l\}$. If $\lambda = \sum_{i=1}^{l} a_i\alpha_i$ is a member of $H^\star$, then $t_\lambda = \sum_{i=1}^{l} a_i t_{\alpha_i}$ and $\eta(j_H(t_\lambda)) = \sum_{i=1}^{l} a_i x_i \in F[x_1, x_2, \ldots, x_l]$. We denote $\eta(j_H(t_\lambda)) = \sum_{i=1}^{l} a_i x_i$ by x_λ. Each $\lambda \in H^\star$ defines a linear function $\hat{\lambda}$ on $S(H) = F[x_1, x_2, \ldots, x_l]$ by putting $\hat{\lambda}(f) = f(\lambda(t_{\alpha_1}), \lambda(t_{\alpha_2}), \ldots, \lambda(t_{\alpha_l}))$.

Further, given $\sigma \in W(\Phi)$, we have an automorphism $\hat{\sigma}$ of $S(H)$ which is given by $\hat{\sigma}(x_i) = x_{\sigma(\alpha_i)}$. Thus $W(\Phi)$ acts on $S(H)$ as a group of automorphisms and the action is given by $\sigma f(x_1, x_2, \ldots, x_l) = f(x_{\sigma(\alpha_1)}, x_{\sigma(\alpha_2)}, \cdots, x_{\sigma(\alpha_l)})$. Evidently, η respects the $W(\Phi)$ actions. Since $\sigma(\psi(z)) = \psi(z)$ for all $\sigma \in W(\Phi)$ and $z \in Z(U(L))$, $\eta(\psi(z)) \in S(H)^{W(\Phi)}$ for all $z \in Z(U(L))$. Thus, $(\eta o \psi)|_{Z(U(L))}$ is a homomorphism $Z(U(L))$ to $S(H)^{W(\Phi)}$. It is clear that $\hat{\lambda}(\eta(u)) = \overline{\lambda}(u)$ for all $u \in U(H)$.

We shall say that a member $\lambda_1 \in H^\star$ is **Linked** to a member $\lambda_2 \in H^\star$ if $\lambda_1+\delta$ is $W(\Phi)$-conjugate to $\lambda_2+\delta$. The notation $\lambda_1 \approx \lambda_2$ stands to say that λ_1 is linked to λ_2. Obviously, $\approx$ is an equivalence relation.

Lemma 3.3.20 *Let* λ_1 *and* λ_2 *be members of* $H^\star$ *which are not linked. Then* $\hat{\lambda}_1|_{S(H)^{W(\Phi)}} \neq \hat{\lambda}_2|_{S(H)^{W(\Phi)}}$.

Proof Suppose that λ_1 and λ_2 are not $W(\Phi)$ conjugate. Then

$$(\sigma(\lambda_1)(t_{\alpha_1}), \sigma(\lambda_1)(t_{\alpha_2}), \ldots, \sigma(\lambda_1)(t_{\alpha_l})) \neq (\tau(\lambda_2)(t_{\alpha_1}), \tau(\lambda_2)(t_{\alpha_2}), \ldots, \tau(\lambda_2)(t_{\alpha_l}))$$

for all $\sigma, \tau \in W(\Phi)$. Again, since $W(\Phi)$ is finite, there is a polynomial $f \in S(H)$ such that

$$f(\lambda_1(t_{\alpha_1}), \lambda_1(t_{\alpha_2}), \ldots, \lambda_1(t_{\alpha_l})) \neq 0,$$

$$f(\sigma(\lambda_1)(t_{\alpha_1}), \sigma(\lambda_1)(t_{\alpha_2}), \ldots, \sigma(\lambda_1)(t_{\alpha_l})) = 0,$$

whenever

$$(\lambda_1(t_{\alpha_1}), \lambda_1(t_{\alpha_2}), \ldots, \lambda_1(t_{\alpha_l})) \neq (\sigma(\lambda_1)(t_{\alpha_1}), \sigma(\lambda_1)(t_{\alpha_2}), \ldots, \sigma(\lambda_1)(t_{\alpha_l})),$$

and also

$$f(\lambda_2(t_{\alpha_1}), \lambda_2(t_{\alpha_2}), \ldots, \lambda_2(t_{\alpha_l})) = 0$$

for all $\sigma \in W(\Phi)$. Take

$$g = \sum_{\sigma \in W(\Phi)} f(x_{\sigma(\alpha_1)}, x_{\sigma(\alpha_2)}, \ldots, x_{\sigma(\alpha_l)}).$$

Evidently, $g \in S(H)^{W(\Phi)}$, $\hat{\lambda}_1(g) \neq 0$, and $\hat{\lambda}_2(g) = 0$. ♯

Proof of Harish-Chandra's Theorem: In the light of Proposition 3.3.19, Lemma 3.3.20, and the fact that $\hat{\lambda}(\eta(u)) = \overline{\lambda}(u)$ for all $u \in Z(U(L))$, it is sufficient to show that the map $(\eta o \psi)|_{Z(U(L))}$ from $Z(U(L))$ to $S(H)^{W(\Phi)}$ is surjective. Further, the isomorphism $\lambda \mapsto t_\lambda$ from $H^\star$ to H induces an isomorphism from $S(H^\star)$ to $S(H)$ which respects the $W(\Phi)$-action, and so, it induces an isomorphism from $S(H^\star)^{W(\Phi)}$ to $S(H)^{W(\Phi)}$. In turn, the epimorphism χ (Chevalley, Proposition 3.3.6) induces an epimorphism $\overline{\chi}$ from $S(L)^G$ to $S(H)^{W(\Phi)}$. Recall Corollary 1.2.5 and the discussions following Corollary 1.2.6. We have an injective linear map $\overline{\nu_m} = \nu o \eta_m^{-1}$ from $S^m(L)$ to $U^m(L)$. This extends to a vector space isomorphism $\overline{\nu}$ from $S(L)$ to $U(L)$ which respects the G-action. Consequently, we have an isomorphism $\hat{\nu}$ from $S(L)^G$ to $U(L)^G = Z(U(L))$ (Lemma 3.3.13), and so, the map $\overline{\chi} o \hat{\nu}^{-1}$ is an epimorphism from $Z(U(L))$ to $S(H)^{W(\Phi)}$. Essentially, $\eta o \psi$ is the map $\eta o \overline{\chi} o \hat{\nu}^{-1}$. As such it is sufficient to show that $\eta|_{S(H)^{W(\Phi)}}$ is surjective from $S(H)^{W(\Phi)}$ to itself. Note that $S(H)^{W(\Phi)}$ is the vector space direct sum $\oplus \sum_{n \in \mathbb{N} \bigcup \{0\}} (S(H)^{W(\Phi)})_n$, where $(S(H)^{W(\Phi)})_n$ is the subspace generated by the set $\{\sum_{\sigma \in W(\Phi)} \prod_{i=1}^d x_{\sigma(\alpha_i)}^{m_i} \mid m_1 + m_2 + \cdots + m_d = n\}$. Thus, it is sufficient to show that for each n, $\sum_{\sigma \in W(\Phi)} \prod_{i=1}^d x_{\sigma(\alpha_i)}^{m_i}$ lies in the image of $\eta|_{S(H)^{W(\Phi)}}$ whenever $m_1 + m_2 + \cdots + m_d = n$. We prove it by the induction on n. It is evident for $n = 0$. Assume the result for n. Consider an element $f = \sum_{\sigma \in W(\Phi)} \prod_{i=1}^d x_{\sigma(\alpha_i)}^{m_i}$, where

$m_1 + m_2 + \cdots + m_d = n+1$. Clearly, $f = \sum_{\sigma \in W(\Phi)} \prod_{i=1}^{d} (x_{\sigma(\alpha_i)} - 1)^{m_i} + g$, where g belongs to $\oplus \sum_{j=0}^{n} (S(H)^{W(\Phi)})_j$. By the induction hypothesis, g is in the image of η. Also, by the definition, $\sum_{\sigma \in W(\Phi)} \prod_{i=1}^{d} (x_{\sigma(\alpha_i)} - 1)^{m_i}$ is in the image of η. It follows that f is in the image of η. This completes the proof of Harish-Chandra's theorem. ♯

Exercises

3.3.1. Consider $sl(2, \mathbb{C})$ with standard basis $\{x, h, y\}$ and root system $\Phi = \{\alpha, -\alpha\}$. Show that $\lambda = \frac{\alpha}{2}$ is the fundamental dominant weight. Show that $< \lambda^2 > = S(H^\star)^{W(\Phi)}$ and $S(H)^{W(\Phi)} = < \frac{h^2}{4} >$. Find a pre-image of $\frac{h^2}{4}$ under the map $\eta o \psi$.

3.3.2. For each of the Lie algebras A_2, B_2, and G_2 with their standard H, Φ and Δ, determine the set $\{\lambda_1, \lambda_2\}$ of fundamental weights. Find generating sets for $S(H^\star)$ and $S(H)$ and their pre-images in $Z(U(L))$.

3.3.3. Let ς be an algebra homomorphism from $Z(U(L))$ to F. Show that there is a $\lambda \in H^\star$ such that $\varsigma = \chi_\lambda$.

3.3.4. Show that the map ψ from $Z(U(L))$ to $S(H)^{W(\Phi)}$ is independent of the choice of Δ.

3.3.5. Let $\lambda \in \Lambda^+$. Suppose that $\lambda + \delta$ is $W(\Phi)$-conjugate to $\mu + \delta$. Show that $\mu < \lambda$ and μ occurs as a weight of $Z(\lambda)$.

3.4 Multiplicity Formulas of Weyl, Kostant, and Steinberg

In this section, we obtain different formulas for the multiplicity $m_\lambda(\mu)$ of a weight μ of $V(\lambda)$.

Let L, H, Φ, and Δ be as usual. For each $\lambda \in H^\star$, let X_λ denote the set $\{\lambda - \sum_{\alpha > 0} a_\alpha \alpha \mid a_\alpha \in \mathbb{N} \bigcup \{0\}\}$. Let $\Sigma(H^\star)$ denote the set of all maps from $H^\star$ to F whose supports lie in a finite union of sets of the form X_λ. Evidently, X_λ is the set of weights appearing in $Z(\lambda)$. Note that the supports of the members of $\Sigma(H^\star)$ may be infinite sets. For example, for $\alpha > 0$, consider the function f_α given by $f_\alpha(-a\alpha) = 1$ for all $a \in \mathbb{N} \bigcup \{0\}$ and $f_\alpha(\lambda) = 0$ otherwise. Evidently, $f_\alpha \in \Sigma(H^\star)$ and its support is an infinite set. $\Sigma(H^\star)$ is a vector space under the obvious point-wise addition and the multiplication by scalars. Clearly, $X_\lambda + X_\mu \subseteq X_{\lambda+\mu}$. For $f, g \in \Sigma(H^\star)$, $(f \star g)(\lambda) = \sum_{\mu+\nu=\lambda} f(\mu) g(\nu)$ makes sense and the function $f \star g$, thus defined, is a member of $\Sigma(H^\star)$. This defines the convolution product $\star$ in $\Sigma(H^\star)$ which makes $\Sigma(H^\star)$ a commutative algebra over F. It is also clear that the formal characters of all standard cyclic modules are members of $\Sigma(H^\star)$. The integral group ring $\mathbb{Z}(\Lambda)$ is also a subset of $\Sigma(H^\star)$. Further, $W(\Phi)$ acts on $\Sigma(H^\star)$ by putting $(\sigma f)(\lambda) = f(\sigma^{-1}(\lambda))$. In particular, $(\sigma e(\lambda))(\mu) = e(\lambda)(\sigma^{-1}(\mu)) = 1$ if $\mu = \sigma(\lambda)$ and 0 otherwise. Thus, $\sigma e(\lambda) = e_{\sigma(\lambda)}$.

Let p denote the function from $H^\star$ to C which is defined by putting

$$p(\lambda) = \mid \{a \in Map(\Phi^+, \mathbb{N}\bigcup\{0\}) \mid \lambda + \sum_{\alpha\in\Delta} a(\alpha)\alpha = 0\} \mid$$

(note that $\mid X \mid$ denotes the cardinality of the set X). Clearly, $p = Ch_{Z(0)}$ belongs to $\Sigma(H^\star)$. Also $p(\lambda) \neq 0$ if and only if $\lambda \in \Lambda^-$. The function p defined above is called the **Kostant function** associated with the triple (H, Φ, Δ).

We have another function q defined by

$$q = \prod_{\alpha>0}\left(e\left(\frac{\alpha}{2}\right) - e\left(\frac{-\alpha}{2}\right)\right),$$

where the product is the convolution product. Evidently, $q \in \Sigma(H^\star)$. The function q is called the **Weyl Function**.

Now, we establish a few identities relating the functions f_α, p, q, and $Ch_{Z(\lambda)}$.

Lemma 3.4.1 *(i)* $p = \prod_{\alpha\in\Phi^+} f_\alpha$, *where the product is the convolution product.*

(ii) $(e(0) - e(-\alpha)) \star f_\alpha = e(0)$.

(iii) $q = (\prod_{\alpha\in\Phi^+}(e(0) - e(-\alpha))) \star e(\delta)$.

(iv) $\sigma(q) = (-1)^{l(\sigma)} q$ *for all* $\sigma \in W(\Phi)$.

(v) $q \star p \star e(-\delta) = e(0)$.

(vi) $Ch_{Z(\lambda)}(\mu) = p(\mu - \lambda) = (p \star e(\lambda))(\mu)$.

(vii) $q \star Ch_{Z(\lambda)} = e(\lambda + \delta)$.

Proof (i) Suppose that $\Phi^+ = \{\beta_1, \beta_2, \ldots, \beta_r\}$. Then by the definition of the convolution product

$$\prod_{\alpha\in\Phi^+} f_\alpha(\lambda) = \sum_{\lambda_1+\lambda_2+\cdots+\lambda_r = \lambda} f_{\beta_1}(\lambda_1) f_{\beta_2}(\lambda_2)\cdots f_{\beta_r}(\lambda_r).$$

In turn, it follows from the definition of f_{β_i} and p that

$$\prod_{\alpha\in\Phi^+} f_\alpha(\lambda) = p(\lambda).$$

This proves (i).

(ii)

$$((e(0)-e(-\alpha))\star f_\alpha)(\lambda) = \sum_{\mu+\nu=\lambda}(e(0)-e(-\alpha))(\mu)f_\alpha(\nu) =$$

$$\sum_{\mu+\nu=\lambda} e(0)(\mu)f_\alpha(\nu) - \sum_{\mu+\nu=\lambda} e(-\alpha)(\mu)f_\alpha(\nu).$$

It follows from the definition of f_α that the RHS is 1 if $\lambda = 0$ and it is zero, otherwise. Thus, $(e(0)-e(-\alpha))\star f_\alpha = e(0)$.

(iii) Clearly, $e(\delta) = e\left(\frac{1}{2}\sum_{\alpha>0}\alpha\right) = \prod_{\alpha>0} e(\frac{\alpha}{2})$. Again, since

$$(e(0)-e(-\alpha))\star e\left(\frac{\alpha}{2}\right) = e\left(\frac{\alpha}{2}\right) - e\left(-\frac{\alpha}{2}\right)$$

and $\Sigma(H^\star)$ is commutative, it follows that

$$\left(\prod_{\alpha>0}(e(0)-e(-\alpha))\right)\star e(\delta) = \prod_{\alpha>0}\left(e\left(\frac{\alpha}{2}\right) - e\left(-\frac{\alpha}{2}\right)\right) = q.$$

(iv) It is sufficient to show that $\sigma_\alpha(q) = -q$ for all $\alpha \in \Delta$. Since σ_α permutes the members of $\Phi^+ - \{\alpha\}$ and takes α to $-\alpha$, it follows that

$$\sigma_\alpha(q) = \left(\prod_{\beta\in\Phi^+-\{\alpha\}}\left(e\left(\frac{\beta}{2}\right) - e\left(-\frac{\beta}{2}\right)\right)\right)\star\left(e\left(-\frac{\alpha}{2}\right) - e\left(\frac{\alpha}{2}\right)\right) = -q.$$

(v) $q\star p\star e(-\delta)$
$= (\prod_{\alpha>0}(e(0)-e(-\alpha)))\star e(\delta)\star p\star e(-\delta)$ (by (iii))
$= (\prod_{\alpha>0}(e(0)-e(-\alpha)))\star p$ (since $e(\delta)\star e(-\delta) = 1$)
$= \prod_{\alpha>0}((e(0)-e(-\alpha))\star f_\alpha) = e(0)$ (by (i)).

(vi) Clearly, $Dim\ Z(\lambda)_\mu = |\{a_\alpha \in \mathbb{N}\bigcup\{0\}, \alpha>0 \mid \lambda-\mu = \sum_{\alpha>0} a_\alpha\alpha\}|$. Hence $Ch_{Z(\lambda)}(\mu) = p(\mu-\lambda) = (p\star e(\lambda))(\mu)$ for all μ.

(vii) By (v) and (vi), $q\star Ch_{Z(\lambda)}$
$= q\star p\star e(\lambda)$
$= q\star p\star e(-\delta)\star e(\delta)\star e(\lambda)$
$= e(\delta)\star e(\lambda)$
$= e(\lambda+\delta)$. ⋆

Our next aim is to establish the following multiplicity formula due to Kostant:

Theorem 3.4.2 (Kostant) *Let $\lambda \in \Lambda^+$. Then the multiplicity $m_\lambda(\mu)$ of μ in $V(\lambda)$ is given by*

$$m_\lambda(\mu) = \sum_{\sigma\in W(\Phi)} sign\sigma\ p(\mu+\delta-\sigma(\lambda+\delta)),$$

where $sign\sigma = (-1)^{l(\sigma)}$.

We need to examine more closely, the module $Z(\lambda)$ and its submodules to prove the theorem.

For $\lambda \in H^\star$, let M_λ denote the class of all L-modules V satisfying the following properties:

(i) V is the direct sum of its weight spaces.
(ii) $zv = \chi_\lambda(z)v$ for all $z \in Z(U(L))$ and $v \in V$.
(iii) The formal character Ch_V of V is a member of $\Sigma(H^\star)$.

Evidently, all submodules and homomorphic images of standard cyclic L-modules of weight λ belong to M_λ. In particular all submodules of $Z(\lambda)$ and their homomorphic images are in M_λ. Indeed, the class M_λ is hereditary in the sense that the submodules of members of M_λ are members of M_λ. It is also divisible in the sense that the quotients of the members of M_λ are also members of M_λ. Harish-Chandra's theorem asserts that $M_\lambda = M_\mu$ if and only if $\lambda + \delta$ is $W(\Phi)$-conjugate to $\mu + \delta$. Let $V \in M_\lambda$. Since $Ch_V \in \Sigma(H^\star)$, there is a weight μ of V such that $\mu + \alpha$ is not a weight of V for each $\alpha > 0$. This means that V_μ has a maximal vector of weight μ.

For $\lambda \in H^\star$, let $\Theta(\lambda)$ denote the set $\{\mu \in H^\star \mid \mu < \lambda,\ and\ \mu \approx \lambda\}$. Obviously, $\Theta(\lambda)$ is a finite set and M is constant on $\Theta(\lambda)$.

Theorem 3.4.3 *Let $\lambda \in H^\star$. Then the following hold:*

(i) $Z(\lambda)$ has a composition series, and the composition factors of $Z(\lambda)$ are of the form $V(\mu)$ for some $\mu \in \Theta(\lambda)$.
(ii) $V(\lambda)$ appears once and only once as composition factor of $Z(\lambda)$.

Proof We first show that $Z(\lambda)$ has a simple submodule U_1 which is isomorphic to $V(\nu_1)$ with $\nu_1 \leq \lambda$ and $\nu_1 \approx \lambda$. If $Z(\lambda)$ is simple, then there is nothing to do. Let V be a nonzero proper submodule of $Z(\lambda)$. Since $Dim(Z(\lambda))_\lambda = 1$, λ cannot be a weight of V. Since $Ch_V \in \Sigma(H^\star)$, for each $\alpha > 0$ and weight $\mu \in \Lambda(V)$, there is a $r \in \mathbb{N}$ such that $\mu + t\alpha \notin \Lambda(V)$ for all $t > r$. Thus, there is a weight μ_1 of V such that $\mu_1 + \alpha$ is not a weight of V for all $\alpha > 0$. In other words, μ_1 is a maximal weight of V. Consequently, V has a nonzero submodule W_1 which is a homomorphic image of $Z(\mu_1)$. As such $\chi_{\mu_1} = \chi_\lambda$ and by Harish-Chandra's theorem, $\mu_1 \approx \lambda$. Thus, $\mu_1 \in \Theta(\lambda)$. If W_1 is simple, then W_1 is isomorphic to $V(\mu_1)$. We can take $U_1 = W_1$, $\nu_1 = \mu_1$, and we are done. If not, then W_1 has a nonzero proper submodule V_2 and repeating the earlier arguments, we obtain a nonzero submodule W_2 of V_2 which is a homomorphic image of $Z(\mu_2)$, where $\mu_2 \in \Theta(\lambda)$. Since W_2 is a proper submodule of W_1, either the number of weights of W_2 is less than that of W_1 or the multiplicity of at least one weight of W_2 is less than that in W_1. If W_2 is simple, then W_2 is isomorphic to $V(\mu_2)$. We can take U_1 to be W_2 and ν_1 to be μ_2, and again, we are done. If not proceed. Evidently, this process terminates giving U_1 and ν_1, since $\Theta(\lambda)$ is finite. If $Z(\lambda)/U_1$ is simple, then, it being standard cyclic, is isomorphic to $V(\nu_2)$ for some $\nu_2 \in \Theta(\lambda)$, and we get a composition series $Z(\lambda) \supset U_1 \supset \{0\}$ with the required properties. If not, then repeating the same arguments, we get a simple submodule U_2/U_1 of $Z(\lambda)/U$ which is isomorphic to $V(\nu_2)$, where $\nu_2 \in \Theta(\lambda)$. If $\mathbb{Z}(\lambda)/U_2$ is simple, then, it being standard cyclic is isomorphic $V(\nu_3)$ for some $\nu_3 \in \Theta(\lambda)$, and

$Z(\lambda) \supset U_2 \supset U_1 \supset \{0\}$ is a composition series of $\mathbb{Z}(\lambda)$ with required property. If not, proceed. This process terminates after finitely many steps giving us a composition series of $Z(\lambda)$ with the required property, since $\Theta(\lambda)$ is finite. This proves (i).

(ii) This follows, since $Dim\ Z(\lambda)_\lambda = 1$. ♯

Corollary 3.4.4 *Let $\lambda \in H^\star$. Express $\Theta(\lambda)$ as an ordered set $\{\mu_1, \mu_2, \ldots, \mu_r\}$ with $\mu_i < \mu_j$ which implies that $i \leq j$. Then we have a $r \times r$ uni-triangular integral matrix $A = [a_{ij}]$ such that*

$$Ch_{V(\mu_j)} = \sum_{i=1}^{j} a_{ij} Ch_{Z(\mu_i)}.$$

Proof It follows from the above theorem that for each j,

$$Ch_{Z(\lambda_j)} = Ch_{V(\lambda_j)} + \sum_{i=1}^{j-1} b_{ij} Ch_{V(\mu_i)},$$

where $b_{ij} \in \mathbb{N}\bigcup\{0\}$ with $b_{ii} = 1$. This gives us a uni-triangular integral matrix $B = [b_{ij}]$ such that

$$[Ch_{V(\mu_1)}, Ch_{V(\mu_2)}, \ldots, Ch_{V(\mu_r)}]B = [Ch_{Z(\mu_1)}, Ch_{Z(\mu_2)}, \ldots, Ch_{Z(\mu_r)}].$$

Since B is a uni-triangular integral matrix, $A = B^{-1} = [a_{ij}]$ is also a uni-triangular integral matrix. The result follows. ♯

Proof of the Theorem 3.4.2 Let $\lambda \in \Lambda^+$. From the Theorem 3.2.13, it follows that $V(\lambda)$ is finite dimensional, σ permutes the set $\Gamma(\lambda)$ of weights of $V(\lambda)$ and $Dim\ V_\mu = Dim\ V_{\sigma(\mu)}$ for all $\mu \in \Gamma(\lambda)$. Thus, $\sigma(Ch_{V(\lambda)}) = Ch_{V(\lambda)}$ for all $\sigma \in W(\Phi)$. From the above corollary,

$$Ch_{V(\lambda)} = \sum_{i=1}^{r} a_{ir} Ch_{Z(\mu_i)},$$

$a_{ir} \in \mathbb{Z}$ and $a_{rr} = 1$. By Lemma 3.4.1 (vii),

$$q \star Ch_{V(\lambda)} = \sum_{i=1}^{r} a_{ir} e(\mu_i + \delta).$$

Again by Lemma 3.4.1 (iv) and the above identity,

$$\sum_{i=1}^{r} a_{ir} e(\sigma^{-1}(\mu_i + \delta)) = \sigma\left(\sum_{i=1}^{r} a_{ir} e\,(\mu_i + \delta)\right) = \sigma(q \star Ch_{V(\lambda)}) =$$
$$\sigma(q) \star \sigma(Ch_{V(\lambda)}) = sign\sigma\ q \star Ch_{V(\lambda)}$$

for all $\sigma \in W(\Phi)$. Now, since σ is a transitive permutation on $\Theta(\lambda)$ and $a_{rr} = 1$, it follows that $a_{ir} = sign\sigma$ whenever $\sigma^{-1}(\mu_i + \delta) = \mu_r + \delta = \lambda + \delta$. Thus,

$$q \star Ch_{V(\lambda)} = \sum_{\sigma \in W(\Phi)} sign\sigma \; e(\sigma(\lambda + \delta)).$$

Now, from Lemma 3.4.1 (v),

$$Ch_{V(\lambda)} = e(0) \star Ch_{V(\lambda)} = q \star p \star e(-\delta) \star Ch_{V(\lambda)} =$$
$$p \star e(-\delta) \star \sum_{\sigma \in W(\Phi)} sign\sigma \; e(\sigma(\lambda + \delta)) =$$
$$p \star \sum_{\sigma \in W(\Phi)} sign\sigma \; e(\sigma(\lambda + \delta) - \delta) = \sum_{\sigma \in W(\Phi)} sign\sigma \; p \star e(\sigma(\lambda + \delta) - \delta).$$

From Lemma 3.4.1 (vi),

$$m_\lambda(\mu) = Ch_{V(\lambda)}(\mu) = \sum_{\sigma \in W(\Phi)} sign\sigma \; p \star e(\sigma(\lambda + \delta) - \delta)(\mu) =$$
$$\sum_{\sigma \in W(\Phi)} sign\sigma \; p(\mu - e(\sigma(\lambda + \delta) - \delta)).$$

This completes the proof. ♯

Remark 3.4.5 If $W(\Phi)$ is small, then the Kostant formula can be used to compute the character $Ch_{V(\lambda)}$, $\lambda \in \Lambda^+$. However, if $W(\Phi)$ is large, then the formula cannot be used to compute it comfortably. In that case, Freudenthal's formula (to be described) gives us a recursive method to compute it.

The following Weyl's character formula allows us to express the character $Ch_{V(\lambda)}$ as a quotient of alternative sums of members of $\mathbb{Z}[\Lambda]$.

Theorem 3.4.6 (Weyl) *Let* $\lambda \in \Lambda^+$. *Then*

$$\sum_{\sigma \in W(\Phi)} sign\sigma \; e(\sigma(\delta)) \star Ch_{V(\lambda)} = \sum_{\sigma \in W(\Phi)} sign\sigma \; e(\sigma(\lambda + \delta)).$$

Proof We have the identity

$$q \star Ch_{V(\lambda)} = \sum_{\sigma \in W(\Phi)} sign\sigma \; e(\sigma(\lambda + \delta))$$

established during the proof of the above Kostant's theorem. Putting $\lambda = 0$ in the above equation and observing that $Ch_{V(0)} = e(0) = 1$, we obtain that

$$q = \sum_{\sigma \in W(\Phi)} sign\sigma e(\sigma(\delta)).$$

Substituting this value of q in the earlier identity, we obtain the desired identity. ♯

Again the Weyl character formula is not so convenient to compute the character as computing a quotient may be quite cumbersome. However, it can be used to compute the $Dim\ V(\lambda)$ as follows.

Let $\Sigma_0(\Lambda)$ denote the subalgebra of $\Sigma(H^\star)$ generated by the set $\{e(\lambda) \mid \lambda \in \Lambda\}$. Evidently, $\mathbb{Z}[\Lambda] \subset \Sigma_0(\Lambda)$. We have the augmentation map ϵ from $\Sigma_0(\Lambda)$ to $\mathbb{C}$ which is given by $\epsilon(\sum_{\lambda\in\Lambda} a_\lambda e(\lambda) = \sum_{\lambda\in\Lambda} a_\lambda$. Evidently, $\epsilon(Ch_{V(\lambda)}) = Dim\ V_\lambda$. We also denote $Dim\ V_\lambda$ by $deg(\lambda)$. Clearly, $deg(\lambda) = \sum_{\mu\in\Lambda(\lambda)} m_\lambda(\mu)$.

For each $\alpha \in \Phi^+$, we have a unique endomorphism ∂_α of $\Sigma_0(\Lambda)$ given by $\partial_\alpha(e(\lambda)) = <\lambda, \alpha> e(\lambda)$. Clearly, $e(\lambda) \star e(\mu) = e(\lambda + \mu)$ and

$$\partial_\alpha(e(\lambda) \star e(\mu)) = \partial_\alpha e(\lambda + \mu) = <\lambda + \mu, \alpha> =$$
$$\partial_\alpha(e(\lambda)) \star e(\mu) + e(\lambda) \star \partial_\alpha(e(\mu)).$$

Thus, ∂_α is a derivation of $\Sigma_0(\Lambda)$. However, $\partial = \prod_{\alpha>0} \partial_\alpha$ is not a derivation, where the product is the usual product of functions. Let ω denote the function from Λ^+ to $\Sigma_0(\Lambda)$ which is given by $\omega(\lambda) = \sum_{\sigma\in W(\Phi)} sign\sigma e(\sigma(\lambda + \delta))$. It can be easily observed that $\omega(0)$ is the Weyl function q and the Weyl formula reads as

$$\omega(0) \star Ch_{V_\lambda}\ \omega(\lambda) = \omega(\lambda).$$

Proposition 3.4.7 $\epsilon(\partial(\omega(\lambda))) = \mid W(\Phi) \mid \prod_{\alpha>0} <\lambda + \delta, \alpha>$ *for all* $\lambda \in \Lambda^+$.

Proof By the definition, $\partial_\alpha(e(\delta)) = <\delta, \alpha> e(\delta)$. Consequently, $\partial(e(\delta)) = \prod_{\alpha>0} <\delta, \alpha> e(\delta)$. Thus, $\epsilon(\partial(e(\delta))) = \prod_{\alpha>0} <\delta, \alpha>$. Similarly,

$$\epsilon(\partial(e(\sigma(\delta)))) = \prod_{\alpha>0} <\sigma(\delta), \alpha> = \prod_{\alpha>0} <\delta, \sigma^{-1}(\alpha)> =$$
$$sign\sigma \prod_{\alpha>0} <\delta, \alpha>,$$

since σ is orthogonal transformation and σ^{-1} takes $l(\sigma^{-1}) = l(\sigma)$ positive roots to negative roots. More generally,

$$\epsilon(\partial(\omega(\lambda))) = \sum_{\sigma\in W(\Phi)} sign\sigma\ \epsilon(\partial(e(\sigma(\lambda + \delta)))) =$$
$$\sum_{\sigma\in W(\Phi)} (sign\sigma)^2 \prod_{\alpha>0} <\lambda + \delta, \alpha> = \mid W(\Phi) \mid \prod_{\alpha>0} <\lambda + \delta, \alpha>. \quad ♯$$

Corollary 3.4.8 $deg(\lambda) = \frac{\prod_{\alpha>0} <\lambda+\delta,\alpha>}{\prod_{\alpha>0} <\delta,\alpha>}$.

Proof From Weyl's formula, $\omega(0) \star Ch_{V(\lambda)} = \omega(\lambda)$. Applying the operator ∂ to both sides and using the Leibniz formula for ∂_α, we obtain that $\epsilon(\partial(\omega(0) \star Ch_{V(\lambda)})) = \epsilon(\partial(\omega(0)))\epsilon(Ch_{V(\lambda)})$. Since $\epsilon(Ch_{V(\lambda)}) = Dim\ V(\lambda) = deg(\lambda)$,

$$deg(\lambda) = \frac{\epsilon(\partial(\omega(\lambda)))}{\epsilon(\partial(\omega(0)))}.$$

The corollary follows from the above proposition. ♯

Example 3.4.9 Consider A_1 with $\Phi = \{\alpha, -\alpha\}$. The fundamental weight $\lambda_1 = \frac{\alpha}{2} = \delta$. Any $\lambda \in \Lambda^+$ is of the form $m\lambda_1, m \geq 0$. Using the above corollary, it is clear that $Dim\ V(\lambda) = deg(\lambda) = m + 1$.

Our next aim is to describe Steinberg's formula to express the tensor product of two standard cyclic simple modules.

Theorem 3.4.10 (Steinberg) *Let λ', λ'' and λ be members of Λ^+. Let $n(\lambda)$ denote the number of times $V(\lambda)$ appears in the decomposition of the tensor product $V(\lambda') \otimes V(\lambda'')$ as the direct sum of simple modules. Then*

$$n(\lambda) = \sum_{\sigma \in W(\Phi)} \sum_{\tau \in W(\Phi)} sign(\sigma\tau)p(\lambda + 2\delta - \sigma(\lambda' + \delta) - \tau(\lambda'' + \delta)).$$

Proof Suppose that

$$Ch_{V(\lambda')\otimes V(\lambda'')} = \sum_{\lambda \in \Lambda^+} n(\lambda)Ch_{V(\lambda)}.$$

In the light of Corollary 3.3.3,

$$Ch_{V(\lambda')} \star Ch_{V(\lambda'')} = \sum_{\lambda \in \Lambda^+} n(\lambda)Ch_{V(\lambda)} \tag{3.2}$$

Multiplying (convolution product) both sides by $\omega(0)$ and using Weyl's formula for $V(\lambda'')$ and $V(\lambda)$, we get

$$Ch_{V(\lambda')} \star \omega(\lambda'') = \sum_{\lambda \in \Lambda^+} n(\lambda)\omega(\lambda) \tag{3.3}$$

Now, by the definition,

$$Ch_{V(\lambda')} = \sum_{\mu \in \Lambda} m_{\lambda'}(\mu)e(\mu) \tag{3.4}$$

Using the Kostant formula for $m_{\lambda'}(\mu)$ and substituting the value of $Ch_{V(\lambda')}$ in (3.3), we obtain that

$$\sum_{\mu\in\Lambda}\sum_{\sigma\in W(\Phi)} sign(\sigma)p(\mu+\delta-\sigma(\lambda'+\delta))e(\mu))\star\omega(\lambda'') = \sum_{\lambda\in\Lambda} n(\lambda)\omega(\lambda).$$

Putting the expressions for $\omega(\lambda'')$ and $\omega(\lambda)$, we get

$$\sum_{\mu\in\Lambda}\sum_{\sigma\in W(\Phi)}\sum_{\tau\in W(\Phi)} sign(\sigma\tau)p(\mu+\delta-\sigma(\lambda'+\delta))e(\tau(\mu+\delta)+\mu) =$$
$$\sum_{\lambda\in\Lambda}\sum_{\sigma\in W(\Phi)} n(\lambda)sign(\sigma)e(\sigma(\lambda+\delta)) \qquad (3.5)$$

(note that $n(\lambda) = 0$ if $V(\lambda)$ does not appear in the decomposition, e.g., if $\lambda\in\Lambda^-$). Putting $\sigma(\lambda+\delta) = \nu+\delta$ (note that as λ varies over Λ, ν also varies over Λ), the RHS of (3.5) becomes

$$\sum_{\nu}\sum_{\sigma\in W(\Phi)} sign(\sigma)n(\sigma^{-1}(\nu+\delta)-\delta)e(\nu+\delta) =$$
$$\sum_{\nu}\sum_{\sigma\in W(\Phi)} sign(\sigma)n(\sigma(\nu+\delta)-\delta)e(\nu+\delta) \qquad (3.6)$$

In the LHS expression, put $\mu=\nu$, where $\tau(\lambda''+\delta)+\mu = \nu+\delta$. We obtain

$$\sum_{\nu}\sum_{\sigma\in W(\Phi)}\sum_{\tau\in W(\Phi)} sign(\sigma\tau)p(\nu+2\delta-\sigma(\lambda'+\delta)-\tau(\lambda''+\delta))e(\nu+\delta) \qquad (3.7)$$

If ν is dominant, then $\sigma(\nu+\delta)-\delta$ can be dominant only if $\sigma = I$. Thus, in Eq. (3.6) $n(\sigma(\nu+\delta)-\delta)\neq 0$ only if $\sigma=I$. Now, using (3.6), (3.7), and (3.5), we get the desired result (with λ replaced by ν). ♯

Remark 3.4.11 The formula is explicit. However, it is convenient for the purpose of computation only when $W(\Phi)$ is small.

Let $\lambda\in\Lambda^+$. The set Γ_λ of weights of $V(\lambda)$ is a saturated set (Remark 3.2.14) as such $\mu\leq\lambda$ and $<\mu+\delta,\mu+\delta>\leq<\lambda+\delta,\lambda+\delta>$ for all $\mu\in\Gamma_\lambda$ (Lemma 2.4.36 and Lemma 2.4.37). Thus, the following Freudenthal multiplicity formula gives us an effective recursive procedure to compute the multiplicity $m_\lambda(\mu)$ starting with $m_\lambda(\lambda) = 1$.

Theorem 3.4.12 (Freudenthal) *Let* $\lambda\in\Lambda^+$. *Then*

$$(<\lambda+\delta,\lambda+\delta> - <\mu+\delta,\mu+\delta>)m_\lambda(\mu) =$$
$$2\sum_{\alpha>0}\sum_{i=1}^{\infty} m_\lambda(\mu+i\alpha) <\mu+i\alpha,\alpha> .$$

Before establishing the theorem, we establish some identities relating the trace and multiplicities.

Recall the definition of the universal Casimir element c_{univ}:

$$c_{univ} = \sum_{i=1}^{l} j_L(h_i)j_L(k_i) + \sum_{\alpha\in\Phi} j_L(x_\alpha)j_L(z_\alpha).$$

If ρ is the irreducible representation afforded by $V(\lambda)$, then the induced homomorphism $\overline{\rho}(c_{univ}) \in End(V(\lambda))$ is the multiplication by a scalar c_λ (say). Thus, for a fixed weight μ of $V(\lambda)$,

$$Tr\overline{\rho}(c_{univ})|_{V(\lambda)_\mu} = c_\lambda Dim V(\lambda)_\mu = c_\lambda m_\lambda(\mu) \tag{3.8}$$

Now,

$$\overline{\rho}(c_{univ}) = \sum_{i=1}^{l} \overline{\rho}(j_L(h_i))\overline{\rho}(j_L(k_i)) + \sum_{\alpha\in\Phi} \overline{\rho}(j_L(x_\alpha))\overline{\rho}(j_L(z_\alpha)) \tag{3.9}$$

Consequently,

$$c_\lambda m_\lambda(\mu) = Tr\overline{\rho}(c_{univ})|_{V(\lambda)_\mu} =$$

$$\sum_{i=1}^{l} Tr(\rho(h_i)\rho(k_i))|_{V(\lambda)_\mu} + \sum_{\alpha\in\Phi} Tr(\rho(x_\alpha)\rho(z_\alpha))|_{V(\lambda)_\mu} \tag{3.10}$$

Since $\rho(h_i)$ on $V(\lambda)_\mu$ is the multiplication by $\mu(h_i)$ and $\rho(k_i)$ on $V(\lambda)_\mu$ is the multiplication by $\mu(k_i)$, it follows that

$$\sum_{i=1}^{l} Tr(\rho(h_i)\rho(k_i))|_{V(\lambda)_\mu} = m_\lambda(\mu)\sum_{i=1}^{l} \mu(h_i)\mu(k_i) \tag{3.11}$$

Note that $\mu(h_i) = \kappa(t_\mu, h_i)$ and $\mu(k_i) = \kappa(t_\mu, k_i)$. Suppose that $t_\mu = \sum_{j=1}^{l} a_j h_j$. Then $\mu(h_i) = \sum_{j=1}^{l} a_j\kappa(h_j, h_i)$ and $\mu(k_i) = \sum_{j=1}^{l} a_j\kappa(h_j, k_i) = a_i$, since $\{k_1, k_2, \ldots, k_l\}$ is dual to $\{h_1, h_2, \ldots, h_l\}$ relative κ. In turn,

$$<\mu, \mu> = \kappa(t_\mu, t_\mu) = \sum_{i,j} a_i a_j \kappa(h_j, h_i) = \sum_{i=1}^{l} \mu(h_i)\mu(k_i) \tag{3.12}$$

Using (3.11) and (3.10), we see that

$$c_\lambda m_\lambda(\mu) = <\mu, \mu> m_\lambda(\mu) + \sum_{\alpha\in\Phi} Tr(\rho(x_\alpha)\rho(z_\alpha))|_{V(\lambda)_\mu} \tag{3.13}$$

Next, we attempt to describe $Tr(\rho(x_\alpha)\rho(z_\alpha))|_{V(\lambda)_\mu}$. Recall Theorem 1.4.28. Replace the standard basis $\{v_0, v_1, \ldots, v_m\}$ by the basis $\{w_0, w_1, \ldots, w_m\}$, where $w_r = r!\frac{<\alpha,\alpha>^r}{2^r}v_r$, $0 \leq r \leq m$. Also replace the standard basis $\{x, h, y\}$ of $sl(2, F)$ by $\{x, t, z\}$, where $t = \frac{<\alpha,\alpha>}{2}h$ and $z = \frac{<\alpha,\alpha>}{2}y$. Then it follows that

(i) $\rho(t)w_r = (m - 2r)\frac{<\alpha,\alpha>}{2}w_r$,
(ii) $\rho(z)w_r = w_{r+1}$, $w_{m+1} = 0$, and
(iii) $\rho(x)w_r = r(m - r + 1)\frac{<\alpha,\alpha>}{2}w_{r-1}$, $\omega_{-1} = 0$.
Consequently,

$$\rho(x)\rho(z)(w_r) = (m - r)(r + 1)\frac{<\alpha, \alpha>}{2}w_r \cdots . \qquad (3.14)$$

Now, let $\lambda \in \Lambda^+$ and ρ denote the representation of L afforded by the module $V(\lambda)$. Consider the basis $B = \{k_1, k_2, \ldots, k_l\}\bigcup\{z_\alpha \mid \alpha \in \Phi\}$ of L as described before Proposition 3.3.6. Let μ be a weight of $V(\lambda)$. Evidently, $\rho(x_\alpha)\rho(z_\alpha)$ is an endomorphism of $V(\lambda)_\mu$ (indeed, $\rho(z_\alpha)$ takes $V(\lambda)_\mu$ to $V(\lambda)_{\mu-\alpha}$ and $\rho(x_\alpha)$ takes $V(\lambda)_{\mu-\alpha}$ back to $V(\lambda)_\mu$). Fix $\alpha > 0$ and consider the restriction $\rho|_{S_\alpha}$ of ρ to S_α, where S_α is the subalgebra of L generated by $\{x_\alpha, h_\alpha, y_\alpha\}$. We look at the non-standard basis $\{x_\alpha, t_\alpha, z_\alpha\}$, where $t_\alpha = \frac{<\alpha,\alpha>}{2}h_\alpha$ and $z_\alpha = \frac{<\alpha,\alpha>}{2}y_\alpha$. Let μ be a weight of L-module $V(\lambda)$ such that $\mu + \alpha$ is not a weight of $V(\lambda)$. Then the α-string through μ is precisely, $\{\mu, \mu - \alpha, \ldots, \mu - q\alpha\}$, where $q = \prec \mu, \alpha \succ$. Evidently,

$$W_\mu = V(\lambda)_\mu \oplus V(\lambda)_{\mu-\alpha} \oplus \cdots \oplus V(\lambda)_{\mu-q\alpha}$$

is a S_α-submodule of $V(\lambda)$. By Weyl's theorem, W_μ is the direct sum of irreducible representations of S_α. Indeed, for a string of weights of S_α which is invariant under σ_α and which involves a given member of a basis of $V(\lambda)_{\mu-i\alpha}$ as a maximal vector, there is one and only one component in the decomposition of W_μ as the direct sum of simple S_α-modules. Thus, if n_i, $0 \leq i \leq \frac{q}{2}$ denotes the number of components with the highest weight $(\mu - i\alpha)(h_\alpha) = q - 2i$ as S_α-module which appears in W_μ, then $m_\lambda(\mu - i\alpha) = DimV(\lambda)_{\mu-i\alpha} = n_0 + n_1 + \cdots + n_i$, where $m_\lambda(\mu - i\alpha) = DimV(\lambda)_{\mu-i\alpha}$. Consequently,

$$n_i = m_\lambda(\mu - i\alpha) - m_\lambda(\mu - (i - 1)\alpha) \cdots . \qquad (3.15)$$

Now, fix a k, $0 \leq k \leq \frac{q}{2}$. We wish to find the trace $Tr(\rho(x_\alpha)\rho(z_\alpha))$ on $V(\lambda)_{\mu-k\alpha}$. Fix r, $0 \leq r \leq k$. Consider the irreducible component of $W_{\mu-k\alpha}$ having the highest weight $q - 2r = (\mu - r\alpha)(h_\alpha)$ and the weight space generated by w_{k-r}. Replacing x by x_α, t by t_α, z by z_α, m by $q - 2r$, and r by $k - r$ in Eq. (3.14), we obtain that

$$\rho(x_\alpha)\rho(z_\alpha)(w_{k-r}) = (q - r - k)(k - r + 1)\frac{<\alpha, \alpha>}{2}w_{k-r} \cdots . \qquad (3.16)$$

We have n_r S_α-simple components of $V(\lambda)_{q-r\alpha}$ of the highest weight $q - 2r$. Hence from (3.16), the matrix of $\rho(x_\alpha)\rho(z_\alpha)|_{V(\lambda)_{\mu-k\alpha}}$ has n_r diagonal entries as given by

Eq. (3.16) relative to a suitable basis of eigenvectors. Thus, varying r from 0 to k, $\rho(x_\alpha)\rho(z_\alpha)|_{V(\lambda)_{\mu-k\alpha}}$ has a diagonal matrix representation with respect to a suitable basis and the trace is given by
$\sum_{r=0}^{k} n_r(q-r-k)(k-r+1)\frac{<\alpha,\alpha>}{2}$
$= \sum_{r=0}^{k}(m_\lambda(\mu-r\alpha)-m_\lambda(\mu-(r-1)\alpha))(q-r-k)(k-r+1)\frac{<\alpha,\alpha>}{2}$ (putting the value of n_r)

$$= \sum_{r=0}^{k} m_\lambda(\mu-r\alpha)(q-2r)\frac{<\alpha,\alpha>}{2} \quad \cdots . \tag{3.17}$$

Since $q = \frac{2<\mu,\alpha>}{<\alpha,\alpha>}$, we obtain

$$Tr\rho(x_\alpha)\rho(z_\alpha)|_{V(\lambda)_{\mu-k\alpha}} = \sum_{r=0}^{k} m_\lambda(\mu-r\alpha) <\mu-r\alpha,\alpha> \quad \cdots . \tag{3.18}$$

Observe that the reflection σ_α maps the set $\{\mu-r\alpha \mid r \leq \frac{q}{2}\}$ to the set $\{\mu-s\alpha \mid \frac{q}{2} < s \leq q\}$ in a one-to-one manner. As such, using a similar argument as above, we see that for $\frac{q}{2} < k \leq q$,

$$Tr\rho(x_\alpha)\rho(z_\alpha)|_{V(\lambda)_{\mu-k\alpha}} = \sum_{r=0}^{q-k} m_\lambda(\mu-r\alpha) <\mu-r\alpha,\alpha> .$$

Since $m_\lambda(\mu-(q-k)\alpha) = 0$, we have

$$Tr\rho(x_\alpha)\rho(z_\alpha)|_{V(\lambda)_{\mu-k\alpha}} = \sum_{r=0}^{q-k-1} m_\lambda(\mu-r\alpha) <\mu-r\alpha,\alpha> \quad \cdots . \tag{3.19}$$

Further, since $q = \frac{2<\mu,\alpha>}{<\alpha,\alpha>}$, for $\frac{q}{2} < r \leq q$,

$$<\mu-r\alpha,\alpha> + <\mu-(q-r)\alpha,\alpha> = 0.$$

Again, since $m_\lambda(\mu-r\alpha) = m_\lambda(\mu-(q-r)\alpha)$, it follows that

$$m_\lambda(\mu-r\alpha) <\mu-r\alpha,\alpha> + m_\lambda(\mu-(q-r)\alpha) <\mu-(q-r)\alpha,\alpha> = 0 \cdots . \tag{3.20}$$

Using (3.18), (3.19), and (3.20), we observe that

$$Tr\rho(x_\alpha)\rho(z_\alpha)|_{V(\lambda)_{\mu-k\alpha}} = \sum_{r=0}^{k} m_\lambda(\mu-r\alpha) <\mu-r\alpha,\alpha> \quad \cdots \tag{3.21}$$

for all k. If we take any weight ν of $V(\lambda)$ and $\nu + k\alpha$ is the last term in the α-string through ν, then taking $\mu = \nu + k\alpha$, and using (3.21), we observe that for all $\mu \in \Gamma(V(\lambda))$

$$Tr\rho(x_\alpha)\rho(z_\alpha)|_{V(\lambda)_\mu} = \sum_{i=0}^{\infty} m_\lambda(\mu + i\alpha) < \mu + i\alpha, \alpha > \cdots . \quad (3.22)$$

Substituting the value of $Tr\rho(x_\alpha)\rho(z_\alpha)|_{V(\lambda)_\mu}$ in (3.7), we obtain the identity

$$c_\lambda m_\lambda(\mu) = < \mu, \mu > m_\lambda(\mu) + \sum_{\alpha\in\Phi}\sum_{i=0}^{\infty} m_\lambda(\mu + i\alpha) < \mu + i\alpha, \alpha > \cdots . \quad (3.23)$$

The above identity is valid for all $\mu \in \Lambda$. For, if $\mu \notin \Gamma_\lambda$, then $m_\lambda(\mu) = 0$ and it also follows (by using an argument similar to the one used to establish (3.20)) that for each α,

$$\sum_{i=1}^{\infty} m_\lambda(\mu + i\alpha) < \mu + i\alpha, \alpha > = 0.$$

Indeed,

$$\sum_{i=-\infty}^{\infty} m_\lambda(\mu + i\alpha) < \mu + i\alpha, \alpha > = 0 \cdots , \quad (3.24)$$

or equivalently,

$$\sum_{i=1}^{\infty} m_\lambda(\mu - i\alpha) < \mu - i\alpha, \alpha > = m_\lambda(\mu) < \mu, \mu > + \sum_{i=1}^{\infty} m_\lambda(\mu + i\alpha) < \mu + i\alpha, \alpha > \cdots . \quad (3.25)$$

In turn,

$$c_\lambda m_\lambda(\mu) = < \mu, \mu > m_\lambda(\mu) + m_\lambda(\mu)\sum_{\alpha>0} < \mu, \alpha > + 2\sum_{\alpha>0}\sum_{i=1}^{\infty} m_\lambda(\mu + i\alpha) < \mu + i\alpha, \alpha > \cdots . \quad (3.26)$$

Since $\sum_{\alpha\in\Delta} \alpha = 2\delta$,

$$c_\lambda m_\lambda(\mu) = < \mu, \mu + 2\delta > m_\lambda(\mu) + 2\sum_{\alpha>0}\sum_{i=1}^{\infty} m_\lambda(\mu + i\alpha) < \mu + i\alpha, \alpha > \cdots . \quad (3.27)$$

Putting $\mu = \lambda$ and observing that $m_\lambda(\lambda + i\alpha) = 0$ for all $\alpha > 0$ and $i \geq 1$, we can solve for c_λ which equals $< \lambda, \lambda + 2\delta > = < \lambda + \delta, \lambda + \delta > - < \delta, \delta >$. Substituting the value of c_λ, we obtain the desired formula. This establishes Freudenthal's multiplicity formula. ♯

Example 3.4.13 Consider the case A_1. $L = sl(2, \mathbb{C})$, $\{x, h, y\}$ the standard basis, $\Phi = \{\alpha, -\alpha\}$, and $\Delta = \{\alpha\}$, where $\alpha(h) = 2$. The fundamental weight $\lambda_1 = a\alpha$, $a \in \mathbb{Q}$ is given by the condition $< \lambda_1, \frac{2\alpha}{<\alpha,\alpha>} > = 1$. Thus, $a = \frac{1}{2}$, and $\lambda_1 = \frac{\alpha}{2} = \delta$. Without any loss, we may assume that $< \alpha, \alpha > = 1$. The weight lattice $\Lambda = \{r\lambda_1 = r\frac{\alpha}{2} \mid r \in \mathbb{Z}\}$ and $\Lambda^+ = \{r\lambda_1 = r\frac{\alpha}{2} \mid r \geq 0\}$. Note that $< r\lambda_1 + \delta, r\lambda_1 + \delta > = (\frac{r+1}{2})^2$. Let $\lambda = n\lambda_1 \in \Lambda^+$. Consider $V(\lambda)$. Then λ is the highest weight in $V(\lambda)$. Indeed, in the earlier sense (see Theorem 1.4.28) $\lambda(h) = n\,\lambda_1(h) = n$ is the highest weight of $V(\lambda)$. Now, $\mu = r\lambda_1 \in \Gamma_\lambda$ only when $\mu \leq \lambda$ and $\sigma_\alpha(\mu) \leq \lambda$. This is equivalent to say that $-m \leq r \leq m$. Already $m_\lambda(\lambda) = 1$. Let $\mu = (n-1)\lambda_1$. The LHS of Freudenthal's formula is

$$(< \lambda + \delta, \lambda + \delta > - < \mu + \delta, \mu + \delta >)m_\lambda(\mu) = \left(\left(\frac{n+1}{2}\right)^2 - \left(\frac{n}{2}\right)^2\right) m_\lambda(\mu).$$

However, for $i \geq 1$, $(n-1)\lambda_1 + i\alpha = (n-1)\lambda_1 + 2i\lambda_1$ does not belong to Γ_λ, $m_\lambda(\mu + i\alpha) = 0$ for all $i \geq 1$. Consequently, the RHS of the Freudenthal formula is 0. This means that $m_\lambda(\lambda - \lambda_1)$ is 0. Now, let $\mu = \lambda - 2\lambda_1$. Then the LHS of the Freudenthal formula gives $nm_\lambda(\mu)$ whereas the RHS becomes n. Consequently, $m_\lambda(\lambda - 2\lambda_1) = 1$. Proceeding recursively, we see that $m_\lambda(\lambda - r\lambda_1) = 1$ if r is even and it is 0 if r is odd. This also justifies Theorem 1.4.28.

In this case, $W(\Phi)$ is small of order 2, hence Kostant and Weyl formulas can also be used conveniently.

Example 3.4.14 Consider the case A_2. $L = sl(3, \mathbb{C})$, with the standard basis, standard root system $\Phi = \{\pm\alpha_1, \pm\alpha_2, \pm(\alpha_1 + \alpha_2)\}$, and $\Delta = \{\alpha_1, \alpha_2\}$. Since ϕ has only one root length, we may assume that $< \alpha_1, \alpha_1 > = 1 = < \alpha_2, \alpha_2 >$. Also, then $< \alpha_1, \alpha_2 > = -\frac{1}{2}$. Let $\{\lambda_1, \lambda_2\}$ be the basis of the lattice Λ of weights consisting of the fundamental weights. Suppose that $\lambda_1 = a\alpha_1 + b\alpha_2$, $\lambda_2 = c\alpha_1 + d\alpha_2$, where $a, b, c, d \in \mathbb{Q}$. Then $1 = < \lambda_1, 2\alpha_1 > = 2a - b$ and $0 = < \lambda_1, 2\alpha_2 > = -\frac{1}{2}a + b$. Solving for a and b, we obtain that $\lambda_1 = \frac{1}{3}(2\alpha_1 + \alpha_2)$ and similarly, $\lambda_2 = \frac{1}{3}(\alpha_1 + 2\alpha_2)$. Solving α_1 and α_2 in terms of λ_1 and λ_2, we find that $\alpha_1 = 2\lambda_1 - \lambda_2)$ and $\alpha_2 = -\lambda_1 + 2\lambda_2)$. Also $\delta = \frac{1}{2}(\alpha_1 + \alpha_2 + (\alpha_1 + \alpha_2)) = \alpha_1 + \alpha_2 = \lambda_1 + \lambda_2$. Further, $< \lambda_1, \alpha_1 > = \frac{1}{2} = < \lambda_2, \alpha_2 >$, and $< \lambda_1, \alpha_2 > = 0 = < \lambda_2, \alpha_1 >$. The Weyl group $W(\Phi) = \{I, \sigma_{\alpha_1}, \sigma_{\alpha_2}, \sigma_{\alpha_1}\sigma_{\alpha_2}, \sigma_{\alpha_2}\sigma_{\alpha_1}, \sigma_{\alpha_1}\sigma_{\alpha_2}\sigma_{\alpha_1}\}$ is the the dihedral group S_3 of order 6. Let $\lambda \in \Lambda^+$. Then $\mu \in \Gamma_\lambda$ if and only if $\sigma(\mu) \leq \lambda$ for all $\sigma \in W(\Phi)$ (Remark 3.2.14). Take $\lambda = \lambda_1 + 3\lambda_2 \in \Lambda^+$. Consider $V(\lambda)$. One can explicitly determine Γ_λ. Already $m_\lambda(\lambda) = 1$. We can check that $< \lambda + \delta, \lambda + \delta > = \frac{28}{3}$. Let $\mu = \lambda - \alpha_1$. Then $< \mu + \delta, \mu + \delta > = \frac{25}{3}$. By Freudenthal's formula

$$(<\lambda+\delta,\lambda+\delta> - <\mu+\delta,\mu+\delta>)m_\lambda(\mu) = 2\Big[\sum_{i=1}^{\infty} m_\lambda(\mu+i\alpha_1) <$$

$$\mu+i\alpha_1,\alpha_1> + \sum_{i=1}^{\infty} m_\lambda(\mu+i\alpha_2) < \mu+i\alpha_2,\alpha_2>$$

$$+\sum_{i=1}^{\infty} m_\lambda(\mu+i\alpha_1+i\alpha_2) < \mu+i\alpha_1+i\alpha_2,\alpha_1+\alpha_2>\Big].$$

Since $\lambda < \mu + i\alpha_1$ for all $i \geq 2$, $\lambda < \mu + i\alpha_2$ for all $i \geq 1$, we see that

$$m_\lambda(\lambda-\alpha_1) = 2 < \lambda,\alpha_1 > = 1.$$

Thus, $m_\lambda(\lambda-\alpha_1) = 1$. Similarly, using that $m_\lambda(\lambda-\alpha_1) = 1$, we can show that $m_\lambda(\lambda-\alpha_1-\alpha_2) = 2$. Indeed, proceeding recursively, we can find all $m_\lambda(\lambda - i\alpha - j\alpha)$.

Exercises

3.4.1. Use Weyl's degree formula to show that $deg(\lambda) = n+1$, where λ_1 is an in Example 3.4.13.

3.4.2. Let ρ be a faithful irreducible representation of L of the smallest positive degree. Show that the highest weight of ρ is λ_i for some i, where $\{\lambda_1, \lambda_2, \ldots, \lambda_l\}$ is the basis of Λ consisting of fundamental weights.

3.4.3. Use Weyl's degree formula to determine the dimension of $V(\lambda)$, where λ is as in Example 3.4.14. Also determine $Ch_{V(\lambda)}$.

3.4.4. Use Kostant's formula to determine $m_\lambda(\lambda-\alpha_1)$, where λ and α_1 are as in Example 3.4.14.

3.4.5. Consider G_2 with the standard root system Φ, $\Delta = \{\alpha_1, \alpha_2\}$, α_1 being short. Let $\{\lambda_1, \lambda_2\}$ denote the basis consisting of fundamental dominant weights. Express λ_1 and λ_2 in terms of α_1 and α_2. Let $\lambda = 2\lambda_1 + e\lambda_2$. Determine $m_\lambda(\lambda-\alpha_1)$ and also $m_\lambda(\lambda-\alpha_1-\alpha_2)$.

3.4.6. Find $Dim V(\lambda) = deg(\lambda)$, where λ is as in Example 3.4.14.

3.4.7. Use the Steinberg formula to decompose $V(\lambda_1) \otimes V(\lambda_2)$ as the direct sum of simple modules, where λ_1 and λ_2 are as in Example 3.4.14.

3.4.7. Decompose $V(\lambda_1+\lambda_2) \otimes V(\lambda_1+2\lambda_2)$ as the direct sum of simple modules, where λ_1 and λ_2 are as in Example 3.4.14.

Chapter 4
Chevalley Groups

The simple complex Lie groups are in natural bijective correspondence with simple Lie algebras over the field of complex numbers. Dickson and Dieudonne introduced and studied the version of some of these classical Lie groups over arbitrary fields. Motivated and tempted by their study of these classical matrix groups, Chevalley, in 1955, unified and introduced groups (bearing his name) by discovering, ingeniously, an integral basis (a basis with respect to which all the structure constants are integers) $\{h_i \mid 1 \leq i \leq l\} \bigcup \{x_\alpha \mid \alpha \in \Phi\}$ of a semi-simple Lie algebra L over $\mathbb{C}$ associated with a root system. Such a basis is also termed as a Chevalley basis. Further, the existence of such a basis allowed to construct analogues of complex simple Lie groups over arbitrary fields. These finite simple groups over finite fields together with their twisted analogues and the alternating groups A_n give the complete list of infinite families of finite simple groups. The $\mathbb{Z}$-span $L(\mathbb{Z})$ of a Chevalley basis has the obvious induced Lie algebra structure over $\mathbb{Z}$. Tensoring with an arbitrary field K gives us a Lie algebra $L(K)$ over K. This allows us to think of $exp(a(ad(x_\alpha)))$ as an automorphism of $L(K)$ for each $a \in K$ and $\alpha \in \Phi$. In turn, we obtain the group $G(K)$ of automorphisms of $L(K)$ generated by $\{exp(a(ad(x_\alpha))) \mid a \in K \ and \ \alpha \in \Phi\}$. Taking K to be finite Galois fields F_q, Chevalley discovered new families (different from A_n) of finite simple groups (termed as finite simple groups of Lie types). Indeed, for each such group $G(F_q)$, there are certain homomorphisms from $SL(2, F_q)$ to $G(F_q)$ and $G(F_q)$ may be viewed as an amalgamation of certain copies of $SL(2, F_q)$ through the members of the Weyl group $W(\Phi)$. In a further development, Tits gave an axiomatic description of these groups with the help of Weyl groups.

The purpose of this chapter is to introduce Chevalley Groups and develop its structure theory. We shall also introduce the twisted groups of Lie types.

R. Lal, *Algebra 4*, Infosys Science Foundation Series,
https://doi.org/10.1007/978-981-16-0475-1_4

4.1 Classical Linear Groups

In this section, we introduce and study (briefly) the classical linear groups over arbitrary fields which provide initial examples of Chevalley groups.

Let K be a field and V be a finite-dimensional vector space of dimension n (say) over K. The general linear group $GL(V) \approx GL(n, K)$ of non-singular linear transformations on V and the special linear group $SL(V) \approx SL(n, K)$ of linear transformations of determinant 1 have already been studied in some detail. We recall some of their basic properties:

(a) $1 \longrightarrow SL(n, K) \xrightarrow{i} GL(n, K) \xrightarrow{Det} K^\star \longrightarrow 1$
is a split exact sequence.

(b) $SL(n, K)$ is generated by the set $\{E_{ij}^{\lambda} \mid i \neq j, \lambda \in K^\star\}$ of transvections and $GL(n, K)$ is generated by the set $\{E_{ij}^{\lambda} \mid i \neq j, \lambda \in K^\star\} \bigcup \{Diag(1, 1, \ldots, 1, d) \mid d \in K^\star\}$.

(c) Let $K = F_q$ be a finite field containing $q = p^m$ elements. Then for $n \geq 3$, $SL(n, F_q)$ is the group generated by the set of transvections subject to the Steinberg relations:

(i) $E_{ij}^{\lambda} E_{ij}^{\mu} = E_{ij}^{\lambda+\mu}$ for all $\lambda, \mu \in K$.
(ii) $[E_{ij}^{\lambda}, E_{kl}^{\mu}] = 1$ for $i \neq l$, J $\neq k$ and for all $\lambda, \mu \in K$.
(iii) $[E_{ij}^{\lambda}, E_{jl}^{\mu}] = E_{il}^{\lambda\mu}$ for $j \neq l$.

The group $GL(n, F_q)$ is also denoted by $GL(n, q)$ and the group $SL(n, F_q)$ is denoted by $SL(n, q)$.

For $n = 2$, $SL(2, q)$ is again generated by the set of transvections subject to the relations (i) and

(iii') $w_{12}^{\lambda} E_{12}^{\mu} w_{12}^{-\lambda} = E_{21}^{-\lambda^2\mu}$ for all $\lambda, \mu \in K$, where

$$w_{12}^{\lambda} = E_{12}^{\lambda} E_{21}^{-\lambda^{-1}} E_{12}^{\lambda} = \begin{bmatrix} 0 & \lambda \\ -\lambda^{-1} & 0 \end{bmatrix}.$$

However, if K is infinite, more relations (coming out of the Schur Multiplier (Chap. 10, Algebra 2)) are needed to present $SL(2, K)$.

(d) $Z(GL(n, K)) = \{aI_n \mid a \in K^\star\}$, $Z(SL(n, K)) = \{aI_n \mid a^n = 1\}$.

(e) $PSL(n, K) = SL(n, K)/Z(SL(n, K))$ is simple except for $n = 2$ with $\mid K \mid \leq 3$.

(f) $\mid GL(n, q) \mid = (q^n - 1)(q^n - q) \cdots (q^n - q^{n-1})$, $\mid SL(n, q) \mid = q^{n-1}(q^n - 1)(q^n - q) \cdots (q^n - q^{n-2})$.

(g) $\mid PSL(n, q) \mid = \frac{q^{n-1}(q^n-1)(q^n-q)\cdots(q^n-q^{n-2})}{(n,q-1)}$.

(h) $PSL(2, 5) \approx SL(2, 4) \approx A_5$: Consider a group G with presentation $< x, y; x^5, y^3, (xy)^2 >$. Using the coset enumeration method of Coxeter and Todd, one finds that the order of G is 60. The elements $p = (12345)$ and $q = (152)$

of A_5 are such that $p^5 = q^3 = (pq)^3 = 1$. Thus, we have a surjective homomorphism from G to A_5 which maps x to p and y to q. Since both the groups are of same order, it is an isomorphism. Similarly, the elements $\overline{E_{12}}$ and $\overline{w_{12}^1}$ of $PSL(2,5)$ are such that $(\overline{E_{12}})^5 = (\overline{w_{12}^1})^3 = (\overline{E_{12}w_{12}^1})^2 = 1$. Again, since $PSL(2,5)$ is also of order 60, it follows that G is also isomorphic to $PSL(2,5)$. This gives an explicit isomorphism from A_5 to $PSL(2,5)$. Use a similar argument to show that $SL(2,4)$ is also isomorphic to A_5.

Further, $PSL(2,9) \approx A_6$. Indeed, it can be shown $PSL(2,9)$ contains a subgroup isomorphic to $PSL(2,5)$. Evidently, this subgroup is of index 6. Thus, $PSL(2,9)$ acts transitively on a set containing 6 elements. Using the fact that $PSL(2,9)$ is simple, it follows that $PSL(2,9)$ is isomorphic to a subgroup of A_6. By order consideration, it follows that $PSL(2,9) \approx A_6$.

Other classical linear groups are obtained with the help of special types of nondegenerate bi-linear forms. Recall that a bi-linear form on a finite-dimensional vector space V over K is a map f from $V \times V$ to K such that

(i) $f(ax + by, z) = af(x, y) + bf(y, z)$, and
(ii) $f(x, ay + bz) = af(x, z) + bf(x, z)$
for all $x, y, z \in V$, and $a, b \in K$.

Thus, a map f from $V \times V$ to K is a bi-linear form if and only if the maps f_x and f^y from V to K defined by $f_x(y) = f(x, y)$ and $f^y(x) = f(x, y)$ are linear functionals. Further, then the maps $x \rightsquigarrow f_x$, and $y \rightsquigarrow f^y$ denoted by L_f and R_f, respectively, are linear transformations from V to $V^\star$.

Let A be a $n \times n$ matrix with entries in K. Then, the map f from $K^n \times K^n$ to K defined by

$$f(\overline{x}, \overline{y}) = \overline{x} A \overline{y}^t$$

is a bi-linear form on K^n, where $\overline{x} = [x_1, x_2, \ldots, x_n]$ denotes a row vector in K^n. Indeed, these are all bi-linear forms on K^n. In fact, given any bi-linear form f on a vector space V of dimension n over K, there exists an isomorphism T from V to K^n(corresponding to each choice of basis), and a matrix A such that $f(x, y) = T(x)^t A T(y)$. Thus, essentially, these are all bi-linear forms on a vector space of dimension n.

Let f and g be bi-linear forms on V and $a, b \in K$. Then, it is easily seen that $af + bg$ defined by $(af + bg)(x, y) = af(x, y) + bg(x, y)$ is a bi-linear form on V. Further, the zero map which takes every thing to 0 is already a bi-linear form. Thus, the set $BL(V, K)$ of bi-linear forms on V is a vector space over K with respect to the operations defined above. Let us fix an ordered basis $\{u_1, u_2, \ldots, u_n\}$ of V. Define a map $M_{u_1, u_2, \ldots, u_n}$ from $BL(V, K)$ to $M_n(K)$ by

$$M_{u_1, u_2, \ldots, u_n}(f) = [a_{ij}],$$

where $a_{ij} = f(u_i, u_j)$. This map is a linear transformation, and it is called the **Matrix representation** map relative to the ordered basis $\{u_1, u_2, \ldots, u_n\}$. The matrix $M_{u_1,u_2,\ldots,u_n}(f)$ is called the **matrix representation** of the bi-linear form f. The matrix representation map $M_{u_1,u_2,\ldots,u_n}$ can easily be seen to be a vector space isomorphism from $BL(V, K)$ to $M_n(K)$. Further, if $\{u_1, u_2, \ldots, u_n\}$ is an ordered basis of V, then $\{f_{ij} = u_i^\star u_j^\star \mid 1 \leq i \leq n, 1 \leq j \leq n\}$ forms a basis of $BL(V, K)$, where $\{u_1^\star, u_2^\star, \ldots, u_n^\star\}$ is the dual basis to $\{u_1, u_2, \ldots, u_n\}$.

We say that a bi-linear form f is **congruent** to a bi-linear form g if there is an isomorphism T from V to V such that $g(x, y) = f(T(x), T(y))$ for all $x, y \in V$. This is equivalent to say that the matrix representation of f and g are congruent with respect to any choice of ordered basis of V. Thus, there is a natural bijective correspondence between equivalence class of congruent bi-linear forms and congruence classes of matrices.

Theorem 4.1.1 *Let f be a bi-linear form on a vector space V of finite dimension n over a field K. Then*

$$\rho(L_f) = \rho(A) = \rho(R_f),$$

where A is a matrix of f corresponding to any choice of basis, and ρ denotes the rank function.

Proof Let us first observe that matrices of f corresponding to different ordered bases are congruent, and so they all have the same rank. Because of the rank-nullity theorem, it is sufficient to show that $\nu(L_f) = \nu(A) = \nu(R_f)$, where ν denotes the nullity. Now,

$$\nu(L_f) = dim(\{x \in V \mid L_f(x) = f_x = 0\}) = dim(\{x \in V \mid f(x, y) = 0 \; for \; all \; y \in V\}).$$

If we fix an ordered basis $\{u_1, u_2, \ldots, u_n\}$ of V, then the map T from V to the vector space K^n, which associates to $x = \Sigma_{i=1}^n x_i u_i$ the row vector $\overline{x} = [x_1, x_2, \ldots, x_n]$ is an isomorphism, and then $f(x, y) = T(x)AT(y)^t = \overline{x}A\overline{y}^t$, where A is the matrix of f with respect to the basis $\{u_1, u_2, \ldots, u_n\}$. This isomorphism takes the subspace $\{x \mid f(x, y) = 0 \; for \; all \; y \in V\}$ of V isomorphically on to the subspace $\{\overline{x} \in K^n \mid \overline{x}A\overline{y}^t = 0 \; for \; all \; \overline{y} \in K^n\}$. Thus, $\nu(L_f) = dim(\{\overline{x} \in K^n \mid \overline{x}A\overline{y}^t = 0 \; for \; all \; y\})$. Next, over any field if C is a row vector such that $C\overline{y}^t = 0$ for all $\overline{y} \in K^n$, then $C = 0$ (verify). Hence, $\{\overline{x} \in K^n \mid \overline{x}A\overline{y}^t = 0 \; for \; all \; \overline{y} \in F^n\} = \{\overline{x} \in F^n \mid \overline{x}A = 0\} = \{\overline{x} \in F^n \mid A^t\overline{x}^t = 0\}$. This shows that $\nu(L_f) = \nu(A^t)$. Since A is a square matrix, $\nu(A^t) = \nu(A)$. This shows that $\rho(L_f) = \rho(A)$. Similarly, we can show that $\rho(R_f) = \rho(A)$. ♯

Definition 4.1.2 Let f be a bi-linear form on V. Then, the common number $\rho(L_f) = \rho(A) = \rho(R_f)$ is called the **rank** of f.

Corollary 4.1.3 *Let f be a bi-linear form on a vector space V of dimension n. Then, the following conditions are equivalent.*

1 *Rank of f is n.*

2 $f_x(y) = 0 \; for \; all \; y \; implies \; that \; x = 0$.

3 $f^y(x) = 0$ *for all* x *implies that* $y = o$.
4 *The matrix representation of* f *with respect to any basis is non-singular.* ♯

A bi-linear form f on V is called **non-degenerate** or **non − singular** bi-linear form if it satisfies any one (and hence all) of the above three conditions in the corollary. We say that a linear endomorphism T of V leaves a bi-linear form f invariant if $f(T(x), T(y)) = f(x, y)$ for all $x, y \in V$.

Proposition 4.1.4 *Let* f *be a nondegenerate bi-linear form on* V. *Let* $G(f)$ *denote the set of all linear endomorphisms of* V *which leave* f *invariant. Then,* $G(f)$ *is a subgroup of* $GL(V)$. *Further, if* f *and* f' *are congruent non-degenerate bi-linear forms on* V, *then* $G(f)$ *and* $G(f')$ *are conjugate in* $GL(V)$. *In particular, they are isomorphic.*

Proof Let T be a linear endomorphism of V such that $f(T(x), T(y)) = f(x, y)$ for all $x, y \in V$. Suppose that $T(x) = 0$. Then $f(x, y) = 0$ for all $y \in V$. Since f is nondegenerate, $x = 0$. Thus, T is an injective endomorphism of V. Since V is finite dimensional $T \in GL(V)$. Thus, $G(f) \subseteq GL(V)$. Clearly, $G(f)$ is a subgroup of $GL(V)$. Suppose that f and f' are congruent. Then, there is an element $T \in GL(V)$ such that $f'(x, y) = f(T(x), T(y))$ for all $x, y \in V$. Clearly, $f(x, y) = f'(T^{-1}(x), T^{-1}(y))$ for all $x, y \in V$. Let $A \in G(f)$. Then

$$f'(T^{-1}AT(x), T^{-1}AT(y)) = f(AT(x), AT(y)) = f(T(x), T(y)) = f'(x, y)$$

for all $x, y \in V$. This shows that $T^{-1}G(f)T \subseteq G(f')$. Similarly, $T^{-1}G(f')T \subseteq G(f)$. ♯

Corollary 4.1.5 *Let* P *be a non-singular* $n \times n$ *matrix with entries in* K. *Let* $G(P)$ *denote the subset* $\{A \in GL(n, K) \mid APA^t = P\}$ *of* $GL(n, K)$. *Then* $G(P)$ *is a subgroup* $GL(n, K)$ *and* $G(P)$ *is conjugate to* $G(Q)$ *whenever* P *and* Q *are congruent.* ♯

A bi-linear form is called a symmetric bi-linear form if $f(x, y) = f(y, x)$ for all $x, y \in V$. This is equivalent to say that the matrix representations of f are symmetric. Assume that the characteristic of K is different from 2. Since every symmetric matrix with entries in a field of characteristic different from 2 is congruent to a diagonal matrix (see Algebra 2), the matrix representation of a symmetric matrix is a diagonal matrix with respect to a basis. Such a basis is called an orthogonal basis. Let f be a nondegenerate symmetric bi-linear form on V. Then $G(f)$ is called an **orthogonal group**. If the field K is such that it contains square root of each of its elements (for example $\mathbb{C}$), then every non-singular symmetric matrix is congruent to the identity matrix (see Theorem 5.6.15, Algebra 2). Thus, in this case, $G(f)$ is isomorphic to the group $\{A \in GL(n, K) \mid AA^t = I\}$. If $K = \mathbb{R}$, then for a non-singular $n \times n$ symmetric matrix there is a unique pair p, q of nonnegative integers p, q with $p + q = n$ such that A is congruent to the diagonal matrix $J_{p,q}$ whose first p diagonal entries are 1 and the rest of the diagonal entries are -1. In this case, the group $G(f)$ is isomorphic to the group $O(p, q) = \{A \in GL(n, \mathbb{R}) \mid AJ_{p,q}A^t = J_{p,q}\}$.

Orthogonal Groups over Finite Fields of Odd Characteristic

Proposition 4.1.6 *Let $K = F_q$ be a finite field of odd characteristic. Consider the set $L = \{a^2 \mid a \in F_q^\star\}$ of squares of nonzero elements of F_q. Then L is a subgroup of $F_q^\star$ of index 2.*

Proof Clearly, L is a subgroup. The map $a \mapsto a^2$ from $K^\star$ to L is a surjective homomorphism whose kernel is the subgroup $\{1, -1\}$ of $K^\star$ (since q is odd, $1 \neq -1$). Indeed, since F_q is finite, $F_q^\star$ is cyclic group of even order generated by a nonzero element ξ (say) of F_q. Thus, $L = <\xi^2>$ of order $\frac{q-1}{2}$. ♯

Proposition 4.1.7 *Let F_q be as in the above proposition. Let $a, b \in F_q^\star$. Then the polynomial $X^2 + aY^2 + b$ in $F_q[X, Y]$ has a zero in F_q^2.*

Proof Clearly $K^\star = L \bigcup \xi L$. Suppose that $a \notin L$. Then $K^\star = L \bigcup aL$. If $-b = c^2 \in L$, then the pair $(c, 0)$ is a solution of the said equation. If $-b = av^2 \notin L$, then the pair $(0, v)$ is a solution. Next, suppose that $a = u^2 \in L$. Again, if $-b = c^2 \in L$, then $(c, 0)$ is a solution. Suppose that $-b \notin L$. Now, $X^2 + aY^2 + b = X^2 + Z^2 + b$, where $Z = uY$. Clearly, if for some c with $-c \notin L$, $X^2 + Z^2 + c = 0$ has a solution, then it has a solution for all d with $-d \notin L$. Thus, if $X^2 + Z^2 + b = 0$ has no solution, then $L \bigcup \{0\}$ will turn out to be a subgroup of the additive group K. But $\mid L \bigcup \{0\} \mid = \frac{q+1}{2}$. This is a contradiction, since $\frac{q+1}{2}$ does not divide q. This completes the proof. ♯

Proposition 4.1.8 *Let f be a nondegenerate symmetric bi-linear form on a vector space V of dimension $n \geq 3$ over F_q, where q is an odd prime power. Then there exists a pair u_1, u_2 of linearly independent elements of V such that $f(u_1, u_1) = 0 = f(u_2, u_2)$ and $f(u_1, u_2) = 1$. Further, if U is the subspace of V generated by $\{u_1, u_2\}$, then $V = U \oplus U^\perp$.*

Proof Since q is odd, we have an ordered orthogonal basis $\{v_1, v_2, \ldots, v_n\}$ of V. Suppose that $f(v_i, v_i) = a_i \neq 0$ and $f(v_i, v_j) = 0$ for all $i \neq j$. From the above proposition, for $a = a_1^{-1}a_2$ and $b = a_1^{-1}a_3$, there is a pair $c, d \in F_q$ such that $c^2 + ad^2 + b = 0$. Take $u_1 = cv_1 + dv_2 + v_3$. Then $f(u_1, u_1) = c^2a_1 + d^2a_2 + a_3 = a_1(c^2 + ad^2 + b) = 0$. Since f is nondegenerate, there is an element $v \in V$ such that $f(u_1, v) \neq 0$. We may assume that $f(u_1, v) = 1$. Now $f(v + xu_1, v + xu_1) = f(v, v) + 2x$. Take $u_2 = v - \frac{f(v,v)}{2}u_1$. Then $f(u_2, u_2) = 0$ and $f(u_1, u_2) = 1$. The last assertion also follows if we observe that $U \bigcap U^\perp = \{0\}$. ♯

Proposition 4.1.9 *Let f be a nondegenerate symmetric bi-linear form on a vector space V of dimension $2m + 1$ over a field F_q, where q is odd. Then there is a basis $\{u_0, u_1, u_2, \ldots, u_{2m}\}$ of V such that $f(u_0, u_0) = a$ for some $a \neq 0$, $f(u_0, u_i) = 0$ for all i, $f(u_i, u_j) = 0$ for all i, j with $\mid i - j \mid \neq m$, and $f(u_i, u_{m+i}) = 1$ for all i, $1 \leq i \leq m$. Further, if f' is another nondegenerate symmetric bi-linear form, then $G(f)$ is isomorphic to $G(f')$.*

Proof We prove the existence of such a basis $\{u_0, u_1, u_2, \ldots, u_{2m}\}$ of V by the induction on m. Suppose that $m = 1$. Then from the above Proposition, we have a linearly independent pair u_1, u_2 of elements of V such that $f(u_1, u_2) = 1 = f(u_2, u_1)$, $f(u_1, u_1) = 0 = f(u_2, u_2)$. Also $V = U \oplus U^{\perp}$. Let u_0 be a nonzero element of $U^{\perp}$. Then, $\{u_0, u_1, u_2\}$ is a basis of V with the required property. Similarly, assuming the assertion for $r < m,\ m \geq 2$, we can establish it for m.

Now, let f' be another nondegenerate symmetric bi-linear form on a vector space V of dimension $2m + 1$ over F_q. Then as before, there is basis $\{v_0, v_1, \ldots, v_{2m+1}\}$ of V such that $f(v_0, v_0) = b$ for some $b \neq 0$, $f(v_0, v_i) = 0$ for all i, $f(v_i, v_j) = 0$ for all i, j with $| i - j | \neq m$, and $f(v_i, v_{m+i}) = 1$ for all i, $1 \leq i \leq m$. Put $c = a^{-1}b$. Consider the automorphism T of V given by $T(u_0) = v_0$, $T(u_i) = v_i$ and $T(u_{m+i}) = cv_{m+i}$ for $1 \leq i \leq m$. Then $f'(x, y) = cf(T^{-1}(x), T^{-1}(y))$ for all $x, y \in V$. Evidently, cf is also a nondegenerate symmetric bi-linear form on V such that $G(f) = G(cf)$. It also follows from the above identity that $G(cf)$ is conjugate to $G(f')$ in $GL(V)$. This shows that $G(f) \approx G(f')$. It particular, it follows that upto isomorphism, there is a unique orthogonal group on any $2m + 1$ dimensional $(m \geq 1)$ vector space over F_q. ♯

Consider the $(2m + 1) \times (2m + 1)$ matrix $J_{2m+1} = [a_{ij}]$, where $a_{11} = 1$, $a_{1i} = 0 = a_{i1}$ for all $i \neq 1$, $a_{jk} = 1 = a_{kj}$ for $| j - k | = m$, $j, k \geq 2$ and all other entries are 0. Then J_{2m+1} is a non-singular symmetric matrix. Consequently, we have a nondegenerate symmetric bi-linear form f on F_q^{2m+1} given by

$$f(\overline{x}, \overline{y}) = \overline{x} J_{2m+1} \overline{y}^t.$$

Thus, the unique (up to isomorphism) orthogonal group on a $2m + 1$ dimensional vector space denoted by $O(2m + 1, q)$ is given by

$$O(2m + 1, q) = \{A \in GL(2m + 1, q) \mid A J_{2m+1} A^t = J_{2m+1}\}.$$

We state few elementary properties of $O(2m + 1, q)$. Evidently, the determinant of each member of $O(2m + 1, q)$ is ± 1. The normal subgroup $SO(2m + 1, q) = \{A \in O(2m + 1, q) \mid Det A = 1$ of $O(2m + 1, q)$ of index 2 is called the **special orthogonal group**. It can be seen $[SO(2m + 1, q), SO(2m + 1, q)] = [O(2m + 1, q), O(2m + 1, q)]$ is a simple subgroup of $SO(2m + 1, q)$ of index 2. This group is denoted by $\Omega(2m + 1, q)$. For $m = 1$, $\Omega(3, q) \approx PSL(2, q)$ (Exercise 4.1.4).

Counting the non-singular matrices satisfying the condition $A J_{2m+1} A^t = J_{2m+1}$, one can establish the following:

Proposition 4.1.10 $| O(2m + 1, q) | = 2q^{m^2}(q^2 - 1)(q^4 - 1) \cdots (q^{2m} - 1)$. ♯

Next, we study orthogonal groups on even-dimensional vector spaces:

Proposition 4.1.11 *Let V be a vector space of dimension $2m \geq 2$ over F_q, where q is odd. Let f be a nondegenerate symmetric bi-linear form on V. Then there is a basis $\{u_1, u_2, \ldots, u_m, u_{m+1}, \ldots, u_{2m}\}$ such that*
1. $f(u_i, u_{m+i}) = 1 = f(u_{m+i}, u_i)$ for all $i \leq m - 1$, $f(u_m, u_{2m}) = 1 = f(u_{2m}, u_m)$, and for the rest of the pair i, j, $f(u_i, u_j) = 0$,

or else
2. $f(u_i, u_{m+i}) = 1 = f(u_{m+i}, u_i)$ *for all* $i \leq m-1$, $f(u_m, u_{2m}) = 0 = f(u_{2m}, u_m)$, $f(u_m, u_m) = -af(u_{2m}, u_{2m}) \neq 0$, *and for the rest of the pair* i, j, $f(u_i, u_j) = 0$, *where* a *is an element of* F_q *which is not a square.*

Proof Using Proposition 4.1.9, and induction, we can find a basis $\{u_1, u_2, \ldots, u_{2m}\}$ which satisfies the condition 1 and 2, except for the conditions to be satisfied by u_m and u_{2m}. Let U denote the subspace of V generated by $\{u_1, u_2, \ldots, u_{m-1}, u_{m+1}, u_{m+2}, \ldots, u_{2m-1}\}$. Then $V = U \oplus U^{\perp}$. Suppose that $U^{\perp}$ contains a nonzero element u such that $f(u, u) = 0$. Since f is nondegenerate, there is vector $v \in U^{\perp}$ such that $\{u, v\}$ generates $U^{\perp}$ and $f(u, v) \neq 0$. Suppose $f(u, v) = b$. If $f(v, v) = 0$, then take $u_m = u$ and $u_{2m} = b^{-1}v$. If $f(v, v) \neq 0$, then $w = u - \frac{2b}{f(v,v)}v$ is such that $f(w, w) = 0$ and $f(u, w) \neq 0$. Take $u_m = u$ and $u_{2m} = f(v, w)^{-1}w$. Thus, in this case, we have a basis satisfying 1. Next, suppose that $f(u, u) \neq 0$ for all nonzero $u \in U^{\perp}$. Take an orthogonal basis $\{u_m, u_{2m}\}$. Then $f(u_m, u_{2m}) = 0 = f(u_{2m}, u_m)$ and $f(u_m, u_m) = -af(u_{2m}, u_{2m}) \neq 0$ for some $a \in F_q$. For any nonzero pair x, y of scalars $xu_m + yu_{2m} \neq 0$. Hence, $f(u_m, u_m)(x^2 - ay^2) = f(xu_m + yu_{2m}, xu_m + yu_{2m}) \neq 0$. This means that a is not a square. ♯

Let f_+ denote a nondegenerate symmetric bi-linear form on a vector space of dimension $2m$ which admits a basis satisfying the condition 1 of the above Proposition and f_- denote a nondegenerate symmetric bi-linear form admitting a basis satisfying the condition 2 of the above Proposition. If f is any nondegenerate symmetric bi-linear form on V, then as in Proposition 4.1.10, $G(f)$ is isomorphic to $G(f_+)$ or to $G(f_-)$. Evidently, $G(f_+)$ is isomorphic to the subgroup

$$\{A \in GL(2m, q) \mid AJ^{+}_{2m}A^{t} = J^{+}_{2m}\}$$

of $GL(2m, q)$, where

$$J^{+}_{2m} = \begin{bmatrix} 0_m & I_m \\ I_m & 0_m \end{bmatrix}$$

is the matrix of f_+ relative to the basis 1 as described in Proposition 4.1.12. We denote this orthogonal group by $O_+(2m, q)$. Counting the number of matrices in $O_+(2m, q)$, we see that

$$\mid O_+(2m, q) \mid = 2q^{m(m-1)}(q^2 - 1)(q^4 - 1)\cdots(q^{2(m-1)} - 1)(q^m - 1).$$

Similarly, $G(f_-)$ is isomorphic to the subgroup

$$\{A \in GL(2m, q) \mid AJ^{-}_{2m}A^{t} = J^{+}_{2m}\}$$

of $GL(2m, q)$, where J^{-}_{2m} is the matrix of f_- relative to the basis 2 as described in Proposition 4.1.12. We denote this orthogonal group by $O_-(2m, q)$. Counting the number of matrices in $O_-(2m, q)$, we see that

$$| O_-(2m, q) | = 2q^{m(m-1)}(q^2 - 1)(q^4 - 1) \cdots (q^{2(m-1)} - 1)(q^m + 1).$$

Consequently, $O_+(2m, q)$ is not isomorphic to $O_-(2m, q)$, and indeed none of them are isomorphic to $O(2m + 1, q)$. This shows that on an odd-dimensional space, there is a unique orthogonal group and on even-dimensional vector space there are two orthogonal groups as described above.

$SO_{\pm}(2m, q)$ is a subgroup of index 2 in $O_{\pm}(2m, q)$. Again the commutator subgroup $[SO_{\pm}(2m, q), SO_{\pm}(2m, q)]$ denoted by $\Omega_{\pm}(2m, q)$ is of index 2 in $SO_{\pm}(2m, q)$. The order of the center $Z(\Omega_{\pm}(2m, q))$ is 2 whenever $q^m \equiv \pm 1 (mod 4)$ and it is 1, otherwise. $P\Omega_{\pm}(2m, q) = \Omega_{\pm}(2m, q)/Z(\Omega_{\pm}(2m, q))$ is a simple group. The reader is referred to "The Geometry of Classical Groups" By J. Dieudonne. For further details.

Symplectic Groups

Recall that a bi-linear form on V is termed as an alternating form if $f(x, x) = 0$ for all $x \in V$. If the characteristic of the field is different from 2, then this is equivalent to say that $f(x, y) = -f(x, y)$ for all $x, y \in V$. This is also equivalent to say that the matrix representation of f with respect to any basis is skew symmetric. Note that over a field of characteristic 2, alternating forms are also symmetric. We are interested in nondegenerate alternating forms.

Proposition 4.1.12 *A vector space V admits a nondegenerate alternating form if and only if $Dim\ V = 2m$ for some m. Further, if f is a non-degenerate alternating form on a vector space V of dimension $2m$, then there is a basis $\{u_1, v_1, u_2, v_2, \ldots, u_m, v_m\}$ such that $f(u_i, u_j) = 0 = f(v_i, v_j)$ for all i, j, $f(u_i, v_j) = 0 = f(v_i, u_j)$ for all $i \neq j$, and $f(u_i, v_i) = 1 = -f(v_i, u_i)$ for all i.*

Proof Given a basis $\{u_1, v_1, u_2, v_2, \ldots, u_m, v_m\}$ of V, we have a unique non degenerate alternating form subject to $f(u_i, u_j) = 0 = f(v_i, v_j)$ for all i, j, $f(u_i, v_j) = 0 = f(v_i, u_j)$ for all $i \neq j$, and $f(u_i, v_i) = 1 = -f(v_i, u_i)$ for all i. Let f be a nondegenerate alternating form on V. Evidently, there is no nondegenerate alternating form on a vector space of dimension 1. Thus, $Dim\ V \geq 2$. Let u_1 be a nonzero element of V. Then $f(u_1, u_1) = 0$. Since f is nondegenerate, there is a vector v such that $f(u_1, v) \neq 0$. Take $v_1 = \frac{1}{f(u_1, v)} v$. Then $f(u_1, v_1) = 1 = -f(v_1, u_1)$. Clearly, $\{u_1, v_1\}$ is linearly independent. If $V = <\{u_1, v_1\}>$, then there is nothing to do. If not, let U be the subspace generated by $\{u_1, v_1\}$. Consider $U^{\perp}$. Let $x = au_1 + bv_1$ be a member of $U^{\perp}$. Then $-a = f(x, u_1) = 0$ and $b = f(x, v_1) = 0$. This means that $U \bigcap U^{\perp} = \{0\}$. Let x be an arbitrary element of V. Put $y = x - f(x, v_1)u_1 + f(x, u_1)v_1$. Clearly, $y \in U^{\perp}$ and $x = f(x, v_1)u_1 - f(x, u_1)v_1 + y$ belongs to $U + U^{\perp}$. This shows that $V = U \oplus U^{\perp}$. Since f restricted to $U^{\perp}$ is also a nondegenerate alternating form, the assertions follow by the induction on $Dim\ V$. ♯

If f is a non degenerate alternating form on an even-dimensional space V, then $G(f)$ is called a **Symplectic Group**. A nondegenerate alternating form is also called a **symplectic form**.

Corollary 4.1.13 *Any symplectic group on a vector space V of dimension $2m$ over a field K is isomorphic to the subgroup*

$$Sp(2m, K) = \{A \in GL(2m, K) \mid APA^t = P\}$$

of $GL(2m, K)$, where

$$P = \begin{bmatrix} 0_m & I_m \\ -I_m & 0_m. \end{bmatrix}$$

Proof Let f be a nondegenerate alternating form on a vector space V of dimension $2m$. From the above Proposition, it follows that the matrix of f with respect to a suitable basis is P as given in the Corollary. Hence $G(f) \approx Sp(2m, K)$. ♯

We state a fact (to follow later) that $PSp(2m, q) = Sp(2m, q)/Z(Sp(2m, q))$ is simple group except when $m = 1, q \leq 3$ or $m = 2$ and $q = 2$.

Let K be a field with an automorphism σ of order 2 (also termed as an involution on K). For example on $\mathbb{C}$, the conjugation map $z \mapsto \overline{z}$ is an automorphism of order 2. Suppose that K be a finite field which admits an automorphism σ of order 2. Then K is quadratic extension of the fixed field K_σ of σ. If $K_\sigma = F_q$, then $K = F_{q^2}$. Evidently, σ given by $\sigma(a) = a^q$ is an automorphism of F_{q^2} of order 2.

Let $a \in F_{q^2}$. Then $(a^{1+q})^q = a^{q+q^2} = a^q a^{q^2} = a^q a = a^{1+q}$. Thus, $a^{1+q} \in F_q$ for all $a \in F_{q^2}$. Consider the map τ from F_{q^2} to F_q defined by $\tau(a) = a^{1+q}$. Clearly τ restricted to $F_{q^2}^\star$ is a homomorphism from $F_{q^2}^\star$ to $F_q^\star$. Since $Ker\tau$ is a cyclic subgroup of $F_{q^2}^\star$ of order $q + 1$, it follows that τ is surjective map. Similarly, $a + a^q \in F_q$ for all $a \in F_{q^2}$ and the map μ from F_{q^2} to F_q defined by $\mu(a) = a + a^q$ is a surjective map.

Let V be a vector space over K with an involution σ. Then a map f from $V \times V$ to K is called a sesqui linear form if the following hold:

(i) $f(ax + by, z) = af(x, z) + bf(y, z)$, and
(ii) $f(x, ay + bz) = \sigma(a)f(x, y) + \sigma(b)f(y, z)$
for all $x, y, z \in V$ and $a, b \in K$. A sesqui linear form f is called a nondegenerate form if $f(x, y) = 0$ for all y implies that $x = 0$ and $f(x, y) = 0$ for all x implies that $y = 0$. If $\{x_1, x_2, \ldots, x_m\}$ is a basis of V, then the matrix $A = [a_{ij} = f(x_i, x_j)]$ is called the matrix of f relative to the given basis.

A sesqui linear form f is called a **Hermitian form** if $f(x, y) = \sigma(f(y, x))$ for all x, y. A sesqui linear form f is a **Hermitian form** if and only if the matrix $A = [a_{ij}]$ associated with f relative to a basis is hermitian matrix in the sense that $A = A^\star = [b_{ij}]$, where $b_{ij} = \sigma(a_{ji})$ for all i, j.

Proposition 4.1.14 *Let f be a nondegenerate hermitian form on a vector space V of dimension m over a finite field F_{q^2}. Then there is a basis $\{u_1, u_2, \ldots, u_m\}$ such that $f(u_i, u_i) = 1$ for all i and $f(u_i, u_j) = 0$ whenever $i \neq j$.*

Proof The proof is by the induction on the dimension of V. Let f be a nondegenerate hermitian form on V. Evidently, $f(V \times V) = F_{q^2}$. Suppose that $f(u, u) = 0$ for all u. Then $f(u + v, u + v) = 0$ for all $u, v \in V$. This implies that $f(u, v) + f(v, u) = 0$ for all $u, v \in V$. Since f is hermitian, $f(v, u) = f(u, v)^q$ for all u, v. In turn, $\mu(f(u, v)) = f(u, v) + f(u, v)^q = 0$ for all u, v. This is a contradiction, since μ is surjective map from F_{q^2} to F_q. Thus, there is a nonzero member u such that $f(u, u) = a \neq 0$. Since f is hermitian, $a^q = a$. Consequently, a is a nonzero member of F_q. Since τ is surjective map, there is an element $b \in F^{\star}_{q^2}$ such that $a = b^{1+q}$. Take $u_1 = b^{-1}u$. Then $f(u_1, u_1) = 1$. The result follows if $DimV = 1$. Assume that the result holds for all vector spaces of dimension less than $DimV$. Then, as above, there is a $u_1 \in V$ such that $f(u_1, u_1) = 1$. Let U denote the subspace generated by u_1. Then as usual, $V = U \oplus U^{\perp}$. Clearly, the $DimU^{\perp} < DimV$ and $f|_{U^{\perp}}$ is nondegenerate hermitian form. By the induction hypothesis, we have a basis $\{u_2, u_3, \ldots, u_m\}$ such that $f(u_i, u_i) = 1$ and $f(u_i, u_j) = 0$ for $i \neq j$. ♯

Corollary 4.1.15 *Let f be a nondegenerate hermitian form on a vector space of dimension m over F_{q^2}, then $G(f)$ is isomorphic to the subgroup*

$$U(m, q) = \{A \in GL(m, q^2) \mid AA^{\star} = I_m\}$$

of $GL(m, q^2)$.

Proof Since the matrix representation of f with respect to a suitable basis is the identity matrix, the result follows. ♯

The group $U(m, q)$ is called a **Unitary Group**. Clearly, $DetA^{\star} = (Det(A))^q = (Det(A))^{-1}$. Hence, $Det(A)^{q+1} = 1$. We have an exact sequence

$$1 \longrightarrow SU(m, q) \xrightarrow{i} U(m, q) \xrightarrow{Det} \mathbb{Z}_{q+1} \longrightarrow 1$$

where $\mathbb{Z}_{q+1} = a \in F^{\star}_{q^2} \mid a^{q+1} = 1$, and $SU(m, q) = SL(m, q^2) \bigcap U(m, q)$. The center $Z(SU(m, q))$ is a cyclic group of order $(m, q + 1)$. The projective special unitary group $PSU(m, q) = SU(m, q)/Z(SU(m, q))$ is a simple group for all $n \geq 3$ except when $n = 3$ and $q = 2$.

Exercises

4.1.1. Compute the order of $U(m, q)$ and also of $PSU(m, q)$.
4.1.2. Determine the order of $Sp(2m, q)$.
4.1.3. Show that $SU(2, q) \approx SL(2, q)$.
4.1.4. Show that $\Omega(3, q) \approx PSL(2, q)$.

4.2 Chevalley Basis

In this section, we shall show the existence of integral bases (bases with structure constants as integers) of L, $U(L)$, and V, where L is a semi-simple Lie algebra over $\mathbb{C}$, $U(L)$ is the universal enveloping algebra of L, and V is a L-module. L, H, Φ, and Δ are as usual.

Proposition 4.2.1 *Let $\alpha, \beta \in \Phi$, $\beta \neq \pm\alpha$. Let $\{\beta - r\alpha, \beta - (r-1)\alpha, \ldots, \beta, \beta + \alpha, \ldots, \beta + q\alpha\}$ be the α-string through β. Then the following hold:*

(i) $\prec \beta, \alpha \succ = r - q = \beta(h_\alpha)$, *where* $h_\alpha = \frac{2t_\alpha}{<\alpha,\alpha>}$.
(ii) At the most two root lengths occur in the root string.
(iii) If $q \neq 0$, *then* $r + 1 = \frac{q<\alpha+\beta,\alpha+\beta>}{<\beta,\beta>}$.

Proof The part (i) has already been established (Propositions 2.1.9, 2.2.8).

(ii) Note that $(\mathbb{Z}\alpha + \mathbb{Z}\beta) \bigcap \Phi = \Phi'$ is also a root system of rank 2. If Φ' is reducible, then $\Phi' = A_1 \times A_1 = \{\pm\alpha, \pm\beta\}$ and there is nothing to do. If Φ' is irreducible, then it is A_2, B_2, or G_2. The result follows in each case.

(iii) If $q \neq 0$, then $\Phi' = A_2$, B_2, or G_2. The identity follows by simple casewise inspection. ♯

Proposition 4.2.2 (Chevalley) *There exists a map x from Φ to $\bigcup_{\alpha\in\Phi} L_\alpha$ with $x(\alpha) = x_\alpha \in L_\alpha$ for each $\alpha \in \Phi$ such that the following hold:*

(i) $[x_\alpha, x_{-\alpha}] = h_\alpha = \frac{2t_\alpha}{<\alpha,\alpha>}$ *for all* $\alpha \in \Phi$.
(ii) For each pair $\alpha, \beta \in \Phi$ such that $\alpha + \beta \in \Phi$, the scalar $\nu_{\alpha,\beta}$ given by $[x_\alpha, x_\beta] = \nu_{\alpha,\beta} x_{\alpha+\beta}$ satisfies the condition $\nu_{\alpha,\beta} = -\nu_{-\alpha,-\beta} = -\nu_{\beta,\alpha}$.
(iii) $\nu_{\alpha,\beta}^2 = q(r+1)\frac{<\alpha+\beta,\alpha+\beta>}{<\beta,\beta>} = (r+1)^2$, *where $\{\beta - r\alpha, \beta - (r-1)\alpha, \ldots, \beta, \beta + \alpha, \ldots, \beta + q\alpha\}$ is the α-string through β.*

Proof (i) We have an automorphism σ of L of order 2 such that $\sigma(L_\alpha) = L_{-\alpha}$ and $\sigma(h) = -h$ for all $h \in H$. For each $\alpha \in \Phi^+$, select a nonzero member u_α of L_α and put $-u_{-\alpha} = \sigma(u_\alpha)$. Since the Killing form κ is nondegenerate, $\kappa(u_\alpha, u_{-\alpha}) \neq 0$. Take $x_\alpha = \sqrt{\frac{2}{<\alpha,\alpha>\kappa(u_\alpha,u_{-\alpha})}}u_\alpha$. Then $\kappa(x_\alpha, x_{-\alpha}) = \frac{2}{<\alpha,\alpha>} = \frac{2}{\kappa(t_\alpha,t_\alpha)}$. From Theorem 2.1.6, it follows that $[x_\alpha, x_{-\alpha}] = h_\alpha$ for each α.

(ii) Since each L_α is one dimensional, for each $\alpha, \beta, \alpha + \beta \in \Phi$, we have a scalar $\nu_{\alpha,\beta}$ such that $[x_\alpha, x_\beta] = \nu_{\alpha,\beta} x_{\alpha+\beta}$ ($\nu_{\alpha,\beta}$ are called structure constants). Further,

$$\begin{aligned}-\nu_{\alpha,\beta} x_{-\alpha-\beta} &= \sigma(\nu_{\alpha,\beta} x_{\alpha+\beta}) = \sigma([x_\alpha, x_\beta]) = [\sigma(x_\alpha), \sigma(x_\beta)] \\ &= [x_{-\alpha}, x_{-\beta}] = \nu_{-\alpha,-\beta} x_{-\alpha-\beta}.\end{aligned}$$

This shows that $\nu_{\alpha,\beta} = -\nu_{-\alpha,-\beta}$. Further,

$$\nu_{\alpha,\beta} x_{\alpha+\beta} = [x_\alpha, x_\beta] = -[x_\beta, x_\alpha] = -\nu_{\beta,\alpha} x_{\beta+\alpha} = -\nu_{\beta,\alpha} x_{\alpha+\beta}.$$

Hence, $\nu_{\alpha,\beta} = -\nu_{\beta,\alpha}$.

(iii) By the Jacobi identity,

$$[[x_\alpha, x_{-\alpha}], x_\beta] + [[x_{-\alpha}, x_\beta], x_\alpha] + [[x_\beta, x_\alpha], x_{-\alpha}] = 0,$$

or

$$[h_\alpha, x_\beta] + \nu_{-\alpha,\beta}\nu_{-\alpha+\beta,\alpha}x_\beta + \nu_{\beta,\alpha}\nu_{\beta+\alpha,-\alpha}x_\beta = 0,$$

or

$$\beta(h_\alpha)x_\beta + \nu_{-\alpha,\beta}\nu_{-\alpha+\beta,\alpha}x_\beta + \nu_{\beta,\alpha}\nu_{\beta+\alpha,-\alpha}x_\beta = 0.$$

This means that

$$\beta(h_\alpha) + \nu_{-\alpha,\beta}\nu_{\beta-\alpha,\alpha} + \nu_{\beta,\alpha}\nu_{\beta+\alpha,-\alpha} = 0. \tag{4.1}$$

Replacing β by $\beta - \alpha$ in the above equation, we obtain

$$(\beta - \alpha)(h_\alpha) + \nu_{-\alpha,\beta-\alpha}\nu_{\beta-2\alpha,\alpha} + \nu_{\beta-\alpha,\alpha}\nu_{\beta,-\alpha} = 0. \tag{4.2}$$

Adding (4.1) and (4.2) and observing that $\nu_{\beta,-\alpha} = -\nu_{-\alpha,\beta}$, we obtain

$$2\beta(h_\alpha) - \alpha(h_\alpha) + \nu_{\beta,\alpha}\nu_{\beta+\alpha,-\alpha} + \nu_{-\alpha,\beta-\alpha}\nu_{\beta-2\alpha,\alpha} = 0. \tag{4.3}$$

Again, replacing β by $\beta - \alpha$ in (4.3) and adding the equation, thus obtained, with (4.3), we obtain

$$3\beta(h_\alpha) - (1+2)\alpha(h_\alpha) + \nu_{\beta,\alpha}\nu_{\beta+\alpha,-\alpha} + \nu_{-\alpha,\beta-2\alpha}\nu_{\beta-3\alpha,\alpha} = 0.$$

Proceeding inductively, we see that

$$(r+1)\beta(h_\alpha) - \frac{r(r+1)}{2}\alpha(h_\alpha) + \nu_{\beta,\alpha}\nu_{\beta+\alpha,-\alpha} + \nu_{-\alpha,\beta-r\alpha}\nu_{\beta-(r+1)\alpha,\alpha} = 0. \tag{4.4}$$

Observing that $\beta(h_\alpha) = (r - q)$, $\alpha(h_\alpha) = 2$ and $\nu_{\beta-(r+1)\alpha,\alpha} = 0$, we get

$$\nu_{\alpha,\beta}\nu_{\beta+\alpha,-\alpha} = -q(r+1). \tag{4.5}$$

Again the Jacobi identity

$$[[x_{-\alpha}, x_{-\beta}], x_{\alpha+\beta}] + [[x_{-\beta}, x_{\alpha+\beta}], x_{-\alpha}] + [[x_{\alpha+\beta}, x_{-\alpha}], x_{-\beta}] = 0$$

gives

$$-\nu_{-\alpha,-\beta}h_{\alpha+\beta} + \nu_{-\beta,\alpha+\beta}h_\alpha + \nu_{\alpha+\beta,-\alpha}h_\beta = 0. \tag{4.6}$$

Since $t_{\alpha+\beta} = t_\alpha + t_\beta$ and $h_\alpha = \frac{2t_\alpha}{<\alpha,\alpha>}$, we obtain

$$-<\alpha+\beta,\alpha+\beta> h_{\alpha+\beta} + <\alpha,\alpha> h_\alpha + <\beta,\beta> h_\beta = 0. \tag{4.7}$$

Since $\{h_\alpha, h_\beta\}$ is linearly independent, the coefficients in (4.6) and (4.7) are proportional. Consequently,

$$\frac{\nu_{-\alpha,-\beta}}{<\alpha+\beta,\alpha+\beta>} = \frac{\nu_{\alpha+\beta,-\alpha}}{<\beta,\beta>} = \frac{\nu_{-\beta,\alpha+\beta}}{<\alpha,\alpha>}.$$

Thus,

$$\frac{\nu_{\alpha+\beta,-\alpha}}{\nu_{-\alpha,-\beta}} = \frac{<\beta,\beta>}{<\alpha+\beta,\alpha+\beta>}, \tag{4.8}$$

and also

$$\frac{\nu_{-\beta,\alpha+\beta}}{\nu_{-\alpha,-\beta}} = \frac{<\alpha,\alpha>}{<\alpha+\beta,\alpha+\beta>}.$$

Using the Eqs. (4.8) and (4.5), we obtain

$$\nu_{\alpha,\beta}\nu_{-\alpha,-\beta} = \frac{\nu_{\alpha,\beta}\nu_{\alpha+\beta,-\alpha} <\alpha+\beta,\alpha+\beta>}{<\beta,\beta>} = q(r+1)\frac{<\alpha+\beta,\alpha+\beta>}{<\beta,\beta>}.$$

Further, using the Proposition 4.2.1 (iii),

$$\nu_{\alpha,\beta}\nu_{-\alpha,-\beta} = -(r+1)^2.$$

Since $\nu_{\alpha,\beta} = -\nu_{-\alpha,-\beta}$, $\nu^2_{\alpha,\beta} = (r+1)^2$ and $\nu_{\alpha,\beta} = \pm(r+1)$. ♯

Corollary 4.2.3 *We have a basis of* $\{h_i \mid 1 \le i \le l\} \bigcup \{x_\alpha \mid \alpha \in \Phi\}$ *such that the structure constants are integers. Indeed:*

(i) $[h_i, h_j] = 0 \le i, j \le l$.
(ii) $[h_i, x_\alpha] = \prec \alpha, \alpha_i \succ x_\alpha$ *for each* i, $1 \le i \le l$ *and* $\alpha \in \Phi$.
(iii) $[x_\alpha, x_{-\alpha}] = h_\alpha$ *is an integral linear combination of* $\{h_i \mid 1 \le i \le l\}$.
(iv) If α *and* β *are independents and* $\{\beta - r\alpha, \beta - (r-1)\alpha, \ldots, \beta, \beta+\alpha, \ldots, \beta + q\alpha\}$ *is a* α*-string through* β*, then* $[x_\alpha, x_\beta] = \pm(r+1)x_{\alpha+\beta}$, $\alpha + \beta \in \Phi$. ♯

A basis described in the above corollary is termed as a **Chevalley Basis** of L.

Remark 4.2.4 Tits, J in 1966 determined the sign of $\nu_{\alpha,\beta} = \pm(r+1)$ for a given Φ which, in turn, gives an alternative construction of semi-simple lie algebra without the use of Serre's theorem.

If $\{x'_\alpha \mid \alpha \in \Phi\}$ is another Chevalley basis and ν' is the associated map from $\Phi \times \Phi$ to $\mathbb{C}$, then there is a map c from Φ to $\mathbb{C}$ such that $x'_\alpha = c_\alpha x_\alpha$, where $c(\alpha) = c_\alpha$. We have the following:

Proposition 4.2.5 *(i)* $c_\alpha c_{-\alpha} = 1$
(ii) $\nu'_{\alpha,\beta}\nu'_{-\alpha,-\beta} = \nu_{\alpha,\beta}\nu_{-\alpha,-\beta}$.

Proof (i) Since $c_\alpha c_{-\alpha}[x_\alpha, x_{-\alpha}] = [x'_\alpha, x'_{-\alpha}] = h_\alpha = [x_\alpha, x_{-\alpha}]$, $c_\alpha c_{-\alpha} = 1$.
(ii) We have

$$c_\alpha c_\beta \nu_{\alpha,\beta} x_{\alpha+\beta} = c_\alpha c_\beta [x_\alpha, x_\beta] = [x'_\alpha, x'_\beta] = \nu'_{\alpha,\beta} x'_{\alpha+\beta} = \nu'_{\alpha,\beta} c_{\alpha+\beta} x_{\alpha+\beta}.$$

for all α, β. It follows that $\nu'_{\alpha,\beta} = \frac{c_\alpha c_\beta}{c_{\alpha+\beta}}\nu_{\alpha,\beta}$ for all α, β. Using (i), we obtain (ii). ♯

Let L be a semi-simple Lie algebra over $\mathbb{C}$ (or over any algebraically closed field of characteristic 0) and $X = \{h_i \mid 1 \leq i \leq l\} \bigcup \{x_\alpha \mid \alpha \in \Phi\}$ be a Chevalley basis of L. Consider, the lattice (abelian subgroup) $L(\mathbb{Z})$ in L generated by X. Note that $L(\mathbb{Z})$ is independent of the basis Δ of Φ. The Lie product in L induces a product in $L(\mathbb{Z})$ and $L(\mathbb{Z})$ is a Lie algebra over $\mathbb{Z}$ with respect to the induced Lie product. Let p be a prime and $F_p = \mathbb{Z}_p$ the prime field of order p. Then F_p is a $\mathbb{Z}$-module. Consider $L(F_p) = L(\mathbb{Z}) \otimes_{\mathbb{Z}} F_p$ which is a vector space over F_p with basis $\{\overline{h_i} = h_i \otimes 1 \mid 1 \leq i \leq l\} \bigcup \{\overline{x_\alpha} = x_\alpha \otimes 1 \mid \alpha \in \Phi\}$. Further, the bracket operation in $L(\mathbb{Z})$ induces a product in $L(F_p)$ and $L(F_p)$ is a Lie algebra over F_p with respect to the induced product. This is just the reduction of $L(\mathbb{Z})$ modulo p. Let K be a field extension of F_p. Then, as before, $L(K) = L(F_p) \otimes_{F_p} K$ is a Lie algebra over K with $\{\overline{h_i} \otimes 1 \mid 1 \leq i \leq l\} \bigcup \{\overline{x_\alpha} \otimes 1 \mid \alpha \in \Phi\}$ as a basis. Indeed, $L(K)$ is the Lie algebra over K obtained by just extending the scalars in $L(F_p)$ to K. For example, if $L = sl(n, \mathbb{C})$, then $L(F_p) = sl(n, F_p)$ and $L(K) = sl(n, K)$. The Lie algebra $L(K)$ may cease to be simple for if p divides n, the scalar matrices in $sl(n, K)$ forms a one dimensional ideal of $sl(n, K)$ which is the center of $sl(n, K)$. $L(K)$ is called the **Chevalley algebra** over K associated with the Lie algebra L. Note that $L(K)$ is isomorphic to $L'(K)$ if L is isomorphic to L'.

Proposition 4.2.6 *$L(\mathbb{Z})$ is invariant under the transformation $\frac{(ad(x_\alpha))^m}{m!}$ for all $\alpha \in \Phi$ and $m \geq 0$.*

Proof Let $X = \{h_i \mid 1 \leq i \leq l\} \bigcup \{x_\alpha \mid \alpha \in \Phi\}$ be a Chevalley basis of L. It is sufficient to show that $\frac{(ad(x_\alpha))^m}{m!}(X) \subseteq L(\mathbb{Z})$ for all $\alpha \in \Phi$ and $m \geq 0$. Clearly, $\frac{(ad(x_\alpha))^0}{0!}(h_i) = h_i$, $\frac{(ad(x_\alpha))^1}{1!}(h_i) = [x_\alpha, h_i] = -\prec \alpha, \alpha_i \succ x_\alpha$, and also $\frac{(ad(x_\alpha))^m}{m!}(h_i) = 0$ for all $m \geq 2$. This shows that $\frac{(ad(x_\alpha))^m}{m!}(h_i)$ belongs to $L(\mathbb{Z})$ for all i and $m \geq 0$. Evidently, $\frac{(ad(x_\alpha))^m}{m!}(x_\alpha) = 0$ for all $m \geq 1$. Next, $\frac{(ad(x_\alpha))^1}{1!}(x_{-\alpha}) = [x_\alpha, x_{-\alpha}] = h_\alpha$ and $\frac{(ad(x_\alpha))^2}{m!}(x_{-\alpha}) = \frac{1}{2}[x_\alpha, h_\alpha] = -x_\alpha$, and $\frac{(ad(x_\alpha))^m}{m!}(x_{-\alpha}) = 0$ for all $m \geq 3$. Finally, let $\beta \in \Phi$ be such that $\beta \neq \pm\alpha$. Let $\{\beta - r\alpha, \beta - (r - 1)\alpha, \ldots, \ldots, \beta, \beta + \alpha, \ldots, \beta + q\alpha\}$ be the α-string through β. From Corollary 4.2.3 (iv), $ad(x_\alpha)(x_\beta) = \pm(r + 1)x_{\beta+\alpha}$. Since $\{(\beta + \alpha) - (r + 1)\alpha, (\beta + \alpha) - (r\alpha, \ldots, \beta + \alpha, (\beta + \alpha) + \alpha, \ldots, (\beta + \alpha) + (q - 1)\alpha\}$ is the α-string through $\beta + \alpha$, $(ad(x_\alpha))^2(x_\beta) = \pm(r + 1)(r + 2)x_{\beta+2\alpha}$. Proceeding inductively, we see that

$$(ad(x_\alpha))^m(x_\beta) = \pm(r + 1)(r + 2)\cdots(r + m)x_{\beta+m\alpha}$$

for all $m \leq q-1$ and it is 0 for all $m \geq q$. Since $\frac{\pm(r+1)(r+2)\cdots(r+m)}{m!} = {}^{(m+r)}C_m$ is an integer for all m, it follows that $\frac{(ad(x_\alpha))^m}{m!}(x_\beta)$ belongs to $L(\mathbb{Z})$ for all $m \geq 0$. ♯

For each $\alpha \in \phi$, $ad(x_\alpha)$ is nilpotent. As such,

$$exp(ad(x_\alpha)) = I + ad(x_\alpha) + \frac{(ad(x_\alpha))^2}{2!} + \cdots \frac{(ad(x_\alpha))^m}{m!} \cdots$$

makes sense. It follows from the above proposition that $exp(ad(x_\alpha))$ acts as an automorphism of the free $\mathbb{Z}$-module $L(\mathbb{Z})$ of rank $n = l+ \mid \Phi \mid$. The matrix representation of $exp(ad(x_\alpha))$ with respect to the chosen Chevalley basis is an integral $n \times n$ matrix of determinant 1. The group $Inn(L)$ generated by the set $\{exp(ad(x_\alpha)) \mid \alpha \in \Phi\}$ can be viewed as a group of $n \times n$ matrices in $SL(n, \mathbb{Z})$. More generally, let T be an indeterminate over $\mathbb{Z}$. Then $L(\mathbb{Z}[T]) = L(\mathbb{Z}) \otimes_{\mathbb{Z}} \mathbb{Z}[T]$ is a free $\mathbb{Z}[T]$-module of rank n with basis $X \otimes 1 = \{h_i \otimes 1 \mid 1 \leq i \leq l\} \bigcup \{x_\alpha \otimes 1 \mid \alpha \in \Phi\}$, where X is the chosen Chevalley basis of L. Evidently,

$$exp(ad(Tx_\alpha)) = I + Tad(x_\alpha) + \frac{T^2(ad(x_\alpha))^2}{2!} + \cdots \frac{T^m(ad(x_\alpha))^m}{m!} \cdots$$

is an automorphism of $L(\mathbb{Z}(T))$ whose matrix representation with respect to the basis $X \otimes 1$ is a $n \times n$ matrix of determinant 1 with entries in $\mathbb{Z}[T]$. Consider the subgroup $G(T)$ of $GL(n, \mathbb{Z}[T])$ generated by $\{exp(ad(Tx_\alpha)) \mid \alpha \in \Phi\}$. Specialize T with elements of a field extension K of F_p to obtain a group $G(K)$ of $n \times n$ matrices over with entries in K. The group $G(K)$ is called the **Chevalley Group of Adjoint Type** over K associated with the semi-simple Lie algebra L. If K is a finite Galois field F_q, then leaving few exceptions, we shall show that $G(F_q)$is a finite simple group. This is how Chevalley discovered several families of finite simple groups.

Still, more generally, our aim is to associate Chevalley Group to an arbitrary finite-dimensional representation ρ of L on V. For the purpose, we look at the subring $U(L)_{\mathbb{Z}}$ with identity (which is also a $\mathbb{Z}$-form on $U(L)$) of $U(L)$ generated by the set $\{\frac{j_L(x_\alpha)^m}{m!} \mid \alpha \in \Phi, m \geq 0\}$ and also a lattice $V_{\mathbb{Z}}$ in V such that $U(L)_{\mathbb{Z}} \cdot V_{\mathbb{Z}} \subseteq V_{\mathbb{Z}}$.

Let A be an associative commutative algebra with identity over the complex field $\mathbb{C}$ (or over any field of characteristic 0). Then for all $a \in A$ and $n \in \mathbb{Z}_+$, $\frac{a(a-1)\cdots(a-n+1)}{n!}$ makes sense and it is denoted by $\binom{a}{n}$. The usual identity "$\binom{a+1}{n} - \binom{a}{n} = \binom{a}{n-1}$" also holds. Here also we adapt the convention that $\binom{a}{0} = 1$ and put $\binom{a}{n} = 0$ for all negative integer n.

Consider $\Phi^+ = \{\alpha_1, \alpha_2, \ldots, \alpha_N\}$ as an ordered set and $\Phi^- = \{-\alpha_1, -\alpha_2, \ldots, -\alpha_N\}$ also as an ordered set. Further, consider an ordered Chevalley basis

$$\{x_{-\alpha_1}, x_{-\alpha_2}, \ldots, x_{-\alpha_N}, h_1, h_2, \ldots, h_l, x_{\alpha_1}, x_{\alpha_2}, \ldots, x_{\alpha_N}\}$$

of L, where $\mid \Phi \mid = 2N$. For each N-tuple $\overline{a} = (a_1, a_2, \ldots, a_N) \in \mathbb{Z}_+^N$, let $e_{\overline{a}}^-$ denote the element

$$\frac{(j_L(x_{-\alpha_1}))^{a_1}}{a_1!} \cdot \frac{(j_L(x_{-\alpha_2}))^{a_2}}{a_2!} \cdots \frac{(j_L(x_{-\alpha_N}))^{a_N}}{a_N!}$$

of $U(L)$, and $e^+_{\overline{a}}$ denote the element

$$\frac{(j_L(x_{\alpha_1}))^{a_1}}{a_1!} \cdot \frac{(j_L(x_{\alpha_2}))^{a_2}}{a_2!} \cdots \frac{(j_L(x_{\alpha_N}))^{a_N}}{a_N!}$$

of $U(L)$. Further, for each l-tuple $\overline{b} = (b_1, b_2, \ldots, b_l)$ in $\mathbb{Z}^l_+$, let $f_{\overline{b}}$ denote the element

$$\binom{j_L(h_1)}{b_1} \cdot \binom{j_L(h_2)}{b_2} \cdots \binom{j_L(h_l)}{b_l}$$

of $U(H) \subseteq U(L)$ (note that $\binom{j_L(h_i)}{b_i}$ makes sense, since $U(H)$ is a associative commutative algebra). We have the following theorem due to Kostant.

Theorem 4.2.7 (Kostant) *The subring* $U(L)_{\mathbb{Z}}$ *of* $U(L)$ *generated by* $\{\frac{(j_L(x_\alpha))^m}{m!} \mid \alpha \in \Phi, m \in \mathbb{Z}_+\}$ *is a lattice in* $U(L)$ *with*

$$\{e^-_{\overline{a}} f_{\overline{b}} e^+_{\overline{c}} \mid \overline{a}, \overline{c} \in \mathbb{Z}^N_+, \text{ and } \overline{b} \in \mathbb{Z}^l_+\}$$

as a $\mathbb{Z}$*-basis.*

We establish a series of lemmas needed for the proof of the theorem.

Lemma 4.2.8 *Let* $f = f(T_1, T_2, \ldots, T_l) \in F[T_1, T_1, \ldots, T_l]$, *where* $T_1, T_1, \ldots, T_l$ *are indeterminates over a field* F *of characteristic 0. Then* $f(\mathbb{Z}^l) \subseteq \mathbb{Z}$ *if and only if* f *can be expressed (of course, uniquely) as a* $\mathbb{Z}$*-linear combination of the set* $X = \{\binom{T_1}{a_1}\binom{T_2}{a_2} \cdots \binom{T_l}{a_l} \mid 0 \leq a_i \leq deg_{T_i} f \text{ for each } i\}$, *where* $deg_{T_i} f$ *denotes the degree of* f *in* T_i.

Proof Evidently, the polynomial $\binom{T_1}{a_1}\binom{T_2}{a_2} \cdots \binom{T_l}{a_l}$ takes integral values on $\mathbb{Z}^l$ for all $(a_1, a_2, \ldots, a_l) \in \mathbb{Z}^l_+$. Hence, any integral linear combination of members of X takes integral values on $\mathbb{Z}^l$.

Conversely, let $f = f(T_1, T_2, \ldots, T_l)$ be a member of $F[T_1, T_2, \ldots, T_l]$ such that $f(\mathbb{Z}^l) \subseteq \mathbb{Z}$. We show that f is integral linear combination of members of X. The proof is by the induction on l. For $l = 0$, there is nothing to do. Suppose that $l = 1$. Let $f(T) \in F[T]$ such that $f(\mathbb{Z}) \subseteq \mathbb{Z}$. If $deg(f) = 0$, then $f = a_0 = a_0\binom{T}{0}$, where a_0 is an integer. If $deg(f) = 1$, then $f(T) = a_0 + a_1 T$. Since $f(\mathbb{Z}) \subseteq \mathbb{Z}$, $f(0) = a_0$ and $f(1) = a_0 + a_1$ are integers. This means $a_o, a_1 \in \mathbb{Z}$ and $f = a_0\binom{T}{0} + a_1\binom{T}{1}$. Assume the result for the polynomials in $F[T]$ satisfying the hypothesis which are of degrees at the most r, $r \geq 1$. Since $\binom{T}{k} = \frac{T^k}{k!} + \sum_{i=0}^{k-1} a_i T^i, a_i \in F$, $T^k = k!\binom{T}{k} + \sum_{i=0}^{k-1} b_i T^i$. By the induction, we observe that $T^k = \sum_{i=1}^{k} c_i \binom{T}{i}$. In turn, it follows that a polynomial $f(T) \in F[T]$ of degree $r + 1$ can be expressed as

$$f(T) = \sum_{j=0}^{r+1} b_j \binom{T}{j}, \tag{4.9}$$

where $b_j \in F$. We need to show that if $f(\mathbb{Z}) \subseteq \mathbb{Z}$, then all b_j are integers. Replacing T by $T+1$ in (4.9), we obtain

$$f(T+1) = \sum_{j=0}^{r+1} b_j \binom{T+1}{j}. \tag{4.10}$$

Subtracting (4.9) from (4.10),

$$f(T+1) - f(T) = \sum_{j=0}^{r+1} b_j \left(\binom{T+1}{j} - \binom{T}{j}\right) = \sum_{j=0}^{r+1} b_j \binom{T}{j-1}. \tag{4.11}$$

Since $f(T+1) - f(T)$ takes integral values on $\mathbb{Z}$, $\sum_{j=0}^{r+1} b_j \binom{T}{j-1}$ takes integral values on $\mathbb{Z}$ and it is of degree r. It follows by the induction hypothesis that $b_j \in \mathbb{Z}$ for all $j \geq 1$. Since $f(0) = b_0 \in \mathbb{Z}$, it follows that all b_j belong to $\mathbb{Z}$. This proves the result for $l = 1$.

Assume the result for l. Let $f(T_1, T_2, \ldots, T_{l+1})$ be a polynomial in $F[T_1, T_2, \ldots, T_{l+1}]$ such that $f(\mathbb{Z}^{l+1}) \subseteq \mathbb{Z}$. If degree of f in T_{l+1} is zero, then the result follows by the induction hypothesis. Assume the result for those polynomials f whose degree in T_{l+1} is $r \geq 0$. Suppose that the degree of f in T_{l+1} is $r+1$. Using the earlier argument used for $l = 1$, we can express f as

$$f(T_1, T_2, \ldots, T_{l+1}) = \sum_{i=0}^{r+1} g_i(T_1, T_2, \ldots, T_l) \binom{T_{l+1}}{i}. \tag{4.12}$$

Again as in case for $l = 1$, replacing T_{l+1} by $T_{l+1} + 1$ in (4.12), we obtain that

$$f(T_1, T_2, \ldots, T_{l+1} + 1) = \sum_{i=0}^{r+1} g_i(T_1, T_2, \ldots, T_l) \binom{T_{l+1}+1}{i}. \tag{4.13}$$

Subtracting (4.12) from (4.13), we get

$$h = f(T_1, T_2, \ldots, T_{l+1} + 1) - f(T_1, T_2, \ldots, T_l) = \\ \sum_{i=0}^{r+1} g_i(T_1, T_2, \ldots, T_l) \binom{T_{l+1}}{i-1}. \tag{4.14}$$

Since h takes integer values on $\mathbb{Z}^{l+1}$, RHS also takes integer values on $\mathbb{Z}^{l+1}$. The degree of h in T_{l+1} is at the most r. By the induction hypothesis h is expressible as unique $\mathbb{Z}$-linear combination of

$$\{\binom{T_1}{a_1}\binom{T_2}{a_2} \cdots \binom{T_{l+1}}{a_{l+1}} \mid 0 \leq a_i \leq deg_{T_i} f \ for \ i \leq l \ and \ a_{l+1} \leq r\}.$$

By the uniqueness consideration, it follows that each $g_i(T_1, T_2, \ldots, T_l)$, $i \geq 1$ is a integral linear combination of elements of the type $\binom{T_1}{a_1}\binom{T_2}{a_2} \cdots \binom{T_l}{a_l}$. Putting $T_{l+1} = 0$, we see that $g_0(T_1, T_2, \ldots, T_l)$ is also Z-linear combination of such elements. The result follows for f. ♯

Lemma 4.2.9 *Consider $sl(2, F)$ with standard basis $\{x, h, y\}$, where F is an algebraically closed field of characteristic 0. Then for each $r, s \in \mathbb{N} \bigcup \{0\}$,*

$$\frac{j_L(x)^r}{r!}\frac{j_L(y)^s}{s!} = \sum_{j=0}^{min(r,s)} \frac{j_L(y)^{s-j}}{(s-j)!}\binom{j_L(h)-s-r+2j}{j}\frac{j_L(x)^{r-j}}{(r-j)!}.$$

Proof We use induction on r to establish the identity. If $r = 0$, then both sides equal $\frac{j_L(y)^s}{s!}$ and there is nothing to do. For $r = 1$, we use induction on s. If $s = 0$, then, as before, both sides equal $\frac{j_L(x)^r}{r!}$. Suppose that $s = 1$. Then

$$\frac{j_L(x)}{1!}\frac{j_L(y)}{1!} = j_L(y)j_L(x) + [j_L(x), j_L(y)] = j_L(y)j_L(x) + j_L([x, y]) =$$
$$j_L(y)j_L(x) + j_L(h) = \sum_{j=0}^{1}\frac{j_L(y)^{s-j}}{(s-j)!}\binom{j_L(h)-1-1-2j}{j}\frac{j_L(x)^{r-j}}{(r-j)!}.$$

Assume the identity for $s \geq 1$. Then

$$\begin{aligned}
&\frac{j_L(x)}{1!}\frac{j_L(y)^{s+1}}{(s+1)!}\\
&= \frac{j_L(x)}{1!}\frac{j_L(y)^s}{s!}\frac{j_L(y)}{s+1}\\
&= \left(\sum_{j=0}^{1}\frac{j_L(y)^{s-j}}{(s-j)!}\binom{j_L(h)-s-1+2j}{j}\frac{j_L(x)^{1-j}}{(1-j)!}\right)\frac{j_L(y)}{s+1} \text{ (by the induction assumption)}\\
&= \frac{j_L(y)^s}{s!}\frac{j_L(x)}{1!}\frac{j_L(y)}{s+1} + \frac{j_L(y)^{s-1}}{(s-1)!}(j_L(h) - s + 1)\frac{j_L(y)}{s+1}\\
&= \frac{j_L(y)^{s+1}}{(s+1)!}\frac{j_L(x)}{1!} + \frac{j_L(y)^s}{(s+1)!}j_L(h) + \frac{j_L(y)^{s-1}}{(s-1)!}(j_L(h) - s + 1)\frac{j_L(y)}{s+1}\\
&= \frac{j_L(y)^{s+1}}{(s+1)!}\frac{j_L(x)}{1!} + \frac{j_L(y)^s}{s!}h \text{ (using the fact that } j_L(h)j_L(y) = -2j_L(y))\\
&= \sum_{j=0}^{1}\frac{j_L(y)^{s+1-j}}{(s+1-j)!}\binom{j_L(h)-(s+1)-1+2j}{j}\frac{j_L(x)^{1-j}}{(1-j)!}.
\end{aligned}$$

This proves the result for $r = 1$. Assuming the result for r and using the similar argument, we can establish the identity for $r + 1$. ♯

Corollary 4.2.10 *Let $L = sl(2, F)$ with the standard basis $\{x, h, y\}$. Then for each $r \in \mathbb{Z}_+$, $\binom{j_L(h)}{r}$ belongs to the subring $U(L)_{\mathbb{Z}}$ of $U(L)$ generated by the set $\{\frac{j_L(x)^r}{r!}, \frac{j_L(y)^s}{s!} \mid r, s \geq 0\}$.*

Proof The proof is by the induction on r. For $r = 0$, $1 = \binom{h}{0}$ belongs to $U(L)_{\mathbb{Z}}$. Assume the result for all $s < r$, $r \geq 1$. From the above lemma

$$\frac{j_L(x)^r}{r!}\frac{j_L(y)^r}{r!} = \binom{j_L(h)}{r} + \sum_{j=0}^{r-1}\frac{j_L(y)^{r-j}}{(r-j)!}\binom{j_L(h)-2r+2j}{j}\frac{j_L(x)^{r-j}}{(r-j)!}$$

belongs to $U(L)_{\mathbb{Z}}$. Since the polynomial $\binom{T-2r+2j}{j}$ takes integral values on $\mathbb{Z}$ for each j, it follows from Lemma 4.2.8 that $\binom{T-2r+2j}{j}$ is integral linear combination of the set $\{\binom{T}{i} \mid i \leq j\}$. Consequently, by the induction assumption, the second term in the RHS, of the above identity belongs to $U(L)_{\mathbb{Z}}$. Since LHS belongs to $U(L)_{\mathbb{Z}}$, $\binom{j_L(h)}{r}$ belongs to $U(L)_{\mathbb{Z}}$. ♯

The following corollary is immediate.

Corollary 4.2.11 *The Kostant theorem holds for $sl(2, F)$.* ♯

Lemma 4.2.12 *Let* $X = \{h_i \mid 1 \le i \le l\} \bigcup \{x_\alpha \mid \alpha \in \Phi\}$ *be a chosen Chevalley basis of* L *and* $L(\mathbb{Z})$ *the lattice with* X *as* $\mathbb{Z}$ *- basis. Then*

$$\underbrace{L(\mathbb{Z}) \otimes_{\mathbb{Z}} L(\mathbb{Z}) \otimes_{\mathbb{Z}} \cdots \otimes_{\mathbb{Z}} L(\mathbb{Z})}_{m}$$

is invariant under $\frac{(ad(x_\alpha))^r}{r!}$ *for each* $r \ge 0$ *and* $\alpha \in \Phi$.

Proof By Proposition 4.2.6, $L(\mathbb{Z})$ is invariant under $\frac{(ad(x_\alpha))^r}{r!}$ for each $r \ge 0$ and $\alpha \in \Phi$. Using induction on m, it is sufficient to show that $L(\mathbb{Z}) \otimes_{\mathbb{Z}} L(\mathbb{Z})$ is invariant under $\frac{(ad(x_\alpha))^r}{r!}$ for each $r \ge 0$ and $\alpha \in \Phi$. Recall that

$$ad(x_\alpha)(u \otimes v) = ad(x_\alpha)(u) \otimes v + u \otimes ad(x_\alpha)(v),$$

and by the induction,

$$\frac{(ad(x_\alpha))^r}{r!}(u \otimes v) = \sum\nolimits_{j=0}^{r} \frac{(ad(x_\alpha))^j}{j!}(u) \otimes \frac{(ad(x_\alpha))^{r-j}}{(r-j)!}(v).$$

It is clear that $\frac{(ad(x_\alpha))^r}{r!}(u \otimes v)$ belongs to $L(\mathbb{Z}) \otimes_{\mathbb{Z}} L(\mathbb{Z})$ for all $u, v \in L(\mathbb{Z})$. ♯

We say that a nonempty subset Ψ of Φ is closed if $\alpha, \beta \in \Psi$ and $\alpha + \beta \in \Phi$ implies that $\alpha + \beta \in \Psi$. For example, Φ, Φ^+, and Φ^- are closed.

Lemma 4.2.13 *Let* $\Psi = \{\alpha_1, \alpha_2, \ldots, \alpha_t\}$ *be an ordered closed subset of* Φ *such that* $\Psi \bigcap -\Psi = \emptyset$. *Let* U_Ψ *denote the subring with identity of* $U(L)$ *which is generated by the set* $\{\frac{j_L(x_{\alpha_i})^r}{r!} \mid 1 \le i \le t$ *and* $r \ge 0\}$. *Then the set* $X = \{\prod_{j=1}^{t} \frac{j_L(x_{\alpha_j})^{m_j}}{m_j!} \mid (m_1, m_2, \ldots, m_t) \in \mathbb{Z}_+^t\}$ *forms a* $\mathbb{Z}$*-basis for* U_Ψ.

Proof Since Ψ is closed, the subspace V of L generated by $\{x_{\alpha_i} \mid 1 \le i \le t\}$ is a Lie sub algebra of L. Evidently, $U_\Psi \subset U(V)$, and hence X forms a F-basis of U_Ψ. We need to show that in the representation of any nonzero element of U_Ψ as unique linear combination of members of X, the coefficient of each member of X is an integer. For $u = \prod_{j=1}^{t} \frac{j_L(x_{\alpha_j})^{m_j}}{m_j!}$, $\sum_{j=1}^{t} m_j = m$ is called the degree of u. Let x be a nonzero member of U_Ψ. Let $c \prod_{j=1}^{t} \frac{j_L(x_{\alpha_j})^{m_j}}{m_j!}$ be a highest degree term appearing in the unique representation of x as linear combination of members of X. Then

$$x = c \prod\nolimits_{j=1}^{t} \frac{j_L(x_{\alpha_j})^{m_j}}{m_j!} + y,$$

where $c \in F^\star$ and the degrees of terms in the linear representation of y are at the most $\sum_{j=1}^{t} m_j$ and if $d \prod_{j=1}^{t} \frac{j_L(x_{\alpha_j})^{n_j}}{n_j!}$ is a term such that $\sum_{j=1}^{t} m_j = \sum_{j=1}^{t} n_j$, then $(m_1, m_2, \ldots, m_t) \ne (n_1, n_2, \ldots, n_t)$. Using the induction on degree and the number of terms appearing in the representations, it suffices to show that $c \in \mathbb{Z}$. Now, $U(L)$

and in particular, U_Ψ acts (the action induced by the adjoint action of L on L) on L, and in turn, it also acts on $\underbrace{L \otimes_{\mathbb{Z}} L \otimes_{\mathbb{Z}} \cdots \otimes_{\mathbb{Z}} L}_{t}$. Consider

$$x \cdot ((\underbrace{x_{-\alpha_1} \otimes x_{-\alpha_1} \otimes \cdots \otimes x_{-\alpha_1}}_{m_1}) \otimes (\underbrace{x_{-\alpha_2} \otimes x_{-\alpha_2} \otimes \cdots \otimes x_{-\alpha_2}}_{m_2}) \otimes \cdots \otimes (\underbrace{x_{-\alpha_t} \otimes x_{-\alpha_t} \otimes \cdots \otimes x_{-\alpha_t}}_{m_t})).$$

Evidently, the first term

$$c \sum_{j=1}^{t} \frac{j_L(x_{\alpha_j})^{m_j}}{m_j!} \cdot ((\underbrace{x_{-\alpha_1} \otimes x_{-\alpha_1} \otimes \cdots \otimes x_{-\alpha_1}}_{m_1}) \otimes (\underbrace{x_{-\alpha_2} \otimes x_{-\alpha_2} \otimes \cdots \otimes x_{-\alpha_2}}_{m_2}) \otimes \cdots \otimes (\underbrace{x_{-\alpha_t} \otimes x_{-\alpha_t} \otimes \cdots \otimes x_{-\alpha_t}}_{m_t}))$$

contributes

$$c((\underbrace{h_{\alpha_1} \otimes h_{\alpha_1} \otimes \cdots \otimes h_{\alpha_1}}_{m_1}) \otimes (\underbrace{h_{\alpha_2} \otimes h_{\alpha_2} \otimes \cdots \otimes h_{\alpha_2}}_{m_2}) \otimes \cdots \otimes (\underbrace{h_{\alpha_t} \otimes h_{\alpha_t} \otimes \cdots \otimes h_{\alpha_t}}_{m_t}))$$

in $\underbrace{H \otimes_{\mathbb{Z}} H \otimes_{\mathbb{Z}} \cdots \otimes_{\mathbb{Z}} H}_{t}$, whereas other terms in x do not contribute any nonzero term in $\underbrace{H \otimes_{\mathbb{Z}} H \otimes_{\mathbb{Z}} \cdots \otimes_{\mathbb{Z}} H}_{t}$, since the other terms of degrees $\sum_{j=1}^{t} m_j$ do not match with the sequence

$$(\underbrace{x_{-\alpha_1} \otimes x_{-\alpha_1} \otimes \cdots \otimes x_{-\alpha_1}}_{m_1}) \otimes (\underbrace{x_{-\alpha_2} \otimes x_{-\alpha_2} \otimes \cdots \otimes x_{-\alpha_2}}_{m_2}) \otimes \cdots \otimes (\underbrace{x_{-\alpha_t} \otimes x_{-\alpha_t} \otimes \cdots \otimes x_{-\alpha_t}}_{m_t}),$$

while the rest of the terms are of smaller degrees. Further, by Lemma 4.2.12, x preserves $L(\mathbb{Z}) \otimes L(\mathbb{Z}) \otimes \cdots \otimes L(\mathbb{Z})$. Since $L(\mathbb{Z})$ is independent of the choice of a basis Δ, and any positive root belong to a basis, we can take a basis so that α_1 is a simple root. Again, since $\{h_\gamma \mid \gamma \in \Delta\}$ form a free $\mathbb{Z}$-basis for $H(\mathbb{Z}) = H \bigcap L(\mathbb{Z})$, it follows that

$$c((\underbrace{h_{\alpha_1}\otimes h_{\alpha_1}\otimes\cdots\otimes h_{\alpha_1}}_{m_1})\otimes(\underbrace{h_{\alpha_2}\otimes h_{\alpha_2}\otimes\cdots\otimes h_{\alpha_2}}_{m_2})\otimes\cdots\otimes$$
$$(\underbrace{h_{\alpha_t}\otimes h_{\alpha_t}\otimes\cdots\otimes h_{\alpha_t}}_{m_t}))$$

belongs to $H(\mathbb{Z})\otimes_{\mathbb{Z}} H(\mathbb{Z})\otimes_{\mathbb{Z}}\cdots\otimes_{\mathbb{Z}} H(\mathbb{Z})$. This means $c\in\mathbb{Z}$. ♯

For simplicity in arguments, the degrees of monomials in $\binom{j_L(h_i)-j}{j}$, $1\le i\le l$, $j\ge 0$ are taken to be 0. Thus, if $x=\prod_{i=1}^{t}\frac{(j_L(x_{\alpha_i}))^{m_i}}{m_i!}\prod_{j=1}^{d}\binom{j_L(h_j)-k_j}{k_j}$, $k_j\ge 0$, then the degree of x is $\sum_{i=1}^{t} m_i$.

Lemma 4.2.14 *Let α,β be members of Φ and $r,s\ge 0$. Then $\frac{(j_L(x_\beta))^r}{r!}\frac{(j_L(x_\alpha))^s}{s!}$ is $\mathbb{Z}$-linear combination of $\frac{(j_L(x_\alpha))^s}{s!}\frac{(j_L(x_\beta))^r}{r!}$ and other monomials of degrees $< r+s$.*

Proof If $\alpha=\beta$, then there is nothing to do. For $\alpha=-\beta$, the result follows from the identity of Lemma 4.2.9. Suppose that $\{\alpha,\beta\}$ is linearly independent. Take $\Psi=\{i\alpha+j\beta\mid i,j\ge 0\}$. Clearly, Ψ is closed. We can order Ψ subject to $\alpha<\beta$. Using Lemma 4.2.13, $\frac{(j_L(x_\beta))^r}{r!}\frac{(j_L(x_\alpha))^s}{s!}$ can be expressed as $c\frac{(j_L(x_\alpha))^s}{s!}\frac{(j_L(x_\beta))^r}{r!}+y$, where $c\in\mathbb{Z}$ and y is $\mathbb{Z}$-linear combination of other monomials. By the PBW theorem, $\frac{(j_L(x_\beta))^r}{r!}\frac{(j_L(x_\alpha))^s}{s!}=c\frac{(j_L(x_\alpha))^s}{s!}\frac{(j_L(x_\beta))^r}{r!}+$ $\mathbb{Z}$-linear combination of basis elements of degrees $< r+s$ with respect to an ordered (Note that we have used a PBW basis in the Lemma 4.2.13). This completes the proof of the Lemma. ♯

Lemma 4.2.15 *Let $\alpha,\beta\in\Phi$ and $f(T)\in F[T]$, where T is indeterminate. Then*

$$(j_L(x_\alpha))^r f(j_L(h_\beta)) = f(j_L(h_\beta)-r\alpha(h_\beta))(j_L(x_\alpha))^r$$

for all $r\ge 0$.

Proof Since $f(T)$ is linear combination of $\{T^m\mid m\ge 0\}$ and multiplication by $(j_L(x_\alpha))^r$ is also linear, it is sufficient to show the result for $f(T)=T^m$, $m\ge 0$. Thus, we need to show that $(j_L(x_\alpha))^r(j_L(h_\beta))^m=(j_L(h_\beta)-r\alpha(h_\beta))^m j_L(x_\alpha)^r$ for all $r,m\ge 0$. We prove it by the induction on m. If $m=0$, then both sides equal to $(j_L(x_\alpha))^r$ and there is nothing to do. For $m=1$, we prove it by induction on r. If $r=0$, then both sides are $j_L(h_\beta)$. Assume the result for r. Then

$$j_L(x_\alpha)^r j_L(h_\beta) = (j_L(h_\beta)-r\alpha(h_\beta))j_L(x_\alpha)^r.$$

Hence, $j_L(x_\alpha)^{r+1}j_L(h_\beta)$
$= j_L(x_\alpha)j_L(x_\alpha)^r j_L(h_\beta)$
$= j_L(x_\alpha)(j_L(h_\beta)-r\alpha(h_\beta))j_L(x_\alpha)^r$
$= j_L(x_\alpha)j_L(h_\beta)(j_L(x_\alpha))^r - r\alpha(h_\beta)(j_L(x_\alpha))^{r+1}$
$= j_L(h_\beta)(j_L(x_\alpha))^{r+1} + [j_L(x_\alpha),j_L(h_\beta)](j_L(x_\alpha))^r - r\alpha(h_\beta)(j_L(x_\alpha))^{r+1}$
$= j_L(h_\beta)(j_L(x_\alpha))^{r+1} - \alpha(h_\beta)(j_L(x_\alpha))^{r+1} - r\alpha(h_\beta)(j_L(x_\alpha))^{r+1}$
$= (j_L(h_\beta)-(r+1)\alpha(h_\beta))j_L(x_\alpha)^{r+1}$.

This proves the result for $m = 1$. A similar calculation may be used to establish the result for $m+1$ by assuming it for m. This proves the Lemma. ♯

Proof of the Kostant Theorem: Let B denote the lattice generated by $\{e^-_{\overline{a}} f_{\overline{b}} e^+_{\overline{c}} \mid \overline{a}, \overline{c} \in \mathbb{Z}^N_+,\ and\ \overline{b} \in \mathbb{Z}^l_+\}$. By Corollary 4.2.10, $\binom{j_L(h_{\alpha_i})}{r_i}$ belongs to $U(L)_{\mathbb{Z}}$ for all i and $r_i \geq 0$. This shows that $B \subseteq U(L)_{\mathbb{Z}}$.

It remains to show that $U(L)_{\mathbb{Z}} \subseteq B$. It suffices to show that each monomial of elements of the set $\{\frac{(j_L(x_\alpha))^m}{m!} \mid \alpha \in \Phi,\ and\ m \geq 0\}$ is integral linear combination of members of the set $\{e^-_{\overline{a}} f_{\overline{b}} e^+_{\overline{c}} \mid \overline{a}, \overline{c} \in \mathbb{Z}^N_+,\ and\ \overline{b} \in \mathbb{Z}^l_+\}$. By Lemmas 4.2.14 and 4.2.15, any monomial in $\{\frac{(j_L(x_\alpha))^m}{m!} \mid \alpha \in \Phi,\ and\ m \geq 0\}$ can be expressed in the order prescribed in the statement of the theorem. Further, $\frac{(j_L(x_\alpha))^r}{r!}\frac{(j_L(x_\alpha))^s}{s!} = \binom{r+s}{s}\frac{(j_L(x_\alpha))^{r+s}}{(r+s)!}$ and the monomials in $\{\binom{j_L(h_i)-a_i}{b_i} \mid 1 \leq i \leq l, a_i \in \mathbb{Z},\ and\ b_i \geq 0\}$ can be expressed as integral linear combinations of monomials in $\{\binom{j_L(h_i)}{m_i} \mid 1 \leq i \leq d, m_i \geq 0\}$ which can be shifted at the required places , thanks to Lemmas 4.2.8 and 4.2.15.(note that the polynomial $\binom{T-a_i}{b_i}$ takes integral values on integers). It follows that $U(L)_{\mathbb{Z}} \subseteq B$. ♯

The following corollary is immediate from the Kostant theorem.

Corollary 4.2.16 *The set* $\{e^-_{\overline{a}} \mid \overline{a} \in \mathbb{Z}^N_+\}$ *is a* $\mathbb{Z}$ *-basis for a lattice* $U(N^-)_{\mathbb{Z}}$ *of* $U(N^-)$, *the set* $\{f_{\overline{b}} \mid \overline{b} \in \mathbb{Z}^l_+\}$ *is a* $\mathbb{Z}$ *-basis for a lattice* $U(H)_{\mathbb{Z}}$ *of* $U(H)$, *and the set* $\{e^+_{\overline{a}} \mid \overline{a} \in \mathbb{Z}^N_+\}$ *is a* $\mathbb{Z}$ *-basis for a lattice* $U(N^+)_{\mathbb{Z}}$ *of* $U(N^+)$, *where* $N^- = \oplus\sum_{\alpha<0} L_\alpha$ *and* $N^+ = \oplus\sum_{\alpha>0} L_\alpha$. ♯

More generally, our aim is to show the existence of a lattice $V_{\mathbb{Z}}$ in a finite-dimensional L-module V such that $U(L)_{\mathbb{Z}} \cdot V_{\mathbb{Z}} \subseteq V_{\mathbb{Z}}$. Such a lattice is called an **admissible lattice**.

Lemma 4.2.17 *Let* $\overline{d} = (d_1, d_2, \ldots, d_l) \in \mathbb{Z}^l$ *and* S *a finite subset of* $\mathbb{Z}^l$ *such that* $\overline{d} \notin S$. *Then there is a polynomial* $f(T_1, T_2, \ldots, T_l) \in F[T_1, T_2, \ldots, T_l]$ *such that* $f(\mathbb{Z}^l) \subseteq \mathbb{Z}$, $f(\overline{d}) = 1$ *and* $f(S) = \{0\}$.

Proof Take k to be sufficiently large so that $S \subseteq D$, where $D = [d_1 - k, d_1 + k] \times [d_2 - k, d_2 + k] \times \cdots \times [d_l - k, d_l + k]$. Consider the polynomial $f(T_1, T_2, \ldots, T_l)$ defined by $\prod_{i=1}^l \binom{T_i - d_i + k}{k} \prod_{i=1}^l \binom{-T_i + d_i + k}{k}$. Clearly $f(\mathbb{Z}^l) \subseteq \mathbb{Z}$, $f(\overline{d}) = 1$, and for each $\overline{a} \in D - \{\overline{d}\}$, $f(\overline{a}) = 0$. Hence, $f(S) = \{0\}$. ♯

Theorem 4.2.18 *Let* V *be a finite-dimensional* L*-module.*
(i) Let A *be a subgroup of* V *such that* $U(L)_{\mathbb{Z}} \cdot A \subseteq A$. *Then* $A = \oplus\sum_{\mu\in\Gamma(V)} V_\mu \bigcap A$, *where* $\Gamma(V)$ *is the set of weights of* V.
(ii) There is a lattice $V_{\mathbb{Z}}$ *in* V *such that* $U(L)_{\mathbb{Z}} \cdot V_{\mathbb{Z}} \subseteq V_{\mathbb{Z}}$.

Proof (i) For each weight $\mu \in \Gamma(V)$, let $\overline{d_\mu} = (\mu(h_1), \mu(h_2), \ldots, \mu(h_l))$. Let $\lambda \in \Gamma(V)$. The set $S = \{\overline{d_\mu} \mid \mu \in \Gamma(V) - \{\lambda\}\}$ is a finite subset of $\mathbb{Z}^l$ and $\overline{d_\lambda} \notin S$. From the above Lemma, there is a polynomial $f(T_1, T_2, \ldots, T_l) \in F[T_1, T_2, \ldots, T_l]$ such that $f(\mathbb{Z}^l) \subseteq \mathbb{Z}$, $f(\overline{d_\lambda}) = 1$ and $f(\overline{d_\mu}) = 0$ for all $\mu \in \Gamma(V) - \{\lambda\}$. Put $h = f(j_L(h_1), j_L(h_2), \ldots, j_L(h_l))$. By Lemma 4.2.8, $h \in U(H)_{\mathbb{Z}})$. Since

$f(\overline{d_\lambda}) = 1$, $v \mapsto h \cdot v$ is a projection of V on V_λ. Since $U(L)_{\mathbb{Z}} \cdot A \subseteq A$, the V_λ component of $h \cdot v$ is in A for each $v \in A$. Since λ is arbitrary member of $\Gamma(V)$, $A = \oplus \sum_{\mu \in \Gamma(V)} A \bigcap V_\mu$.

(ii) By Weyl's theorem, V is direct sum of simple L-modules. It is sufficient, therefore, to assume that $V = V(\lambda)$ is a simple L-module with highest weight λ. Let v^+ be a maximal vector with highest weight λ. Put $V_{\mathbb{Z}} = U(N^-)_Z \cdot v^+$. Since $j_L(x_\alpha) \cdot v^+ = 0$ for all $\alpha > 0$ and $\{e^+_{\overline{a}} \mid \overline{a} \in \mathbb{Z}^N_+\}$ is a $\mathbb{Z}$-basis of $U(N^+)_Z$, $U(N^+)_Z \cdot v^+ = \mathbb{Z}v^+$. Since

$$\binom{j_L(h_i)}{b_i} \cdot v^+ = \frac{j_L(h_i)(j_L(h_i)-1)\cdots(j_L(h_i)-b_i+1)}{b_i!}(v^+) = \frac{\lambda(h_i)(\lambda(h_i)-1)\cdots(\lambda(h_i)-b_i+1)}{b_i!}(v^+)$$

belongs $\mathbb{Z}v^+$, $U(H)_{\mathbb{Z}} \cdot v^+ = \mathbb{Z}v^+$. Thus,

$$U(L)_{\mathbb{Z}} \cdot v^+ = U(N^-)U(H)_{\mathbb{Z}}U(N^+)_{\mathbb{Z}}v^+ = U(N^-)\mathbb{Z}v^+ = U(N^-)_{\mathbb{Z}}v^+ = V_{\mathbb{Z}}.$$

This means that $U(L)_{\mathbb{Z}}V_{\mathbb{Z}} \subseteq V_{\mathbb{Z}}$. It is also clear that $V_{\mathbb{Z}} \bigcap V_\lambda = \mathbb{Z}v^+$. Further, $\{e^-_{\overline{a}}v^+ \mid \overline{a} \in \mathbb{Z}^N_+\}$ is finite and so $V_{\mathbb{Z}}$ is finitely generated. Since $U(N^-)_{\mathbb{Z}}$, being a lattice in $U(N^-)$ contains a basis of $U(N^-)$, $V_{\mathbb{Z}} = U(N^-)_{\mathbb{Z}}v^+$ generates V as a F-space. Thus, to show that $V_{\mathbb{Z}}$ is a lattice, we need to show that the $rank V_{\mathbb{Z}}$ is at the most $Dim V$. Suppose not. Then $rank V_{\mathbb{Z}} > Dim V$. Let r be the smallest number such that there is a set $\{v_1, v_2, \ldots, v_r\}$ of $\mathbb{Z}$-linearly independent vectors in $V_{\mathbb{Z}}$ which is linearly dependent over F. Suppose that $\sum_{i=1}^r a_iv_i = 0$, $a_i \in F^\star$, $v_i \neq 0$. If all elements of $U(L)_{\mathbb{Z}}v_1$ has zero $V(\lambda)_\lambda$ component in the decomposition of $V(\lambda)$ as direct sum of its weight spaces, then the subspace of $V(\lambda)$ generated by $U(L)_{\mathbb{Z}}v_1$ will be a nonzero proper sub module of $V(\lambda)$. This is a contradiction, since $V(\lambda)$ is simple. Hence, there is a nonzero element $u \in U(L)_{\mathbb{Z}}$ such that $V(\lambda)_\lambda$ component of uv_1 is nonzero. Further, by part (i) of the Lemma, $V(\lambda)_\lambda$-component of uv_i lies in $V_{\mathbb{Z}}$ for each i. Thus, $V(\lambda)_\lambda$-component of of uv_i is m_iv_i for some $m_i \in \mathbb{Z}$. This means that $\sum_{i=1}^r m_iv_i = 0$, where $m_1 \neq 0$. Now,

$$m_1 \sum_{i=1}^r a_iv_i - \left(\sum_{i=1}^r a_im_i\right)v_1 = 0.$$

Hence,

$$\sum_{i=2}^r a_i(m_1v_i - m_iv_1) = 0.$$

Clearly, $\{m_1v_i - m_iv_1 \mid 2 \leq i \leq r\}$ is $\mathbb{Z}$-linearly independent subset of $V_{\mathbb{Z}}$ which is linearly dependent over F $(a_i \neq 0)$, a contradiction to the minimality of r. This shows that $V_{\mathbb{Z}}$ is an admissible lattice in V. ♯

Let V be a L-module and $L' = \{x \in L \mid x \cdot v = 0 \ for \ all \ v \in V\}$. Then L' is an ideal of L and V is faithful L/L'-module. Indeed, L' is sum of those simple ideals of L which act trivially on V. There is not much loss in generality in assuming that V is a faithful L-module. So, we assume that V is a faithful L-module. We have already seen that

$$\Lambda_r \subseteq \Lambda(V) \subseteq \Lambda_{univ}.$$

Let $L_V = \{x \in L \mid x \cdot v \in V_{\mathbb{Z}} \ for\ all\ v \in V_{\mathbb{Z}}\}$ denote the **stabilizer** of the admissible lattice $V_{\mathbb{Z}}$. Evidently, the lattice $L(\mathbb{Z})$ of L generated by the chosen Chevalley basis is contained in L_V. Also, L_V is closed under the Lie product.

Proposition 4.2.19 *L_V is an admissible lattice for the L-module L and it is independent of the chosen admissible lattice $V_{\mathbb{Z}}$ of V.*

We need the following two Lemmas to prove the Proposition.

Lemma 4.2.20 *$H_V = L_V \bigcap H$ is a lattice in H such that $H(\mathbb{Z}) \subseteq H_V \subseteq H_0$, where $H_0 = \{h \in H \mid \lambda(h) \in \mathbb{Z} \ for\ all\ \lambda \in \Lambda_r\}$ and $H(\mathbb{Z}) = H \bigcap L(\mathbb{Z})$ is the $\mathbb{Z}$-span of $\{h_\alpha \mid \alpha \in \Phi\}$.*

Proof Let $h \in H$. It follows from Theorem 4.2.18 that $h \cdot V_{\mathbb{Z}} \subseteq V_{\mathbb{Z}}$ if and only if $\lambda(h) \in \mathbb{Z}$ for all $\lambda \in \Lambda(V)$. Further, since $\Lambda_r \subseteq \Lambda(V) \subseteq \Lambda_{univ}$,

$$\{h \in H \mid \lambda(h) \in \mathbb{Z} \ for\ all\ \lambda \in \Lambda_{univ}\} \subseteq \{h \in H \mid \lambda(h) \in \mathbb{Z} \ for\ all\ \lambda \in \Lambda(V)\} \subseteq \{h \in H \mid \lambda(h) \in \mathbb{Z} \ for\ all\ \lambda \in \Lambda_r\}.$$

Consequently,

$$H(\mathbb{Z}) \subseteq H_V \subseteq H_0,$$

where $H(\mathbb{Z})$ is the $\mathbb{Z}$-span of $\{h_\alpha \mid \alpha \in \Phi\}$. This shows that H_V is a lattice in H. ♯

Lemma 4.2.21 *For each $x \in L$, $u \in U(L)$, and $n \in \mathbb{N}\bigcup\{0\}$,*

$$\frac{(adj_L(x))^n}{n!}(u) = \sum\nolimits_{i=0}^{n}(-1)^i \frac{(j_L(x))^{n-i}}{(n-i)!} u \frac{(j_L(x))^i}{i!}.$$

Proof The proof is by the induction on n. For $n = 0$, both sides are u. For $n = 1$, $\frac{j_L(x)}{1!}u = J_L(x)u - uj_L(x)$. Thus, the identity holds for 0 and 1. Assume the identity for n. Then

$$\begin{aligned}
&\frac{(adj_L(x))^{n+1}}{(n+1)!}(u) \\
&= \frac{ad(j_L(x))}{(n+1)}\left(\frac{(adj_L(x))^n}{n!}(u)\right) \\
&= \frac{ad(j_L(x))}{(n+1)}[\sum\nolimits_{i=0}^{n}(-1)^i \frac{(j_L(x))^{n-i}}{(n-i)!} u \frac{(j_L(x))^i}{i!}] \\
&= \sum\nolimits_{i=0}^{n}(-1)^i \frac{(j_L(x))^{n+1-i}}{(n-i)!(n+1)} u \frac{(j_L(x))^i}{i!} - \sum\nolimits_{i=0}^{n}(-1)^i \frac{(j_L(x))^{n-i}}{(n-i)!} u \frac{(j_L(x))^{i+1}}{i!(n+1)}.
\end{aligned}$$

Putting $i = j - 1$ in the second term, it becomes

$$-\sum\nolimits_{j=1}^{n+1}(-1)^j \frac{(j_L(x))^{n+1-j}}{(n+1-j)!} u \frac{(j_L(x))^j}{(j-1)!(n+1)}.$$

Substituting and arranging, we obtain that

$$\frac{(adj_L(x))^{n+1}}{(n+1)!}(u) = \sum_{j=0}^{n+1}(-1)^i \frac{(j_L(x))^{n+1-j}}{(n+1-j)!}u\frac{(j_L(x))^j}{j!}. \ \sharp$$

Proof of Proposition 4.2.19. Since $L(\mathbb{Z})$ is the lattice generated by the chosen Chevalley basis,

$$L(\mathbb{Z}) = H(\mathbb{Z}) + \oplus\sum_{\alpha\in\Phi}\mathbb{Z}x_\alpha.$$

From the above lemma $\frac{(adj_L(x_\alpha))^n}{n!}(j_L(u)) \cdot v$ belongs to $V_\mathbb{Z}$ for all $u \in L_V$ and $v \in V_\mathbb{Z}$. This means that L_V is invariant under $\frac{(adj_L(x_\alpha))^n}{n!}$ for all $\alpha \in \Phi$ and $m \geq 0$. Again by Theorem 4.2.18 (i),

$$L_V = H_V \oplus (\oplus\sum_{\alpha\in\Phi}(L_V \bigcap L_\alpha)).$$

Since $L(\mathbb{Z}) \subseteq L_V$, $\mathbb{Z}x_\alpha \subseteq (L_V \bigcap L_\alpha)$. Again, since H_V is a lattice in H, it suffices to show that $\mathbb{Z}x_\alpha = (L_V \bigcap L_\alpha)$ for each $\alpha \in \Phi$. Consider the map $ad(x_{-\alpha})|_{L_\alpha}$ from L_α to H. This is injective and the image is the subspace Fh_α of H, since L_α is one dimensional. Clearly, $ad(x_{-\alpha})(L_V \bigcap L_\alpha) \subseteq H_V = L_V \bigcap H$. Since H_V is a lattice in H, $ad(x_{-\alpha})(L_V \bigcap L_\alpha)$ is infinite cyclic. Thus, $L_V \bigcap L_\alpha$ is infinite cyclic generated by $\frac{1}{n}x_\alpha$ for some $n \in \mathbb{N}$, as x_α has to be an integral multiple of the generator. In turn,

$$\frac{(ad(x_{-\alpha}))^2}{2!}\left(\frac{x_\alpha}{n}\right) = \frac{x_{-\alpha}}{n}$$

belongs to L_V, thanks to the above Lemma. Again, since L_V is closed under Lie product,

$$-\left(\frac{ad(x_\alpha)}{n}\right)^2\left(\frac{x_{-\alpha}}{n}\right) = \frac{2}{n^3}x_\alpha$$

belongs to L_V. This means that $\frac{2}{n^3}x_\alpha$ is an integral multiple of $\frac{1}{n}x_\alpha$. Hence, $n = 1$ and $L_V \bigcap L_\alpha = \mathbb{Z}x_\alpha$. $\sharp$

Example 4.2.22 Consider $L = sl(2, F)$ with standard basis $\{h, y, x\}$, where F is as usual an algebraically closed field of characteristic 0. Evidently, $\{h, y, x\}$ is a Chevalley basis of L. Consider L as L-module through adjoint action. Clearly, L is a faithful simple L-module with x as a maximal vector of maximal weight α, where $\alpha(h) = 2$ (note that $[h, x] = hx - xh = 2x$). $L(\mathbb{Z}) = \mathbb{Z}h + \mathbb{Z}y + \mathbb{Z}x$. $U(L)_\mathbb{Z}$ is the lattice with $\{\frac{j_L(x_{-\alpha})^m}{m!}\binom{j_L(h)}{b}\frac{j_L(x_\alpha)^r}{r!} \mid m, b, r \geq 0\}$ as $\mathbb{Z}$-basis. Evidently, $L(\mathbb{Z})$ is an admissible lattice and then the lattice L_L of $L(\mathbb{Z})$ is given by $L_L = \mathbb{Z}\frac{h}{2} + \mathbb{Z}y + \mathbb{Z}x$.

We have another faithful L-module F^2 with matrix multiplication action. It can be easily seen that $e_1 = (1, 1)^t$ is a maximal vector with maximal weight $\lambda = \frac{\alpha}{2}$ given by $\lambda(h) = 1$, and $\mathbb{Z}e_1 + \mathbb{Z}e_2$ is an admissible lattice. The stabilizing admissible lattice L_{F^2} is given by $L_{F^2} = \mathbb{Z}h + \mathbb{Z}y + \mathbb{Z}x = L(\mathbb{Z})$. Note that respective weight lattices are Λ_{univ} and Λ_r (check it).

Exercises

4.2.1. Show that any two different choice of Chevalley basis of a Lie algebra determine isomorphic $\mathbb{Z}$-Lie algebras.
4.2.2. Determine two Chevalley bases of $sl(2, F)$.
4.2.3. Determine a Chevalley base of each of the following: (i) A_l, (ii) B_l, and (iii) G_2.
4.2.4. Let $L = sl(2, F)$ and F be a field of characteristic different from 2. Describe the Chevalley algebra $L(K)$ of adjoint type and also the Chevalley group of adjoint type.
4.2.5. Consider the standard two-dimensional L-module F^2, where $L = sl(2, F)$. Find an admissible lattice, and the corresponding stabilizer lattice.
4.2.6. Determine an admissible lattice for the standard representation of $sl(3, F)$ on F^3 and also the stabilizer lattice.
4.2.7. Let V and W be L-modules with admissible lattices A and B. Show that $A \otimes B$ is admissible lattice for the module $V \otimes W$.

4.3 Chevalley Groups

In this section, we shall introduce Chevalley groups $G(V, K)$ associated with a triple (L, V, K), where L is a simple Lie algebra over an algebraically closed field F of characteristic 0, and K is a field. There is no loss in assuming that $F = \mathbb{C}$. We also study the structure of $G(V, K)$.

Let V be a L-module and $V_{\mathbb{Z}}$ an admissiblle lattice in V. Let K be an arbitrary field. We can treat K as a $\mathbb{Z}$-module. Put $V^K = V_{\mathbb{Z}} \otimes_{\mathbb{Z}} K$, $L^K = L(\mathbb{Z}) \otimes_{\mathbb{Z}} K$, $H^K = H(\mathbb{Z}) \otimes_{\mathbb{Z}} K$, $V_\mu^K = (V_{\mathbb{Z}})_\mu \otimes_{\mathbb{Z}} K$, and $Kx_\alpha^K = \mathbb{Z}x_\alpha \otimes_{\mathbb{Z}} K$, where $(V_{\mathbb{Z}})_\mu = V_{\mathbb{Z}} \bigcap V_\mu$, μ being a weight of V. The following corollary is immediate from Theorem 4.2.18.

Corollary 4.3.1 *(i)* $V^K = \oplus \sum_{\mu \in \Lambda(V)} V_\mu^K$, $Dim_K V_\mu^K = Dim_F V_\mu$.
(ii) $x_\alpha^K \neq 0$ *for all* $\alpha \in \Phi$, $L^K = H^K \oplus \sum_{\alpha \in \phi} Kx_\alpha^K$, $Dim_K H^K = Dim_F H$, *and* $Dim_K L^K = Dim_F L$. ♯

For each $\alpha \in \Phi$ and $m \in \mathbb{Z}_+$, $\frac{j_L(x_\alpha)^m}{m!} \in U(L)_{\mathbb{Z}}$, and it acts on the admissible lattice $V_{\mathbb{Z}}$. In turn, for $T \in \mathbb{Z}[T]$, $\frac{j_L(x_\alpha)^m}{m!} \otimes T^m$ acts on $V_{\mathbb{Z}} \otimes_{\mathbb{Z}} \mathbb{Z}[T]$. We denote $\frac{j_L(x_\alpha)^m}{m!} \otimes T^m$ by $\frac{T^m j_L(x_\alpha)^m}{m!}$. Since for large m the action of $j_L(x_\alpha)^m$ on V is 0, the action of $\frac{T^m j_L(x_\alpha)^m}{m!}$ on $V_{\mathbb{Z}} \otimes_{\mathbb{Z}} \mathbb{Z}[T]$ is zero for large m. Thus, $exp(Tj_L(x_\alpha)) = \sum_{m=0}^{\infty} \frac{T^m j_L(x_\alpha)^m}{m!}$ makes sense. and it acts as an automorphism on $V_{\mathbb{Z}} \otimes_{\mathbb{Z}} \mathbb{Z}[T]$. Indeed, $(exp(Tj_L(x_\alpha)))^{-1} = exp(-Tj_L(x_\alpha))$. Let $a \in K$. The specialization map $T \mapsto a$ induces a surjective homomorphism η_a from $(V_{\mathbb{Z}} \otimes_{\mathbb{Z}} \mathbb{Z}[T]) \otimes_{\mathbb{Z}} K$ to $V_{\mathbb{Z}} \otimes_{\mathbb{Z}} K = V^K$. In turn, $exp(aj_L(x_\alpha)) = \sum_{m=0}^{\infty} \frac{a^m j_L(x_\alpha)^m}{m!}$ acts as an automorphism on V^K. We denote this automorphism by $x_\alpha^V(a)$. Since $aj_L(x_\alpha)$ and $bj_L(x_\alpha)$ commute,

$$x_\alpha^V(a + b) = x_\alpha^V(a)x_\alpha^V(b)$$

for all $a, b \in K$. The subgroup $\{x_\alpha^V(a) \mid a \in K\}$ of $GL(V^K)$ is denoted by $X_\alpha^V(K)$, and the subgroup of $GL(V^K)$ generated by $\bigcup_{\alpha\in\Phi} X_\alpha^V(K)$ is denoted by $G(V, K)$. We shall see soon that the group $G(V, K)$ depends only on the module V (weights of V) and not on the choice of the admissible lattice. So, the notation is unambiguous. The group $G(V, K)$ is called **Chevalley Group** associated with (L, V, K).

Our next aim is to look at the structure of $G(V, K)$ and also to look at the presentation of $G(V, K)$ in terms of the generators $x_\alpha^V(a)$.

Theorem 4.3.2 *Let $\alpha, \beta \in \Phi$ such that $\alpha + \beta \in \Phi$ or $\alpha + \beta \neq 0$. Fix an ordering in the set $\{i\alpha + J\beta \mid i, j \geq 1\}$, lexicographic ordering (say) with $\alpha + \beta$ the smallest element. Consider the power series ring $U(L)_{\mathbb{Z}}[[s, t]]$ over $U(L)_{\mathbb{Z}}$ in two variables s and t. Then*

$$[exp(sj_L(x_\alpha)), exp(tj_L(x_\beta))] = \prod_{i,j\geq 1} exp(c_{ij}s^it^j j_L(x_{i\alpha+j\beta})),$$

where, as usual, $[u, v]$ denotes the commutator $uvu^{-1}v^{-1}$, and in the RHS, the product is taken in the chosen order. Further, $c_{11} = \nu_{\alpha,\beta}$, c_{ij} are all integers which depend only on the chosen ordering and not on s or t.

Proof Recall that if T_1 and T_2 are commuting linear transformations on a vector space, then $exp(T_1 + T_2) = exp(T_1)exp(T_2)$. In particular $exp(-T) = (exp(T))^{-1}$. If A is an associative algebra and $a \in A$, then $ad(a) = L_a - R_a$, where L_a is left multiplication by a and R_a is right multiplication by a and they commute because of the associativity. Consequently,

$$exp(ad(a)) = L_{exp(a)}R_{exp(-a)}$$

for all $a \in A$ with $ad(a)$ nilpotent. Thus,

$$exp(ad(a))(b) = exp(a)bexp(-a) \tag{4.15}$$

for all $a, b \in A$ with $ad(a)$ nilpotent. Consider the element

$$f(s,t) = [exp(sj_L(x_\alpha)), exp(tj_L(x_\beta))]\prod_{i,j\geq 1} exp(-c_{ij}s^it^j j_L(x_{i\alpha+j\beta})),$$

where the product in the RHS is in the reversed order to the chosen order of $\{i\alpha + j\beta \mid i, j \geq 1\}$. It is sufficient to show that for a choice of c_{ij} with $c_{11} = \nu_{\alpha,\beta}$, $f(s,t) = 1$. Observe that

$$s\frac{d}{ds}(exp(sj_L(x_\alpha))) = sj_L(x_\alpha)exp(sj_L(x_\alpha)). \tag{4.16}$$

Now, using the product rule of derivation,

$$
\begin{array}{c}
s\frac{\partial}{\partial s}(f(s,t)) = sj_L(x_\alpha)f(s,t)+ \\
exp(sj_L(x_\alpha))exp(tj_L(x_\beta))(-sj_L(x_\alpha))exp(-j_L(x_\alpha)exp(-tj_L(x_\beta))\cdot \\
\prod_{i,j\geq 1} exp(-c_{ij}s^it^j j_L(x_{i\alpha+j\beta}) + [exp(sj_L(x_\alpha)), exp(tj_L(x_\beta))]\cdot \\
\sum_{k,l}\Big(\Big(\prod_{i\alpha+j\beta>k\alpha+l\beta} exp(-c_{ij}s^it^j j_L(x_{i\alpha+j\beta})\cdot \\
(-c_{kl}ks^kt^l j_L(x_{k\alpha+l\beta})exp(-s^kt^l c_{kl} j_L(x_{k\alpha+l\beta})))\cdot \\
\prod_{i\alpha+j\beta j\leq k\alpha+l\beta} exp(-c_{ij}s^it^j j_L(x_{i\alpha+j\beta}))))\Big)\Big).
\end{array}
\tag{4.17}
$$

Further, from (4.15)

$$
(exp(ad(tj_L(x_\beta)))(-sj_L(x_\alpha)) = exp(tj_L(x_\beta))(-sj_L(x_\alpha))(exp(-tj_L(x_\beta))). \tag{4.18}
$$

Also,

$$
exp(ad(tj_L(x_\beta)))(-sj_L(x_\alpha)) = -sj_L(x_\alpha) - \nu_{\alpha,\beta}stj_L(x_{\alpha+\beta}) - \cdots. \tag{4.19}
$$

Using (4.18) and (4.19), we can express the RHS of (4.17) as $Af(s,t)$, where A is a polynomial in s^i, t^j, and $j_L(x_{i\alpha+j\beta})$, $i, j \geq 1$. Observe that in the expression for $f(s,t)$ is homogeneous of degree 0 relative to the grading $s \to -\alpha, t \to -\beta$, and $j_L(x_\gamma) \to \gamma$. Consequently, $s\frac{\partial}{\partial s}(f(s,t))$ and so A also has the same property. It follows from the degree considerations that

$$
A = \sum_{k,l\geq 1}(-c_{kl} + p_{kl})s^kt^l j_L(x_{k\alpha+l\beta}),
$$

where p_{kl} is a polynomial in c_{ij} with $i+j < k+l$. Inductively, we can choose c_{ij} so that $A = 0$. But, then $s\frac{\partial}{\partial s}(f(s,t)) = 0$. This means that $f(s,t) = f(0,t) = 1$.

Finally, we need to to show that all c_{ij} are integers with $c_{11} = \nu_{\alpha,\beta}$. Clearly, the coefficient of s^it^j in the power series of $f(s,t)$ is $-c_{ij}$ + the terms coming from the exponentials of $j_L(x_{k\alpha+l\beta})$ with $k+l < i+j$. It follows by the induction that $c_{ij}j_L(x_{i\alpha+j\beta})$ belongs to $U(L)_{\mathbb{Z}}$. This means that $c_{ij} \in \mathbb{Z}$. Also for $i = 1 = j$, the coefficient of $j_L(x_{\alpha+\beta})$ is $-c_{11} + \nu_{\alpha,\beta}$. So the initial choice is $c_{11} = \nu_{\alpha,\beta}$ to make $A = 0$. This completes the proof of the Theorem. ♯

Corollary 4.3.3 *Consider the Chevalley group $G(V, K)$. If $\alpha, \beta \in \Phi$ such that $\alpha + \beta \in \Phi$. Then there exists integers c_{ij}, $i, j \geq 1$ with $c_{11} = \nu_{\alpha,\beta}$ depending upon a chosen ordering of the set $\{i\alpha + j\beta \mid i, j \geq\}$ with smallest element $\alpha + \beta$ such that*

$$
[x^V_\alpha(a), x^V_\beta(b)] = \prod_{i,j\geq 1} x^V_{i\alpha+j\beta}(c_{ij}a^ib^j)
$$

for all $a, b \in K$, where the product is taken in the chosen order. ♯

Corollary 4.3.4 *If $\alpha, \beta \in \Phi$ and $\alpha + \beta \notin \Phi$, then $[x^V_\alpha(a), x^V_\beta(b)] = 1$ for all $a, b \in K$. Suppose that if $i, j \geq 1$ and $i\alpha + j\beta \in \Phi$, then $i = 1 = j$. Then $\nu_{\alpha,\beta} = \pm(r+1)$ and the RHS of the expression in Theorem 4.3.2 is $exp(\nu_{\alpha,\beta}j_L(x_{\alpha+\beta}))$.*

Proof Since $\alpha + \beta \notin \Phi$, $[x_\alpha, x_\beta] = 0$. Using the Jacobi identity, we observe that $ad(x_\alpha)$ and $ad(x_\beta)$ commute. Thus, the multiplication by $j_L(x_\alpha)$ and the multiplication by $j_L(x_\beta)$ on $U(L)$ commute. It follows $exp(aj_L(x_\alpha))$ and $exp(bj_L(x_\beta))$ commute. Hence $[x_\alpha^v(a), x_\beta^v(b)] = 1$ for all $a, b \in K$. The rest of the assertion follows from the Theorem 4.3.2. ♯

Example 4.3.5 Consider $L = sl(l+1, F)$, $l \geq 2$. We have a Chevalley basis $\{e_{ij} \mid i \neq j, 1 \leq i, j \leq l+1\} \bigcup \{e_{ii} - e_{i+1\, i+1} \mid 1 \leq i \leq l\}$ of L. $\Phi = \{(i, j) \mid i \neq j\}$, $x_{(i,j)} = e_{ij}$. Note that $[e_{ij}, e_{ji}] = e_{ii} - e_{jj}$. We have the faithful L-module $V = F^{l+1}$ with the obvious action. It can be easily observed that $x_{ij}^V(a) = exp(aj_L(e_{ij})) = E_{ij}^a$. The relation described in Theorem 4.3.2 is the usual Steinberg relation.

Recall that a subset Ψ of Φ is called a closed set of roots if $\alpha, \beta \in \Psi$ and $\alpha + \beta \in \Phi$ implies that $\alpha + \beta \in \Psi$. Thus, Φ^+, $\Phi^+ - \{\alpha \in \Delta\}$, $\{\alpha\}$, and $\Phi_r = \{\alpha \in \Phi \mid ht(\alpha) \geq r \geq 1\}$ are all closed sets. Let Ψ be a closed subset of Φ. A subset ζ of Ψ is called an ideal of Ψ if $\alpha \in \zeta$, $\beta \in \Psi$ and $\alpha + \beta \in \Psi$ implies that $\alpha + \beta \in \zeta$. Clearly, Φ^+, $\Phi^+ - \{\alpha \in \Delta\}$, and $\Phi_r = \{\alpha \in \Phi \mid ht(\alpha) \geq r \geq 1\}$ are all ideals of Φ^+.

Proposition 4.3.6 *Let ζ be an ideal of a closed subset Ψ of Φ. Let $X_\zeta^V(K)$ and $X_\Psi^V(K)$ denote the subgroups of $G(V, K)$ generated by $\bigcup_{\alpha \in \zeta} X_\alpha^V(K)$ and $\bigcup_{\alpha \in \Psi} X_\alpha^V(K)$, respectively. Suppose that $\alpha \in \Psi$ implies that $-\alpha \notin \Psi$. Then $X_\zeta^V(K)$ is a normal subgroup of $X_\Psi^V(K)$.*

Proof It is sufficient to show that $x_\alpha^V(a)x_\beta^V(b)x_\alpha^V(a)^{-1}$ belongs to $X_\zeta^V(K)$ for all $\alpha \in \Psi$ and $\beta \in \zeta$, $a, b \in K$. Since $\gamma \in \Psi$ implies that $-\gamma \notin \Psi$, $\alpha + \beta \neq 0$. It follows from Corollary 4.3.3 that $[x_\alpha^V(a), x_\beta^V(b)]$ belongs to $X_\zeta^V(K)$. Hence, $x_\alpha^V(a)x_\beta^V(b)x_\alpha^V(a)^{-1}$ belongs to $X_\zeta(K)$ for all $\alpha \in \Psi$, $\beta \in \zeta$, and $a, b \in K$. ♯

Proposition 4.3.7 *Let Ψ be closed set of roots such that $\alpha \in \Psi$ implies that $-\alpha \notin \Psi$. Let $\leq$ be an order in Ψ such that $ht(\alpha) < ht(\beta)$ implies that $\alpha < \beta$. Then every element of $X_\Psi^V(K)$ is uniquely expressible as $\prod_{\alpha \in \Psi} x_\alpha^V(a_\alpha)$, where $a_\alpha \in K$ and the product is taken in the chosen order of Ψ. Indeed, the result holds for an arbitrary ordering in Ψ.*

Proof We first use the induction on $\mid \Psi \mid$ to show that every element of $X_\Psi^V(K)$ is expressible as $\prod_{\alpha \in \Psi} x_\alpha^V(a_\alpha)$, where $a_\alpha \in K$ and the product is taken in the chosen order of Ψ. It follows trivially, if $\mid \Psi \mid = 1$. Assume the induction hypothesis. Let α_1 be the first element of Ψ. Then $\Psi - \{\alpha_1\}$ is an ideal of Ψ. By the above Proposition $X_{\Psi - \{\alpha_1\}}^V(K)$ is a normal subgroup of $X_\Psi^V(K)$ and so $X_\Psi^V(K) = X_{\alpha_1}^V(K)X_{\Psi - \{\alpha_1\}}^V(K)$. The assertion follows by the induction hypothesis.

Next, we show the uniqueness of the expression, again, by the induction on $\mid \Psi \mid$. Suppose that $\mid \Psi \mid = 1$. Then $\Psi = \{\alpha_1\}$ for some $\alpha_1 \in \Phi$. Since $X_{\alpha_1}^V(K) \neq \{0\}$, there is a weight λ of V and a vector $v \in (V_{\mathbb{Z}})_\lambda$ such that $j_L(x_{\alpha_1})v \neq 0$. Now, suppose that $x_{\alpha_1}^v(a_{\alpha_1}) = x_{\alpha_1}^v(b_{\alpha_1})$. Then

$$x^v_{\alpha_1}(a_{\alpha_1}) \cdot v = v + a_{\alpha_1} j_L(x_{\alpha_1}) \cdot v + z$$

and

$$x^v_{\alpha_1}(b_{\alpha_1}) \cdot v = v + b_{\alpha_1} j_L(x_{\alpha_1}) \cdot v + z',$$

where $a_{\alpha_1} j_L(x_{\alpha_1} \cdot v)$ and $b_{\alpha_1} j_L(x_{\alpha_1} \cdot v)$ belong to $V_{\lambda+\alpha_1}$ while z and z' are sums of elements in other weight spaces. This shows that $a_{\alpha_1} = b_{\alpha_1}$. Assume the induction hypothesis. Let $u = \prod_{\alpha\in\Psi} x^V_\alpha(a_\alpha) = \prod_{\alpha\in\Psi} exp(a_\alpha j_L(x_\alpha))$ be a nonzero member of $X^V_\Psi(K)$. Clearly, $x^V_{\alpha_1}(a_{\alpha_1})^{-1}u$ belongs to $X^V_{\Psi-\{\alpha_1\}}(K)$. Using the induction hypothesis, we see that the representation is unique.

The last assertion follows from the following abstract group theoretic result. If G is a group with subgroups $G_1, G_2, \ldots, G_r$ such that (i) every element $g \in G$ has a unique representation as $g = g_1 g_2 \cdots g_r$, $g_i \in G_i$, and (ii) for each i, $G_i G_{i+1} \cdots G_r$ is a normal subgroup of G, then given any permutation $p \in S_r$, every element g of G is uniquely expressible as $g = h_1 h_2 \cdots h_r$, where $h_i \in G_{p(i)}$. ♯

Corollary 4.3.8 *The group $X^V_\alpha(K)$ is isomorphic to the additive group $(K, +)$.*

Proof It follows from the above Proposition that the map $a \mapsto x^V_\alpha(a)$ is bijective homomorphism. ♯

Corollary 4.3.9 *Fix an order of ϕ^+. The subgroup $U(V, K) = X^V_{\Phi^+}(K) = \prod_{\alpha\in\Phi^+} X^V_\alpha(K)$ of $G(V, K)$ is a unipotent subgroup in the sense that the eigen value of each member of $U(V, K)$ is 1. Further, we can choose a basis of V^K with respect to which $U(V, K)$ is the group of uni-upper triangular matrices. Similarly, the subgroup $U(V, K)_- = X^V_{\Phi^-}(K) = \prod_{\alpha\in\Phi^-} X^V_\alpha(K)$ of $G(V, K)$ is a unipotent subgroup and it is the group of uni-lower triangular matrices with respect to a suitable basis.*

Proof Let B be a basis of weight vectors. Take an ordering $\leq$ in B subject to the condition that if "$\lambda - \mu$ is sum of positive roots, then $\lambda < \mu$". It follows from proof of the Proposition 4.3.7 that the matrix representation of each member of U is uni-upper triangular. The rest of the assertion follows, similarly. ♯

The following Corollaries follow immediately from the preceding results.

Corollary 4.3.10 *Fix an ordering $\leq$ in Φ^+ subject to the condition that "$ht(\alpha) < ht(\beta)$" implies that $\alpha < \beta$. Let $U(V, K)_r$ denote the subgroup $X^V_{\Phi_r}(K)$ of $U(V, K)$. Then the following hold:*

(i) For each r, $U(V, K)_r$ is a normal subgroup of $U(V, K)$.
(ii) For each r, $[U(V, K), U(V, K)_r] \subseteq U(V, K)_{r+1}$.
(iii) $U(V, K)$ is a nilpotent subgroup of $G(V, K)$. ♯

Corollary 4.3.11 *Let Ψ and ζ be disjoint closed subsets of Φ^+ such that $\Phi^+ = \Psi \bigcup \zeta$. Then $U(V, K) = X^V_\Psi(K) X^V_\zeta(K)$. In particular, if α is a simple root, then $U(V, K) = X^V_\alpha(K) X^V_{\Phi^+ - \{\alpha\}}(K)$.* ♯

Example 4.3.12 If $L = A_l = sl(l+1, F)$ and $V = F^{l+1}$ the usual L-module, then $U(V, K)$ is the subgroup of $SL(l+1, K)$ consisting of uni-upper triangular matrices. $U(V, K)_2$ is the subgroup of uni-upper triangular matrices whose just above diagonal entries are 0 and so on.

We fix up some more notations to develop the structure theory of Chevalley groups. Denote the element $x^V_\alpha(a)x^V_{-\alpha}(-a^{-1})x^V_\alpha(a)$ of $G(V, K)$ by $w^V_\alpha(a)$, and the element $w^V_\alpha(a)(w^V_\alpha(1))^{-1}$ by $h^V_\alpha(a)$, $a \in K^\star$. Thus, in case $L = sl(n, F)$, $n \geq 2$, $V = F^n$, we have $w^V_{ij}(a) = E^a_{ij}E^{-a^{-1}}_{ji}E^a_{ij} = I - e_{ii} - e_{jj} + ae_{ij} - a^{-1}e_{ji}$, $i > j$ is the matrix $[b_{kl}]$, where $b_{kk} = 1$ for $i \neq k \neq j$, $b_{ii} = 0 = b_{jj}$, $b_{ij} = a$, $b_{ji} = -a^{-1}$, and all other entries are 0. Also $h^V_{ij}(a)$ is the diagonal matrix all of whose diagonal entries are 1, except the i_{th} and the j_{th} diagonal entries which are a and a^{-1}, respectively.

Recall the action of $W(\Phi)$ on H which is given by $\sigma_\alpha(t_\beta) = t_{\sigma_\alpha(\beta)}$. The action is extended to whole of H by the linearity. In particular, $\sigma_\alpha(h_\beta) = h_{\sigma_\alpha(\beta)}$.

Proposition 4.3.13 *(i)* $w^V_\alpha(a)j_L(h)w^V_\alpha(a)^{-1} = j_L(\sigma_\alpha(h))$.

(ii) For each $\mu \in \Gamma(V)$, *let* $v \in V^K_\mu$. *Then, there is an element* $v' \in V^K_{\sigma_\alpha(\mu)}$ *such that* $w^V_\alpha(a)v = a^{-\langle\mu,\alpha\rangle}v'$ *for each* $a \in K$.

(iii) $w^V_\alpha(a)j_L(x_\beta)w^V_\alpha(a)^{-1} = c(\alpha, \beta)a^{-\beta(h_\alpha)}j_L(x_{\sigma_\alpha(\beta)})$, *where* $c(\alpha, \beta) = \pm 1 = c(\alpha, -\beta)$ *is independent of* a, *the* L*-module* V *(note that* $j_L(x_\beta)$ *acts on* V^K*) and* K.

(iv) $h_\alpha(a)v = a^{\langle\mu,\alpha\rangle}v$ *for each* $v \in V^K_\mu$.

Proof (i) Suppose that $\alpha(h) = 0$. Then $[x_{\pm\alpha}, h] = 0$. Consequently, $j_L(x_{\pm\alpha})j_L(h) = j_L(h)j_L(x_{\pm\alpha})$. In turn, $x^V_{\pm\alpha}(a)j_L(h) = j_L(h)x^V_{\pm\alpha}(a)$ for all a. Thus, in this case, both sides of (i) are equal to $j_L(h)$. Now, every element of H is uniquely expressible as $ah_\alpha + h$, where $a \in \mathbb{C}$ and $\alpha(h) = 0$ (any $k \in H$ is expressible as $k = \frac{\alpha(k)}{2}h_\alpha + (k - \frac{\alpha(k)}{2}h_\alpha)$). Since both sides are linear in h, it suffices to establish the identity for h_α. Clearly, in this case, both sides depend on the Lie algebra $S_\alpha = <\{h_\alpha, x_\alpha, y_\alpha\}>$ and not on the representation. Thus, it is sufficient to establish the identity for $L = sl(2, F)$ with the usual two-dimensional representation. Clearly,

$$w^{F^2}_\alpha(a) = exp(ae_{12})exp(-a^{-1}e_{21})exp(ae_{12}) = E^a_{12}E^{-a^{-1}}_{21}E^a_{12} = \begin{bmatrix} 0 & a \\ -a^{-1} & 0 \end{bmatrix},$$

and

$$j_L(h_\alpha) = j_L(h_{11}) = \begin{bmatrix} 1 & 0 \\ 0 & -1 \end{bmatrix}.$$

Thus,

$$w^{F^2}_{12}(a)j_L(h_{11})w^{F^2}_{12}(a)^{-1} = \begin{bmatrix} -1 & 0 \\ 0 & 1 \end{bmatrix} = j_L(\sigma_\alpha(h_{11})).$$

This proves (i).

(ii) Let $v \in V^K_\mu$. It follows from the definition of the action of $x^V_\alpha(a)$ and $w^V_\alpha(a)$ that

$$v' = w_\alpha^V(a)v = \sum_{i=-\infty}^{\infty} a^i v_i,$$

where $v_i \in V_{\mu+i\alpha}^K$ (the sum is finite as $\Gamma(V)$ is finite). If $h \in H$, then

$$j_L(h)v' = j_L(h)w_\alpha^V(a)v = w_\alpha^V(a)w_\alpha^V(a)^{-1}j_L(h)w_\alpha^V(a)v = w_\alpha^V(a)j_L(\sigma_\alpha(h))v = \sigma_\alpha(\mu)(h)v',$$

by(i) (note that $w_\alpha^V(a)^{-1} = w_\alpha^V(-a)$ and the identity in (i) is independent of a) and the supposition that $v \in V_\mu^K$. This shows that $v' \in V_{\sigma_\alpha(\mu)}^K$ and hence the only nonzero term in the summation representing v' is the i_{th} term, where $i = - \prec \mu, \alpha \succ$. This proves (ii).

(iii) Applying (ii) to the adjoint representation of L on L, for $v = x_\beta \in L_\beta$, we have an element $v' = c(\alpha,\beta)x_{\sigma_\alpha(\beta)} \in L_{\sigma_\alpha(\beta)}^K$ which is independent of a, and it is such that

$$w_\alpha^L(a)j_L(x_\beta)w_\alpha^L(a)^{-1} = c(\alpha,\beta)a^{-\prec\beta,\alpha\succ}j_L(x_{\sigma_\alpha(\beta)}).$$

Further, since $w_\alpha^L(1)$ is an automorphism of $L(\mathbb{Z})$ and x_α is a primitive element for $L(\mathbb{Z})$ for each α, $c(\alpha,\beta) = \pm 1$. Next, using (i) and the above identity,

$$j_L(h_{\sigma_\alpha(\beta)}) = w_\alpha^L(1)j_L(h_\beta)w_\alpha^L(1)^{-1} = [w_\alpha^L(1)j_L(x_\beta)w_\alpha^L(1)^{-1}, w_\alpha^L(1)j_L(x_{-\beta})w_\alpha^L(1)^{-1}] = c(\alpha,\beta)c(\alpha,-\beta)j_L(h_{\sigma_\alpha(\beta)}).$$

Thus, $c(\alpha,\beta)c(\alpha,-\beta) = 1$. Hence, $c(\alpha,\beta) = c(\alpha,-\beta)$.

(iv) Since $w_\alpha^V(a)^{-1} = w_\alpha^V(-a)$, $h_\alpha^V(a) = w_\alpha^V(-a)^{-1}w_\alpha^V(-1)$. By (ii), $w_\alpha^V(-a)v = (-a)^{-\prec\mu,\alpha\succ}v'$ and also $w_\alpha^V(-1)v = (-1)^{-\prec\mu,\alpha\succ}v'$. Hence, $h_\alpha^V(a)v = w_\alpha^V(-a)^{-1}w_\alpha^V(-1)v = w_\alpha^V(-a)^{-1}(-1)^{-\prec\mu,\alpha\succ}v' = (-1)^{-\prec\mu,\alpha\succ}(-a)^{\prec\mu,\alpha\succ}v = a^{\prec\mu,\alpha\succ}v$ for each $v \in V_\mu^K$. ♯

Proposition 4.3.14 *(i)* $w_\alpha^V(1)h_\beta^V(a)w_\alpha^V(1)^{-1} = h_{\sigma_\alpha(\beta)}^V(a)$.

(ii) $w_\alpha^V(a)x_\beta^V(b)w_\alpha^V(1)^{-1} = x_{\sigma_\alpha(\beta)}^V(ca^{-\prec\beta,\alpha\succ}b)$, *where* $c = c(\alpha,\beta)$ *is as in the above Proposition. In particular,* $w_\alpha^V(1)x_\beta^V(b)w_\alpha^V(a)^{-1} = x_{\sigma_\alpha(\beta)}^V(cb)$.

(iii) $h_\alpha^V(a)x_\beta^V(b)h_\alpha^V(a)^{-1} = x_\beta^V(a^{\prec\beta,\alpha\succ}b)$.

Proof (i) Let $v \in V_\mu^K$. Then $w_\alpha^V(1)^{-1}v \in V_{\sigma_\alpha(\mu)}^K$ (Proposition 4.3.13 (ii)). Again, by Proposition 4.3.13 (iv),

$$w_\alpha^V(1)h_\beta^V(a)w_\alpha^V(1)^{-1}v = w_\alpha^V(1)a^{\prec\sigma_\alpha(\mu),\beta\succ}w_\alpha^V(1)^{-1}v = a^{\prec\sigma_\alpha(\mu),\beta\succ}v = a^{\prec\mu,\sigma_\alpha(\beta)\succ}v = h_{\sigma_\alpha(\beta)}^V(a)v.$$

Since μ and v are arbitrary, $w_\alpha^V(1)h_\beta^V(a)w_\alpha^V(1)^{-1} = h_{\sigma_\alpha(\beta)}^V(a)$.

(ii) By Proposition 4.3.13 (iii),

$$w_\alpha^V(a)bj_L(x_\beta)w_\alpha^V(a)^{-1} = ca^{-\prec\beta,\alpha\succ}bj_L(x_{\sigma_\alpha(\beta)}).$$

Exponentiating, we get the desired result.

(iii) By applying Proposition 4.3.13 (iv) to the adjoint representation, we obtain

$$h^V_\alpha(a)bj_L(x_\beta)h^V_\alpha(a)^{-1} = a^{\prec\beta,\alpha\succ}bj_L(x_\beta).$$

Exponentiating again, we get the desired identity. ♯

Corollary 4.3.15 *Let $H(V, K)$ denote the subgroup of $G(V, K)$ generated by the set $\{h^V_\alpha(a) \mid \alpha \in \Phi$ and $a \in K\}$ and $B(V, K)$ denote the subgroup generated by $H(V, K) \bigcup U(V, K)$. Then the following hold:*

(i) *$U(V, K)$ is a normal subgroup of $B(V, K)$.*
(ii) $B(V, K) = U(V, K)H(V, K)$.
(iii) *$H(V, K) \bigcap U(V, K) = \{1\}$. In particular, every element b of $B(V, K)$ is uniquely expressible as uh, where $u \in U(V, K)$ and $h \in H(V, K)$.*

Proof (i) By Proposition 4.3.14 (iii), $h^V_\alpha(a)x^V_\beta(b)h^V_\alpha(a)^{-1} = x^V_\beta(a^{\prec\beta,\alpha\succ}b)$ belongs to $U(V, K)$ for all $\beta > 0$. This shows that $U(V, K)$ is normal subgroup of $B(V, K)$.

(ii) Follows from (i).

(iii) We have a basis of V^K such that the matrix of each element of $U(V, K)$ is uni-upper triangular (Corollary 4.3.9) and the matrix of each element of $H(V, K)$ is diagonal (Proposition 4.3.13 (iv)). This means that $H(V, K) \bigcap U(V, K) = \{1\}$. The rest is evident. ♯

Lemma 4.3.16 $X^V_\alpha(K) = X^V_\beta(K)$ *implies that* $\alpha = \beta$.

Proof Clearly, $X^V_\alpha(K) \neq \{1\}$ for each $\alpha \in \Phi$. If α and β are in Φ^+ (or in Φ^-), then the result follows from Proposition 4.3.7. If they are of opposite sign, one represents uni-upper triangular and the other represents uni-lower triangular. ♯

Proposition 4.3.17 *Let $N(V, K)$ denote the subgroup of $G(V, K)$ generated by the set $\{w^V_\alpha(a) \mid \alpha \in \Phi$ and $a \in K\}$. Then the following hold:*

(i) *$H(V, K)$ is a normal subgroup of $N(V, K)$.*
(ii) *The association $\sigma_\alpha \mapsto H(V, K)w^V_\alpha(1) = H(V, K)w^V_\alpha(a)$ induces a homomorphism η from $W(\Phi)$ to $N(V, K)/H(V, K)$.*
(iii) *η is an isomorphism.*

Proof (i) Since $w^V_\alpha(1)h^V_\beta(a)w^V_\alpha(1)^{-1} = h^V_{\sigma_\alpha(\beta}(a)$ (Proposition 8.3.14 (i)),

$$w^V_\alpha(b)h^V_\beta(a)w^V_\alpha(b)^{-1} = h^V_\beta(b)w^V_\alpha(1)h^V_\beta(a)w^V_\alpha(1)^{-1}h^V_\alpha(b)^{-1}$$

belongs to $H(V, K)$. Hence, $H(V, K)$ is a normal in $N(V, K)$.

(ii) Since $w^V_\alpha(a)w^V_\alpha(1)^{-1} = h^V_\alpha(a)$ belongs to $H(V, K)$, $H(V, K)w^V_\alpha(1) = H(V, K)w^V_\alpha(a)$ for all $a \in K$. Recall that $W(\Phi)$ is generated by the set $\{\sigma_\alpha \mid \alpha \in \Phi\}$ subject to the relations $\{\sigma_\alpha^2 = I \mid \alpha \in \Phi\}$ and $\{\sigma_\alpha\sigma_\beta\sigma_\alpha^{-1} = \sigma_{\sigma_\alpha(\beta)} \mid \alpha, \beta \in \Phi\}$. Thus, it suffices to show that

$$(H(V, K)w^V_\alpha(1))^2 = H(V, K), \tag{4.20}$$

and

$$H(V,K)w_\alpha^V(1)H(V,K)w_\beta^V(1)H(V,K)w_\beta^V(1)^{-1} = H(V,K)w_{\sigma_\alpha(\beta)}^V(1). \quad (4.21)$$

Now,

$$(H(V,K)w_\alpha^V(1))^2 = H(V,K)w_\alpha^V(1)H(V,K)w_\alpha^V(1) =$$
$$H(V,K)w_\alpha^V(1)H(V,K)w_\alpha^V(-1) = H(V,K)w_\alpha^V(1)w_\alpha^V(-1) = H(V,K).$$

This establishes (4.20).
Next,

$$\begin{aligned}
& w_\alpha^V(1)w_\beta^V(1)w_\alpha^V(1)^{-1} \\
&= w_\alpha^V(1)x_\beta^V(1)x_{-\beta}^V(-1)x_\beta^V(1)w_\alpha^V(1)^{-1} \text{ (by definition of } w_\beta^V(1)) \\
&= w_\alpha^V(1)x_\beta^V(1)w_\alpha^V(1)^{-1}w_\alpha^V(1)x_{-\beta}^V(-1)w_\alpha^V(1)^{-1}w_\alpha^V(1)x_\beta^V(1)w_\alpha^V(1)^{-1} \\
&= x_{\sigma_\alpha(\beta)}^V(c)x_{-\sigma_\alpha(\beta)}^V(-c)x_{\sigma_\alpha(\beta)}^V(c) \\
&= w_{\sigma_\alpha(\beta)}^V(c) \text{ (note that } c = \pm 1).
\end{aligned}$$

This proves (4.21). Thus, η is a homomorphism from $W(\Phi)$ to $N(V,K)/H(V,K)$.

(iii) Since $N(V,K)$ is generated by $\{w_\alpha^V(a) \mid \alpha \in \Phi \textit{ and } a \in K\}$, it follows that η is surjective. Let $\sigma = \sigma_{\alpha_1}\sigma_{\alpha_2}\cdots\sigma_{\alpha_t}$ be a member of $Ker\eta$. Then

$$\eta(\sigma) = H(V,K)w_{\alpha_1}^V(1)w_{\alpha_2}^V(1)\cdots w_{\alpha_t}^V(1) = H(V,K).$$

In other words $w_{\alpha_1}^V(1)w_{\alpha_2}^V(1)\cdots w_{\alpha_t}^V(1) = \hat{h}$ belongs to $H(V,K)$. Now,

$$w_{\alpha_1}^V(1)w_{\alpha_2}^V(1)\cdots w_{\alpha_t}^V(1)X_\alpha^V(K)w_{\alpha_t}^V(1)^{-1}w_{\alpha_{t-1}}^V(1)^{-1}\cdots w_{\alpha_1}^V(1)^{-1} = X_{\sigma(\alpha)}^V(K).$$

In turn, by Proposition 4.3.14 (iii), $X_{\sigma(\alpha)}^V(K) = \hat{h}X_\alpha^V(K)\hat{h}^{-1} = X_\alpha^V(K)$. By Lemma 4.3.16, $\sigma(\alpha) = \alpha$. Since α is arbitrary, $\sigma = I$. ♯

Convention: If $n_1H(V,K) = H(V,K)n_1 = H(V,K)n_2 = n_2H(V,K)$, then $n_1n_2^{-1},\ n_1^{-1}n_2 \in H(V,K) \subseteq B(V,K)$ and so $B(V,K)n_1 = B(V,K)n_2$ and $n_1B(V,K) = n_2B(V,K)$. Thus, for $\sigma \in W(\Phi)$, we have a unique $B(V,K)n \in G(V,K)/^rB(V,K)$, and unique $nB(V,K)$ such that $\eta(\sigma) = B(V,K)n$. We denote this unique $B(V,K)n$ by $B(V,K)\sigma$ and $nB(V,K)$ by $\sigma B(V,K)$ also.

Proposition 4.3.18 *Let $\alpha \in \Delta$ be a simple root. Then the following hold:*

(i) $G_\alpha = B(V,K)\bigcup(B(V,K)w_\alpha^V(1)B(V,K)) = B(V,K)\bigcup(B(V,K)\sigma_\alpha B(V,K))$ *is a subgroup of* $G(V,K)$.
(ii) If $\sigma \in W(\Phi)$ *is such that* $\sigma(\alpha) \in \Phi^+$ *(i.e.,* $l(\sigma\sigma_\alpha) = l(\sigma)+1$*), then*

$$\sigma B(V,K)\sigma_\alpha \subseteq B(V,K)\sigma\sigma_\alpha B(V,K), \textit{ and in turn,}$$
$$(B(V,K)\sigma B(V,K))(B(V,K)\sigma_\alpha B(V,K)) \subseteq B(V,K)\sigma\sigma_\alpha B(V,K).$$

Further, $(B(V,K)\sigma B(V,K))(B(V,K)\sigma_\alpha B(V,K)) = B(V,K)\sigma\sigma_\alpha B(V,K)$.

(iii) If $\sigma \in W(\Phi)$ *is such that* $\sigma(\alpha) \in \Phi^-$, *then*

$$\sigma B(V,K)\sigma_\alpha \subseteq (B(V,K)\sigma\sigma_\alpha B(V,K)) \bigcup (B(V,K)\sigma B(V,K)),\ \textit{and in turn,}$$
$$(B(V,K)\sigma B(V,K))(B(V,K)\sigma_\alpha B(V,K)) \subseteq$$
$$(B(V,K)\sigma\sigma_\alpha B(V,K)) \bigcup (B(V,K)\sigma B(V,K)).$$

(iv) For all $\sigma \in W(\Phi)$,

$$\sigma B(V,K)\sigma_\alpha \subseteq (B(V,K)\sigma\sigma_\alpha B(V,K)) \bigcup B(V,K)\sigma B(V,K),\ \textit{and in turn,}$$
$$(B(V,K)\sigma B(V,K))(B(V,K)\sigma_\alpha B(V,K)) \subseteq$$
$$(B(V,K)\sigma\sigma_\alpha B(V,K)) \bigcup (B(V,K)\sigma B(V,K)).$$

(v) $B(V,K) \bigcap U(V,K)_- = \{1\}$.
(vi) $B(V,K) \bigcap N(V,K) = H(V,K)$.

Proof (i) Since $B(V,K)$ is a subgroup and $\eta(\sigma)^{-1} = \eta(\sigma)$, it follows that inverse of each element in G_α is in G_α. Thus, it is sufficient to show that $G_\alpha G_\alpha \subseteq G_\alpha$. Now, $G_\alpha G_\alpha$ is contained in

$$B(V,K)(B(V,K) \bigcup B(V,K)\sigma_\alpha B(V,K)) \bigcup$$
$$((B(V,K)\sigma_\alpha B(V,K))(B(V,K) \bigcup B(V,K)\sigma_\alpha B(V,K))).$$

In turn,
$$G_\alpha G_\alpha \subseteq G_\alpha \bigcup (B(V,K)\sigma_\alpha B(V,K)\sigma_\alpha B(V,K)).$$

Since $B(V,K)G_\alpha B(V,K) \subseteq G_\alpha$, it suffices to show that $\sigma_\alpha B(V,K)\sigma_\alpha \subseteq G_\alpha$. Further, $x^V_{-\alpha}(a) = x^V_\alpha(a^{-1})w^V_\alpha(-a^{-1})x^V_\alpha(a^{-1})$ belongs to $B(V,K)w^V_\alpha(1)B(V,K) = B(V,K)\sigma_\alpha B(V,K)$ for each α. Hence, $X^V_{-\alpha}(K) \subseteq G_\alpha$. Next,

$\sigma_\alpha B(V,K)\sigma_\alpha$
$= w^V_\alpha(1)B(V,K)w^V_\alpha(-1)$
$= w^V_\alpha(1)B(V,K)w^V_\alpha(1)^{-1}$
$= w^V_\alpha(1)X^V_\alpha(K)X^V_{\Phi^+-\{\alpha\}}(K)H(V,K)w^V_\alpha(1)^{-1}$
$= w^V_\alpha(1)X^V_\alpha(K)w^V_\alpha(1)^{-1}w^V_\alpha(1)X^V_{\Phi^+-\{\alpha\}}(K)w^V_\alpha(1)^{-1}w_\alpha(1)H(V,K)w^V_\alpha(1)^{-1}$
$= X^V_{-\alpha}(K)X^V_{\Phi^+-\{\alpha\}}(K)H(V,K)$ (since $\sigma_\alpha(\Phi^+ - \{\alpha\}) \subseteq (\Phi^+ - \{\alpha\})$ (check it))
$\subseteq G_\alpha B(V,K) = G_\alpha$. This proves (i)

(ii) $\sigma B(V,K)\sigma_\alpha$
$= \sigma X^V_\alpha(K)X^V_{\Phi^+-\{\alpha\}}(K)H(V,K)\sigma_\alpha$
$= \sigma X^V_\alpha(K)\sigma^{-1}\sigma\sigma_\alpha\sigma_\alpha^{-1}X^V_{\Phi^+-\{\alpha\}}(K)\sigma_\alpha\sigma_\alpha^{-1}H(V,K)\sigma_\alpha$
$\subseteq B(V,K)\sigma\sigma_\alpha B(V,K)$ (for $\sigma X^V_\alpha(K)\sigma^{-1} \subseteq B(V,K)$, $\sigma_\alpha^{-1}X^V_{\Phi^+-\{\alpha\}}(K)$
$\subseteq B(V,K)$, *and* $\sigma_\alpha^{-1}H(V,K)\sigma_\alpha \subseteq B(V,K)$).

Again, since $B(V, K)^2 = B(V, K)$,

$$(B(V, K)\sigma B(V, K))(B(V, K)\sigma_\alpha B(V, K)) = B(V, K)\sigma B(V, K)\sigma_\alpha B(V, K)$$
$$\subseteq B(V, K)(B(V, K)\sigma\sigma_\alpha B(V, K))B(V, K) = B(V, K)\sigma\sigma_\alpha B(V, K).$$

(iii) Suppose that $\sigma(\alpha) \in \Phi^-$. Put $\sigma' = \sigma\sigma_\alpha$. Then $\sigma'(\alpha) \in \Phi^+$ and $\sigma = \sigma'\sigma_\alpha$. Now,
$\sigma B(V, K)\sigma_\alpha$
$= \sigma'(\sigma_\alpha B(V, K)\sigma_\alpha)$
$\subseteq \sigma'(B(V, K) \bigcup B(V, K)\sigma_\alpha B(V, K))$ (from the proof of (i))
$= \sigma' B(V, K) \bigcup \sigma' B(V, K)\sigma_\alpha B(V, K)$
$\subseteq (\sigma' B(V, K)\sigma'^{-1})\sigma' \bigcup (\sigma' B(V, K)\sigma'^{-1}\sigma'\sigma_\alpha B(V, K)$
$\subseteq (B(V, K)\sigma\sigma_\alpha B(V, K)) \bigcup (B(V, K)\sigma B(V, K))$.
Again,

$$(B(V, K)\sigma B(V, K))(B(V, K)\sigma_\alpha B(V, K)) = B(V, K)\sigma B(V, K)\sigma_\alpha B(V, K)$$
$$\subseteq B(V, K)((B(V, K)\sigma\sigma_\alpha B(V, K)) \bigcup (B(V, K)\sigma B(V, K)).$$

(iv) Follows from (ii) and (iii).

(v) We have a basis with respect to which the matrices of the members of $B(V, K)$ are non-singular upper triangular and the matrices of members of $U(V, K)_-$ are lower uni-triangular. This shows that $B(V, K) \bigcap U(V, K)_- = \{I\}$.

(vi) Obviously $H(V, K) \subseteq B(V, K) \bigcap N(V, K)$. We show that $B(V, K) \bigcap N(V, K) \subseteq H(V, K)$. Suppose not. Then there is a $n \in B(V, K) \bigcap N(V, K)$ such that $H(V, K)n \neq H(V, K)$. Consequently, there is a $\sigma \neq I$ such that $\eta(\sigma) = H(V, K)n$. Since $\sigma \neq I$, there is a $\alpha \in \Phi^+$ such that $\sigma(\alpha) < 0$ and hence $nX^V_\alpha(K)n^{-1} \subseteq B(V, K) \bigcap U(V, K)_-$. This is a contradiction. ♯

Corollary 4.3.19 *Let $\sigma \in W(\Phi)$ and $\sigma = \sigma_{\alpha_1}\sigma_{\alpha_2}\cdots\sigma_{\alpha_r}$ be a minimal representation of σ as product of simple reflections. Then*

$$B(V, K)\sigma B(V, K) = B(V, K)\sigma_{\alpha_1} B(V, K)\sigma_{\alpha_2} B(V, K)\cdots B(V, K)\sigma_{\alpha_r} B(V, K).$$

Proof Under the hypothesis, $\sigma_{\alpha_1}\sigma_{\alpha_2}\cdots\sigma_{\alpha_i}(\alpha_i) < 0$ for all i. The result follows from the (ii) part of the above Proposition. ♯

Proposition 4.3.20 *Let $G(V, K) = < \bigcup_{\alpha\in\Phi} X^V_\alpha(K) >$ be a Chevalley group. Then $G(V, K) = < S >$, where $S = (\bigcup_{\alpha\in\Delta} X^V_\alpha(K)) \bigcup \{w^V_\alpha(1) \mid \alpha \in \Delta\}$.*

Proof Since $W(\Phi)$ is generated by simple reflections and every root is $W(\Phi)$-conjugate to a simple root, the result follows from the following identity:

$$w^V_\alpha(1)X^V_\beta(K)w^V_\alpha(1)^{-1} = X^V_{\sigma_\alpha(\beta)}(K). \ \sharp$$

Theorem 4.3.21 (Bruhat Decomposition)
(i) $G(V, K) = \bigcup_{\sigma\in W(\Phi)} B(V, K)\sigma B(V, K)$.

(ii) $B(V,K)\sigma B(V,K) = B(V,K)\sigma' B(V,K)$ *if and only if* $\sigma = \sigma'$. *In other words a system of coset representative system of* $N(V,K)$ *modulo* $H(V,K)$ *is also a double coset representative system for* $B(V,K)|G(V,K)|B(V,K)$.

Proof (i) By Proposition 4.3.20, the set S of generaters of $G(V,K)$ is contained in $\bigcup_{\sigma\in W(\Phi)} B(V,K)\sigma B(V,K)$. Further, by Proposition 4.3.18, $u(\bigcup_{\sigma\in W(\Phi)} B(V,K)\sigma B(V,K)) \subseteq \bigcup_{\sigma\in W(\Phi)} B(V,K)\sigma B(V,K)$ for all $u \in S\bigcup S^{-1}$. Hence, $G(V,K) = \bigcup_{\sigma\in W(\Phi)} B(V,K)\sigma B(V,K)$.

(ii) Suppose that $B(V,K)\sigma B(V,K) = B(V,K)\sigma' B(V,K)$. We have to show that $\sigma' = \sigma$. The proof is by the induction on $l(\sigma)$. Suppose that $l(\sigma) = 0$. Then $\sigma = I$. In this case, $B(V,K)\sigma' B(V,K) = B(V,K)$. Hence if $n \in N(V,K)$ is such that $\eta(\sigma') = H(V,K)n$, then $n \in B(V,K)$. This means that $\eta(\sigma')B(V,K)\eta(\sigma')^{-1} = B(V,K)$. Consequently, $\sigma'(\Phi^+)\bigcap\Phi^- = \emptyset$ and so $\sigma'(\Phi^+) = \Phi^+$. This shows that $\sigma' = I$. Now, assume the result for all $\tau \in W(\Phi)$ for which $l(\tau) < l(\sigma)$, $l(\sigma) \geq 1$. Suppose that $B(V,K)\sigma B(V,K) = B(V,K)\sigma' B(V,K)$. Let $\alpha \in \Delta$ be such that $l(\sigma\sigma_\alpha) < l(\sigma)$. Now,

$$\eta(\sigma\sigma_\alpha) \subseteq B(V,K)\sigma' B(V,K)B(V,K)\sigma_\alpha B(V,K) \subseteq$$
$$(B(V,K)\sigma' B(V,K))\bigcup(B(V,K)\sigma'\sigma_\alpha B(V,K)) =$$
$$(B(V,K)\sigma B(V,K))\bigcup(B(V,K)\sigma'\sigma_\alpha B(V,K)).$$

By the induction hypothesis, $\sigma\sigma_\alpha = \sigma$ or $\sigma\sigma_\alpha = \sigma'\sigma_\alpha$. Now, $\sigma\sigma_\alpha \neq \sigma$. for $\sigma_\alpha \neq I$. Hence, $\sigma\sigma_\alpha = \sigma'\sigma_\alpha$ and so $\sigma = \sigma'$. ♯

Theorem 4.3.22 *Let* $\sigma \in \Phi$ *and* $n_\sigma \in N(V,K)$ *be such that* $\eta(\sigma) = H(V,K)n_\sigma$. *Let* S *denote the set* $\Phi^+\bigcap\sigma^{-1}(\Phi^-)$ *and* T *denote the set* $\Phi^+\bigcap\sigma^{-1}(\Phi^+)$. *Put* $U(V,K)^-_\sigma = X^V_S(K)$ *and* $U(V,K)^+_\sigma = X^K_T$. *Then,*

(i) $U(V,K) = U(V,K)^+_\sigma U(V,K)^-_\sigma$.
(ii) $U(V,K)^+_\sigma \bigcap U(V,K)^-_\sigma = \{I\}$.
(iii) $B(V,K)\sigma B(V,K) = B(V,K)n_\sigma U(V,K)^-_\sigma$.
(iv) Every element of $B(V,K)\sigma B(V,K)$ *has a unique representation as* $bn_\sigma v$, *where* $b \in B(V,K)$ *and* $v \in U(V,K)^-_\sigma$.
(v) Every element of $G(V,K)$ *has a unique representation as* $uhn_\sigma v$, *where* $u \in U(V,K)$, $h \in H(V,K)$, *and* $v \in U(V,K)^-_\sigma$.

Proof (i) Since $S\bigcup T = \Phi^+$, $U(V,K) = U(V,K)^+_\sigma U(V,K)^-_\sigma$.

(ii) Since $S\bigcap T = \emptyset$, the result follows.

(iii) $B(V,K)\sigma B(V,K) = B(V,K)\sigma U(V,K)H(V,K)$
$= B(V,K)\sigma X^V_T(K)X^V_S(K)H(V,K)$
$= B(V,K)\sigma X^V_T(K)\sigma^{-1}\sigma X^V_S(K)H(V,K)$
$= B(V,K)\sigma X^V_S(K)H(V,K)$ (since $U(V,K) \triangleleft B(V,K)$ and $X^V_T(K) \subseteq U(V,K)$)
$= B(V,K)n_\sigma X^V_S(K)H(V,K) = B(V,K)n_\sigma U(V,K)^-_\sigma$.

(iv) Suppose that $bn_\sigma v = b'n_\sigma v'$. Then $b^{-1}b' = n_\sigma vv'^{-1}n_\sigma^{-1}$. But, we can get a basis with respect to which LHS is upper triangular, where as the RHS is lower triangular. This means that $b = b'$ and $v = v'$.

(v) Follows from (iv), if we observe that an element $b \in B(V, K)$ has a unique representation as $b = uh$, where $u \in U(V, K)$ and $h \in H(V, K)$ $(U(V, K) \bigcap H(V, K) = \{I\})$. ♯

Propositions 4.3.18, 4.3.20, and Theorem 4.3.21 assert that a Chevalley Group admits a (B, N) $(B = B(V, K)$ and $N = N(V, K))$ pair (or a Tits system) in the sense of the following definition:

Definition 4.3.23 A Group G is said to admit a **Tits System** (or a (B, N)**-Pair**) if it has a pair (B, N) of subgroups such that the following hold:

(i) $G = < B \bigcup N >$.
(ii) $H = B \bigcap N$ is a normal subgroup of N.
(iii) N/H is generated by a set $S = \{s_i \mid i \in I\}$ of involutions.
(iv) If $n_i \in N$, $i \in I$ are such that $s_i = Nn_i$, then $n_i B n_i \neq B$.
(v) For each $n \in N$, $n_i Bn \subseteq (Bn_i nB) \bigcup (BnB)$.

Remark 4.3.24 The concept of (B, N) pair was introduced by J.Tits (Ann of Math, 1964) as an axiomatic system in order to develop its theory and to unify certain family of groups. Chevalley Groups, Reductive algebraic groups over local fields, Doubly transitive groups on sets containing more than 2 elements are certain examples. For details see Bourbaki, "Groupes et al.ġbres de Lie", 1968.

Example 4.3.25 Consider $sl(2, F)$ and the corresponding Chevalley group $SL(2, K)$ relative to the usual representation of $sl(2, F)$ on F^2. Clearly, $U(F^2, K) = \{E_{12}^a \mid a \in K\}$, $\Phi = \{\alpha, -\alpha\}$, $W(\Phi) = \{I, \sigma_\alpha\}$, $x_\alpha^{F^2}(a) = E_{12}^a, a \in K$,

$$w_\alpha^{F^2}(a) = w_{12}^{F^2}(a) = E_{12}^a E_{21}^{-a^{-1}} E_{12}^a = \begin{bmatrix} 0 & a \\ -a^{-1} & 0 \end{bmatrix},$$

$$h_\alpha^{F^2}(a) = h_{12}(a) = \begin{bmatrix} a & 0 \\ 0 & a^{-1} \end{bmatrix}.$$

Thus,

$$H(F^2, K) = \left\{ \begin{bmatrix} a & 0 \\ 0 & a^{-1} \end{bmatrix} \mid a \in K^\star \right\},$$

and

$$N(F^2, K) = < \{ \begin{bmatrix} 0 & a \\ -a^{-1} & 0 \end{bmatrix} \mid a \in K^\star \} > .$$

Evidently, $N(F^2, K) = N_{SL(2,K)}(H(F^2, K)) = H(F^2, K) \bigcup \left\{ \begin{bmatrix} 0 & 1 \\ -1 & 0 \end{bmatrix} H(F^2, K) \right\}$. Describe all these for $sl(n, F)$ with usual representation on F^n as exercise.

Proposition 4.3.26 *The center $Z(G(V, K)) \subseteq H(V, K)$.*

Proof Let $x \in Z(G(V, K))$. From Theorem 4.3.22 (v), x has a unique representation as $x = uhn_\sigma v$, where $u \in U(V, K)$, $h \in H(V, K)$, and $v \in U(V, K)_\sigma^-$, $\sigma \in W(\Phi)$. If $\sigma \neq I$, then there is a $\alpha \in \Phi^+$ such that $\sigma(\alpha) \in \Phi^-$. But $xx_\alpha^V(1) = x_\alpha^V(1)x$. This is a contradiction (Theorem 4.3.22). Thus, $\sigma = I$ and $x = uh$. We have a unique element $\sigma_0 \in W(\Phi)$ such that $\sigma_0(\Phi^+) = \Phi^-$. Consequently, $x = w_{\sigma_0}^V(1)xw_{\sigma_0}^V(1)^{-1}$ is upper triangular as well as lower triangular with respect to a suitable basis. Since h is diagonal and u is uni-upper triangular, $u = 1$. This shows that $x = h \in H(V, K)$. ♯

Theorem 4.3.27 *Let $G^\star$ denote the group generated by the set $\{x'_\alpha(a) \mid \alpha \in \Phi$, and $a \in K\}$ of symbols subject to the following relations:*

(i) $x'_\alpha(a+b) = x'_\alpha(a)x'_\alpha(b)$ *for all* $\alpha \in \Phi$ *and* $a, b \in K$.
(ii) For all $\alpha, \beta \in \Phi$ *with* $\alpha + \beta \neq 0$,

$$[x'_\alpha, x'_\beta] = \prod_{i,j\geq 1} x'_{i\alpha+j\beta}(c_{ij}a^ib^j),$$

where c_{ij} *are as in the Theorem 4.3.2.*
(iii) For all $\alpha, \beta \in \Phi$ *and* $a \in K$, $w'_\alpha(1)h'_\beta(a)w'_\alpha(1)^{-1} = h'_{\sigma_\alpha(\beta)}(a)$, *where* $w'_\alpha(a) = x'_\alpha(a)x'_{-\alpha}(-a^{-1})x'_\alpha(a)$ *and* $h'_\alpha(a) = w'_\alpha(a)w'_\alpha(1)^{-1}$.
(iv) For all $\alpha, \beta \in \Phi$ *and* $a \in K$, $w'_\alpha(1)x'_\beta(a)w'_\alpha(1)^{-1} = x'_{\sigma_\alpha(\beta)}(ca)$, *where c is as in Proposition 4.3.14.*
(v) For all $\alpha, \beta \in \Phi$ *and* $a, b \in K$, $h'_\alpha(a)x'_\beta(b)h'_\alpha(a)^{-1} = x'_\beta(a^{\langle\beta,\alpha\rangle}b)$.

Let $U^\star$ *denote the subgroup of* $G^\star$ *generated by the set* $\{x'_\alpha(a) \mid \alpha \in \Phi^+$ *and* $a \in K\}$, $H^\star$ *denote the subgroup generated by the set* $\{h'_\alpha(a) \mid \alpha \in \Phi$ *and* $a \in K\}$, $B^\star$ *denote the subgroup generated by* $U^\star \bigcup H^\star$, *and* $U_\sigma^{-\star}$ *denote the subgroup generated by the set* $\{x'_\alpha(a) \mid \alpha \in \Phi^+ \bigcap \sigma^{-1}(\Phi^-)$ *and* $a \in K\}$. *Then the following hold.*

(a) If $G(V, K)$ *is the Chevalley group associated with the representation space* V, *then there is a unique surjective homomorphism* η^V *from* $G^\star$ *to* $G(V, K)$ *given by* $\eta^V(x'_\alpha(a)) = x_\alpha^V(a), \alpha \in \Phi, a \in K$.

(b) Every element of $U^\star$ *can be uniquely expressed as* $\prod_{\alpha\in\Phi^+} x'_\alpha(a_\alpha)$.

(c) If $\sigma \in W(\Phi)$ *is product* $\sigma_{\alpha_1}\sigma_{\alpha_2}\cdots\sigma_{\alpha_r}$ *of simple reflections, then we write* $w'_\sigma(1) = w'_{\alpha_1}(1)w'_{\alpha_2}(1)\cdots w'_{\alpha_r}(1)$. *Every element of* $G^\star$ *can be uniquely expressed as* $u^\star h^\star w'_\sigma(1)v^\star$, *where* $u^\star \in U^\star$, $h^\star \in H^\star$, $\sigma \in W(\Phi)$, *and* $v^\star \in U_\sigma^{-\star}$.

(d) $Ker\ \eta^V \subseteq Z(G^\star) \subseteq H^\star$.

Proof (a). It follows from the earlier results that $G(V, K)$ is generated by the set $\{x_\alpha^V(a) \mid \alpha \in \Phi$, *and* $a \in K\}$ which satisfy the relations (i) – (v) with $x'_\alpha(a)$ replaced by $x_\alpha^V(a)$. Hence, there is a unique surjective homomorphism η^V from $G^\star$ to $G(V, K)$ given by $\eta^V(x'_\alpha(a)) = x_\alpha^V(a), \alpha \in \Phi, a \in K$.

(b) Suppose that $\prod_{\alpha\in\Phi^+} x'_\alpha(a_\alpha) = \prod_{\alpha\in\Phi^+} x'_\alpha(b_\alpha)$. Then $\prod_{\alpha\in\Phi^+} x_\alpha^V(a_\alpha) = \eta^V\left(\prod_{\alpha\in\Phi^+} x'_\alpha(a_\alpha)\right) = \eta^V\left(\prod_{\alpha\in\Phi^+} x'_\alpha(b_\alpha)\right) = \prod_{\alpha\in\Phi^+} x_\alpha^V(b_\alpha)$. It follows from Proposition 4.3.7, $a_\alpha = b_\alpha$ for each $\alpha \in \Phi^+$. This proves (b).

(c) From (b), it also follows that $\eta^V|_{U^\star}$ is an isomorphism from $U^\star$ to $U(V, K)$. Suppose that $u^\star h^\star w'_\sigma(1)v^\star = u'^\star h'^\star w'_\tau(1)v'^\star$. Applying the homomorphism η^V

again and using Theorems 4.3.21 and 4.3.22, it follows that $\eta^V(u^\star) = \eta^V(u'_\star)$, *and* $\eta^V(v^\star) = \eta^V(v'^\star)$. Since $\eta^V|_{U^\star}$ is an isomorphism, $u^\star = u'^\star$ and $v^\star = v'^\star$. The rest is immediate.

(d) Let $g = u^\star h^\star w'_\sigma(1)v^\star$ belong to the $Ker\ \eta^V$, where $u^\star \in U^\star$, $h^\star \in H^\star$, $\sigma \in W(\Phi)$, and $v^\star \in U^\star_\sigma$. Then $\eta^V(u^\star)\eta^V(h^\star)\eta^V(w'_\sigma(1))\eta^V(v^\star) = 1$. It follows from Example 4.2.22, that $1 = \eta^V(u^\star) = \eta^V(h^\star) = \eta^V(w'_\sigma(1)) = \eta^V(v^\star)$. Since $\eta^V|_{U^\star}$ is an isomorphism, $u^\star = 1 = v^\star$. Since $\eta^V(w'_\sigma(1)) = n_\sigma = 1, \sigma = I$ and so $w'_\sigma(1) = 1$. This means that $g = h^\star = \prod_{\alpha\in\Phi} h'_\alpha(a_\alpha)$. It follows from the defining relations of $G^\star$ that

$$gx'_\beta(b)g^{-1} = x'_\beta\left(\prod_{\alpha\in\Phi} a_\alpha^{\prec\beta,\alpha\succ} b\right).$$

Applying η^V, $x^V_\beta(b) = x^V_\beta\left(\prod_{\alpha\in\Phi} a_\alpha^{\prec\beta,\alpha\succ} b\right)$. Consequently, $b = \prod_{\alpha\in\Phi} a_\alpha^{\prec\beta,\alpha\succ} b$ and so $\prod_{\alpha\in\Phi} a_\alpha^{\prec\beta,\alpha\succ} = 1$. Thus g commutes with $x'_\beta(b)$ for all $\beta \in \Phi$ and $b \in K$. This shows that $g \in Z(G^\star)$. Finally, it remains to show that $Z(G^\star) \subseteq H^\star$. Since $Ker\ \eta^V$ is contained in $H^\star$ and $\eta^V(Z(G^\star)) \subseteq Z(G(V, K))$, it suffices to show that $Z(G(V, K)) \subseteq H(V, K)$. In fact, this is Proposition 4.3.26. ♯

Corollary 4.3.28 *Let S be a set of words in $\{h^V_\alpha(a) \mid \alpha \in \Phi, a \in K\}$ which forms a set of defining relations for $H(V, K)$. Then the set S together with relations $(i) - (v)$ in the above theorem with $x'_\alpha(a)$ replaced by $x^V_\alpha(a)$ forms a set of defining relations for $G(V, K)$ in terms of the generators $x^V_\alpha(a)$.*

Proof Follows from the fact that $Ker\ \eta^V$ is contained in $H^\star$. ♯

Corollary 4.3.29 *Let $G(V, K)$ and $G(V', K)$ be Chevalley groups associated with L-modules V and V', respectively. Then there is a homomorphism $\eta^{V',V}$ from $G(V', K)$ to $G(V, K)$ with $\eta^{V',V}(x^{V'}_\alpha(a)) = x^V_\alpha(a)$ for each $\alpha \in \Phi$ and $a \in K$ if and only if there is a homomorphism χ from $H(V', K)$ to $H(V, K)$ such that $\chi(h^{V'}_\alpha(a)) = h^V_\alpha(a)$ for all $\alpha \in \Phi$ and $a \in K$.*

Proof Suppose that we have a homomorphism $\eta^{V',V}$ from $G(V', K)$ to $G(V, K)$ such that $\eta^{V',V}(x^{V'}_\alpha(a)) = x^V_\alpha(a)$ for each $\alpha \in \Phi$ and $a \in K$. Then $\eta^V = \eta^{V',V} o \eta^{V'}$. Clearly, $\chi = \eta^{V',V}|_{H(V',K)}$ has the required property. Conversely, χ be a homomorphism from $H(V', K)$ to $H(V, K)$ such that $\chi(h^{V'}_\alpha(a)) = h^V_\alpha(a)$ for each $\alpha \in \Phi$ and $a \in K$. It follows from the above corollary that $Ker\ \eta^{V'} \subseteq Ker\ \eta^V$. This ensures the existence of a homomorphism $\eta^{V',V}$ with the required property. ♯

It becomes essential to look at the structure of $H(V, K)$. Indeed, the presentation depends only on the weight lattice $\Lambda(V)$ of V.

Theorem 4.3.30 *The group $H(V, K)$ is an abelian group generated by the set $\{h^V_\alpha \mid \alpha \in \Delta\}$ such that the following hold:*

(i) $h^V_\alpha(ab) = h^V_\alpha(a)h^V_\alpha(b)$ *for all* $\alpha \in \Phi$ *and* $a, b \in K^\star$.
(ii) $\prod_{i=1}^l h^V_{\alpha_i}(a_i) = 1$ *if and only if* $\prod_{i=1}^l a_i^{\prec\mu,\alpha_i\succ} = 1$ *for all* $\mu \in \Lambda(V)$.

(iii) $Z(G(V,K)) = \{\prod_{i=1}^{l} h_{\alpha_i}^{V}(a_i) \mid \prod_{i=1}^{l} a_i^{\langle\beta,\alpha_i\rangle} = 1\ \forall \beta \in \Lambda_r\}$.

Proof Evidently, $H(V,K)$ is abelian group generated by the set $\{h_\alpha^V \mid \alpha \in \Delta\}$.

(i) Follows from Proposition 4.3.13 (iv).

(ii) $\prod_{i=1}^{l} h_{\alpha_i}^{V}(a_i) = 1$ if and only if $\prod_{i=1}^{l} h_{\alpha_i}^{V}(a_i)v = v$ for all $v \in V_\mu$ and for all $\mu \in \Lambda(V)$. The assertion follows from the Proposition 4.3.13 (iv).

(iii) This, again, follows from Proposition 4.3.13 (iv). ♯

Let us denote the L-module L with adjoint action by V_0 and the L-module $V(\lambda_1) \oplus V(\lambda_2) \oplus \cdots \oplus V(\lambda_l)$ by V_1, where $\{\lambda_1, \lambda_2, \ldots, \lambda_l\}$ is the basis of Λ_{univ} consisting of fundamental weights. Thus, $\Lambda(V_0) = \Lambda_r$ and $\Lambda(V_1) = \Lambda_{univ}$.

Corollary 4.3.31 *(i) Every element h of $H(V_1,K)$ has a unique representation as*

$$h = \prod_{i=1}^{l} h_i^{V_1}(a_i),\ a_i \in K^\star.$$

(ii) $Z(G(V_0,K)) = \{1\}$.

Proof (i) Suppose that $\prod_{i=1}^{l} h_i^{V_1}(a_i) = 1$. We need to show that $a_j = 1$ for each j. Since λ_j is a weight of V_1, it follows from Theorem 4.3.30 (ii) that $\prod_{i=1}^{l} a_i^{\langle\lambda_j,\alpha_i\rangle} = 1$. This shows that $a_j = 1$ for.

(ii) From Theorem 4.3.30 (iii), $h = \prod_{i=1}^{l} h_i^{V_0}(a_i)$ belongs to $Z(G(V_0,K))$ if and only if $\prod_{i=1}^{l} a_i^{\langle\beta,\alpha_i\rangle} = 1$ for each $\beta \in \Lambda_r$. Since $\Lambda(V_0) = \Lambda_r$, it follows from Theorem 4.3.30 (ii) that $h = 1$. ♯

Corollary 4.3.32 *Suppose that $\Lambda(V) \subseteq \Lambda(V')$. Then there is a homomorphism $\eta^{V',V}$ from $G(V',K)$ to $G(V,K)$ such that $\eta^{V',V}(x_\alpha^{V'}(a)) = x_\alpha^V(a)$ for all α and a. Also $Ker\,\eta \subseteq Z(G(V',K))$. If $\Lambda(V) = \Lambda(V')$, then $\eta^{V',V}$ is an isomorphism.*

Proof Let $h_\alpha^V(a) = \prod_{i=1}^{l} h_{\alpha_i}^{V}(a_i)$ be a member of $H(V,K)$ and $h_\alpha^{V'}(a) = \prod_{i=1}^{l} h_{\alpha_i}^{V'}(a_i)$ be the corresponding member of $H(V',K)$. Suppose that $\prod_{i=1}^{l} h_{\alpha_i}^{V'}(a_i) = 1$ in $H(V',K)$. Then by Theorem 4.3.30 (ii), $\prod_{i=1}^{l} a_i^{\langle\mu,\alpha_i\rangle} = 1$ for all $\mu \in \Lambda(V')$. Since $\Lambda(V) \subseteq \Lambda(V')$, $\prod_{i=1}^{l} a_i^{\langle\mu,\alpha_i\rangle} = 1$ for all $\mu \in \Lambda(V)$. Again by Theorem 4.3.30 (ii), $\prod_{i=1}^{l} h_{\alpha_i}^{V}(a_i) = 1$ in $H(V,K)$. Thus, we have a homomorphism χ from $H(V',K)$ to $H(V,K)$ such that $\chi(h_\alpha^{V'})(a)) = h_\alpha^V(a)$ for all $\alpha \in \Phi$ and $a \in K$. From Corollary 4.3.29, we have a surjective homomorphism $\eta^{V',V}$ from $G(V',K)$ to $G(V,K)$ such that $\eta^{V',V}(x_\alpha^{V'}(a)) = x_\alpha^V(a))$ for all α and a and $Z(G(V',K)) \subseteq Ker\eta^{V',V}$. Finally, if $\Lambda(V) = \Lambda(V')$, then we have a homomorphism $\eta^{V',V}$ from $G(V,K)$ to $G(V',K)$ such that $\eta^{V,V'}(x_\alpha^V(a)) = x_\alpha^{V'}(a))$ for all α and a. Clearly, $\eta^{V',V} o \eta^{V,V'} = I_{G(V',K)}$ and $\eta^{V,V'} o \eta^{V',V} = I_{G(V,K)}$. ♯

Definition 4.3.33 The Chevalley group $G(V_0,K)$ is called the **Adjoint Chevalley Group** or **Chevalley group of adjoint type** associated with the pair (L,K). We shall also denote it by G_{adj}, where (L,K) is already understood to be there. The group $G(V_1,K)$ is called the **Universal Chevalley Group**. We shall also denote it by G_{univ}.

The following Corollary gives us a presentation of the universal Chevalley group $G_{univ} = G(V_1, K)$. The Corollary follows immediately from Theorem 4.3.27 and the corollaries following it.

Corollary 4.3.34 *Let L be a simple Lie algebra. Let $\tilde{G}$ denote the group generated by the set $\{x'_\alpha \mid \alpha \in \Phi$ and $a \in K\}$ of symbols subject to the relations (i) - (v) in Theorem 4.3.27 together with*

(vi) $h'_\alpha(ab) = h'_\alpha(a)h'_\alpha(b)$ for all $\alpha \in \Phi$ and $a, b \in K^\star$, where $h'_\alpha(a) = w'_\alpha(a)w'_\alpha(1)^{-1}$. Then, we have a unique isomorphism η from $\tilde{G}$ to $G(V_1, K)$ such that $\eta(x'_\alpha(a) = x^{V_1}_\alpha(a)$ for all α and a. ♯

Remark 4.3.35 The group $G(V, \mathbb{C})$ is a Lie group. Indeed, the universal Chevalley group $G(V_1, \mathbb{C})$ is a simply connected Lie group and the map $\eta^{V_1, V}$ is a covering map. Thus, $Ker\eta^{V_1,V}$ is the fundamental group of $G(V, \mathbb{C})$. Following the analogy, we term $Ker\eta^{V_1,V}$ to be the fundamental group of the Chevalley group $G(V, K)$. It can be shown as an exercise that the fundamental group of $G(V, K)$ is isomorphic to $Hom(\Lambda_{univ}/\Lambda(V), K^\star)$. Consequently, the fundamental group $\pi_1(G(V, \mathbb{C}))$ is isomorphic to $\Lambda_{univ}/\Lambda(V)$.

Remark 4.3.36 Given any faithful L-module V, we have two important associated central extensions $G(V_1, K) \stackrel{\eta^{V^1,V}}{\to} G(V, K)$ and $G(V, K) \stackrel{\eta^{V,V_0}}{\to} G(V_0, K)$.

Corollary 4.3.37 *Let $K = F_q$ be a finite field containing $q = p^n$ elements and V be a L-module. Then, we have the following:*

(i) $\mid U(V, F_q) \mid = q^{|\Phi^+|}$.
(ii) $\mid U(V, F_q)^-_\sigma \mid = q^{l(\sigma)}$.
(iii) $\mid G(V, F_q) \mid = q^{|\Phi^+|} \mid H(V, F_q) \mid (\sum_{\sigma \in W(\Phi)} q^{l(\sigma)})$.
(iv) $\mid G(V_1, F_q) \mid = q^{|\Phi^+|}(q-1)^l(\sum_{\sigma \in W(\Phi)} q^{l(\sigma)})$.
(v) $U(V, F_q)$ is a Sylow p-subgroup of $G(V, F_q)$ and $H(V, F_q)$ is an abelian p'-subgroup of $G(V, F_q)$.

Proof (i) Follows from the definition of $U(V, F_q)$ and Proposition 4.3.7.

(ii) Again follows from the definition of $U(V, F_q)^-_\sigma$ and Proposition 4.3.7.

(iii) Follows from (i), (ii), and Theorem 4.3.22.

(iv) Follows from (i), (ii), (iii), and Corollary 4.3.31.

(v) This is evident from the order consideration. ♯

Example 4.3.38 For $A_1 = sl(2, \mathbb{C})$, $\Phi = \{\alpha, -\alpha\}$, $\lambda_{univ} = \mathbb{Z}\frac{\alpha}{2}$ and $\Lambda_r = \mathbb{Z}\alpha$. If $Char K \neq 2$, then $G(V, K) \approx SL(2, K)$, whereas, $G(V_0, K) \approx PSL(2, K)$. If $Char K = 2$, then $G(V_1, K) = G(V_0, K) \approx SL(2, K)$. Indeed, the map η from $SL(2, K)$ to $G(V_1, K)$ given by $\eta(E^a_{12}) = x^{V_1}_\alpha(a)$ and $\eta(E^a_{21}) = x^{V_1}_{-\alpha}(a)$ induces an isomorphism. In case characteristic K different from 2, $Z(SL(2, K)) = \{I, -I\}$. Consequently, $G(V_0, K) \approx PSL(2, K)$. Similarly, for $A_2 = sl(3, \mathbb{C})$, if the characteristic of K is different from 3, then $G(V_1, K) = SL(3, K)$ and $G(V_0, K) = PSL(3, K$. If the characteristic of K is 3, then $G(V_1, K) = SL(3, K) = G(V_0, K)$.

More generally, we list the information about universal Chevalley group $G(V_1, K)$, the adjoint Chevalley group $G(V_0, K)$, the intermediary Chevalley groups $G(V, K)$, and also Λ_{univ}/Λ_r for all 9 types of semi-simple Lie algebras:

Type of L	Λ_{univ}/Λ_r	$G(V_1, K)$	$G(V, K)$	$G(V_0, K)$	$\mid Z(G(V_1, F_q)) \mid$
A_l	$\mathbb{Z}_{l+1}$	$SL(l+1, K)$	-	$PSL(l+1, K)$	$(l+1, q-1)$
B_l	$\mathbb{Z}_2$	$Spin(2l+1, K)$	-	$SO(2l+1, K)$	$(2, q-1)$
C_l	$\mathbb{Z}_2$	$Sp(2l, K)$	-	$PSp(2l, K)$	$(2, q-1)$
D_{2l+1}	$\mathbb{Z}_4$	$Spin(4l+2, K)$	$SO(4l+2, K)$	$PSO(4l+2, K)$	$(4, q-1)$
D_{2l}	$\mathbb{Z}_2 \times \mathbb{Z}_2$	$Spin(4l, K)$	$SO(4l, K)$	$PSO(4l, K)$	$(4, q-1)$
E_6	$\mathbb{Z}_3$	–	-	–	$(3, q-1)$
E_7	$\mathbb{Z}_2$	–	-	–	$(2, q-1)$
E_8	1	$G_{univ} = G_{adj}$	-	$G_{univ} = G_{adj}$	1
F_4	1	$G_{univ} = G_{adj}$	-	$G_{univ} = G_{adj}$	1
G_2	1	$G_{univ} = G_{adj}$	-	$G_{univ} = G_{adj}$	1

The orders of the finite exceptional Chevalley groups $E_6(q)$, $E_7(q)$, $E_8(q)$, $F_4(q)$, and $G_2(q)$ of adjoint type over a field F_q are given as follows:

1. $\mid E_6(q) \mid = \frac{q^{36}(q^2-1)(q^5-1)(q^6-1)(q^8-1)(q^9-1)(q^{12}-1)}{(q-1,3)}$.
2. $\mid E_7(q) \mid = \frac{q^{63}(q^2-1)(q^6-1)(q^8-1)(q^{10}-1)(q^{12}-1)(q^{12}-1)(q^{14}-1)(q^{18}-1)}{(q-1,3)}$.
3. $\mid E_8(q) \mid = q^{120}(q^2-1)(q^8-1)(q^{12}-1)(q^{14}-1)(q^{18}-1)(q^{20}-1)(q^{24}-1)(q^{30}-1)$.
4. $\mid F_4(q) \mid = q^{24}(q^2-1)(q^6-1)(q^8-1)(q^{12}-1)$.
5. $\mid G_2(q) \mid = q^6(q^2-1)(q^6-1)$.

Our next aim is to establish the following theorem about the simplicity of Chevalley groups of adjoint type.

Theorem 4.3.39 (Chevalley–Dickson) *Adjoint Chevalley group $G(V_0, K)$ associated with a simple Lie algebra L over $\mathbb{C}$ is always simple except in the following cases:*

(i) L is of the type A_1 and $\mid K \mid = 2$ or $\mid K \mid = 3$.
(ii) L is of the type B_2 and $\mid K \mid = 2$.
(iii) L is G_2 and $\mid K \mid = 2$.

Let us first observe that in all the three excluded cases the adjoint Chevalley groups are not simple. If $L = sl(2, \mathbb{C})$ is of the type A_1, then $G(V_0, F_2) \approx SL(2, F_2) \approx S_3$ has a normal subgroup isomorphic to A_3, and $G(V_0, F_3) \approx PSL(2, F_3)$ contains a normal subgroup isomorphic to A_4. If L is of type B_2, then $G(V_0, K) \approx PSO(3, F_3)$ contains a subgroup of index 2 which is isomorphic to A_6. If L is of the type G_2, then $G(V_0, F_2)$ contains $SU(3, 3)$ as a subgroup of index 2.

The idea of the proof of Chevalley–Dickson theorem is as follows. First observe that $B(V, K)$ is a solvable subgroup of $G(V, K)$ ($U(V, K)$ is a nilpotent normal subgroup such that $B(V, K)/U(V, K) \approx H(V, K)$ is abelian). We shall show that if A is a nontrivial normal subgroup of $G(V, K)$, then $G(V, K) = AB(V, K)$. Finally, if we also show that $G(V, K)$ is perfect in the sense that $[G(V, K), G(V, K)] =$

$G(V, K)$, then $G(V, K)/A \approx B(V, K)/(B(V, K) \bigcap A)$ becomes perfect, a contradiction to the solvability of $B(V, K)$. We prove all these in several steps in the form of Lemmas.

Lemma 4.3.40 *Let $G(V, K)$ be a Chevalley group and $\sigma \in W(\Phi)$. Suppose that $\sigma = \sigma_{\alpha_{i_1}}\sigma_{\alpha_{i_2}} \cdots \sigma_{\alpha_{i_r}}$ is a minimal representation of σ as product of simple reflections. Then for each j, $w^V_{\alpha_{i_j}}(1)$ belongs to the subgroup generated by $B(V, K) \bigcup (w^V_\sigma(1)$ $B(V, K)w^V_\sigma(1)^{-1})$*

Proof The proof is by the induction on $l(\sigma)$. Suppose that $l(\sigma) = 1$. Then $\sigma = \sigma_\alpha$ is a simple reflection and $X^V_{-\alpha}(K) = X^V_{\sigma_\alpha(\alpha)}(K) = w^V_\sigma(1)X^V_\alpha(K)w^V_\sigma(1)^{-1}$ is contained in $B(V, K) \bigcup w^V_\sigma(1)B(V, K)w^V_\sigma(1)^{-1}$. Assume the result for all τ with $l(\tau) < r$, $r \geq 2$. Suppose that $l(\sigma) = r$ and $\sigma = \sigma_{\alpha_{i_1}}\sigma_{\alpha_{i_2}} \cdots \sigma_{\alpha_{i_r}}$ is a minimal representation of σ. Put $\beta = -\sigma^{-1}(\alpha_{i_1})$. Then as above $X^V_{-\alpha_{i_1}}(K) = X^V_{\sigma(\beta)}(K) = w^V_{\sigma_{\alpha_{i_1}}}(1)X^V_\beta w^V_{\sigma_{\alpha_{i_1}}}(1)^{-1}$ is a subset of $B(V, K) \bigcup w^V_\sigma(1)B(V, K)w^V_\sigma(1)^{-1}$. This shows that $w^V_{\sigma_{\alpha_{i_1}}}(1)$ belongs to $B(V, K) \bigcup w^V_\sigma(1)B(V, K)w^V_\sigma(1)^{-1}$. Thus, $w^V_{\sigma_{\alpha_{i_1}}}(1)w^V_\sigma(1)$ $B(V, K)w^V_\sigma(1)^{-1}w^V_{\sigma_{\alpha_{i_1}}}(1)^{-1}$ is contained in $B(V, K) \bigcup w^V_\sigma(1)B(V, K)w^V_\sigma(1)^{-1}$. Since $l(\sigma_{\alpha_1}\sigma) < l(\sigma)$, the result follows by the induction hypothesis. ♯

Lemma 4.3.41 *Let $G(V, K)$ be Chevalley group. For each subset π of Δ, let W_π denote the subgroup of $W(\Phi)$ generated by the set $\{\sigma_\alpha \mid \alpha \in \pi\}$, and G_π denote $\bigcup_{\sigma \in W_\pi} B(V, K)\sigma B(V, K)$. Then, we have the following:*

(i) G_π is a subgroup containing $G(V, K)$.
(ii) $G_\pi = G_{\pi'}$ implies that $\pi = \pi'$.
(iii) Any subgroup of $G(V, K)$ which contains $B(V, K)$ is of the form G_π for a unique π.

Proof (i) Clearly, G_π is closed under inversion. Again, by Proposition 4.3.18 (iii),

$$(B(V, K)\sigma B(V, K))(B(V, K)\sigma_\alpha B(V, K) \subseteq$$
$$B(V, K)\sigma B(V, K)) \bigcup (B(V, K)\sigma\sigma_\alpha B(V, K)).$$

This shows that G_π is a subgroup.

(ii) Suppose that $\pi \neq \pi'$. Let $\alpha \in \pi$ such that $\alpha \notin \pi'$. Then if $\sigma \in W_{\pi'}$, then $\sigma(\alpha) = \alpha + \sum_{\beta \in \pi'} a_\beta \beta$ but $\sigma_\alpha(\alpha) = -\alpha$. Since Δ is linearly independent, $\sigma_\alpha \notin W_{\pi'}$. It follows from the Bruhat decomposition (Theorem 4.3.21) that $G_\pi \neq G_{\pi'}$.

(iii) Let P be a subgroup containing $B(V, K)$. Consider the subset $\pi = \{\alpha \in \Delta \mid w^V_{\sigma_\alpha}(1) \in P\}$ (note that $w^V_{\sigma_\alpha}(1)$ is same as $w^V_\alpha(1)$). Clearly, $G_\pi \subseteq P$. Since $B(V, K) \subseteq P$ and $G(V, K) = \bigcup_{\sigma \in W(\Phi)} B(V, K)\sigma B(V, K)$, it is sufficient to show that $w^V_\sigma(1) \in P$ implies that $w^V_\sigma(1) \in G_\pi$. Let $w^V_\sigma(1) \in P$. Suppose that $\sigma = \sigma_{\alpha_{i_1}}\sigma_{\alpha_{i_2}} \cdots \sigma_{\alpha_{i_r}}$ be a minimal representation of σ as product of simple reflections. From the previous Lemma, $w^V_{\sigma_{\alpha_{i_j}}}(1) \in P$ for each j. This means that $\alpha_j \in \pi$ for each j. Hence $\sigma \in W_\pi$. Consequently, $w^V_\sigma(1) \in G_\pi$. ♯

Corollary 4.3.42 *Let σ be a member of $W(\Phi)$. Then*

$$B(V,K)\bigcup(B(V,K)w_{\sigma}^{V}(1)B(V,K)) \text{ is a subgroup if and only if } \sigma = I \text{ or } \sigma = \sigma_{\alpha} \text{ is a simple reflection. } \sharp$$

Example 4.3.43 Consider the Lie algebra $sl(l+1,\mathbb{C})$ of the type A_l and $V = \mathbb{C}^{l+1}$ the obvious L-module. Then $G(V_1,K) = SL(l+1,K)$ and $B(V_1,K)$ is the group $T^0(l+1,K)$ of upper triangular matrices of determinant 1. Let α_i denote the simple root associated with $(i,i+1)$ (see Theorem 2.3.2). Then $B(V_1,K)\bigcup(B(V_1,K)w_{\sigma_i}^{V_1}(1)B(V_1,K)$, where $w_{\sigma_i}^{V_1} = E_{i\,i+1}^{1}E_{i+1\,i}^{-1}E_{i\,i+1}^{1}$ is a subgroup of $SL(l+1,K)$. Interpret this subgroup. More generally, let $\{i_1,i_2,\ldots,i_r\}$ be a subset of $\{1,2,\ldots,l\}$, where $i_1 < i_2 < \cdots < i_r \leq l$. Let $\pi = \{\alpha_{i_1},\alpha_{i_2},\ldots,\alpha_{i_r}\}$ where α_{i_j} is the simple root associated with (i_j,i_j+1). Interpret the group G_{π}. How many such subgroups are there?

Definition 4.3.44 For each subset π of Δ, G_{π} is called a **Principal Parabolic Subgroup** of $G(V,K)$. Conjugates of principal parabolic subgroups are called the **Parabolic Subgroups** of $G(V,K)$. In particular, $B(V,K)$ is a parabolic subgroup. We term the conjugates of $B(V,K)$ as **Borel Subgroup** of $G(V,K)$. Indeed, it is a maximal solvable subgroup.

Proposition 4.3.45 *(i) G_{π} is conjugate to $G_{\pi'}$ if and only if $\pi = \pi'$.*

(ii) $N_{G(V,K)}(G_{\pi}) = G_{\pi}$ for all subsets π of Δ. In particular, $N_{G(V,K)}(B(V,K)) = B(V,K)$.

(iii) $G_{\pi}\bigcap G_{\pi'} = G_{\pi\bigcap\pi'}$.

(iv) We have a natural bijection from the set of conjugates of G_{π} to $G(V,K)/^{r}G_{\pi}$ given by $gG_{\pi}g^{-1} \mapsto G_{\pi}g$.

Proof (i) Suppose that $G_{\pi} = gG_{\pi'}g^{-1}$. It is sufficient to show that $w_{\alpha}^{V}(1) \in G_{\pi'}$ implies that $w_{\alpha}^{V}(1) \in G_{\pi}$. By Theorem 4.3.22 (v), g has a unique representation as $g = bw_{\sigma}^{V}(1)x$, where $b \in B(V,K)$, and $x \in U_{\sigma}^{-}(V,K)$. Since $gw_{\alpha}^{V}(1)g^{-1} \in G_{\pi}$, it follows from Lemma 4.3.40 that $w_{\alpha}^{V}(1) \in G_{\pi}$. This shows that $G_{\pi'} \subseteq G_{\pi}$. Similarly, $G_{\pi} \subseteq G_{\pi'}$.

(ii) Suppose that $gG_{\pi}g^{-1} = G_{\pi}$. Let $g = bw_{\sigma}^{V}(1)x$ be the unique representation of g. Then $gw_{\alpha}^{V}(1)g^{-1} \in G_{\pi}$ for each $\alpha \in \pi$. Again using Lemma 4.3.40, $w_{\sigma}^{V}(1) \in G_{\pi}$. Hence $g \in G_{\pi}$.

(iii) The arguments used in (i) and (ii) imply that $w_{\alpha}^{V}(1) \in G_{\pi}\bigcap G_{\pi'}$ if and only if $\alpha \in \pi\bigcap\pi'$. The result follows.

(iv) Immediate from (ii). $\sharp$

Remark 4.3.46 With the obvious modifications, the above results are true for arbitrary Groups with (B,N) pairs.

Lemma 4.3.47 *Let L be a simple Lie algebra over $\mathbb{C}$. Then $Core_{G(V_0,K)}(B(V_0,K)) = \{1\}$.*

Proof Suppose not. Let A be a nontrivial normal subgroup of $G(V_0, K)$ which is contained in $B(V_0, K)$. Let $x \in A, x \neq 1$. Since $A \subseteq B(V_0, K), x = uh$, where $u \in U(V_0, K)$ and $h \in H(V_0, K)$. Suppose that $u \neq 1$. But, then there is a $\sigma \in W(\phi)$ such that $w_\sigma^{V_0}(1)xw_\sigma^{V_0}(1)^{-1}$ does not belong to $B(V_0, K)$. Thus, $u = 1$ and $h \neq 1$. Since $Z(G(V_0, K)) = \{1\}$, there is a root α and distinct elements $a, b \in K^\star$ such that $hx_\alpha^{V_0}(a)h^{-1} = x_\alpha^{V_0}(b)$. But, then $1 \neq x_\alpha^{V_0}(b - a) = [h, x_\alpha^{V_0}(a)]$ belongs to A. This is impossible because of the previous argument. ♯

Lemma 4.3.48 *Let L be a simple Lie algebra over $\mathbb{C}$ and A be a nontrivial normal subgroup of $G(V_0, K)$. Then $G(V_0, K) = AB(V_0, K)$.*

Proof Since A is a nontrivial normal subgroup of $G(V_0, K)$, it follows from the above Lemma that $A \nsubseteq B(V_0, K)$ and $AB(V_0, K)$ is a subgroup of $G(V_0, K)$ containing $B(V_0, K)$ as a proper subgroup. From Lemma 4.3.41 (iii), there is a nonempty subset π of Δ such that $AB(V_0, K) = G_\pi$. We need to show that $\pi \supseteq \Delta$. Suppose that Δ is not contained in π. Since L is simple, Φ and so Δ is irreducible. Thus, there is a pair α, β of simple roots such that $\alpha \in \pi$, $\beta \notin \pi$ and $< \alpha, \beta > \neq 0$. Since $\alpha \in \pi$, $B(V_0, K)w_\alpha^{V_0}(1)B(V_0, K) \subseteq G_\pi$. Let $b_1w_\alpha^{V_0}(1)b_2$ belong to A. Then $b_2b_1w_\alpha^{V_0}(1) = b_2b_1w_\alpha^{V_0}(1)b_2b_2^{-1}$ belongs to A (for A is normal subgroup). In turn, using Proposition 4.3.18, we see that $w_\beta^{V_0}(1)b_2b_1w_\alpha^{V_0}(1)w_\beta^{V_0}(1)^{-1}$ belongs to

$$(A\bigcap((B(V_0, K)w_\alpha^{V_0}(1)w_\beta^{V_0}(1)B(V_0, K))\bigcup(B(V_0, K)w_\beta^{V_0}(1)w_\alpha^{V_0}(1)))w_\beta^{V}(1)) \subseteq G_\pi.$$

Thus, $\sigma_\alpha\sigma_\beta \in W_\pi$ or $\sigma_\beta\sigma_\alpha\sigma_\beta \in W_\pi$. Clearly, $\sigma_\beta\sigma_\alpha\sigma_\beta = \sigma_{\sigma_\beta(\alpha)}$. Since $\prec \alpha, \beta \succ \neq 0$, $\sigma_\beta(\alpha)$ is not a simple root. Hence $l(\sigma_\beta\sigma_\alpha\sigma_\beta) \neq 1$. Further, $l(\sigma_\beta\sigma_\alpha\sigma_\beta) \neq 2$ for $det(\sigma_\beta\sigma_\alpha\sigma_\beta) = -1 = (-1)^{l(\sigma_\beta\sigma_\alpha\sigma_\beta)}$. Thus, $l(\sigma_\beta\sigma_\alpha\sigma_\beta) \geq 3$. This means that $\sigma_\alpha\sigma_\beta$ and $\sigma_\beta\sigma_\alpha\sigma_\beta$ are minimal representations in terms of simple reflections. Hence by Lemma 4.3.40, $\sigma_\beta \in W_\pi$. This is a contradiction, since $\beta \notin \pi$. It follows that $\pi \supseteq \Delta$ and so $AB(V_0, K) = G_\Delta = G(V_0, K)$. ♯

Finally, it remains to establish the following Lemma.

Lemma 4.3.49 *Under the hypothesis of Theorem 4.3.39, $G(V_0, K)$ is perfect in the sense that $G(V_0, K)' = [G(V_0, K), G(V_0, K)] = G(V_0, K)$.*

Proof It is sufficient to show that $x_\alpha^{V_0}(a) \in G(V_0, K)'$ for all $\alpha \in \Phi$ and $a \in K^\star$. Suppose that $| K | \geq 4$. Then, there is an element $a \in K^\star$ such that $a^2 - 1 \neq 0$. By Proposition 4.3.14 (iii) with $\alpha = \beta$,

$$[h_\alpha^V(a), x_\alpha^V(b)] = x_\alpha^V(a^{\prec\alpha,\alpha\succ}b)x_\alpha^V(-b) = x_\alpha^V((a^2 - 1)b)$$

belongs to $G(V_0, K)'$. Since $a^2 - 1 \neq 0$, $x_\alpha^V(c) \in G(V_0, K)'$ for all $\alpha \in \Phi$ and $c \in K^\star$. Thus, it is sufficient to prove the result when $rankL \geq 2$ and $| K | = 2$ or $| K | = 3$. We prove it by casewise analysis.

(i) Suppose that L is of the type A_l, D_l or E_l (see their Dynkin diagrams). Then all the roots are of same length. Consequently, any two roots are congruent under a member of the Weyl group $W(\Phi)$. For each root α, we can find a basis Δ with $\beta, \gamma \in \Delta$ and $\alpha = \beta + \gamma$ and no other positive integral linear combination of β, γ are roots. From Corollary 4.3.3,

$$x_\alpha^V(\nu_{\beta,\gamma}ab) = [x_\beta^V(a), x_\gamma^V(b)].$$

Since $\nu_{\beta,\gamma} \neq 0$, $x_\alpha^V(c) \in G(V_0, K)'$ for all $c \in K^\star$. Since α is arbitrary, it follows that $G(V_0, K)$ is perfect.

(ii) Suppose that L is of the type B_l, $l \geq 2$. If α is a long root, then we can find β, γ such that $\alpha = \beta + \gamma$ and no other positive integral linear combination of β, γ is a root. As before $x_\alpha^{V_0}(c) \in G(V_0, K)'$. If α is short, we can find β, γ such that $\alpha = \beta + \gamma$, $\nu_{\beta,\gamma} \neq 0$ and all other positive linear combinations of β and γ are long. As such, $\prod_{(i,j)\neq(1,1)} x_{i\beta+j\gamma}^{V_0}(c_{ij}a^i b^j)$ belongs to $G(V_0, K)'$. Again from Corollary 4.3.3,

$$[x_\beta^{V_0}(a), x_\gamma^{V_0}(B)] = x_\alpha^{V_0}(\nu_{\beta,\gamma}ab)\prod_{(i,j)\neq(1,1)} x_{i\beta+j\gamma}^{V_0}(c_{ij}a^i b^j).$$

It follows that $x_\alpha^{V_0}(\nu_{\beta,\gamma}ab)$ belongs to $G(V_0, K)'$. Since $\nu_{\alpha,\beta} \neq 0$ and a, b are arbitrary, it follows that $x_\alpha^{V_0}(c) \in G(V_0, K)'$ for all c. The proof is complete in this case.

(iii) The earlier arguments can be used to complete the proof in case L is of the type G_2 or F_4.

(iv) Finally consider the case when L is of the type C_l. If α is short, the the earlier argument shows that $x_\alpha^{V_0}(a) \in G(V, K)'$. If α is long but $\mid K \mid = 3$, then also the result follows. Suppose that $l \geq 3$ and α is long and $\mid K \mid = 2$. We can chose β, γ with β long and γ short and $\alpha = \beta + 2\gamma$. Since $\beta + \gamma$ is short, $x_{\beta+\gamma}^{V_0}(a) \in G(V_0, K)'$ for all a. The result follows if we show that

$$[x_\beta^{V_0}(a), x_\gamma^{V_0}(b)] = x_{\beta+\gamma}^{V_0}(\pm ab)x_{\beta+2\gamma}^{V_0}(\pm ab^2).$$

This follows from Corollary 4.3.3, since $\nu_{\gamma,\beta} = \pm 1$ and $\nu_{\gamma,\beta+\gamma} = \pm 2$. ♯

The following Corollary is an important observation:

Corollary 4.3.50 *If* $\mid K \mid\geq 4$, *then every solvable normal subgroup of* $G(V, K)$ *is is contained in the center. Further, it is finite.*

Proof Let $G(V, K)$ be a Chevalley group. The finiteness follows from the first statement of the Corollary, since $Z(G(V, K))$ is finite. From Corollary 4.3.32, we have a surjective homomorphism η^{V,V_0} from $G(V, K)$ to $G(V_0, K)$ such that $Ker\eta^{V,V_0} \subseteq Z(G(V, K))$. Let A be a solvable normal subgroup of $G(V, K)$. Then $AKer\eta^{V,V_0}$ is solvable normal subgroup. Hence, $\eta^{V,V_0}(AKer\eta^{V,V_0})$ is a normal subgroup of $G(V_0, K)$. Since $G(V_0, K)$ is simple, $A \subseteq Ker^{V,V_0} \subseteq Z(G(V, K))$. ♯

Presentation of a Universal Chevalley Group

For constructing certain important outer automorphisms of a universal Chevalley group needed to introduce the twisted groups of Lie types (to be introduced in the next section), we describe a convenient and useful minimal presentation of a universal Chevalley group. The relation of $G^\star$ in Corollary 4.3.34 is derivable from a smaller set of relations. More explicitly, we have the following theorem.

Theorem 4.3.51 *Let L be a simple Lie algebra (so Φ is irreducible) over $\mathbb{C}$. Consider the set $S = \{\hat{x}_\alpha(a) \mid \alpha \in \Phi$ and $a \in K\}$ of symbols. Let $\hat{G}$ be the group generated by S subject to the relations A, B, and C given below in case $rank L \geq 2$ and subject to the relations A, B', and C given below in case $rank L = 1$.*

A. $\hat{x}_\alpha(a+b) = \hat{x}_\alpha(a)\hat{x}_\alpha(b)$ for all $\alpha \in \Phi$ and $a, b \in K$.

B. For each $\alpha, \beta \in \Phi$ with $\alpha + \beta \neq 0$,

$$[\hat{x}_\alpha(a), \hat{x}_\beta(b)] = \prod_{i,j\geq 1} \hat{x}_{i\alpha+j\beta}(c_{ij}a^i b^j,$$

where c_{ij} are as in the Theorem 4.3.2.

B'. $\hat{w}_\alpha(a)\hat{x}_\alpha(b)\hat{w}_\alpha(a)^{-1} = \hat{x}_{-\alpha}(-a^2b)$ for all $a \in K^\star$, $b \in K$, $\alpha \in \Phi$, where $\hat{w}_\alpha(a) = \hat{x}_\alpha(a)\hat{x}_{-\alpha}(-a^{-1})\hat{x}_\alpha(a)$.

C. $\hat{h}_\alpha(ab) = \hat{h}_\alpha(a)\hat{h}_\alpha(b)$ for all $\alpha \in \Phi$, $a, b \in K^\star$, where $\hat{h}_\alpha(a) = \hat{w}_\alpha(a)w_\alpha(1)^{-1}$.

Then there is a unique isomorphism ζ from $\hat{G}$ to G_{univ} such that $\zeta(\hat{x}_\alpha(a) = x_\alpha^{V_1}(a)$ for all α and a.

Proof The relations A, B, and C are already included in the set of relations in Corollary 4.3.34 with $\hat{x}$ replaced by x', $\hat{w}$ replaced by w' , where as B' is derivable from (iv), and (v) of Corollary 4.3.34 (see Exercise 4.3.6). Thus, we get a surjective homomorphism $\hat{\eta}$ from $\hat{G}$ to $\tilde{G}$ such that $\hat{\eta}(\hat{x}_\alpha(a)) = x'_\alpha(a)$ for all α and a. Thus, $\zeta = \hat{\eta} o \eta$ is a surjective homomorphism from $\hat{G}$ to G_{univ} such that $\zeta(\hat{x}_\alpha(a)) = x_\alpha^{V_1}(a)$ for all α and a. It suffices to show that ζ is injective. The following two lemmas establish the injectivity of ζ.

Lemma 4.3.52 *Let Ψ be a closed subset of Φ such that $\Psi \bigcap -\Psi = \emptyset$ Let $\hat{X}_\Psi(K)$ denote the subgroup of $\hat{G}$ generated by the set $\{\hat{x}_\alpha(a) \mid \alpha \in \Psi$ and $a \in K\}$. Then $\zeta|_{\hat{X}_\Psi(K)}$ is an isomorphism from $\hat{X}_\Psi(K)$ to $X_\Psi^{V_1}(K)$.*

Proof Clearly, $\zeta|_{\hat{X}_\Psi(K)}$ is surjective homomorphism $\hat{X}_\Psi(K)$ to $X_\Psi^{V_1}(K)$. Using A and B and the fact that Ψ is a closed subset of Φ satisfying $\Psi \bigcap -\Psi = \emptyset$, we observe that $\{\prod_{\alpha\in\Psi} \hat{x}_\alpha(a_\alpha) \mid a_\alpha \in K\}$ forms a subgroup of $\hat{G}$. Thus, $\hat{X}_\Psi(K) = \{\prod_{\alpha\in\Psi} \hat{x}_\alpha(a_\alpha) \mid a_\alpha \in K\}$. Let $\prod_{\alpha\in\Psi} \hat{x}_\alpha(a_\alpha)$ be such that $\zeta\left(\prod_{\alpha\in\Psi} \hat{x}_\alpha(a_\alpha)\right) = \prod_{\alpha\in\Psi} x_\alpha^{V_1}(a_\alpha) = 1$. From Proposition 4.3.7, it follows that $a_\alpha = 0$ for all $\alpha \in \Psi$. From A, $\hat{x}_\alpha(0) = 1$ and so $\prod_{\alpha\in\Psi} \hat{x}_\alpha(a_\alpha) = 1$. ♯

Lemma 4.3.53 *The following relations are derivable from A and B in case $rank L \geq 2$ and from A and B' in case $rank L = 1$.*

(i) $\hat{w}_\alpha(a)\hat{x}_\beta(b)\hat{w}_\alpha(-a) = \hat{x}_{\sigma_\alpha(\beta)}(ca^{-<\beta,\alpha>}b)$,

(ii) $\hat{w}_\alpha(a)\hat{w}_\beta(b)\hat{w}_\alpha(-a) = \hat{w}_\alpha(ca^{-<\beta,\alpha>}b)$,

(iii) $\hat{w}_\alpha(a)\hat{h}_\beta(b)\hat{w}_\alpha(-a) = \hat{h}_{\sigma_\alpha(\beta)}(ca^{-<\beta,\alpha>}b)\hat{h}_{\sigma_\alpha(\beta)}(ca^{-<\beta,\alpha>})^{-1}$,
(iv) $\hat{h}_\alpha(a)\hat{x}_\beta(b)\hat{h}_\alpha(a)^{-1} = \hat{x}_\beta(a^{<\beta,\alpha>}b)$,
(v) $\hat{h}_\alpha(a)\hat{w}_\beta(b)\hat{h}_\alpha(a)^{-1} = \hat{w}_\beta(a^{<\beta,\alpha>}b)$,
(vi) $\hat{h}_\alpha(a)\hat{h}_\beta(b)\hat{h}_\alpha(a)^{-1} = \hat{h}_\beta(a^{<\beta,\alpha>}b)\hat{h}_\beta(a^{<\beta,\alpha>})^{-1}$,
for all $\alpha, \beta \in \Phi$, $a, b \in K$, *and* $c = c_{\alpha,\beta} = \pm 1$ *as usual.*

Proof The relations (ii)–(vi) can easily be seen to be derivable from (i) and the definitions of $\hat{w}_\alpha(a)$ and $\hat{w}_\alpha(a)$. Thus, it is sufficient to establish (i). Suppose that $\alpha \neq \pm\beta$. Consider the set $S = \{i\alpha + j\beta \mid j \geq 1\}$. Then $\beta \in S$. The relation B implies that $\hat{x}_\alpha(a)\hat{X}_S(K)\hat{x}_\alpha^{-1}, \hat{x}_{-\alpha}(a)\hat{X}_S(K)\hat{x}_{-\alpha}(a)^{-1}$, and in turn, $\hat{w}_\alpha(a)\hat{X}_S(K)\hat{w}_\alpha^{-1}$ are contained in $\hat{X}_S(K)$. In particular, $\hat{w}_\alpha(a)\hat{x}_\beta(b)\hat{w}_\alpha(-a) \in \hat{X}_S(K)$. Since $\hat{\eta}|_{\hat{X}_S(K)}$ is injective (above Lemma), it is sufficient to show that $w_\alpha^{V_1}(a)x_\beta^{V_1}(b)w_\alpha^{V_1}(-a) = x_{\sigma_\alpha(\beta)}^{V_1}(ca^{-<\beta,\alpha>}b)$. This, of course, follows from Proposition 4.3.14 (ii).

Next, assume that $\alpha = \beta$ and $rankL \geq 2$. Then, we have roots γ and δ together with a positive integer n such that $\alpha = \gamma + n\delta$ with $c_{1,n} \neq 0$ (see Exercise 4.3.7) and

$$[\hat{x}_\gamma(a), \hat{x}_\delta(b)] = \prod\nolimits_{k,l \geq 1} \hat{x}_{k\gamma+l\delta}(c_{k,l}a^k b^l).$$

Consider the set $\hat{S} = \{k\sigma_\alpha(\gamma) + l\sigma_\alpha(\delta) \mid k, l \geq 1\}$. Conjugating the above equation by $\hat{w}_\alpha(a)$ and using the earlier case, we observe that
$\hat{w}_\alpha(a)\hat{x}_{k\gamma+l\delta}(c_{k,l}a^k b^l)\hat{w}_\alpha(a)^{-1}$ belongs to $\hat{X}_{\hat{S}}(K)$ for all k, l except, perhaps , $\hat{w}_\alpha(a)\hat{x}_{\gamma+n\delta}(c_{1,n}ab^n)\hat{w}_\alpha(a)^{-1}$. But $\hat{w}_\alpha(a)[\hat{x}_\gamma(a), \hat{x}_\delta(b)]\hat{w}_\alpha(a)^{-1}$ belongs to $\hat{X}_{\hat{S}}(K)$. Hence
$\hat{\omega}_\alpha(a)\hat{x}_{\gamma+n\delta}(C_{1,n}ab^n)\hat{\omega}_\alpha(a)^{-1} \in \hat{X}_{\hat{S}}(K)$. This proves (i) in case $rankL \geq 2$. Next suppose that $rankL = 1$ and $\alpha = \beta$, then the relation (i) is precisely B'. The case $\alpha \neq \beta$ follows from the earlier case. ♯

Corollary 4.3.54 *The relations (i)–(v) in Theorem 4.3.27 are derivable from A and B in case* $rankL \geq 2$ *and they are derivable from A and* B' *in case* $rankL = 1$. ♯

Exercises

4.3.1. Show that $N_{G(V,K)}(U(V,K)) = B(V,K)$.
4.3.2. For all types of simple Lie algebras, determine $\mid G(V_0, F_q) \mid$.
4.3.3. Verify the table following Example 4.3.38.
4.3.4. Show that $Z(G(V_1, K)) \approx Hom(\Lambda_{univ}/\Lambda_r, K^\star)$.
4.3.5. Give a presentation of $SL(2, F_q)$.
4.3.6. Show that the relation B' of Theorem 4.3.51 is derivable from the relations (iii), (iv), and (v) of Theorem 4.3.27.
4.3.7. Let L be a simple Lie algebra with $rankL \geq 2$. Let $\alpha \in \Phi$. Show that there are roots β and γ and a positive integer n such that $\alpha = \beta + n\gamma$ with $[x_\beta^V(a), x_\gamma^V(b)] = \prod_{k,l \geq 1} x_{k\beta+l\gamma}^V(c_{k,l}a^k b^l)$.
4.3.8. Establish the formula for the order of $G_2(q)$.
8.3.9. Suppose that $l(\sigma\tau) = l(\sigma) + l(\tau)$. Show that

$$(B(V,K)\bigcap(n_0n_\sigma)^{-1}B(V,K)n_0n_\sigma)(B(V,K)\bigcap(n_0n_\tau)^{-1}B(V,K)n_0n_\tau) = B(V,K),$$

where $\nu(n_0) = \sigma_0$, $\nu(n_\sigma) = \sigma$, $\nu(n_\tau) = \tau$, ν the quotient map from $N(V,K)$ to $N(V,K)/H(V,K) \approx W(\Phi)$, and σ_0 is the unique member of $W(\Phi)$ which takes Φ^+ to Φ^-.

4.4 Twisted Groups

In this section, we introduce the Twisted Groups of Lie types which again gives new families of finite simple groups.

Let G be a group and σ an automorphism of G. Then the set $G^\sigma = \{g \in G \mid \sigma(g) = g\}$ of fixed points of σ forms a subgroup. For example, centralizers of elements of G are such groups. At this point, let me inform the readers that the centralizers of involutions in finite simple groups have played very significant role in the classification of finite simple groups. Let us have few more examples. For a fixed n, consider the matrix

$$J_{2n} = \begin{bmatrix} 0_n & I_n \\ -I_n & 0_n \end{bmatrix},$$

where 0_n is the $n \times n$ zero matrix I_n is the $n \times n$ identity matrix. The map σ from $SL(2n, K)$ to itself given by $\sigma(A) = J_{2n}(A^t)^{-1}J_{2n}^{-1}$ is an automorphism of $SL(2n, K)$, and $SL(2n, K)^\sigma = Sp(n, K)$ is the symplectic group. If σ is a map from $SL(n, K)$ to itself given by $\sigma(A) = (A^t)^{-1}$, then $SL(n, K)^\sigma = SO(n, K)$. Let τ be a non identity involutory automorphism of K (for example, $z \mapsto \overline{z}$ in $\mathbb{C}$), then the map σ from $SL(n, K)$ to it self given by $\sigma(A) = (A^\star)^{-1}$ is an automorphism, where $A^\star$ is the tranjugate of A. Then $SL(n, K)^\sigma = SU(n, K)$.

Consider the universal Chevalley group $G(V_1, K)$. We shall describe some special types of automorphisms of $G(V_1, K)$ (the graph, the diagonal, and the field automorphisms), and thereby introduce the groups $G(V_1, K)^\sigma$, and this further allows us discover other family of finite simple groups termed as twisted groups of Lie types. As per classification of finite simple groups, the adjoint finite Chevalley groups, twisted finite simple groups of Lie types, and A_n constitute all infinite families of finite simple groups (leaving prime cyclic groups). The rest of the finite simple groups are exactly 26 in number, and they are called the Sporadic simple groups.

Field Automorphisms: Consider $SL(n, K)$. Let τ be an automorphism of K. Then $E_{ij}^a \mapsto E_{ij}^{\tau(a)}$ respects the Steinberg identities and it induces an automorphism $\hat{\tau}$ from $SL(n, K)$ to itself. Indeed, $\hat{\tau}([a_{ij}]) = [b_{ij}]$, where $b_{ij} = \tau(a_{ij})$. More generally, consider the set $S = \{\hat{x}_\alpha(a) \mid \alpha \in \Phi \textit{ and } a \in K\}$ (Theorem 4.3.51) of symbols. The automorphism τ induces a bijective map $\hat{\tau}$ from S to S given by $\hat{\tau}(\hat{x}_\alpha(a)) = \hat{x}_\alpha(\tau(a))$. Evidently, $\hat{\tau}$ and $\widehat{\tau^{-1}}$ respect the relations A, B, B', and C of Theorem 4.3.51. Thus, τ induces an automorphism $\hat{\tau}$ of $G(V_1, K)$ which is given by $\hat{\tau}(x_\alpha^{V_1}(a)) = x_\alpha^{V_1}(\tau(a))$. Such an automorphism of $G(V_1, K)$ is called a **Field automorphism**.

Diagonal Automorphisms: Let χ be a map from Δ to $K^{\star}$. Then χ induces a homomorphism $\hat{\chi}$ from Λ_r to $K^{\star}$ (as Λ_r is a free abelian group with basis Δ). Again the map $\hat{x}_{\alpha}(a) \mapsto \hat{x}_{\alpha}(\hat{\chi}(\alpha)a)$ from S to S respects the relation A, B, B', and C. Thus, it induces an endomorphism $\hat{d}$ of $G(V_1, K)$ given by $\hat{d}(x_{\alpha}^{V_1}(a)) = x_{\alpha}^{V_1}(\hat{\chi}(\alpha)a)$. An automorphism of this type is called a **Diagonal automorphism**.

Graph Automorphisms: These are the automorphisms of Chevalley groups associated with automorphisms (symmetries) of the corresponding Dynkin diagrams. An automorphism of a Dynkin diagram is a bijective map from the set $S = \{v_i \mid 1 \leq i \leq d\}$ of nodes of the diagram to itself which preserves the number of bonds joining the nodes. Thus, if $n_{v(i)v(j)}$ denotes the number of bonds joining v_i and v_j, then a permutation τ of S is an automorphism if $n_{v_iv_j} = n_{\tau(v_i)\tau(v_j)}$ for all i, j. An automorphism τ of a Dynkin diagram associated with a simple Lie algebra over $\mathbb{C}$ is described in the following figure:

One Root Length

$A_l, l \geq 1$:

$\tau = (v_1, v_l)(v_2, v_{l-1})\ldots, \tau^2 = I$

$D_l, l \geq 4$:
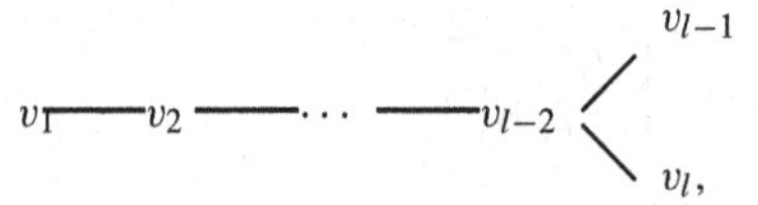

$\tau = (v_{l-1}, v_l), \tau^2 = I$

D_4 :
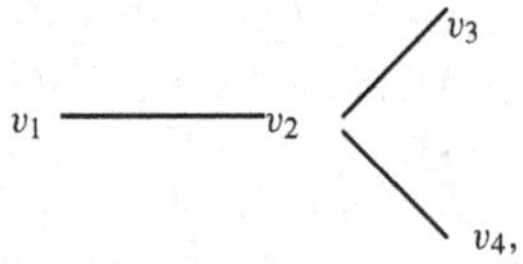

$\tau = (v_1, v_3, v_4), \tau^3 = I$ or $\tau = (v_3, v_4)$

E_6 :
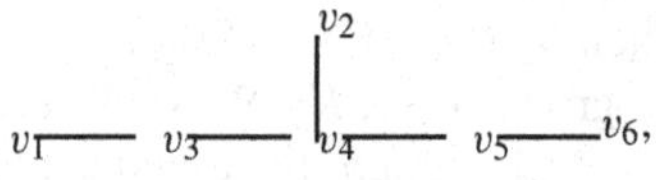

$\tau = (v_1, v_6)(v_3, v_5), \tau^2 = I$

Two root lengths

B_2 :

$\tau = (v_1, v_2), \sigma^2 = I$

F_4 :

$\tau = (v_1, v_4)(v_2, v_3), \tau^2 = I$

G_2 :

$\tau = (1, 2), \tau^2 = I$

The rest of the Dynkin diagrams (associated with simple Lie algebras) have no nontrivial automorphisms.

Proposition 4.4.1 *Let Φ be an irreducible root system with $\Delta = \{\alpha_1, \alpha_2, \ldots, \alpha_l\}$ as a base. Suppose that Φ has only one root length (associated with simple Lie* *algebras of types A_l, D_l, or E_l). Let τ be an automorphism of the corresponding Dynkin diagram. Then τ induces a permutation on Δ and the linear automorphism $\overline{\tau}$ on the Euclidean space $E = <\Phi>$ induced by τ is an isometry such that $\overline{\tau}(\Phi) = \Phi$.*

Proof Let $n_{v_i v_j}$ denote the number of bonds joining the nodes v_i and v_j, where v_i and v_j are the nodes associated with α_i and α_j, respectively. Since τ is an automorphism of the associated Dynkin diagram,

$$\frac{4 < \alpha_i, \alpha_j >^2}{|| \alpha_i ||^2 || \alpha_j ||^2} = n_{v_i v_j} = n_{\tau(v_i)\tau(v_j)} = \frac{4 < \tau(\alpha_i), \tau(\alpha_j) >^2}{|| \tau(\alpha_i) ||^2 || \tau(\alpha_j) ||^2}.$$

Since all the roots are of same length and $< \alpha_k, \alpha_l > \leq 0$ for all k, l, $< \alpha_i, \alpha_j > = < \tau(\alpha_i), \tau(\alpha_j) >$ for all i, j. Consequently, the induced linear automorphism $\overline{\tau}$ is an isometry. Now $\overline{\tau}\sigma_{\alpha_i}\overline{\tau}^{-1} = \sigma_{\tau(\alpha_i)} \in W(\Phi)$. Since $W(\Phi)$ is generated by simple reflections, $\overline{\tau}\sigma\overline{\tau}^{-1} \in W(\Phi)$ for all $\sigma \in W(\Phi)$. Let $\alpha \in \Phi$. Then $\alpha = \sigma(\alpha_i)$ for some $\sigma \in W(\Phi)$ and i and therefore $\overline{\tau}(\alpha) = \overline{\tau}(\sigma)(\alpha_i) = \overline{\tau}\sigma\overline{\tau}^{-1}\overline{\tau}(\alpha_i)$ belongs to $\{\sigma(\alpha_i) \mid \sigma \in W(\Phi), \alpha_i \in \Delta\} = \Phi$. ♯

Let Φ be an irreducible root system (associated with L). Consider the dual root system $\check{\Phi} = \{\check{\alpha} = \frac{2\alpha}{<\alpha,\alpha>} \mid \alpha \in \Phi\}$ of Φ. Note that $\check{\check{\Phi}} = \Phi$. So Φ and $\check{\Phi}$ are dual to each other. If Δ is a base of Φ, then $\check{\Delta} = \{\check{\alpha} \mid \alpha \in \Delta\}$ is a base of $\check{\Phi}$ (verify). Further, $\frac{<\alpha,\beta>}{||\alpha||||\beta||} = \frac{<\check{\alpha},\check{\beta}>}{||\check{\alpha}||||\check{\beta}||}$ for all $\alpha, \beta \in \Phi$. Thus, the map $\alpha \mapsto \check{\alpha}$ from Φ to $\check{\Phi}$ preserves angles. Also $|| \check{\alpha} || = \frac{2}{||\alpha||}$ for all $\alpha \in \Phi$. If Φ has one root length (A_l, D_l, E_6, E_7, and E_8), then $\check{\Phi}$ also has one root length. If Φ has two root lengths, then $\check{\Phi}$ also has two root lengths. However, the map $\alpha \mapsto \check{\alpha}$ reverses the lengths of the roots. Also note that $\prec \check{\alpha}, \check{\beta} \succ = \prec \beta, \alpha \succ$ (verify) for all $\alpha, \beta \in \Phi$. If Φ has two root lengths, then we associate the number $p_\Phi = \frac{||\alpha_0||^2}{||\beta_0||^2}$ to Φ, where α_0 is a long root and β_0 is a short root. Thus, for Φ associated with B_l, C_l, or F_4, $p_\Phi = 2$ and for Φ associated with G_2, $p_\Phi = 3$. The rest of the irreducible root systems are of root length 1. Note that $p_\Phi = p_{\check{\Phi}}$.

Example 4.4.2 (i) Let Φ be an irreducible root system of root length 1. Then Φ is self dual in the sense that Φ and $\check{\Phi}$ are root systems associated with the Lie algebra of same type (A_l, $D_l (l \geq 4)$, E_6, E_7, *or* E_8).

(ii) Let Φ be an irreducible root system of rank ≥ 3 with two root lengths such that $p_\Phi = 2$. If Φ is associated with a Lie algebra of type B_l, the $\check{\Phi}$ is associated with a Lie algebra of type C_l and vice-versa. If Φ is associated with B_2 or to F_4, then it is self dual. Let $\Phi = \{\alpha, \beta, \alpha + \beta, 2\alpha + \beta, -\alpha, -\beta, -(\alpha + \beta), -(2\alpha + \beta)\}$ be a root system associated with B_2 with $\{\alpha, \beta\}$ as basis, α short root and β a long root. Without any loss we may assume that $< \alpha, \alpha > = 1$. Then $p_\Phi = \frac{<\beta,\beta>}{<\alpha,\alpha>} = 2$, $< \beta, \beta > = 2$, $< \alpha, \beta > = -1$. Further, $\check{\alpha} = \frac{2\alpha}{<\alpha,\alpha>} = 2\alpha$, $\check{\beta} = \frac{2\beta}{<\beta,\beta>} = \beta$, $(\alpha \check{+} \beta) = 2(\alpha + \beta) = \check{\alpha} + 2\check{\beta}$, and $(2\alpha \check{+} \beta) = 2\alpha + \beta = \check{\alpha} + \check{\beta}$. Geometrically,

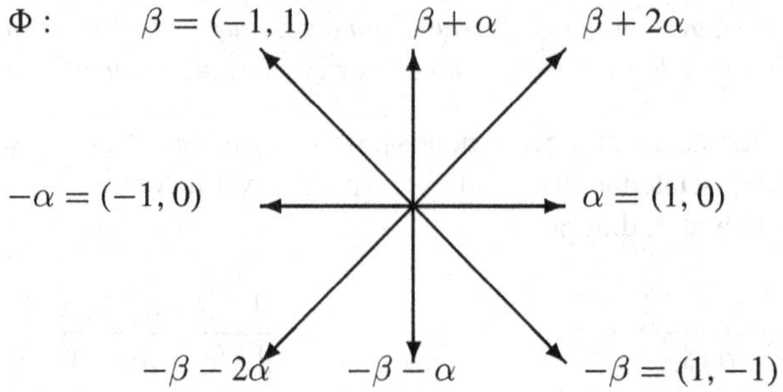

,

and

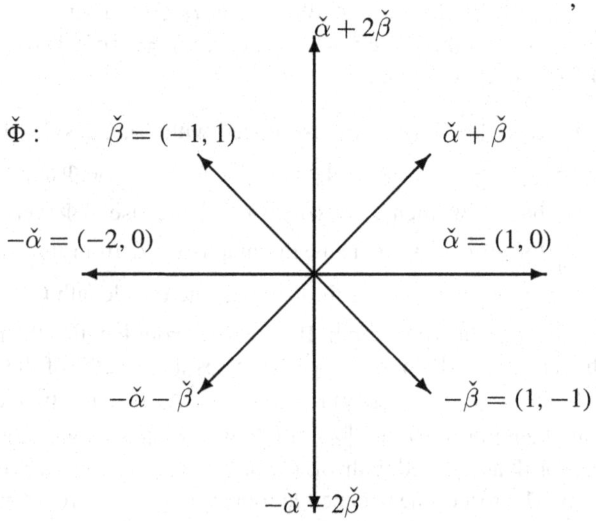

Evidently, by adjusting the length and reflecting about the line bisecting the vectors $\check{\alpha}$ and $\check{\beta}$ we obtain an automorphism of ϕ given by $\alpha \mapsto \beta$, $\beta \mapsto \alpha$, $(\alpha + \beta) \mapsto 2\alpha + \beta$, and $(\beta + 2\alpha) \mapsto \alpha + \beta$. Indeed, this is the automorphism associated with the corresponding Dynkin diagram.

Similarly, a root system Φ associated F_4 or to G_2 (with $p_\Phi = 3$) is self dual, and we can obtain automorphisms in these cases also which are associated with the corresponding Dynkin diagrams.

Theorem 4.4.3 *Let Φ be an irreducible root system with $\Delta = \{\alpha_1, \alpha_2, \ldots, \alpha_l\}$ as a base. Suppose that Φ has one root length. Let τ be an automorphism of the corresponding Dynkin diagram. Let K be a field. Then there is a map ϵ from Φ to $\{1, -1\}$ together with an automorphism g_τ of the universal Chevalley group $G(V_1, K)$ such that $\epsilon(\Delta \bigcup -\Delta) = \{1\}$ and $g_\tau(x_\alpha^{V_1}(a) = x_{\tau(\alpha)}^{V_1}(\epsilon(\alpha)a))$ for all $\alpha \in \Phi$ and $a \in K$.*

Proof We first define a map ϵ from Φ to $\{1, -1\}$ as follows. For $\alpha \in \Delta \bigcup -\Delta$, put $\epsilon(\alpha) = 1$. For $\alpha \in \Phi^+$, we define $\epsilon(\alpha)$ by the induction on $l(\alpha)$. If $l(\alpha) = 1$, $\epsilon(\alpha)$ is already 1. Assume that $\epsilon(\beta)$ is already defined, for all β for which $l(\beta) < l(\alpha)$, where $l(\alpha) \geq 2$. Suppose that $\alpha = \beta + \gamma$, $\beta, \gamma \in \Phi^+$. Then $l(\beta) < l(\alpha)$ and $l(\gamma) < l(\alpha)$. Further, since $\overline{\tau}$ is an automorphism, $\nu_{\overline{\tau}(\beta),\overline{\tau}(\gamma)} = \pm\nu_{\beta,\gamma}$. Put

$$\epsilon(\alpha) = \frac{\epsilon(\beta)\epsilon(\gamma)\nu_{\overline{\tau}(\beta),\overline{\tau}(\gamma)}}{\nu_{\beta,\gamma}} = \pm 1.$$

If $\alpha = -\beta$ is a negative root, then the $[x_\beta, x_{-\beta}] = h_\beta$ implies that $\epsilon(\beta)\epsilon(-\beta) = 1$. This defines $\epsilon(\alpha) = \epsilon(\beta) = \pm 1$. It can be easily seen that the map $\check{x}_\alpha(a) \mapsto \check{x}_{\overline{\tau}(\alpha)}(\epsilon(\alpha)a)$, $\check{h}_\alpha(a) \mapsto \check{h}_{\overline{\tau}(\alpha)}(\epsilon(\alpha)a)$ respects the relations A, B, and C of Theorem 4.3.51. Hence, it induces an automorphism g_τ on the universal Chevalley group. ♯

Theorem 4.4.4 *Suppose that Φ has two root lengths with $p_\Phi = 2$. Let $G(V_1, K)$ and $\check{G}(V_1, K)$ denote the universal Chevalley groups associated with Φ and $\check{\Phi}$, respectively, where K is a field of characteristic p_Φ. Put $\xi(\alpha) = p_\Phi$ if α is long and put $\xi(\alpha) = 1$ if α is short. Then, we have a homomorphism η from $G(V_1, K)$ to $\check{G}(V_1, K)$ such that $\eta(x_\alpha(a)) = x_{\check{\alpha}}(\epsilon_\alpha a^{\xi(\alpha)})$. Further, if K is perfect field (e.g., finite field), then η is an isomorphism.*

Proof Suppose first that $p_\Phi = 2$. Then Φ is associated with B_l, C_l, or F_4. Since K is of characteristic 2, $\epsilon_\alpha = 1$ for all α. We need to show that η respects the relations A, B, and C. If α is short, then $\eta(x_\alpha(a+b)) = x_{\check{\alpha}}(a+b) = x_{\check{\alpha}}(a) + x_{\check{\alpha}}(b) = \eta(x_\alpha(a)) + \eta(x_\alpha(b))$. If α is long, then $\eta(x_\alpha(a+b)) = x_{\check{\alpha}}(a+b)^2 = x_{\check{\alpha}}(a^2 + b^2) = x_{\check{\alpha}}(a^2) + x_{\check{\alpha}}(b^2) = \eta(x_\alpha(a)) + \eta(x_\alpha(b))$. This shows that η respects the relation A. Similarly, using the definitions of $h_\alpha(a)$ and $h_{\check{\alpha}}(a)$, it can be easily seen that η respects the relation C. Finally, using Example 4.4.2, Exercise 4.3.7, and the fact that the characteristic of K is 2, it can be observed that η respects the relation B as well. This proves the existence of the homomorphism as required. If K is perfect, then every element b of K is uniquely expressed as a^2. Consequently, η is an isomorphism. ♯

The following corollary is immediate, since image of the center under an isomorphism is center and $G(V_1, K)/Z(G(V_1, K)) \approx G(V_0, K)$.

Corollary 4.4.5 *Under the hypothesis of the above theorem with K as perfect field, η induces an isomorphism from the adjoint Chevallety group $G(V_0, K)$ to $\check{G}(V_0, K)$.* ♯

Corollary 4.4.6 *If K is a perfect field of characteristic 2, then $Spin_{2n+1}(K) \approx Sp_{2n}(K)$ (see the table following Example 4.3.38).* ♯

From now onward, for convenience, let us denote the universal Chevalley groups associated with $A_l, B_l, C_l, D_l, E_6, E_7, E_8, F_4$, and G_2 over the field K by $A_l^u(K), B_l^u(K), C_l^u(K), D_l^u(K), E_6^u(K), E_7^u(K), E_8^u(K), F_4^u(K)$, and $G_2^u(K)$, respectively, and the corresponding adjoint Chevalley groups by

$A_l(K)$, $B_l(K)$, $C_l(K)$, $D_l(K)$, $E_6(K)$, $E_7(K)$, $E_8(K)$, $F_4(K)$, and $G_2(K)$, respectively. Thus, $A_l(K) \approx A_l^u(K)/Z(A_l^u(K))$ and so on.

The following corollary is immediate from Theorem 4.4.4 (see Example 4.4.2).

Corollary 4.4.7 *We have an automorphism η of $B_2^u(K)$ $(B_2(K))$ given by $\eta(x_{\pm\alpha}(a)) = x_{\pm\beta}(a)$, $\eta(x_{\pm\beta}(a)) = x_{\pm\alpha}(a^2)$, $x_{\pm(\alpha+\beta)}(a) = x_{\pm(2\alpha+\beta)}(a)$, $x_{\pm(\beta+2\alpha)}(a) = x_{\pm(\alpha+\beta)}(a^2)$, where K is a field of characteristic 2. Similarly, we have an automorphism η of $F_4^u(K)$ such that $\eta(x_{\alpha_1}(a)) = x_{\alpha_4}(a)$, $\eta(x_{\alpha_4}(a)) = x_{\alpha_1}(a^2)$, $\eta(x_{\alpha_2}(a)) = x_{\alpha_3}(a)$, and $\eta(x_{\alpha_3}(a)) = x_{\alpha_2}(a^2)$, where $\Delta = \{\alpha_1, \alpha_2, \alpha_3, \alpha_4\}$, α_1, α_2 short and the other two long.* ♯

The following theorem for $p_\Phi = 2$ is precisely Corollary 4.4.7. The proof for the case when $p_\Phi = 3$ is a little involved. The idea of the proof is again to check that the map g_τ described in the theorem respects the relations A, B, and C. The reader can refer to the Yale University Lecture notes on Chevalley groups by Steinberg or to the "Simple Groups of Lie types" by Carter for details.

Theorem 4.4.8 *Let Φ be an irreducible root system. Let K be a perfect field of characteristic p_Φ. Let τ be the nontrivial automorphism of the corresponding Dynkin diagram. Put $\xi(\alpha) = 1$ if α is long and put $\xi(\alpha) = p_\Phi$ if α is short. Then, we have an automorphism g_τ of $G_{adj}(K)$ and a map χ from Φ to $\{1, -1\}$ with $\chi(\pm\alpha) = 1$ for $\alpha \in \Delta$ such that $g_\tau(x_\alpha^\star(a)) = x_{\overline{\tau}(\alpha)}^\star(\chi(\alpha)a^{\xi(\alpha)})$.* ♯

The automorphisms g_τ described in the preceding theorems are called the graph automorphisms associated with the symmetry τ of the Dynkin diagram.

Let Φ be an irreducible root system associated with a simple Lie algebra over $\mathbb{C}$. Let Δ be a base of Φ. Suppose that the associated Dynkin diagram admits a nontrivial symmetry(automorphism) τ. Thus, L is of type A_l, $D_l, l \geq 4$, E_6, B_2, F_4, G_2, or $D_l, l \geq 4$. Clearly τ induces a permutation on Δ which is again denoted by τ. We introduce an isometry $\hat{\tau}$ on $E = <\Phi> = <\Delta>$ in each of the cases as follows: (i) If Φ has one root length (A_l, D_l, or E_6), then define $\hat{\tau}$ by putting $\hat{\tau}(\alpha) = \tau(\alpha)$ for each α in Δ. Thus, if $v = \sum_{i=1}^{l} a_i\alpha_i$, then $\hat{\tau}(v) = \sum_{i=1}^{l} a_i\tau(\alpha_i)$. (ii) If Φ has two root lengths with $p_\Phi = 2$ (B_2, or F_4), then define $\hat{\tau}$ by putting $\hat{\tau}(\alpha) = \frac{\tau(\alpha)}{\sqrt{2}}$ if α is long, and $\hat{\tau}(\alpha) = \sqrt{2}\tau(\alpha)$ if α is short. (iii) If $p_\Phi = 3$ (of the type G_2), then define $\hat{\tau}$ by putting $\hat{\tau}(\alpha) = \frac{\tau(\alpha)}{\sqrt{3}}$ if α is long and $\hat{\tau}(\alpha) = \sqrt{3}\tau(\alpha)$ if α is short. Clearly, the order of $\hat{\tau}$ is same as that of τ and $\hat{\tau}$ preserves the sign.

Let E_0 denote the subspace $\{v \in E \mid \hat{\tau}(v) = v\}$ of E. The projection p_0 of E on to E_0 is given by

$$p_0(v) = v_0 = \frac{1}{m}\sum_{i=0}^{m-1} \hat{\tau}^i(v),$$

where $\hat{\tau}$ is of order m. The projection $p_0(v)$ of v will be denoted by v_0. Thus, $v \in E_0$ if and only if $v_0 = v$.

Proposition 4.4.9 $\hat{\tau}W(\Phi)\hat{\tau}^{-1} = W(\Phi)$.

Proof Let $\alpha \in \Delta$. Then $\hat{\tau}(\alpha) = a\tau(\alpha)$ for some $a > 0$. Then
$\hat{\tau}\sigma_{\alpha}\hat{\tau}^{-1}(v)$
$= \hat{\tau}(\hat{\tau}^{-1}(v) - 2\frac{<\hat{\tau}^{-1}(v),\alpha>\alpha}{<\alpha,\alpha>})$
$= v - 2\frac{<v,\hat{\tau}(\alpha)>\tau(\alpha)}{<\tau(\alpha),\tau(\alpha)>}$
$= \sigma_{\tau(\alpha)}(v)$
belongs to $W(\Phi)$. Since simple reflections generate $W(\Phi)$, $\hat{\tau}W(\Phi)\hat{\tau}^{-1} = W(\Phi)$. ♯

Let $W_0(\Phi)$ denote the subgroup $\{\sigma \in W(\Phi) \mid \hat{\tau}\sigma\hat{\tau}^{-1} = \sigma\} = W(\Phi) \bigcap C_{\hat{\tau}}(Isom(E))$ of $W(\Phi)$. Now, we describe some properties of $W_0(\Phi)$ and its action on E_0.

Proposition 4.4.10 *$W_0(\Phi)$ acts faithfully on E_0.*

Proof Let $v_0 \in E_0$ and $\sigma \in W_0(\Phi)$. Then $\hat{\tau}(\sigma(v_0)) = \sigma(\hat{\tau}(v_0)) = \sigma(v_0)$. This means that $\sigma(v_0) \in E_0$ for each $\sigma \in W_0(\Phi)$ and $v_0 \in E_0$. Thus $W_0(\Phi)$ acts on E_0. Let σ be a non identity element of $W_0(\Phi)$. Then, there is an element $\alpha \in \Phi^+$ such that $\sigma(\alpha) \in \Phi^-$. Evidently, $\hat{\tau}^i(\alpha)$ is positive for each i in the sense that it is positive linear combination of members of Δ. But $\sigma(\hat{\tau}(\alpha)) = \hat{\tau}(\sigma(\alpha)) < 0$. This means σ takes some positive member of E_0 to to a negative member E_0. Hence, there is an element v_0 of E_0 such that $\sigma(v_0) \neq v_0$. ♯

Proposition 4.4.11 *Let J be an orbit of the action of τ as permutation on Δ. Let W_J be the subgroup of $W(\Phi)$ generated by the set $\{\sigma_\alpha \mid \alpha \in J\}$. Let σ_0^J denote the unique element of W_J which maps the positive elements in ϕ_J to negative roots. Then σ_0^J belongs to $W_0(\Phi)$, and $W_0(\Phi)$ is generated by the set $S = \{\sigma_0^J \mid J\ is\ an\ orbit\ of\ \tau\}$.*

Proof First observe that σ_0^J is the unique element of W_J such that $\sigma_0^J(\Phi_J^+) = \Phi_J^-$. Since $\tau(\alpha) \in J$ for each $\alpha \in J$ and $\hat{\tau}\sigma_\alpha\hat{\tau}^{-1} = \sigma_{\tau(\alpha)}$, it follows that $\hat{\tau}W_J\hat{\tau}^{-1} = W_J$. In particular, $\hat{\tau}\sigma_0^J\hat{\tau}^{-1} \in W_J$. Again, since $\hat{\tau}$ preserves sign of roots, $\hat{\tau}\sigma_0^J\hat{\tau}^{-1}(\Phi_J^+) = \Phi_J^-$. From the initial observation, $\hat{\tau}\sigma_0^J\hat{\tau}^{-1} = \sigma_0^J$. This shows that σ_0^J belongs to $W_0(\Phi)$.

Finally, we show that S generates $W_0(\Phi)$. Already, $< S > \subseteq W_0(\Phi)$. By the induction on $l(\sigma)$, we show that $\sigma \in < S >$ for each $\sigma \in W_0(\Phi)$. If $l(\sigma) = 1$, then $\sigma = \sigma_\alpha$ for some $\alpha \in \Delta$. Since $\sigma_\alpha \in W_0(\Phi)$, $\hat{\tau}\sigma_\alpha\hat{\tau}^{-1} = \sigma_\alpha$. This means that $\sigma_\alpha = \sigma_{\tau(\alpha)}$. It follows that $\tau(\alpha) = \alpha$. This means that $\{\alpha\}$ is an orbit of τ and $\sigma_\alpha = \sigma_0^{\{\alpha\}}$ belongs to S. Assume that the result holds for all those members of $W_0(\Phi)$ whose length is less than $l(\sigma)$, where $\sigma \in W_0(\Phi)$, $l(\sigma) \geq 2$. Since $\sigma \neq I$, there is a root $\alpha \in \Delta$ such that $\sigma(\alpha) \in \Phi^-$. Let J denote the τ orbit of α. Since τ preserves the sign of the roots, it follows that $\sigma(\beta) \in \Phi^-$ for each $\beta \in J$. Clearly, σ_0^J maps Φ_J^+ to Φ_J^- but it does not change sign of any element of $\Phi - \Phi_J$. Thus, $l(\sigma\sigma_0^J) = l(\sigma) - l(\sigma_0^J)$. By the induction hypothesis $\sigma\sigma_0^J$ belongs to $< S >$. Already, $\sigma_0^J \in S$. This shows that $\sigma \in < S >$. ♯

Proposition 4.4.12 *Let J be an orbit of the action of τ on Δ. Let $\alpha \in J$. Then $\sigma_0^J|_{E_0}$ is the reflection $\sigma_{p_0(\alpha)}$ on E_0 for each $\alpha \in J$.*

Proof If $\alpha, \beta \in J$, then $\tau^i(\alpha) = \beta$ for some i. Consequently, $p_0(\alpha)$ is a positive multiple of $p_0(\beta)$. In turn, it also follows that $p_0(\gamma)$ and $p_0(\delta)$ are positive multiples of each other whenever $\gamma, \delta \in \Phi_J^+$. Since $\sigma_0^J \in W_0(\Phi)$, $\hat{\tau}\sigma_0^J = \sigma_0^J\hat{\tau}$. Consequently, $\sigma_0^J(p_0(\gamma)) = p_0(\sigma_0^J(\gamma))$ for all $\gamma \in \Phi_J^+$. Also, by the definition, $\sigma_0^J(\gamma) \in \Phi_J^-$ for all $\gamma \in \Phi_J^+$. This means that $\sigma_0^J(p_0(\gamma))$ is a negative multiple of $p_0(\gamma)$. Since σ_0^J is an isometry, $\sigma_0^J(p_0(\gamma)) = -ap_0(\gamma)$ for some $a > 0$. Let $\alpha \in J$ and $v \in E_0$ be such that $< v, p_0(\alpha) > = 0$. Since $p_0(\beta)$ is a positive multiple of $p_0(\alpha)$ for each $\beta \in J$, it follows that $< v, \beta > = 0$ for all $\beta \in J$. This means that $\sigma_0^J(v) = v$ for all v in the plane perpendicular to $p_0(\alpha)$ for all $\alpha \in J$. This shows that $\sigma_0^J|_{E_0}$ is the reflection $\sigma_{p_0(\alpha)}$ for all $\alpha \in J$. ♯

The following corollary is immediate.

Corollary 4.4.13 *$W_0(\Phi)$ treated as group of isometries of E_0 is generated by the set $\{\sigma_{p_0(\alpha)} \mid \alpha \in \Delta\}$.* ♯

Let Φ_0 denote the set $\{p_0(\alpha) \mid \alpha \in \Phi\}$, Δ_0 denote the set $\{p_0(\alpha) \mid \alpha \in \Delta\}$. Denote $W_0(\Phi)$ treated as subgroup of isometries of E_0 by $W_0(\Phi_0)$. We shall see that Φ_0 satisfies all the defining conditions of a root system on E_0 with base Δ_0 except the condition that $a\alpha \in \Phi_0$ (Δ_0) for $\alpha \in \Phi_0$ (Δ_0) implies that $a = \pm 1$.

Proposition 4.4.14 *The set $\wp = \{\sigma(\Phi_J^+) \mid \sigma \in W_0(\Phi)$ and J is an orbit of $\tau\}$ is a partition of Φ. Further, α and β belongs to the same member of the partition if and only if $p_0(\alpha)$ is a positive multiple of $p_0(\beta)$.*

Proof Let $\alpha \in \Phi^+$. Consider the unique element $\sigma_0 \in W(\Phi)$ given by $\sigma_0(\Phi^+) = \Phi^-$. Since $\hat{\tau}$ preserves sign of each root in Φ, $\hat{\tau}\sigma_0\hat{\tau}^{-1}(\Phi^+) = \Phi^-$. This means that $\hat{\tau}\sigma_0\hat{\tau}^{-1} = \sigma_0$ and so $\sigma_0 \in W_0(\Phi)$. From Proposition 4.4.11,

$$\sigma_0 = \sigma_0^{J_1}\sigma_0^{J_2}\cdots\sigma_0^{J_r}$$

for some orbits $J_1, J_2, \ldots, J_r$ of τ. Since $\alpha \in \Phi^+$, $\sigma_0(\alpha) \in \Phi^-$. Thus, there is index i such that $\sigma_0^{J_{i+1}}\sigma_0^{J_{i+2}}\cdots\sigma_0^{J_r}(\alpha)$ belongs to Φ^+ where as $\sigma_0^{J_i}\sigma_0^{J_{i+1}}\cdots\sigma_0^{J_r}(\alpha)$ belongs to Φ^-. Since $\sigma_0^{J_i}(\Phi_{J_i}^+) \subseteq \Phi^-$ and $\sigma_0^{J_i}(\Phi^+ - \Phi_{J_i}^+) \subseteq \Phi^+$, it follows that $\sigma_0^{J_{i+1}}\sigma_0^{J_{i+2}}\cdots\sigma_0^{J_r}(\alpha)$ belongs to $\Phi_{J_i}^+$. This means that α belongs $\sigma(\Phi_{J_i}^+)$, where $\sigma = \sigma_0^{J_r}\sigma_0^{J_{r-1}}\cdots\sigma_0^{J_{i+1}}$ is an element of $W_0(\Phi)$. Also $-\alpha$ belongs to $\sigma\sigma_0^{J_i}(\Phi_{J_i}^+)$. This shows that the union of the members of $\wp$ is Φ.

Next, we show that α, β belong to the same member of $\wp$ if and only if $p_0(\alpha)$ is a positive multiple of $p_0(\beta)$. This, of course, will also show that the distinct members of $\wp$ are disjoint. Suppose that α, β belong to $\sigma(\Phi_J^+)$ for some $\sigma \in W_0(\Phi)$ and an orbit J of τ. Then $\alpha = \sigma(\gamma)$ and $\beta = \sigma(\delta)$, where γ, δ belong to Φ_J^+. Clearly, $p_0(\gamma) = ap_0(\delta)$ for some $a > 0$. Since σ is an isometry, $\sigma(p_0(\gamma)) = a\sigma(p_0(\delta))$. Again, since $\sigma\hat{\tau} = \hat{\tau}\sigma$, we see that $\sigma p_0 = p_0\sigma$. Consequently, $p_0(\alpha) = p_0(\sigma(\gamma)) = ap_0(\sigma(\delta)) = ap_0(\beta)$, where $a > 0$. Conversely, suppose that $p_0(\alpha) = ap_0(\beta)$, $\alpha, \beta \in \Phi$ and $a > 0$. Since union of members of $\wp$ is Φ, $\alpha \in \sigma(\Phi_J^+)$ for some orbit J of τ and $\sigma \in W_0(\Phi)$. This means $\sigma^{-1}(\alpha) \in \Phi_J^+$ for some $\sigma \in W_0(\Phi)$ and an orbit

J of τ. Thus, $\sigma^{-1}(\alpha)$ is a positive linear combination of some members of J. Again, since $p_0\sigma^{-1} = \sigma^{-1}p_0$,

$$p_0(\sigma^{-1}(\alpha)) = \sigma^{-1}(p_0(\alpha)) = \sigma^{-1}(ap_0\beta) = a\sigma^{-1}(p_0(\beta)) = ap_0(\sigma^{-1}(\beta)),$$

where $a > 0$. This means that $p_0(\sigma^{-1}(\beta))$ is also a positive linear combination of members of J. Since J is an orbit of τ, $\sigma^{-1}(\beta)$ is also a positive linear combination of members of J and so β also belongs to $\sigma(\Phi_J^+)$. It follows that α and β belong to the same member of $\wp$. The proof is complete. ♯

Theorem 4.4.15 *(i)* Φ_0 *generates* E_0.

(ii) Any element of Φ_0 *is a nonnegative linear combination or it is a nonpositive linear combination of members* Δ_0.

(iii) Define an equivalence relation $\approx$ *on* Δ_0 *by putting* $\alpha \approx \beta$ *if* α *is a positive multiple of* β. *Then, we obtain a basis* $\hat{\Delta}_0$ *by choosing one and only one member from each equivalence class.*

(iv) For each $p_\alpha \in \Phi_0$, *there is a* $\sigma \in W_0(\Phi)$ *such that* $\sigma|_{E_0} = \sigma_{p_0(\alpha)}$.

(v) $p_0(\alpha) \in \Phi_0$ *implies that* $\sigma_{p_0(\alpha)}(p_0(\beta))$ *belongs to* Φ_0 *for each* $p_0(\beta) \in \Phi_0$.

Proof (i) Since $<\Phi> = E$, $<\Phi_0> = p_0(E) = E_0$.

(ii) Since each member of Φ is a nonnegative or a nonpositive linear combination of members of Δ and p_0 takes positive elements to positive elements, it follows that each member of $\Phi_0 = p_0(\Phi)$ is a nonnegative or a nonpositive linear combination of members of $p_0(\Delta) = \Delta_0$.

(iii) Let $\{J_1, J_2, \ldots, J_r\}$ be the set of distinct orbits of τ. Selecting one and only one member $p_0(\alpha_i)$ from $p_0(\Phi_{J_i}^+)$, we obtain a linearly independent subset of E_0 which is clearly a basis for E_0.

(iv) Let $p_0(\alpha) \in \Phi_0$. By Proposition 4.4.14, $\alpha \in \sigma(\Phi_J^+)$ for some $\sigma \in W_0(\Phi)$ and an orbit J. Thus, $\alpha = \sigma(\beta)$, for some $\beta \in J$. Now, $p_0(\alpha) = p_0(\sigma(\beta)) = \sigma(p_0(\beta))$. By Proposition 4.4.12, $\sigma_{p_0(\beta)}$ coincides with an element of $W_0(\Phi)$ treated as group of isometris of E_0. Thus, $\sigma_{p_0(\alpha)} = \sigma\sigma_\beta\sigma^{-1}$ agrees with an element of $W_0(\Phi)$ on E_0. This proves (iv).

(v) Let $p_0(\alpha)$ and $p_0(\beta)$ belong to Φ_0. From the earlier observation, there is an element $\sigma \in W_0(\Phi)$ which coincides with $\sigma_{p_0(\alpha)}$ on E_0. Consequently, $\sigma_{p_0(\alpha)}(p_0(\beta)) = \sigma(p_0(\beta)) = p_0(\sigma(\beta))$ belongs to Φ_0. This proves (v). ♯

Example 4.4.16 Consider the root system Φ of the type A_l, $l = 2m-1$ with basis $\Delta = \{\alpha_1, \alpha_2, \ldots, \alpha_{2m-1}\}$. The permutation τ of Δ associated with the nontrivial symmetry of the Dynkin diagram is given by $\tau = (\alpha_1, \alpha_{2m-1})(\alpha_2, \alpha_{2m-2})\cdots(\alpha_{m-1}, \alpha_{m+1})$. The orbits are $J_1, J_2, \ldots, J_m$, where $J_i = \{\alpha_i, \alpha_{2m-i}\}$ for $i \leq m-1$ and $J_m = \{\{\alpha_m\}\}$. $\hat{\tau}$ is the isometry of E which agrees with τ on Δ. $E_0 = \{v \in E \mid \hat{\tau}(v) = v\} = \{\sum_{i=1}^{2m-1} a_i\alpha_i \mid a_i = a_{2m-i}\ for\ i \leq m-1\}$. The projection map p_0 is given by $p_0(v) = \frac{1}{2}(v + \hat{\tau}(v))$. $\Phi_0 = \{\frac{1}{2}(\alpha + \hat{\tau}(\alpha)) \mid \alpha \in \Phi\}$. $\Delta_0 = \{p_0(\alpha_1), p_0(\alpha_2), \ldots, p_0(\alpha_m)\}$ is a basis of Φ_0, $p_0(\alpha_i) = \frac{1}{2}(\alpha_i + \alpha_{2m-i}) = p_0(\alpha_{2m-i})$ for $i \leq m-1$ and $p_0(\alpha_m) = \alpha_m$. It can be easily checked that $\prec p_0(\alpha_i), p_0(\alpha_{i+1}) \succ = 1$ for all $i \leq m-2$, whereas it is $\sqrt{2}$ for $i = m-1$. Also

$\| p_0(\alpha_i) \| = \frac{1}{2}$ for all $i \leq m-1$, whereas $\| p_0(\alpha_m) \| = 1$. This means that Φ_0 in this case is of the type C_m and $W_0(\Phi)$ is isomorphic to $W(C_m)$. Similarly, if $l = 2m$, then it can be seen that Φ_0 is of the type B_m and $W_0(\Phi) \approx W(B_m)$.

Suppose that Φ is of type D_4 with basis $\Delta = \{\alpha_1, \alpha_2, \alpha_3, \alpha_4\}$. The associated Dynkin diagram is

D_4 :

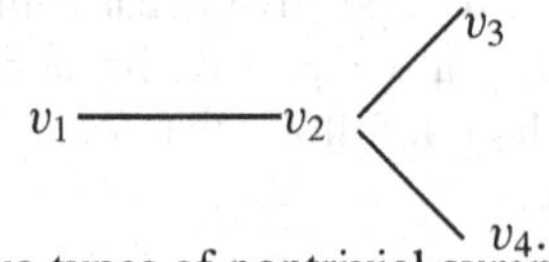

We have two types of nontrivial symmetries given by a transposition (v_3, v_4) and a cycle (v_1, v_3, v_4). These induce the transposition (α_3, α_4) and the cycle $(\alpha_1, \alpha_3, \alpha_4)$. It can be seen that Φ_0 associated with (α_3, α_4) is of the type B_3, and that associated with the cycle $(\alpha_1, \alpha_3, \alpha_4)$ is of the type G_2.

If Φ is of the type $D_l, l > 4$, then the associated Φ_0 is of the type B_{l-1}. If Φ is of the type E_6, then it can be seen that Φ_0 is of the type F_4. If Φ is of the type B_2, or of the type G_2, then it is evident that Φ_0 is of the type A_1.

Finally, we describe Φ_0 when Φ is of the type F_4. Let $\Delta = \{\alpha_1, \alpha_2, \alpha_3, \alpha_4\}$ be a base of Φ, where α_1, α_2 are short and α_3, α_4 are long. Note that $\prec \alpha_1, \alpha_2 \succ \prec \alpha_1, \alpha_2 \succ = 1 = \prec \alpha_3, \alpha_4 \succ \prec \alpha_4, \alpha_3 \succ$, whereas $\prec \alpha_2, \alpha_3 \succ \prec \alpha_3, \alpha_2 \succ = 2$ (see the Dynkin diagram for F_4). Evidently, $\tau = (\alpha_1, \alpha_4)(\alpha_2, \alpha_4)$, and $\Delta_0 = \{\frac{1}{2}(\sqrt{2}\alpha_1 + \alpha_4), \frac{1}{2}(\sqrt{2}\alpha_2 + \alpha_3)\}$. It can be seen that the angle between the two members of Δ_0 is $\frac{7\pi}{8}$. Thus, Φ_0 is different from all irreducible root systems of rank 2. In this case, $W_0(\Phi)$ is generated by two reflections about planes inclined at $\frac{7\pi}{8}$. Determine the corresponding graph.

For convenience, in case the the triple (L, V, K) is understood to be there, we denote the Chevalley group $G(V, K)$ by $G(K)$, the universal Chevaley group $G(V_1, K)$ by $G_{univ}(K)$, and the adjoint Chevalley group $G(V_0, K)$ by $G_{adj}(K)$. The diagonal subgroup $H(V, K)$ is denoted by $H(K)$, the unipotent subgroups $U^+(V, K)$ and $U^-(V, K)$ by $U^+(K)$ and $U^-(K)$, the Borel subgroup $B(V, K)$ by $B(K)$, and the monomial subgroup $N(V, K)$ by $N(K)$, respectively.

It is clear from the above discussions that only a graph automorphism of a Chevalley group $G_{adj}(K)$ of adjoint type defined over a field K does not give new simple groups except, perhaps, in case of F_4. We look at composites of field and graph automorphisms. Let $\hat{f}$ be a field automorphism of $G_{adj}(K)$ associated with an automorphism f of K. In case of B_2 and also in case of F_4, assume that K is a perfect field of characteristic 2. In case of G_2 assume that K is perfect field of characteristic 3.

Proposition 4.4.17 *Graph automorphisms and field automorphisms commute with each other. If all the roots are of same length, g_τ and τ have same order (2 or 3). If there are two root lengths, then g_τ^2 is the field automorphism associated with the automorphism $a \mapsto a^{p_\Phi}$.*

Proof Consider the graph automorphism g_τ of $G_{adj}(K)$ associated with a nontrivial symmetry τ of the associated Dynkin diagram and the field automorphism $\hat{f}$ associated with an automorphism f of K. Suppose that all the roots are of same length

(A_l, D_l, E_6). Then $g_\tau(x_\alpha(a)) = x_{\tau(\alpha)}(a)$ and $\hat{f}(x_\alpha(a)) = x_\alpha(f(a))$ for all $\alpha \in \Delta$ and $a \in K$. Clearly,

$$\hat{f} g_\tau(x_\alpha(a)) = \hat{f}(x_{\tau(\alpha)}(a)) = x_{\tau(\alpha)}(f(a)) = g_\tau \hat{f}(x_\alpha(a))$$

for all $\alpha \in \Delta$ and $a \in K$.

Next, suppose that Φ has two root lengths (B_2, F_4, or G_2). Then

$$\hat{f} g_\tau(x_\alpha(a)) = \hat{f}(x_{\tau(\alpha)}(a^{\xi(\alpha)}) = x_{\tau(\alpha)}(f(a^{\xi(\alpha)})) = g_\tau \hat{f}(x_\alpha(a))$$

for all $\alpha \in \Delta$ and $a \in K$. Thus, $\hat{f}$ and g_τ commute.

Next, if all the roots are of same length, then $g_\tau(x_\alpha(a)) = x_{\tau(\alpha)}(a)$. Evidently, order of g_τ is same as that of τ. Suppose that there are two root lengths. Then $g_\tau^2(x_\alpha) = g_\tau(x_{\tau(\alpha)}(a^{\xi(\alpha)}) = x_{\tau(\tau(\alpha))}(a^{\xi(\alpha)\xi(\tau(\alpha))}) = x_\alpha(a^{p_\Phi})$. This means that g_τ^2 is the field automorphism associated with the automorphism $a \mapsto a^{p_\Phi}$ of K. ♯

Proposition 4.4.18 *Let n be the order of a nontrivial symmetry τ of the associated Dynkin diagram ($n = 2$ for A_l, D_l, E_6, F_4, G_2 and $n = 3$ for D_4). Then $\hat{f} g_\tau$ is of order n if and only if (i) $f^n = I$ in case all roots are of same length and (ii) $f^2(a)^{p_\Phi} = a$ for all $a \in K$ in case of two root lengths.*

Proof Suppose that all roots are of same length. Suppose further, that $(g_\tau \hat{f})^n = I$, where f is a nontrivial automorphism of K. Clearly, $(g_\tau \hat{f})^n(x_\alpha(a)) = x_{\tau^n(\alpha)}(f^n(a)) = x_\alpha(a)$ for all $\alpha \in \Delta$ and $a \in K$ if and only if $f^n(a) = a$ for all a.

Next suppose that there are two root lengths. Then $n = 2$ and $(g_\tau \hat{f})^2(x_\alpha(a)) = g_\tau^2 \hat{f}^2(x_\alpha(a)) = g_\tau^2(x_\alpha(f^2(a))) = x_{\tau^2(\alpha)}(f^2(a^{\xi(\tau(\alpha))\xi(\alpha)})) = x_\alpha(f^2(a)^{p_\Phi})$ for all $\alpha \in \Delta$ and $a \in K$. This shows that $(g_\tau \hat{f})^n = I$ if and only if $f^2(a)^{p_\Phi} = a$ for all $a \in K$. ♯

Corollary 4.4.19 *(i) In case of $A_l(K)$, $E_6(K)$, and $D_l(K), l \geq 4$ with a symmetry τ of order 2, an automorphism $g_\tau \hat{f}$ of order 2 exists if and only if $K \approx F_{q^2}$ for some prime power q.*

(ii) In case of $D_4(K)$ with a symmetry τ of order 3, an automorphism $g_\tau \hat{f}$ of order 3 exists if and only if $K \approx F_{q^3}$ for some prime power q.

(iii) In case of $B_2(K)$, $F_4(K)$, the characteristic of the field has to be 2, and such an automorphism $g_\tau \hat{f}$ exists if and only if $K \approx F_{2^{2m+1}}$ for some m.

(iv) In case of $G_2(K)$, the characteristic of the field has to be 3 and such an automorphism $g_\tau \hat{f}$ exists if and only if $K \approx F_{3^{2m+1}}$ for some $m \geq 0$.

Proof A finite field K admits an automorphism f of order 2 if and only if $K \approx F_{q^2}, q = p^m$ (the automorphism f is given by $f(a) = a^q$). It admits an automorphism of order 3 if and only if $K \approx F_{q^3}, q = p^m$. Again, an automorphism f of K satisfying $f^2(a)^2 = a$ for all $a \in K$ exists if an only if $K \approx F_{p^{2m+1}}$ for some prime p. The results follow from the above observations and the preceding Proposition. ♯

Proposition 4.4.20 *Let $G_{adj}(K)$ be a Chevalley group of adjoint type over a field K together with a nontrivial symmetry τ of the associated Dynkin diagram. Assume that K is a perfect field of characteristic 2 in case of B_2 or F_4 and that K is a perfect field of characteristic 3 in case of G_2. Let $\hat{f}$ be a nontrivial field automorphism such that $\eta = g_\tau \hat{f}$ is of order n, where n is the order of τ. Then $\eta(U^+_{adj}(K)) = U^+_{adj}(K)$, $\eta(U^-_{adj}(K)) = U^-_{adj}(K)$, $\eta(B^+_{adj}(K)) = B^+_{adj}(K)$, $\eta(H_{adj}(K)) = H_{adj}(K)$, and $\eta(N_{adj}(K)) = N_{adj}(K)$. Further, η induces an automorphism $\hat{\eta}$ on $W(\Phi) \approx N_{adj}(K)/H_{adj}(K)$ which is given by $\hat{\eta}(\sigma_\alpha) = \sigma_{\tau(\alpha)}$, $\alpha \in \Delta$.*

Proof For each $\alpha \in \Delta$, $\eta(x_\alpha(a)) = g_\tau(\hat{f}(x_\alpha(a)) = g_\tau(x_\alpha(f(a))) = x_{\tau(\alpha)}(f(a)^{\xi(\alpha)})$ belongs to $U^+_{adj}(K)$. It follows that $\eta(U^+_{adj}(K)) = U^+_{adj}(K)$. Similarly, $\eta(U^-_{adj}(K)) = U^-_{adj}(K)$. Since $U^+_{adj}(K)H_{adj}(K) = B^+_{adj}(K)$ is the normalizer of $U^+_{adj}(K)$, it follows that $\eta(B^+_{adj}(K)) = B^+_{adj}(K)$. Similarly, $\eta(B^-_{adj}(K)) = B^-_{adj}(K)$. Since $B^+_{adj}(K) \bigcap B^-_{adj}(K) = H_{adj}(K)$ and η is bijective, $\eta(H_{adj}(K)) = H_{adj}(K)$.

For each $\alpha \in \Delta$, recall the element $n_\alpha = w_\alpha(1) = x_\alpha(1)x_{-\alpha}(-1)x_\alpha(1) \in N_{adj}(K)$ such that $\sigma_\alpha \mapsto n_\alpha$ induces an isomorphism from $W(\Phi)$ to N_{adj}/H_{adj}, where $n_\alpha = w_\alpha(1) = x_\alpha(1)x_{-\alpha}(-1)x_\alpha(1)$. Clearly, $\{n_\alpha \mid \alpha \in \Delta\}$ together with $H_{adj}(K)$ generate $N_{adj}(K)$. Further,

$$\eta(n_\alpha) = g_\tau \hat{f}(w_\alpha(1)) = w_{\tau(\alpha)}(1) = n_{\tau(\alpha)}.$$

for each $\alpha \in \Delta$. This shows that $\eta(N_{adj}(K)) = N_{adj}(K)$. Evidently, we have unique automorphism $\hat{\eta}$ such that $\hat{\eta}(\sigma_\alpha) = \sigma_{\tau(\alpha)}$ for all $\alpha \in \Delta$. ♯

Let $U^+_\eta(K)$ denote the subgroup $\{u \in U^+(K) \mid \eta(u) = u\}$ of $U^+(K)$, $U^-_\eta(K)$ denote the subgroup $\{u \in U^-(K) \mid \eta(u) = u\}$ of $U^-(K)$ and G_η the subgroup of $G_{adj}(K)$ generated by $U^+_\eta \bigcup U^-_\eta$. Further, let $H_\eta(K)$ denote the subgroup $G_\eta(K) \bigcap H_{adj}(K)$ and $N_\eta(K)$ denote the subgroup $G_\eta(K) \bigcap N_{adj}(K)$. The group $G_\eta(K)$ is called a **Twisted Group**. Leaving very few exceptions the twisted group $G_\eta(K)$ are simple groups. They are called the twisted simple groups of Lie types.

The subgroups $U^+_\eta(K)$, $U^-_\eta(K)$, $H_\eta(K)$, and $N_\eta(K)$ bear the similar relationship with $G_\eta(K)$ as the subgroups $U^+_{adj}(K)$, $U^-_{adj}(K)$, $H_{adj}(K)$, and $N_{adj}(K)$ bear with $G_{adj}(K)$. We state few results without proof. The following result is the analogue of Theorem 4.3.22.

Proposition 4.4.21 *Every element $g \in G_\eta(K)$is uniquely expressible as $g = uhn_\sigma v$, where $u \in U^+_\eta(K)$, $h \in H_\eta(K)$, and $\sigma \in W_\eta(\Phi)$, $n_\sigma \in N_\eta(K)$, and $v \in (U^-_\sigma)_\eta(K)$.* ♯

Theorem 4.4.22 *The group $G_\eta(K)$ is a group with $(B_\eta(K), N_\eta(K))$ as (B, N)-pair.* ♯

Notations: We fix up some standard notations for the twisted groups over finite fields:

1. ${}^2A_l(q^2)$: Consider the Chevalley group $A_l(K)$. The nontrivial symmetry τ of the associated Dynkin diagram is of order 2. As such, an associated twisted group exists if and only if $K \approx F_{q^2}$ for some prime power p. We denote this twisted group by ${}^2A_l(q^2)$. Observe that this group is $PSU(q)$. These are all simple groups except when $q = 2$
2. ${}^2D_l(q^2)$: Consider the Chevalley group $D_l(K)$, $l \geq 4$ together with the nontrivial symmetry τ of order 2 (note that in case of $D_4(K)$, there is a symmetry of order 3 also). An associated twisted subgroup exists if and only if $K \approx F_{q^2}$. We denote this twisted group by ${}^2D_l(q^2)$. These are all simple groups.
3. ${}^2E_6(q^2)$: As above, a twisted subgroup associated with $E_6(K)$ exists if and only if $K \approx F_{q^2}$. We denote this twisted group by ${}^2E_6(q^2)$. These are all simple groups.
4. ${}^3D_4(q^3)$: A twisted group associated with $D_4(K)$ together with a symmetry of order 3 exists if and only if $K \approx F_{q^3}$. We denote this twisted group by ${}^3D_4(q^3)$. These are all simple groups.
5. ${}^2B_2(2^{2m+1})$: Similarly, using Corollary 4.4.19, we see that a twisted subgroup associated with $B_2(K)$ exists if and only if $K \approx F_{2^{2m+1}}$. We denote this twisted group by ${}^2B_2(2^{2m+1})$.These are all simple groups except ${}^2B_2(2)$.
6. ${}^2F_4(2^{2m+1})$: Again, a twisted subgroup associated with $F_4(K)$ exists if and only if $K \approx F_{2^{2m+1}}$. We denote this twisted group by ${}^2F_4(2^{2m+1})$. All these groups are also simple groups except ${}^2F_4(2)$.
7. ${}^2G_2(3^{2m+1})$: A twisted subgroup associated with $G_2(K)$ exists if and only if $K \approx F_{3^{2m+1}}$. We denote this twisted group by ${}^2G_2(3^{2m+1})$. All these groups are simple except ${}^2G_2(3)$.

Thus, all twisted groups are simple except ${}^2A_2(4)$, ${}^2B_2(2)$, ${}^2F_4(2)$, and ${}^2G_2(3)$.

The series ${}^2B_2(2^{2m+1})$ of finite simple groups was discovered by Suzuki ("A new type of simple groups of finite order", Proc. Nat. Acad . Sci (46) 868–870). These groups are also called the **Suzuki Groups**, and they are also denoted by $Sz(2^{2m+1})$. These groups are the only finite simple groups whose orders are not divisible by 3. The groups ${}^2G_2(3^{2m=1})$ and ${}^2F_4(2^{2m+1})$ were discovered by Ree in 1961. These are called the **Ree Groups**. The groups ${}^2E_6(q)$ and ${}^3D_4(q)$ are called **Steinberg Groups**

The orders of twisted simple groups can be obtained by using the Proposition 4.4.21 and they are given as follows:

1. $|{}^3\ D_4(q^3)\ | = q^{12}(q^2-1)(q^6-1)(q^8+q^4+1)$.
2. $|{}^2\ E_6(q^2)\ | = \frac{q^{36}(q^2-1)(q^5+1)(q^6-1)(q^8-1)(q^9+1)(q^{12}-1)}{(q+1,3)}$.
3. $|{}^2\ B_2(q)\ | = q^2(q^2+1)(q-1),\ q = 2^{2m+1}$.
4. $|{}^2\ F_4(q)\ | = q^{12}(q-1)(q^3+1)(q^4-1)(q^6+1),\ q = 2^{2m+1}$.
5. $|{}^2\ G_2(q)\ | = q^3(q^3+1)(q-1),\ q = 3^{2m+1}$.

List of All Non-abelian Finite Simple Groups with Their Schur Multipliers

1. $A_n, n \geq 5$. The Schur Multipliers of A_5 and $A_n, n \geq 8$ are of order 2. For A_6 and A_7 it is a cyclic group of order 6.
2. Adjoint Chevalley groups $A_l(q)$, where $l \geq 2$ or $l = 1$ and $q \geq 4$. The Schur Multiplier of $A_1(4)$, $A_2(2)$, and $A_3(2)$ are of order 2. The Schur Multiplier of

$A_1(9)$ is of order 6, and that of $A_2(4)$ is $\mathbb{Z}_3 \times \mathbb{Z}_4 \times \mathbb{Z}_4$. The Schur Multiplier of the rest of $A_l(q)$ is a cyclic group of order $(l+1, q-1)$.

3. Adjoint Chevalley Groups $B_l(q)$, where $l \geq 3$ or $l = 2$ and $q \geq 3$. The Schur Multiplier of $B_3(2)$ is of order 2, and that of $B_3(3)$ is of order 6. The Schur Multiplier of the rest of $B_l(q)$ are of order $(q-1, 2)$.
4. Adjoint Chevalley Groups $C_l(q), l \geq 3$. The Schur Multiplier of $C_3(2)$ is of order 2 and for the rest of $C_l(q)$, it is of order $(q-1, 2)$.
5. Adjoint Chevalley Groups $D_l(q), l \geq 4$. The Schur Multiplier of $D_4(2)$ is $\mathbb{Z}_2 \times \mathbb{Z}_2$, the Schur multiplier of $D_l(q)$ for odd l is a cyclic group of order $(q^l - 1, 4)$ and it is elementary abelian of order $(q^l - 1, 4)$ for even l.
6. Exceptional Chevalley groups $E_6(q)$. The Schur Multiplier of $E_6(q)$ is of order $(q-1, 3)$.
7. Exceptional Chevalley groups $E_7(q)$. The Schur Multiplier of $E_7(q)$ is of order $(q-1, 2)$.
8. Exceptional Chevalley groups $E_8(q)$. The Schur Multiplier of $E_8(q)$ is trivial.
9. Exceptional Chevalley groups $F_4(q)$. The Schur Multiplier of $F_4(q)$ is trivial except for $F_4(2)$ it is of order 2.
10. Exceptional Chevalley groups $G_2(q), q \geq 3$. The Schur Multiplier of $G_2(q)$ is trivial except for $G_3(3)$ it is of order 3 and for $G_2(4)$ it is of order 2.
11. Twisted simple groups ${}^2A_l(q^2) l \geq 2$. The Schur Multiplier are all cyclic groups of order $(q+1, l+1)$ except for ${}^2A_3(2^2)$ it is $\mathbb{Z}_2$, for ${}^2A_3(3^2)$ it is $\mathbb{Z}_3 \times \mathbb{Z}_3 \times \mathbb{Z}_4$, for ${}^2A_5(2^2)$ it is $\mathbb{Z}_2 \times \mathbb{Z}_2 \times \mathbb{Z}_3$.
12. Twisted Simple Groups ${}^2D_l(q^2) l \geq 4$. The Schur Multiplier of ${}^2D_l(q^2) l \geq 4$ is cyclic group of order $(q^{l+1}, 4)$.
13. Twisted Simple Groups ${}^3D_4(q^3)$. The Schur Multiplier of ${}^3D_4(q^4)$ is trivial.
14. Twisted Simple Groups ${}^2E_6(q^2)$. The Schur Multiplier of ${}^2E_6(q^2)$ is of order $(q+1, 3)$ except for ${}^2E_6(2^2)$ it is $\mathbb{Z}_2 \times \mathbb{Z}_2 \times \mathbb{Z}_3$.

The Groups 11–14 are also termed as Steinberg groups.

15. Twisted Groups ${}^2B_2(2^{2m+1}), m \geq 1$(termed as Suzuki groups). The Schur Multiplier is trivial except for ${}^2B_2(2^3)$ it is $\mathbb{Z}_2 \times \mathbb{Z}_2$.
16. Twisted Group ${}^2F_4(2^{2m+1})$ (termed as a Ree Group of type 1). The Schur Multiplier is trivial.
17. Twisted Group ${}^2G_2(3^{2m+1})$ (termed as Ree groups of Type2). The Schur Multiplier is trivial.

The 26 sporadic simple groups as given below.

18. The 5 Mathew groups:

    (i) The Mathew group M_{11} of order $2^4 \cdot 3^2 \cdot 5 \cdot 11$ and with trivial Schur Multiplier.
    (ii) The Mathew group M_{12} of order $2^6 \cdot 3^3 \cdot 5 \cdot 11$ and with Schur Multiplier of order 2.
    (iii) The Mathew group M_{22} of order $2^7 \cdot 3^2 \cdot 5 \cdot 7 \cdot 11$ and with $\mathbb{Z}_{12}$ as Schur Multiplier.

(iv) The Mathew group M_{23} of order $2^7 \cdot 3^2 \cdot 5 \cdot 11 \cdot 23$ and with trivial Schur Multiplier.
(v) The Mathew group M_{24} of order $2^{10} \cdot 3^3 \cdot 5 \cdot 11 \cdot 23$ and with trivial Schur Multipler.

19. The 4 Janko Groups:

    (i) The Janko group J_1 of order $2^3 \cdot 3 \cdot 5 \cdot\ 7 \cdot 11 \cdot 19$ and with trivial Schur Multiplier.
    (ii) The Janko group J_2 of order $2^7 \cdot 3^3 \cdot 5^2 \cdot\ 7$ and with Schur Multiplier of order 2.
    (iii) The Janko group J_3 of order $2^7 \cdot 3^5 \cdot 5 \cdot\ 17 \cdot 19$ and with Schur Multiplier of order 3.
    (iv) The Janko group J_4 of order $2^{21} \cdot 3^3 \cdot 5 \cdot\ 7 \cdot 11^3 \cdot 23 \cdot 29 \cdot 31 \cdot 37 \cdot 43$ and with trivial Schur Multiplier.

20. The 3 Conway Groups:

    (i) The Conway group Co_1 of order $2^{21} \cdot 3^9 \cdot 5^4 \cdot 7^2 \cdot 11 \cdot 13 \cdot 23$ and with Schur Multiplier of order 2.
    (ii) The Conway group Co_2 of order $2^{18} \cdot 3^6 \cdot 5^3 \cdot 7 \cdot 11 \cdot 23$ and with trivial Schur Multiplier.
    (iii) The Conway group Co_3 of order $2^{10} \cdot 3^7 \cdot 5^3 \cdot 7 \cdot 11 \cdot 23$ and with trivial Schur Multiplier.

21. The 3 Fischer Groups:

    (i) The Fischer group Fi_{22} of order $2^{17} \cdot 3^9 \cdot 5^2 \cdot 7 \cdot 11 \cdot 13$ and with Schur Multiplier of order 6.
    (ii) The Fischer group Fi_{23} of order $2^{18} \cdot 3^{13} \cdot 5^2 \cdot 7 \cdot 11 \cdot 13 \cdot 17 \cdot 23$ and with trivial Schur Multiplier.
    (iii) The Fischer group Fi_{24} of order $2^{21} \cdot 3^{16} \cdot 5^2 \cdot 7^3 \cdot 11 \cdot 13 \cdot 17 \cdot 23 \cdot 29$ and with Schur Multiplier of order 3.

22. The Higman–Sims Group HS of order $2^9 \cdot 3^2 \cdot 5^3 \cdot 7 \cdot 11$ and with Schur Multiplier of order 2.
23. The Suzuki Sporadic Simple Group Suz of order $2^{13} \cdot 3^7 \cdot 5^2 \cdot 7 \cdot 11 \cdot 13$ and with Schur Multiplier of order 6.
24. The McLaughlin Group Mc of order $2^7 \cdot 3^6 \cdot 5^3 \cdot 7 \cdot 11$ and with Schur Multiplier of order 3.
25. The Held Group He of order $2^{10} \cdot 3^3 \cdot 5^2 \cdot 7^3 \cdot 17$ and with trivial Schur Multiplier.
26. Lyons Group Ly of order $2^8 \cdot 3^7 \cdot 5^6 \cdot 7 \cdot 11 \cdot 31 \cdot 37 \cdot 67$ and with trivial Schur Multiplier.
27. The Rudavalis Group Ru of order $2^{14} \cdot 3^3 \cdot 5^3 \cdot 7 \cdot 13 \cdot 29$ and with Schur Multiplier of order 2.
28. The O'Non Group $O'N$ of order $2^9 \cdot 3^4 \cdot 5 \cdot 7^3 \cdot 11 \cdot 19 \cdot 31$ and with Multiplier of order 3.

29. The Thompson Group Th of order $2^{15} \cdot 3^{10} \cdot 5^3 \cdot 7^2 \cdot 13 \cdot 19 \cdot 31$ and with trivial Schur Multiplier.
30. The Harada–Norton Group HN of order $2^{14} \cdot 3^6 \cdot 5^6 \cdot 7 \cdot 11 \cdot 19$ and with trivial Schur Multiplier.
31. The Baby Monster Group B of order $2^{41} \cdot 3^{13} \cdot 5^6 \cdot 7^2 \cdot 11 \cdot 13 \cdot 17 \cdot 19 \cdot 23 \cdot 31 \cdot 47$ and with Schur Multiplier of order 2.
32. The Fischer–Griess Monster Group M of order $2^{46} \cdot 3^{20} \cdot 5^9 \cdot 7^6 \cdot 11^2 \cdot 13^3 \cdot 17 \cdot 19 \cdot 23 \cdot 29 \cdot 31 \cdot 41 \cdot 47 \cdot 59 \cdot 71$ and with trivial Schur Multiplier.

Chevalley Groups as Algebraic Groups

All algebraic groups considered are linear algebraic groups. We state the following Theorem without proof (see Lecture notes on Chevalley Groups by Steinberg for a proof) for their convenient use in the following chapter:

Theorem 4.4.23 *(i) A Chevalley group $G(V, K)$ over an algebraically closed field K is a semi-simple algebraic group.*

(ii) $B(V, K)$ is a Borel subgroup of $G(V, K)$ considered as semi-simple algebraic group.

(iii) $H(V, K)$ is maximal Torus.

(iv) The groups $G(V, K)$, $B(V, K)$, $H(V, K)$, and $N(V, K)$ are all realizable over the prime subfield K_0 of K. ♯

In the above Theorem, assume that K is an algebraically closed field of prime characteristic p. Fix a $q = p^m$. Consider the Frobenius map F on $G(V, K)$ induced by $a \mapsto a^q$. Then the Lang map L (see Sect. 4.4 of Algebra 3) from $G(V, K)$ to itself is given by $L(g) = g^{-1}F(g)$. By Lang's theorem, L is surjective map of varieties. The group $G(V, K)^F = L^{-1}(1) = \{g \in G(V, K) \mid F(g) = g\}$ is the corresponding Chevalley group $G(V, F_q)$ defined over F_q. A maximal torus subgroup of $G(V, F_q)$ is a subgroup of the form $T^F = \{t \in T \mid F(t) = t\}$, where T is a F-stable maximal torus of $G(V, K)$. In general a F-stable torus need not be contained in a F-stable Borel subgroup of $G(V, K)$. Thus, a maximal torus of $G(V, F_q) = G(V, K)^F$ need not lie in a Borel subgroup of $G(V, F_q)$. A F-stable maximal torus of $G(V, K)$ is called a **maximally split torus** if it is contained in F-stable Borel subgroup. A maximal torus of $G(V, F_q)$ is called maximally split if it is of the form T^F, where T is a maximally split torus of $G(V, K)$. In turn, it follows that any two maximal tori contained in B^F are conjugate in B^F and any two maximally split torus of $G(V, F_q)$ are conjugate in $G(V, F_q)$. In general, two maximal tori of $G(V, F_q)$ need not be conjugate in $G(V, F_q)$.

Exercises

4.4.1 Let Φ be an irreducible root system with two root lengths and $\Delta = \{\alpha_1, \alpha_2, \ldots, \alpha_l\}$ be a basis of Φ. Show that $\alpha = \sum_{i=1}^{l} n_i\alpha_i$ is a long root if and only if p_Φ divides n_i whenever α_i is short root.
Hint. Use induction and observe it when $l = 2$.

4.4.2. Describe maximally split F-stable torus subgroups of $GL(2, q)$ and also of $SL(2, q)$. Also determine maximally non split F-stable total subgroups of $GL(2, q)$ and also of $SL(2, q)$.

Chapter 5
Representation Theory of Chevalley Groups

This chapter is an introduction to the basic representation theory followed by the representation theory of Chevalley groups including the Steinberg representations, the principal series representations, the discrete series representations, and Deligne–Lusztig virtual characters.

5.1 Language of Representation Theory

In this section, we shall develop the language of representation theory together with the Schur theory for compact and finite groups with special emphasis on Weyl groups (Symmetric groups) and linear groups $GL(2, F_q)$. The reader may also refer to the Chap. 9 of Algebra 2 for the basic representation theory. However, here, our approach is somewhat different.

Let G be a group and V be a complex vector space. A homomorphism ρ from G to $GL(V)$ is called a representation of G. The dimension of V, if finite, is called the degree of the representation. If G is a topological group and V is a Banach space (Hilbert space), then $GL(V)$ is replaced by the group of invertible bounded operators (Uninary operators), and ρ is assumed to be continuous. If both sides are equipped with analytical structures and ρ is holomorphic, then we term this representation as a **holomorphic representation**.

Representations $G \stackrel{\rho}{\to} GL(V)$ and $G \stackrel{\eta}{\to} GL(W)$ are said to be **equivalent** if there is an isomorphism T from V to W such that $To\rho(g) = \eta(g)oT$ for all $g \in G$. A homomorphism T with this property is called an **intertwining operator**. Let $G \stackrel{\rho}{\to} GL(V)$ be a representation. If W is a subspace of V such that $\rho(g)(w) \in W$ for each $g \in G$ and $w \in W$, then $G \stackrel{\rho|_W}{\to} GL(W)$ given by $\rho|_W(g) = \rho(g)|_W$ is called a **subrepresentation** of ρ. By the abuse of language, we also term the subspace W as

R. Lal, *Algebra 4*, Infosys Science Foundation Series,
https://doi.org/10.1007/978-981-16-0475-1_5

a subrepresentation. A representation ρ is said to be an **irreducible representation** if it has no nontrivial proper subrepresentation.

Proposition 5.1.1 *Every finite-dimensional irreducible representation of an abelian group G is of degree 1.*

Proof Let G be an abelian group and $\rho : G \longrightarrow GL(V)$ be an irreducible representation, where V is a finite-dimensional vector space over $\mathbb{C}$. Let $g \in G$. Let λ_g be an eigenvalue of $\rho(g)$, and V_{λ_g} be the corresponding eigenspace. Let $v \in V_{\lambda_g}$. Since G is abelian, $\rho(g)(\rho(h)(v)) = \rho(h)(\rho(g)(v)) = \lambda_g \rho_h(v)$. This means that $\rho(h)(v) \in V_{\lambda_g}$ for all $h \in G$ and $v \in V_{\lambda_g}$. Thus, V_{λ_g} affords a nontrivial subrepresentation of ρ. Since ρ is irreducible, $V_{\lambda_g} = V$. Let v be a nonzero member of V. Then, the subspace $< v >$ generated by v is also invariant under G and it affords a nontrivial subrepresentation of ρ. Hence $V = < v >$. ♯

Operations on Representations

Let $G \stackrel{\rho}{\to} GL(V)$ and $G \stackrel{\eta}{\to} GL(W)$ be representations of G. Then, the representation $G \stackrel{\rho\oplus\eta}{\to} GL(V \oplus W)$ defined by $(\rho \oplus \eta)(g)(v \oplus w) = \rho(g)(v) \oplus \eta(g)(w)$ is called the **direct sum** of ρ and η.

The representation $G \stackrel{\rho\otimes\eta}{\to} GL(V \otimes W)$ defined by $(\rho \otimes \eta)(g)(v \otimes w) = \rho(g)(v) \otimes \eta(g)(w)$ is called the **tensor product** of ρ and η.

The representation $G \stackrel{\bigwedge^r \rho}{\to} GL(\bigwedge^r V)$ defined by $(\bigwedge^r \rho)(g)(v_1 \wedge v_2 \wedge \cdots \wedge v_r) = \rho(g)(v_1) \wedge \rho(g)(v_2) \wedge \cdots \wedge \rho(g)(v_r)$ is called the rth **exterior power** of ρ.

The representation $G \stackrel{S^r(\rho)}{\to} GL(S^r(V))$ defined by $S^r(\rho)(g)(\overline{v_1 \otimes v_2 \otimes \cdots \otimes v_r}) = \overline{\rho(g)(v_1) \otimes \rho(g)(v_2) \otimes \cdots \otimes \rho(g)(v_r)}$ is called the rth **symmetric power** of ρ.

We have a representation $G \stackrel{Hom(\rho,\eta)}{\to} GL(Hom(V, W)) \approx GL(V^\star \otimes W)$ given by $Hom(\rho, \eta)(g)(\phi)(v) = \eta(g)(\phi(\rho(g^{-1})(v)))$ (we have an isomorphism μ from $V^\star \otimes W$ to $Hom(V, W)$ given by $\mu(f \otimes w)(v) = f(v)w$). In particular, we have a representation $G \stackrel{Hom(\rho,1)}{\to} GL(V^\star)$, where 1 is the trivial representation. This representation is called the **dual representation** of ρ and it is denoted by $\rho^\star$. Thus, $\rho^\star(g)(f)(v) = f(\rho(g^{-1})(v))$.

A subrepresentation η of ρ is called a **direct summand** of ρ if there is a subrepresentation μ of ρ such that $\rho = \eta \oplus \mu$. ρ is said to be **indecomposable** if it cannot be expressed as direct sum of two nontrivial proper subrepresentations. By the Remak–Krull–Schmidt Theorem, every representation is direct sum of indecomposable representations and this representation is unique in an obvious sense. Note that an indecomposable representation need not be irreducible. For example, the representation ρ of $\mathbb{R}$ on $\mathbb{C}^2$ given by $\rho(a)(x, y) = ax + y$ is indecomposable but it is not irreducible. It has a subrepresentation $\mathbb{C} \times \{0\}$ which is not a direct summand.

A representation ρ is said to be **completely reducible** also **semi-simple** if it can be expressed as direct sum of irreducible representations. Thus, the representation ρ of $\mathbb{R}$ on $\mathbb{C}^2$ given by $\rho(a)(x, y) = ax + y$ is not completely reducible. For all representations to be completely reducible, the group has to be of special type, e.g., compact group (in particular finite group) on which invariant integral exists. In fact, in Cartan's classification of semi-simple Lie algebras, there was a gap in the argument which was rectified by the Weyl's unitarian trick: "The restriction defines a bijective correspondence from the set of all holomorphic representations of a complex linear Lie group $G \subseteq GL(n, \mathbb{C})$ to the set of all representations of the compact subgroup $K = U(n) \bigcap G$. Further, this correspondence respects equivalence and irreducibility" (Theorem 5.1.47).

Let G be a compact (in particular finite) group. Let $C(G)$ denote the space of all continuous real valued functions on G. Then, there is a unique invariant integral $\int_G$ on $C(G)$ in the sense that the following hold:

(i) $\int_G(\alpha f + \beta f')dg = \alpha \int_G(f) + \beta \int_G(f')$.
(ii) $f \geq 0$ implies that $\int_G f dg \geq 0$.
(iii) $\int_G 1 dg = 1$, where 1 in the LHS is the constant function which takes the value 1 on G.
(iv) $\int_G L_a(f)dg = \int_G f dg = \int_G R_a(f)dg$ for all $a \in G$, where L_a is the left translation given by $L_a(f)(x) = f(ax)$ and R_a is the right translation given by $R_a(f)(x) = f(xa^{-1})$.

The integral introduced above is called the **Haar Integral** on G. A proof for the existence and uniqueness of Haar integral can be found in "Topological Groups" by Pontryagin. For Lie groups, the existence and uniqueness is easy to establish. However, here, we shall give the concrete descriptions of Haar integrals in our examples.

The Haar integral induces a metric d on $C(G)$ which is given by

$$d(f, f') = \int_G \mid f - f' \mid dg.$$

Using Danniell's process, the integral can be extended to a larger class $L^1(G)$ of all integrable functions and $L^1(G)$ is a Banach algebra with respect to the convolution product $\star$ given by

$$(f \star f')(x) = \int_G f(xg^{-1})f'(g)dg.$$

The technique is to complete the metric and interpret the members of the completion as functions. The integral, being continuous linear functional, can be extended to the completion. The integral $\int_G$ on $C(G)$ can be extended to the space of complex valued continuous functions by putting $\int_G(f + if')dg = \int_G f dg + i \int_G f' dg$.

Example 5.1.2 Let G be a finite group (so a compact group). Then, $C(G)$ is the space of all real valued functions and the integral $\int_G$ is given by

$$\int_G f dg = \frac{1}{\mid G \mid} \sum_{g \in G} f(g).$$

Example 5.1.3 Consider $S^1 = \{e^{i\theta} \mid 0 \leq \theta < 2\pi\}$. A continuous map f from S^1 to $\mathbb{R}$ can be realized as a continuous map from $[0, 2\pi]$ to $\mathbb{R}$ such that $f(0) = f(2\pi)$. Define $\int_{S^1}$ by putting $\int_{S^1} f(\theta)dg = \frac{1}{2\pi}\int_0^{2\pi} f(\theta)d\theta$.

Example 5.1.4 Consider the group $SU(2)$ of 2×2 unitary matrices of determinant 1. Then $SU(2)$ can be identified with the group S^3 of unit quaternions. The isomorphism being

$$(x_0, x_1, x_2, x_3) \mapsto \begin{bmatrix} x_0 + ix_1 & x_2 + ix_3 \\ -x_2 + ix_3 & x_0 - ix_1 \end{bmatrix},$$

with $x_0^2 + x_1^2 + x_2^2 + x_3^2 = 1$. We have a coordinatization of S^3 which is given by

$$x_0 = cos\theta, \ \ 0 \leq \theta \leq \pi,$$

$$x_1 = sin\theta cos\phi, \ \ 0 \leq \phi \leq \pi,$$

$$x_2 = sin\theta sin\phi cos\psi, \ \ 0 \leq \psi \leq 2\pi,$$

$$x_3 = sin\theta sin\phi sin\psi, \ \ 0 \leq \psi \leq 2\pi.$$

This coordinatization is not global. The set $\{(cos\theta, \pm sin\theta, 0, 0) \mid 0 \leq \theta \leq \pi\}$ of exceptional points is of Lebesgue measure 0. If we determine the Riemannian metric induced by the Euclidean metric, then the Haar integral can be given by

$$\int_{S^3} f(g)dg = \frac{1}{2\pi^2}\int_0^{\pi}\int_0^{\pi}\int_0^{2\pi} f(\theta, \phi, \psi)sin^2\theta sin\phi d\theta d\phi d\psi.$$

The verification is left to the reader.

The following is the first basic result in the representation theory.

Theorem 5.1.5 *(Maschke's Theorem)Let $G \xrightarrow{\rho} GL(V)$ be a representation of a compact group G. Then every subrepresentation of ρ is a direct summand of ρ.*

Proof Let $G \xrightarrow{\eta} GL(W)$ be a subrepresentation of ρ. Then W is a subspace of V and $\eta(g)(w) = \rho(g)(w) \in W$ for each $g \in G$ and $w \in W$. Let $<, >$ be an inner inner product on V. Define a map $<, >'$ from $V \times V$ to $\mathbb{C}$ by

$$< v, \ w >' = \int_G < \rho(g)(v), \rho(g)(w) > dg.$$

It can be checked that $<, >'$ is an inner product on V such that $< \rho(g)(v), \ \rho(g)(w) >' = < v, \ w >'$ for all $g \in G$. Clearly, $\rho(g)$ is unitary with respect to this inner product. Let $W^{\perp}$ be the orthogonal complement of W with respect to this inner product. Let $v \in W^{\perp}$. Then

$$< w, \ \rho(g)(v) >' = < \rho(g^{-1})(w), \ v >' = 0$$

for all $w \in W$. Hence $\rho(g)(v) \in W^{\perp}$. This gives us a subrepresentation $G \xrightarrow{\mu} GL(W^{\perp})$ of ρ such that $\rho = \eta \oplus \mu$. ♯

Th following Corollary is also immediate:

Corollary 5.1.6 *Every representation of a compact group is equivalent to a uniatary representation.* ♯

Another Proof: Let $V \xrightarrow{p} W$ be a vector space projection. Define a map $\overline{p}$ on V by

$$\overline{p}(v) = \int_G \rho(g)p(\rho(g^{-1})(v))dg.$$

It can be checked that $\overline{p}$ is a linear transformation such that $\overline{p}o\rho(h) = \rho(h)o\overline{p}$ for each $h \in G$. Thus, $Ker\overline{p}$ determines a subrepresentation μ of ρ such that $\rho = \eta \oplus \mu$. ♯

Using the induction on the dimension of V, we get the following Corollary:

Corollary 5.1.7 *Every representation of a compact group is completely reducible in the sense that it is direct sum of irreducible representations.* ♯

Remark 5.1.8 1. The conclusion of Maschke's Theorem is true for representations of finite groups over finite fields whose characteristics do not divide the order of the group. The second proof can be carried out.

2. The conclusion of the Maschke's Theorem is not true for arbitrary noncompact groups or for the representations of finite groups on vector spaces over fields whose characteristic divides the order of the group. For example, the representation ρ of $\mathbb{R}$ on $\mathbb{C}^2$ given by $\rho(a)(x, y) = x + ay$ has a subrepresentation which is not direct summand. However, on some noncompact group, the result can be extended using Weyl's unitary trick.

3. It is a fact that all irreducible representations of a compact group are finite dimensional.

Following are some guiding problems in the representation theory of groups:

1. Classify all indecomposable and irreducible representations of a group.
2. To decompose a representation as direct sum of irreducible representations.
3. The set of equivalence classes of representations of a group G form an algebraic structure under the operations like $\oplus$, $\otimes$, $\wedge$, and S^r. To determine the presentation and the structure of such an algebraic structure (Adam's operations, in particular, are useful in K-Theory).

Example 5.1.9 1. Any homomorphism ρ from G to $GL(1, \mathbb{C})$ is an irreducible representation of degree 1.

2. We have the standard representation (the matrix multiplication) of $GL(n, \mathbb{C})$ on $\mathbb{C}^n$. The Det and the powers of Det give one-dimensional irreducible representations of $GL(n, \mathbb{C})$. In fact $\bigwedge^n \rho = Det$, where ρ is the standard representation. We shall

see as to how the representations of S_n operating on the standard representation of $GL(n, \mathbb{C})$ determine irreducible representations via Schur functors.

3. Consider S_n. The map ρ from S_n to $GL(n, \mathbb{C})$ given by $\rho(p)(e_i) = e_{p(i)}$ gives us a representation. Let $V = \{\sum_{i=1}^{n} a_i e_i \mid \sum_{i=1}^{n} a_i = 0\}$. Then V determines a subrepresentation of ρ. Clearly, $\mathbb{C}^n = V \oplus \mathbb{C}(e_1 + e_2 + \cdots + e_n)$. We shall see that the representation on V and their exterior powers are all irreducible. The trivial representation 1 and the $sign$ representation ($sign(p) = 1$ if p is even and $sign(p) = -1$ if p is odd) are the only two degree 1 representations of S_n. The representation ρ_s of S_n afforded by the subspace V described above is called the **Standard representation** of S_n. Note that V has a nice basis $\{e_1 - e_2, e_2 - e_3, \ldots, e_{n-1} - e_n\}$.

4. The group $SU(2)$ has natural standard representation ρ on $\mathbb{C}^2$ of degree 2 which is given by $\rho(g)([\alpha, \beta]) = [\alpha, \beta]g$. Let $\wp_n$ denote the space consisting of homogeneous polynomials of degree n in $\mathbb{C}[X_1, X_2]$ together with 0. Clearly, $B = \{X_1^r X_2^{n-r} \mid 0 \leq r \leq n\}$ is a basis of $\wp_n$ and so the dimension of $\wp_n$ is $n+1$. Define a representation ρ_n of $SU(2)$ on $\wp_n$ by putting $\rho_n(g)(f(X_1, X_2)) = f([X_1, X_2]g)$. We shall see that all ρ_n are irreducible and any irreducible representation of $SU(2)$ is equivalent to ρ_n for a unique n.

All representations are assumed to be finite dimensional unless stated otherwise.

Now, let us assume that G is finite. Let $\mathbb{C}(G)$ denote the complex group algebra. We have a representation ρ_{reg} of G on $\mathbb{C}(G)$ which is given by $\rho_{reg}(g)(\sum_{h \in G} \alpha_h h) = \sum_{h \in G} \alpha_h gh$. This representation is called the **regular representation** of G. Indeed, $\mathbb{C}(G)$ is the $\mathbb{C}(G)$-module affording the regular representation. The members of G treated as members of $\mathbb{C}(G)$ form a basis of $\mathbb{C}(G)$. Thus, the degree of the regular representation is $\mid G \mid$.

Let ρ be an irreducible representation of G on V. Then V is a simple $\mathbb{C}(G)$-module affording ρ. Let $v_0 \in V$, $v_0 \neq 0$. Then $\mathbb{C}(G)v_0$. is a nontrivial submodule of V. Hence $V = \mathbb{C}(G)v_0$. The map ϕ given by $\phi(a) = av_0$ is a surjective $\mathbb{C}(G)$-module homomorphism from $\mathbb{C}(G)$ to V. By the Maschke's Theorem $\mathbb{C}(G) = Ker\phi \oplus V$. This shows that every irreducible representation of G appears in the regular representation ρ_{reg} of G as a direct summand, and so there are only finitely many irreducible representations $\rho_1, \rho_2, \ldots, \rho_r$ (up to equivalence) of G, each appearing in ρ_{reg} as direct summand. We may assume that $\rho_i \approx \rho_j$ only when $i = j$. Suppose that ρ_i appears n_i times in ρ_{reg}. Then

$$\mathbb{C}(G) = n_1 V_1 \oplus n_2 V_2 \oplus \cdots \oplus n_r V_r, \tag{5.1}$$

where V_i affords the representation ρ_i. Each V_i is simple $\mathbb{C}(G)$-module.

The following Lemma of Schur is the most fundamental Lemma in the representation theory.

Lemma 5.1.10 (Schur) *Let $G \overset{\rho_1}{\to} GL(V_1)$ and $G \overset{\rho_2}{\to} GL(V_2)$ be irreducible represetations of G, where G is a compact group not necessarily finite. Let $V_1 \overset{T}{\to} V_2$ be a linear transformation such that $To\rho_1(g) = \rho_2(g)oT$ for all $g \in G$. Then $T = 0$ or T is an isomorphism. Further, if $V_1 = V_2$ ($\rho_1 = \rho_2$), then there is a $\lambda \in \mathbb{C}$ such that $T(v) = \lambda v$ for all $v \in V_1$.*

Proof Under the hypothesis of the Lemma, $KerT$ affords a subrepresentation of ρ_1 and $ImageT$ affords a subrepresentation of ρ_2. Since ρ_2 is irreducible, $ImageT = 0$ or else $ImageT = V_2$. Thus $T = 0$ or T is surjective. Suppose that $T \neq 0$. Then $KerT \neq V_1$. Since ρ_1 is irreducible, $KerT = 0$. This means that T is an isomorphism.

Finally, suppose that $V_1 = V_2 = V$. Let λ be an eigenvalue of T. Then $Ker(T - \lambda I)$ is a nontrivial subrepresentation of ρ, and since ρ is irreducible, $Ker(T - \lambda I) = V$. This means that $T = \lambda I$. ♯

Remark 5.1.11 Schur Lemma is valid for all unitary representations on Hilbert spaces not necessarily finite dimensional.

Corollary 5.1.12 *If V is a simple $\mathbb{C}(G)$- module, then $End_{\mathbb{C}(G)}(V)$ is a field isomorphic to $\mathbb{C}$.* ♯

If G is a finite group, then using the Schur Lemma and the Eq. 5.1, we see that

$$End_{\mathbb{C}(G)}(\mathbb{C}(G)) \approx M_{n_1}(\mathbb{C}) \times M_{n_2}(\mathbb{C}) \times \cdots \times M_{n_r}(\mathbb{C}). \tag{5.2}$$

Next, given a ring R with identity, the map ϕ from R to $End_R(R, +)$ given by $\phi(\alpha) = f_\alpha$ is an anti-isomorphism of rings, where f_α is given by $f_\alpha(a) = \alpha a$. (any member f of $End_R(R, +)$ is of the form f_α, where $\alpha = f(1)$). Thus, $\mathbb{C}(G)$ is anti-isomorphic to $M_{n_1}(\mathbb{C}) \times M_{n_2}(\mathbb{C}) \times \cdots \times M_{n_r}(\mathbb{C})$. Since $A \mapsto A^t$ is an anti-isomorphism from $M_n(\mathbb{C})$ to itself, $M_{n_1}(\mathbb{C}) \times M_{n_2}(\mathbb{C}) \times \cdots \times M_{n_r}(\mathbb{C})$ is anti-isomorphic to itself. Consequently,

$$\mathbb{C}(G) \approx M_{n_1}(\mathbb{C}) \times M_{n_2}(\mathbb{C}) \times \cdots \times M_{n_r}(\mathbb{C}), \tag{5.3}$$

where r denotes the number of equivalence classes of irreducible representations of G, and n_i is the multiplicity of ρ_i in ρ_{reg}. Comparing the dimensions, we get

$$\mid G \mid = n_1^2 + n_2^2 + \cdots + n_r^2. \tag{5.4}$$

Note that the space W_i of column matrices with n_i rows is a simple $M_{n_i}(\mathbb{C})$-module in obvious manner. In turn, it is also a simple $\mathbb{C}(G)$-module which is isomorphic to V_i. Indeed, $M_{n_i}(\mathbb{C}) = n_i W_i \approx n_i V_i$. Thus, n_i is also the degree of the irreducible representation ρ_i. We may assume $n_1 = 1$ corresponding to the trivial irreducible representation of G.

Let us compare the centers of both sides of 4. Since $Z(M_n(\mathbb{C}))$ is the space of scalar matrices, the dimension of the center of the RHS is r. Consider the center $Z(\mathbb{C}(G))$ of $\mathbb{C}(G)$. An element $\sum_{g \in G} \alpha_g g$ belongs to $Z(\mathbb{C}(G))$ if and only if $h \sum_{g \in G} \alpha_g g h^{-1} = \sum_{g \in G} \alpha_g g$ for all $h \in G$. This amounts to say that $\alpha_g = \alpha_{hgh^{-1}}$ for all $g, h \in G$. In other words, $\alpha_g = \alpha_h$ if and only if g and h are in the same conjugacy class. Let $\{C_1, C_2, \ldots, C_t\}$ be the set of all conjugacy classes of G and $c_i = \sum_{g \in C_i} g$. Then $c_i \in Z(\mathbb{C}(G))$ and all members of $Z(\mathbb{C}(G))$ are linear combinations of $\{c_1, c_2, \ldots, c_t\}$. Since $C_1, C_2, \ldots, C_t$ are pairwise disjoint, $\{c_1, c_2, \ldots, c_t\}$ is linearly independent. Thus, the dimension of the center of LHS is t. This establishes the following Proposition.

Proposition 5.1.13 *The number of irreducible representations of a finite group G over $\mathbb{C}$ is the number of conjugacy classes of G (the number of conjugacy class of G is called the class number of G).* ♯

Example 5.1.14 Consider S_3. The class number of S_3 is 3. Thus, there are 3 irreducible representations of S_3 over $\mathbb{C}$. Let ρ_1, ρ_2, and ρ_3 be irreducible representations of S_3 of degrees $n_1 = 1$, n_2, and n_3, respectively. Then from (5.4), $6 = | S_3 | = 1 + n_2^2 + n_3^2$. This means that $n_2 = 1$ and $n_3 = 2$. The representation ρ_1 is the trivial representation, the representation ρ_2 is the $sign$ representation, and $\rho_3 = \rho_s$ is the standard representation (see Example 5.1.9 (3)).

The number of conjugacy classes of S_n is the number $p(n)$ of partitions of n, as such, there are $p(n)$ irreducible representations of S_n of which the two one-dimensional representations are the trivial and the $sign$ representations.

The Quaternion group Q_8 has 5 conjugacy classes and so it has 5 irreducible representations. The one degree representations are just homomorphisms from Q_8 to $\mathbb{C}^\star$. There are 4 homomorphisms from Q_8 to $\mathbb{C}^\star$. Since $8 = 1 + 1 + 1 + 1 + 2^2$, the 5th representation is of degree 2. The irreducible representation ρ of degree 2 is given by

$$\rho(i) = \begin{bmatrix} i & 0 \\ 0 & -i \end{bmatrix}, \rho(j) = \begin{bmatrix} 0 & 1 \\ -1 & 0 \end{bmatrix}, \rho(k) = \begin{bmatrix} 0 & i \\ i & 0 \end{bmatrix}.$$

We shall show soon that the degrees of irreducible representations divide the order of the group.

Characters of Representations

Let G be a group (not necessarily finite) and $\rho : G \longrightarrow GL(V)$ be a representation of G. Then, the map χ_ρ from G to $\mathbb{C}$ defined by $\chi_\rho(g) = Tr(\rho(g))$ is called a **Character** of G afforded by the representation ρ. If ρ is irreducible, then it is called an irreducible character of G.

1. Characters are class functions in the sense that they are constant on each conjugacy class. This is because similar transformations have same trace. Thus, characters considered as members of $\mathbb{C}(G)$ are members of the center of $\mathbb{C}(G)$.
2. Equivalent representations have same characters: If $\rho : G \longrightarrow GL(V)$ and $\eta : G \longrightarrow GL(W)$ are equivalent representations, then there is an isomorphism T from V to W such that $To\rho(g) = \eta(g)oT$ for all $g \in G$. This means that $Tr(\rho(g)) = Tr(\eta(g))$ for all g.

Following are few examples:

3. If ρ is a representation of G on V, then $\chi_\rho(1) = tr(\rho(1)) = Tr(I_V) = DimV = deg(\rho)$.
4. Consider the regular representation ρ_{reg} of G on $\mathbb{C}(G)$. Let g be a nonidentity element of G. Since $\rho_{reg}(g)(h) \neq h$ for all $h \in G$ and G forms a basis of $\mathbb{C}(G)$, it follows that $\chi_{\rho_{reg}}(g) = 0$ for all $g \neq 1$ and $\chi_{\rho_{reg}}(1) = | G |$.

5. Consider the standard representation ρ_s of S_3. Here, V is the subspace $\{\alpha_1 e_1 + \alpha_2 e_2 + \alpha_3 e_3 \mid \alpha_1 + \alpha_2 + \alpha_3 = 0\}$ of $\mathbb{C}^3$. Clearly, $\{e_1 - e_2, e_2 - e_3\}$ is a basis of V. Now, $\chi_{\rho_s}(I) = deg(\rho_s) = 2$. Since $\rho_s(1,2)(e_1 - e_2) = -(e_1 - e_2)$ and $\rho_s(1,2)(e_2 - e_3) = e_1 - e_3 = (e_1 - e_2) + (e_2 - e_3)$, it follows that $\chi_{\rho_s}(1,2) = 0 = \chi_{\rho_s}(2,3) = \chi_{\rho_s}(1,3)$. Similarly, $\chi_{\rho_s}(1,2,3) = -1 = \chi_{\rho_s}(1,3,2)$.

The following identities can be easily established by looking at traces of the matrix representations of the representations with respect to suitable bases:

6. $\chi_{\rho_1 \oplus \rho_2} = \chi_{\rho_1} + \chi_{\rho_2}$.
7. $\chi_{\rho_1 \otimes \rho_2} = \chi_{\rho_1}\chi_{\rho_2}$.
8. $\chi_{Hom(\rho_1,\rho_2)}(g) = \overline{\chi_{\rho_1}(g)}\chi_{\rho_2}(g)$. In particular, $\chi_{\rho^\star} = \overline{\chi_\rho}$.
9. $\chi_{\wedge^2\rho}(g) = \frac{1}{2}[\chi_\rho(g)^2 - \chi_\rho(g^2)]$.
10. Since $V \otimes V = S^V \oplus \bigwedge^2 V$, $\chi_{\rho\otimes\rho} = \chi_{S^\rho} + \chi_{\wedge^2\rho}$. Thus, $\chi_{S^2\rho}(g) = \frac{1}{2}[\chi_\rho(g)^2 + \chi_\rho(g^2)]$.

Schur Orthogonality Relation

Consider the algebra $\mathbb{C}(G)$ of complex valued continuous function on G. We have an inner product $<, >$ on $\mathbb{C}(G)$ given by given by

$$< f,\ f' > = \int_G f(g)\overline{f'(g)}dg.$$

We may assume, without any loss, that the representation ρ is unitary (Corollary 5.1.6), and as such $\rho(g^{-1}) = \overline{\rho(g)}^t$. Thus, $\chi_\rho(g^{-1}) = \overline{\chi_\rho(g)}$.

Theorem 5.1.15 (Schur) *Let G be a compact group (in particular finite group). Let $\rho : G \longrightarrow GL(V)$ and $\eta : G \longrightarrow GL(W)$ be finite-dimensional unitary representations (indeed, any irreducible unitary representation of G is finite dimensional). Let T be a linear transformation from V to W. Then, the map $\overline{T}$ from V to W given by*

$$\overline{T}(v) = \int_G \eta(g)T(\rho(g^{-1})(v))dg$$

is an intertwining operator in the sense that $\overline{T}o\rho(g) = \eta(g)o\overline{T}$ for all g (it is a G-homomorphism).

Proof $\overline{T}o\rho(h)(v)$
$= \int_G \eta(g)T(\rho(g^{-1})(\rho(h)(v)))dg$
$= \int_G \eta(g)T(\rho(g^{-1}h)(v))dg$
$= \int_G \eta(hk)T(\rho(k^{-1})(v))dk$
$= \eta(h)o\overline{T}(v)$
for all v. This shows that $\overline{T}$ is a G-map. ♯

Corollary 5.1.16 *Let G be a compact group (in particular a finite group). Let $\rho : G \longrightarrow GL(V)$ and $\eta : G \longrightarrow GL(W)$ be finite-dimensional irreducible unitary representations. Let T be a linear transformation from V to W. Then*

(i) $\overline{T} = 0$ *if* $\rho \not\approx \eta$,
(ii) $\overline{T} = \frac{TrT}{DimV} I_V$ *if* $\rho = \eta$.

Proof (i) This is evident from the Schur Lemma and the above result.
(ii) Suppose that $\rho = \eta$ with $V = W$. Then $\overline{T} = \alpha I$ for some α, thanks to Lemma 5.1.10. Consequently,

$$\begin{aligned}\alpha DimV &= Tr\overline{T} = Tr(\int_G \rho(g)T((\rho(g^{-1})))dg = \int_G Tr((\rho(g)T\rho(g)^{-1}))\\ &= \int_G TrTdg = TrT\int_G dg = TrT.\end{aligned}$$

Thus, $\alpha = \frac{TrT}{DimV}$. ♯

Let $\rho : G \longrightarrow GL(V)$ be a representation of V, where V is a vector space over $\mathbb{C}$. Fix a basis $\{x_1, x_2, \ldots, x_n\}$ of V. Let $[\rho_{ij}(g)]$ denote the matrix representation of $\rho(g)$ with respect to this basis. More explicitly $\rho(g)(x_j) = \sum_{i=1}^{n} \rho_{ij}(g)x_i$. The functions ρ_{ij} from G to $\mathbb{C}$, thus obtained, are called the matrix representation functions. Since ρ is unitary, $\rho(g^{-1}) = \rho(g)^{-1} = \overline{\rho(g)}^t$ for each $g \in G$. Consequently, $\rho_{ij}(g^{-1}) = \overline{\rho_{ji}(g)}$ for all i, j and $g \in G$.

Theorem 5.1.17 (Schur Orthogonality Relation) *Let* $\rho : G \longrightarrow GL(V)$ *and* $\eta : G \longrightarrow GL(W)$ *be irreducible representations of a compact group* G *over* $\mathbb{C}$. *Then*
(i) $< \rho_{ji}, \eta_{lk} > = 0$ *if* $\rho \not\approx \eta$ *or* $(j, i) \neq (l, k)$, *and*
(ii) $< \rho_{ji}, \rho_{ji} > = \frac{1}{DimV}$,
where ρ_{ij} *and* η_{kl} *are corresponding matrix representation functions with respect to some choice of bases.*

Proof Let $\{x_1, x_2, \ldots, x_n\}$ be a basis of V and $\{y_1, y_2, \ldots, y_m\}$ be a basis of W. Consider the linear transformation T_{ki} which takes x_i to y_k and the rest of the x_j to 0. Consider the average

$$\begin{aligned}&\overline{T_{ki}}(x_j)\\ &= \int_G \eta(g)(T_{ki}(\rho(g^{-1})(x_j)))dg\\ &= \int_G \eta(g)(T_{ki}(\textstyle\sum_{m=1}^{n} \rho_{mj}(g^{-1})(x_m)))dg\\ &= \int_G \eta(g)(\rho_{ij}(g^{-1})(y_k))dg\\ &= \int_G \rho_{ij}(g^{-1})(\textstyle\sum_{l=1}^{m} \eta_{lk}(g)y_l)dg\\ &= \textstyle\sum_{l=1}^{m}(\int_G \rho_{ij}(g^{-1})\eta_{lk}(g)dg)y_l\\ &= \textstyle\sum_{l=1}^{m}(\int_G \overline{\rho_{ji}(g)}\eta_{lk}(g)dg)y_l\\ &= \textstyle\sum_{l=1}^{m} < \rho_{ji}, \eta_{lk} > y_l \quad \cdots (\star).\end{aligned}$$

Since both the representations are irreducible, if $\rho \not\approx \eta$, then $\overline{T_{ki}} = 0$ and so the coefficient of each y_l is 0. This shows that

$$< \rho_{ji}, \eta_{lk} > = 0$$

for all i, j, l, k.

Now suppose that $\rho = \eta$ and $y_i = x_i$. Then from ($\star$),

$$\overline{T_{ki}}(x_j) = \sum_{l=1}^{n} \overline{< \rho_{ji}, \rho_{lk} >} x_l = \alpha_{ki} x_j$$

for some α_{ki}. Thus, for $l \neq j$,

$$< \rho_{ji}, \ \rho_{lk} > = 0.$$

Similarly, $< \rho_{ji}, \ \rho_{lk} > = 0$ for $i \neq k$.

Now, let us take $k = i$. Then

$$\overline{T_{ii}}(x_j) = \sum_{l=1}^{n} (\int_G \rho_{ij}(g^{-1}) \rho_{li}(g) dg) x_l = \sum_{l=1}^{n} \overline{< \rho_{ji}, \rho_{li} >} x_l = \alpha_{ii} x_j$$

for some α_{ii}. Thus, for $l = j$,

$$\alpha_{ii} = < \rho_{ji}, \rho_{ji} > = \alpha_{jj}.$$

Consequently,

$$n\alpha_{ii} = \sum_{j=1}^{n} \int_G \overline{\rho}_{ji}(g) \rho_{ji}(g) dg = \int_G \rho_{ii}(g^{-1}g) = \int_G 1 dg = 1.$$

This means that

$$< \rho_{ji}, \ \rho_{ji} > = \alpha_{ii} = \frac{1}{DimV}.$$

♯

Corollary 5.1.18 *If ρ is irreducible representation of G over $\mathbb{C}$, then $< \chi_\rho, \ \chi_\rho > = 1$, and if $\rho \not\approx \eta$, then $< \chi_\rho, \ \chi_\eta > = 0$.*

Proof Suppose that $\rho \not\approx \eta$. Then from the above Theorem,

$$< \chi_\rho, \ \chi_\eta > = \int_G \chi_\rho(g) \chi_\eta(g^{-1}) dg = \int_G (\sum_{i=1}^{n} \rho_{ii}(g) \sum_{k=1}^{m} \eta_{kk}(g^{-1})) dg = 0.$$

Further,

$$< \chi_\rho, \ \chi_\rho > = \int_G \chi_\rho(g) \chi_\rho(g^{-1}) dg = \int_G (\sum_{i=1}^{n} \rho_{ii}(g) \sum_{k=1}^{n} \rho_{kk}(g^{-1})) dg = \frac{n}{n} \int_G 1 dg = 1.$$

Corollary 5.1.19 *Two representations are equivalent if and only if they have same characters.*

Proof Let ρ and η be two representations of G. We can find pairwise nonequivalent irredicible representations $\{\rho_1, \rho_2, \ldots, \rho_t\}$ of G and nonnegative integers $a_1, a_2, \ldots, a_t$ and $b_1, b_2, \ldots, b_t$ such that

$$\rho = a_1\rho_1 \oplus a_2\rho_2 \oplus \cdots a_t\rho_t$$

and

$$\eta = b_1\rho_1 \oplus b_2\rho_2 \oplus \cdots \oplus b_t\rho_t.$$

Then

$$\chi_\rho = a_1\chi_{\rho_1} + a_2\chi_{\rho_2} + \cdots + a_t\chi_{\rho_t},$$

and

$$\chi_\eta = b_1\chi_{\rho_1} + b_2\chi_{\rho_2} + \cdots + b_t\chi_{\rho_t},$$

Suppose that $\chi_\rho = \chi_\eta$. Then by the above Corollary, $a_i = <\chi_\rho, \chi_{\rho_i}> = <\chi_\eta, \chi_{\rho_i}> = b_i$ for each i. This shows that $\rho \approx \eta$. The converse is evident. ♯

Corollary 5.1.20 *A representation ρ is irreducible if and only if $<\chi_\rho, \chi_\rho> = 1$.*

Proof Let $\rho = a_1\rho_1 + a_2\rho_2 + \cdots + a_t\rho_t$ be a representation, where $\{\rho_1, \rho_2, \ldots, \rho_t\}$ is a set of pairwise nonequivalent irreducible representations of G. Then $<\chi_\rho, \chi_\rho> = a_1^2 + a_2^2 + \cdots + a_t^2$. Hence $<\chi_\rho, \chi_\rho> = 1$ if and only if for some i, $a_i = 1$ and for $j \neq i$, $a_j = 0$. ♯

The standard ρ_s of S_3 is irreducible, since $<\chi_{\rho_s}, \chi_{\rho_s}> = 1$. Show that the standard representations of S_n are irreducible for each n.

Since the dimension of the space of the class functions on a finite group G is the number of irreducible characters, the following Corollary is immediate.

Corollary 5.1.21 *The set of irreducible characters of G form an orthonormal basis for the space of class functions.* ♯

Let ρ be a irreducible representation of a finite group G over $\mathbb{C}$. Then V is a simple $\mathbb{C}(G)$—module, and we have a homomorphism $\overline{\rho}$ from the ring $\mathbb{C}(G)$ to $End_{\mathbb{C}}(V)$ defined by $\overline{\rho}(\Sigma_{g\in G}\alpha_g g) = \Sigma_{g\in G}\alpha_g\rho(g)$. If $\Sigma_{g\in G}\alpha_g g$ is in the center of $\mathbb{C}(G)$, then $\overline{\rho}(\Sigma_{g\in G}\alpha_g g)$ commutes with $\rho(g)$ for all $g \in G$, and so it belongs to $End_{\mathbb{C}(G)}(V)$. Since V is a simple $\mathbb{C}(G)$—module, members of $End_{\mathbb{C}(G)}(V)$ are multiplications by scalars. Let $\{C_1, C_2, \cdots, C_r\}$ be the set of distinct conjugacy classes of G. Let $u_i = \Sigma_{g\in C_i} g$. Then as observed, $\{u_1, u_2, \cdots, u_r\}$ form a basis for the center of $\mathbb{C}(G)$. From the previous observation $\overline{\rho}(u_i)$ is multiplication by a scalar α_i (say).

Proposition 5.1.22 *Let G be a finite group and ρ be an irreducible representation of G over $\mathbb{C}$. Then, the scalars α_i described in the above paragraph are algebraic integers.*

Proof Let $u_i = \Sigma_{g\in C_i} g$, where $\{C_1, C_2, \cdots, C_r\}$ is the set of all distinct conjugacy classes of G. Let $v \in C_k$ and a_{ij}^k denote the cardinality of the set $X_{ij}^v = \{(g, h) \in C_i \times C_j \mid gh = v\}$. If $w \in C_k$, then there is $x \in G$ such that $w = xvx^{-1}$. The map $(g, h) \rightsquigarrow (xgx^{-1}, xhx^{-1})$ is clearly a bijective map from X_{ij}^{v} to X_{ij}^{w}. Thus, the integer a_{ij}^{k} depends only on i, j, k, and not on the choice of $v \in C_k$. This also shows that

$$u_i u_j = \Sigma_{k=1}^{r} a_{ij}^{k} u_k.$$

Thus,

$$\overline{\rho}(u_i)\overline{\rho}(u_j) = \overline{\rho}(u_i u_j) = \Sigma_{k=1}^{r} a_{ij}^{k} \overline{\rho}(u_k).$$

The left-hand side is multiplication by $\alpha_i \alpha_j$, and the R.H.S. is multiplication by $\Sigma_{k=1}^{r} a_{ij}^{k} \alpha_k$. This shows that

$$\alpha_i \alpha_j = \Sigma_{k=1}^{r} a_{ij}^{k} \alpha_k$$

for all i, j and k. We can take C_1 to be the conjugacy class $\{e\}$, and so $u_1 = e$. Since $\overline{\rho}(u_1) = \rho(e) = I_V$, $\alpha_1 = 1 \neq 0$. The above equation shows that the column vector

$$\begin{bmatrix} \alpha_1 \\ \alpha_2 \\ \cdot \\ \cdot \\ \cdot \\ \alpha_r \end{bmatrix}$$

is an eigenvector of the matrix $[b_{jk}]$, where $b_{jk} = a_{ij}^{k}$, and the corresponding eigenvalue is α_i. Thus, α_i is a root of the monic polynomial $det(xI_r - [b_{jk}])$ whose coefficients are all integers (note that $b_{jk} = a_{ij}^{k}$ are all nonnegative integers). This shows that each α_i is an algebraic integer. ♯

Corollary 5.1.23 *Let G be a finite group and ρ be an irreducible representation over $\mathbb{C}$ of degree n. Let $g \in G$. Let $m = [G : C_G(g)]$ be the number of conjugates to g. Then $\frac{m\chi_\rho(g)}{n}$ is an algebraic integer(observe that every algebraically closed field of characteristic 0 contains the field of algebraic numbers).*

Proof Let C_i be the conjugacy class determined by g. Then, the trace of $\rho(g) =$ the trace of $\rho(x)$ for each $x \in C_i$. Thus, $trace\overline{\rho}(u_i) = m \cdot trace\rho(g) = m \cdot \chi_\rho(g)$. Since $\overline{\rho}$ is multiplication by α_i, and the degree of ρ is n, it follows that $trace\overline{\rho}(u_i) = n \cdot \alpha_i$. Thus, $\alpha_i = \frac{m\chi_\rho(g)}{n}$. The result follows from the above theorem. ♯

Corollary 5.1.24 *The degree of an irreducible representation of a finite group G divides the order of the group.*

Proof Let G be a finite group and ρ be an irreducible representation of G. Let $\{C_1, C_2, \cdots, C_r\}$ be the set of conjugacy classes of G. Put $u_i = \sum_{x \in C_i} x$. Note that ρ induces a ring homomorphism $\overline{\rho}$ from $\mathbb{C}(G)$ to $End(V)$ by putting $\overline{\rho}(\sum_{g \in G} \alpha_g g) = \sum_{g \in G} \alpha_g \rho(g)$. Then $\overline{\rho}(u_i)$ commutes with each $\rho(g)$, since $u_i \in Z(\mathbb{C}(G))$. Since $\mathbb{C}$ is algebraically closed, $\overline{\rho}(u_i)$ is multiplication by a scalar α_i (say). Then

$$m_i \chi_i = Tr\overline{\rho}(u_i) = n\alpha_i,$$

where m_i is the number of elements in C_i, χ_i is the value of the character χ on C_i, and n is the degree of the representation. By the orthogonality relation

$$1 =< \chi_\rho, \chi_\rho > = \frac{1}{|G|}\sum_{g\in G}\chi_\rho(g)\chi_\rho(g^{-1}) = \frac{1}{|G|}\sum_{i=1}^{r} m_i\chi_i\chi_i^\star.$$

Thus, $\frac{|G|}{n} = \sum_{i=1}^{r}\frac{m_i\chi_i}{n}\chi_i^\star$. By Corollary 5.1.23, each $\frac{m_i\chi_i}{n}$ is an algebraic integer. Already $\chi_i^\star$ is an algebraic integer. Since sums and products of algebraic integers are algebraic integers, it follows that $\frac{|G|}{n}$ is an algebraic integer. Since a rational number is an algebraic integer if and only if it is an integer, n divides $|G|$. ♯

A compact group may have infinitely many irreducible representations. However, we have the following important Theorem of Peter–Weyl for an arbitrary compact groups. We use some analysis to prove the Theorem.

Theorem 5.1.25 (Peter-Weyl)*Let G be a compact linear group. Let $\hat{G}$ denote the set obtained by selecting one and only one member from each equivalence class of irreducible representations of G. Then, the set*

$$\Omega = \{\sqrt{deg(\rho)}\rho_{ij} \mid \rho \in \hat{G}, i, j \leq deg(\rho)\}$$

is an orthonormal basis of the Hilbert space $L^2(G)$ of complex valued square integrable functions functions on G, where the inner product on $L^2(G)$ is given by

$$< \alpha, \beta > = \int_G \alpha(g)\overline{\beta(g)}dg.$$

Proof By the Schur orthogonality relation, Ω forms an orthogonal set. Again, the space $\mathbb{C}(G)$ of all continuous functions is dense in the space $L^2(G)$ of square integrable functions, and hence it is sufficient to show that the subspace generated by Ω is dense in $\mathbb{C}(G)$. Since G is a linear group, we can talk of the space $\wp(G)$ of polynomial functions on G. By the Stone-Weirstrass Theorem, $\wp(G)$ is dense in $\mathbb{C}(G)$ with respect to sup norm topology. But sup norm topology is finer than the topology induced by inner product. It follows that $\wp(G)$ is dense in $\mathbb{C}(G)$ with inner product topology. Thus, it is sufficient to show that $\wp(G)$ is contained in the subspace generated by Ω. Let $f \in \wp(G)$. Suppose that the degree of f is m. Let $\wp_m$ denote the space of all polynomials on G of degree at the most m. We have a representation ρ of G on $\wp_m$ which is given by

$$\rho(g)(\phi)([x_{ij}]) = \phi([x_{ij}]\cdot g).$$

Now, ρ is direct sum of finitely many irreducible representations in $\hat{G}$. Further, the map $\phi \mapsto \phi(I)$ is a linear function on $\wp_m(G)$. By the Riesz representation Theorem, there is a $\eta \in \wp_m(G)$ such that $< \phi, \eta > = \phi(I)$. Now, $< \rho(g)(f), \eta > = \rho(g)(f)(I) = f(g)$ for all g. This shows that f is a linear combination of members of Ω. ♯

Remark 5.1.26 The Peter-Weyl Theorem is valid for any compact group but the proof requires the deeper analysis of spectral representations.

$SL(2,\mathbb{C})$, $SU(2)$, and Weyl groups (in particular symmetric groups) are building blocks in the theory of Lie groups and their representations. Indeed, in any semi-simple group, $SL_(2,\mathbb{C})$ and $SU(2)$ are interwoven with Weyl groups in a suitable manner. A little later, we shall study the representations of Weyl groups and S_n, in particular.

Example 5.1.27 Representations of S^1: Since S^1 is commutative all irreducible representations of S^1 are one dimensional and hence it is a continuous homomorphism ρ from S^1 to $\mathbb{C}^\star$. Since S^1 is compact $\rho(S^1)$ is bounded. In particular, $\{\mid \rho(z) \mid^n \mid n \in \mathbb{Z}\}$ is bounded for each $z \in S^1$ This means that ρ is a continuous homomorphism from S^1 to S^1 such that $\mid \rho(z) \mid = 1$ for each $z \in S^1$ and. Continuous homomorphisms from S^1 to S^1 are of the form $z \mapsto z^n$, $n \in \mathbb{Z}$. A consequence of the Peter-Weyl Theorem is the classical theorem of Fourier which asserts that the set of all periodic square integrable functions on $\mathbb{R}$ have Fourier series representations.

Example 5.1.28 Representations of $SU(2)$: Consider the space V_n consisting of 0 and all homogeneous polynomials in $\mathbb{C}[X, Y]$ of degree n. Clearly, $\{X^iY^{n-i} \mid 0 \leq i \leq n\}$ is a basis of V_n. Thus, V_n is of dimension $n + 1$. Define a map ρ_n from $SU(2)$ to $GL(V_n)$ by putting $\rho_n(g)(f)(X, Y) = f([X, Y] \cdot g)$. Then ρ_n is a representation of $SU(2)$ of degree $n + 1$. Let us calculate the character χ_{ρ_n}. Since χ_{ρ_n} is a class function, it is sufficient to calculate its values on representatives of conjugacy classes. Since we can decompose $\mathbb{C}^2$ as direct sum of eigenspaces of g, and the eigenvalues of g are of the form $e^{i\theta}$, g is conjugate to a diagonal matrix

$$\begin{bmatrix} e^{i\theta} & 0 \\ 0 & e^{-i\theta} \end{bmatrix}$$

for some θ. Note that $\begin{bmatrix} e^{i\theta} & 0 \\ 0 & e^{-i\theta} \end{bmatrix}$ is similar to $\begin{bmatrix} e^{-i\theta} & 0 \\ 0 & e^{i\theta} \end{bmatrix}$, since

$$\begin{bmatrix} 0 & -1 \\ 1 & 0 \end{bmatrix} \cdot \begin{bmatrix} e^{i\theta} & 0 \\ 0 & e^{-i\theta} \end{bmatrix} \cdot \begin{bmatrix} 0 & 1 \\ -1 & 0 \end{bmatrix} = \begin{bmatrix} e^{-\theta} & 0 \\ 0 & e^{i\theta} \end{bmatrix}.$$

Thus, a character χ_ρ is determined by an even and 2π periodic function $\chi_\rho(\theta) = Tr(\rho(\begin{bmatrix} e^{i\theta} & 0 \\ 0 & e^{-i\theta} \end{bmatrix}))$.

Indeed, any $g \in SU(2)$ treating as a member of S^3 has a parametric representation as

$$\begin{bmatrix} cos\theta + isin\theta cos\phi & sin\theta sin\phi cos\psi + isin\theta sin\phi sin\psi \\ -sin\theta sin\phi cos\psi + isin\theta sin\phi sin\psi & cos\theta - isin\theta cos\phi \end{bmatrix}.$$

The characteristic equation of g given above is $\lambda^2 - 2cos\theta\lambda + 1 = 0$, and as such the eigenvalues of $g \in SU(2)$ are $\pm e^{i\theta}$. More generally, it also follows that any class function on $SU(2)$ depends only on the parameter θ which is even and 2π periodic. The integral of a class function on $SU(2)$ is given by

$$\int_{SU(2)} f(g)dg = \frac{1}{2\pi^2}\int_0^{2\pi}\int_0^{\pi}\int_0^{\pi} f(\theta,0,0)sin^2\theta sin\phi d\theta d\phi d\psi = \frac{2}{\pi}\int_0^{\pi} f(\theta)sin^2\theta d\theta = \frac{1}{\pi}\int_0^{2\pi} f(\theta)sin^2\theta d\theta.$$

This is the **Weyl Integral Formula** for $SU(2)$.

Thus, the character χ_ρ of a representation ρ is given by

$$\chi_\rho(g) = \frac{2}{\pi}\int_0^{\pi}\chi_\rho(\theta)sin^2\theta d\theta, \text{ where } g \text{ is given as above.}$$

This is called the **Weyl character formula** for $SU(2)$. Consider the representation ρ_n of $SU(2)$ and the corresponding character χ_{ρ_n} given by

$$\chi_{\rho_n}(\begin{bmatrix} e^{i\theta} & 0 \\ 0 & e^{-i\theta}\end{bmatrix} = Tr(\rho_n(\begin{bmatrix} e^{i\theta} & 0 \\ 0 & e^{-i\theta}\end{bmatrix})).$$

Further,

$$[X,Y]\begin{bmatrix} e^{i\theta} & 0 \\ 0 & e^{-i\theta}\end{bmatrix} = [Xe^{i\theta}, Ye^{-i\theta}].$$

Hence

$$\rho_n(\begin{bmatrix} e^{i\theta} & 0 \\ 0 & e^{-i\theta}\end{bmatrix})(X^mY^{n-m}) = (Xe^{i\theta})^m(Ye^{-i\theta})^{n-m} = e^{im\theta}X^me^{-i(n-m)\theta}Y^{n-m} = e^{i(2m-n)\theta}X^mY^{n-m}.$$

Thus,

$$Tr(\rho_n(\begin{bmatrix} e^{i\theta} & 0 \\ 0 & e^{-i\theta}\end{bmatrix})) = \sum_{m=0}^{n} e^{i(2m-n)\theta} = e^{-in\theta}\frac{1-e^{2i(n+1)\theta}}{1-e^{2i\theta}} = \frac{sin(n+1)\theta}{sin\theta}.$$

Consequently,

$$< \chi_{\rho_n}, \chi_{\rho_n} > = \frac{2}{\pi}\int_0^{\pi}\chi_{\rho_n}(\theta)\overline{\chi_{\rho_n}(\theta)}sin^2\theta d\theta = \frac{2}{\pi}\int_0^{\pi} sin^2(n+1)\theta d\theta = 1.$$

This shows that the representation ρ_n is an irreducible representation of $SU(2)$ for each n.

Conversely, we show that these are all irreducible representations of $SU(2)$. Let ρ be an irreducible representation of $SU(2)$ and χ_ρ the corresponding character. Then
$1 =< \chi_\rho, \chi_\rho >$ (by the orthogonality relation)
$= \frac{2}{\pi}\int_0^{\pi}(\chi_\rho(\theta))^2 sin^2\theta d\theta$ (for χ_ρ is a class function)
$= \frac{1}{\pi}\int_0^{2\pi}(\chi_\rho(\theta)sin\theta)^2 d\theta$
$= 2 \parallel \chi_\rho(\theta)sin\theta \parallel^2_{S^1}$

$= 2\sum_{r=-\infty}^{\infty} \mid < \chi_\rho(\theta) sin\theta,\ e^{ir\theta} >_{S^1} \mid^2$ (by Peter- weyl theorem for S^1 and Parswalls formula)
$= 2\sum_{r=-\infty}^{\infty} \mid \frac{1}{2\pi}\int_0^{2\pi} \chi_\rho(\theta) sin\theta e^{-ir\theta} d\theta \mid^2$
$= 2\sum_{r=-\infty}^{\infty} \mid \frac{i}{2\pi}\int_0^{2\pi} \chi_\rho(\theta) sin\theta sinr\theta d\theta \mid^2$
$= 4\sum_{r=1}^{\infty} \mid \frac{1}{4\pi^2}\int_0^{2\pi} \chi_\rho(\theta)\chi_{\rho_{n-1}}(\theta) sin^2\theta d\theta \mid^2$
$= \sum_{r=1}^{\infty} \mid \frac{2}{\pi}\int_0^{\pi} \chi_\rho(\theta)\chi_{\rho_{n-1}}(\theta) sin^2\theta d\theta \mid^2$
$= \sum_{r=0}^{\infty} \mid < \chi_\rho,\ \chi_{\rho_n} > \mid^2$.
Since χ_ρ is irreducible, $\chi_\rho = \chi_{\rho_n}$ for some n. ♯

Next, we shall quickly give relationship between representations of Lie groups and the representations of corresponding Lie algebras. We restrict ourself to linear Lie groups, viz., closed subgroups of general linear groups over $\mathbb{C}$. First, we describe exponential maps and logarithmic maps.

Consider $M_n(\mathbb{C})$. This is a complex Hilbert space with inner product $<,\ >$ given by

$$< A,\ B > = \sum_{i,j} a_{ij}\overline{b_{ij}} = Tr(AB^\star).$$

$\| A \|^2 = \sum_{i,j} \mid a_{ij} \mid^2$, and $\| AB \| \leq \| A \|\ \| B \|$. In other words, $M_n(\mathbb{C})$ is a complex Banach algebra. Define a map *exp* from $M_n(\mathbb{C})$ to $M_n(\mathbb{C})$ by putting

$$exp(A) = I + A + \frac{A^2}{2!} + \cdots = Lim_{n\to\infty}\sum_{r=0}^{n}\frac{A^r}{r!}.$$

By the Weirstrass M-test, the series in the RHS is uniformly absolutely convergent in any bounded domain in $M_n(\mathbb{C})$.

Proposition 5.1.29 *If* $AB = BA$, *then* $exp(A + B) = exp(A)exp(B)$.

Proof Note that

$$\sum_{r=0}^{n}\frac{A^r}{r!}\sum_{r=0}^{n}\frac{B^r}{r!} = \sum_{s=0}^{2n}\frac{(A+B)^s}{s!} - \sum_{i>n\ or\ j>n}\frac{A^i}{i!}\frac{B^j}{j!}.$$

Suppose that $M = max(\| A \|,\ \| B \|)$. Then

$$\| \sum_{i>n\ or\ j>n, i+j\leq 2n}\frac{A^i}{i!}\frac{B^j}{j!} \| \leq \frac{n(n+1)M^{2n}}{n!} = \frac{(M^2)^{n-2}}{(n-2)!}\frac{n+1}{n-1}M^2.$$

Clearly, the RHS tends to 0 as n tends to ∞. This proves that $exp(A + B) = exp(A)exp(B)$. ♯

Corollary 5.1.30 *$exp(A)$ belongs to $GL(n, \mathbb{C})$ for each $A \in M_n(\mathbb{C})$ and $(exp(A))^{-1} = exp(-A)$.*

Proof Since A and $-A$ commute, $I = exp(o) = exp(A + (-A)) = exp(A)exp(-A)$. This shows that $(exp(A))^{-1} = exp(-A)$. ♯

Next, consider the series

$$(A - I) - \frac{(A - I)^2}{2} + \frac{(A_I)^3}{3} - \cdots = \sum_{n=1}^{\infty}(-1)^{n+1}\frac{(A - I)^n}{n}.$$

By the Leibnitz test, the above series is convergent in the domain $D = \{A \mid \| A - I \| < 1\}$. This defines a map Log from D to $M_n(\mathbb{C})$. Clearly, $Log(I) = 0$. The map exp and Log are local inverses to each other. The following Propositions can be easily established by using simple analysis:

Proposition 5.1.31 *(i) $exp(Log(A)) = A$, if $\| A - I \| < 1$.*
(ii) $Log(exp(A)) = A$, if $\| A \| < Log2$.
(iii) $exp(BAB^{-1}) = Bexp(A)B^{-1}$.
(iv) Eigenvalues of A are of the form e^{λ}, where λ is an eigenvalue of A.
(v) $exp(tA)exp(tB) = exp(t(A + B) + \frac{t^2}{2}[A, B] + O(t^2))$, where $[A, B] = AB - BA$.
(vi) $exp(tA)\exp(tB)exp(-tA) = exp(tB + t^2[A, B] + O(t^3))$.
(vii) $[exp(tA), exp(tB)] = exp(t^2[A, B] + O(t^2))$, where $[exp(A), exp(B)] = exp(A)exp(B)exp(A)^{-1}exp(B)^{-1}$. ♯

Corollary 5.1.32 *(i) $Lim_{n\to\infty}(exp(\frac{A}{n})exp(\frac{B}{n}))^n = exp(A + B)$.*
(ii) $Lim_{n\to\infty}[exp(\frac{A}{n}), exp(\frac{B}{n})]^{n^2} = exp[A, B]$.

Proof (i) $Lim_{n\to\infty}(exp(\frac{A}{n})exp(\frac{B}{n}))^n$
$= Lim_{n\to\infty}(exp(\frac{A+B}{n}) + O(\frac{1}{n^2}))^n$ (by Proposition 5.1.31(v))
$= Lim_{n\to\infty}(exp((A + B) + O(\frac{1}{n}))$
$= exp(A + B)$.
(ii) $Lim_{n\to\infty}[exp(\frac{A}{n}), exp(\frac{B}{n})]^{n^2}$
$= Lim_{n\to\infty}(exp(\frac{A+B}{n^2} + O(\frac{1}{n^3})))^{n^2}$ (by Proposition 5.3.31(vii))
$= exp([A, B])$. ♯

The following Theorem is immediate from the above Corollary.

Theorem 5.1.33 *Let G be a linear Lie group contained in $GL(n, \mathbb{C})$. Let $L(G)$ denote the set $\{A \in M_n(\mathbb{C}) \mid exp(tA) \in G\ \forall t \in \mathbb{R}\}$. Then $L(G)$ is a Lie algebra over $\mathbb{R}$.* ♯

Example 5.1.34 $L(G)$ need not be a Lie algebra over $\mathbb{C}$. Consider $U(n) = \{A \mid A^{\star}A = I = AA^{\star}\}$. Thus, $L(U(n)) = \{A \in M_n\mathbb{C}) \mid (exp(A))^{\star} = (exp(A))^{-1} = exp(-A)\}$. Since $(exp(A))^{\star} = exp(A^{\star})$, it follows that $L(U(n)) = \{A \mid A^{\star} = -A\}$. Clearly, $L(U(n))$ is not Lie algebra over $\mathbb{C}$ as it is not even a vector space over $\mathbb{C}$. $L(U(n))$ is denoted by $u_n(\mathbb{C})$ which is real Lie algebra of skew hermitian matrices.

Example 5.1.35 $L(GL(n, \mathbb{C})) = M_n(\mathbb{C}) = gl(n, \mathbb{C})$. This is because $A \in M_n(\mathbb{C})$ implies that $exp(tA) \in GL(n, \mathbb{C})$. $GL(n, \mathbb{R})$ is also a closed subgroup of $GL(n, \mathbb{C})$ with $L(GL(n, \mathbb{R})) = M_n(\mathbb{R}) = gl(n, \mathbb{R})$. $L(SL(n, \mathbb{C})) = sl(n, \mathbb{C})$. $L(O(n, \mathbb{C})) = o(n, \mathbb{C}) = \{A \in M_n(\mathbb{C}) \mid A^t = -A\}$ is the Lie algebra of skew symmetric matrices. $L(Sp(2n, \mathbb{C})) = sp(2n, \mathbb{C})$. Similarly, we can determine Lie algebras of all classical groups.

Let G be a linear Lie group. Then, for all $A \in L(G)$, $exp(t(gAg^{-1})) = gexp(tA)g^{-1}$ belongs to G for all $g \in G$ and $t \in \mathbb{R}$. Thus, $gAg^{-1} \in L(G)$ for all $g \in G$ and $A \in L(G)$. In turn, we get a continuous representation Ad of G on $L(G)$ which is given by $Ad(g)(A) = gAg^{-1}$. This representation is called the **Adjoint representation** of G on its Lie algebra.

Differential Representations

Proposition 5.1.36 *Let G be a linear Lie group and $A \in L(G)$. Then, the map η_A from $\mathbb{R}$ to G given by $\eta_A(t) = exp(tA)$ is a continuous homomorphism from $\mathbb{R}$ to G. Indeed, η_A is also differentiable whose derivative $\eta'_A(0)$ at 0 is A. Conversely, any continuous homomorphism from $\mathbb{R}$ to G is of the form η_A for some $A \in L(G)$.*

Proof Clearly, η_A is a homomorphism. Since $(\mathbb{R}, +)$ and G are topological groups in which the group operations are differentiable, it is sufficient to show that η_A is differentiable at 0 and the derivative at 0 is A. This is evident, since

$$Lim_{n\to\infty}\frac{\eta_A(t) - \eta_A(0)}{t} = Lim_{n\to\infty}\frac{\eta_A(t) - I}{t} = A.$$

Conversely, let ϕ be a continuous homomorphism from $(\mathbb{R}, +)$ to G. Then $\phi(0) = I$. Since ϕ is continuous, there is a $\delta > 0$ such that

$$\| \phi(s) - I \| < 1$$

for all s, $| s | \leq \delta$. Consider

$$B = \int_0^{\delta} \phi(t)dt.$$

We show that $B \in GL(n, \mathbb{C})$. Now,

$$\| B - \delta I \| = \| \int_0^{\delta} (\phi(s) - I)ds \| \leq \int_0^{\delta} \| (\phi(s) - I) \| ds < \delta.$$

Thus, $\| \frac{B}{\delta} - I \| < 1$. This shows that $\frac{B}{\delta}$ is invertible and hence B is invertible. Now,

$$\int_0^{\delta} \phi(s + t)dt = \int_s^{s+\delta} \phi(u)du = \phi(s)\int_0^{\delta} \phi(t)dt = \phi(s)B.$$

The map $s \mapsto \int_s^{s+\delta} \phi(u)du$ is differentiable, since ϕ is continuous (Fundamental Theorem of integral calculus). Thus, the map ϕ is also differentiable, and

$$\phi'(s) = Lim_{t\to 0}\frac{\phi(s+t)-\phi(s)}{t} = \phi(s)A,$$

where

$$A = Lim_{t\to 0}\frac{\phi(t)-\phi(0)}{t} = \phi'(0).$$

This means that $\phi(s) = e^{sA} \in G$ for all s and so $\phi = \eta_A$, $A \in L(G)$. ♯

Now, we describe as to how to realize representations of Lie groups with the help of the representations of their Lie algebras and vice-versa.

Theorem 5.1.37 *Let ρ be a continuous homomorphism from a linear Lie group G to a linear Lie group G'. Then, ρ is differentiable and the differential ρ' induces a Lie algebra homomorphism from $L(G)$ to $L(G')$.*

Proof Let ρ be a continuous homomorphism from a linear Lie group $G \subseteq GL(n, \mathbb{C})$ to a linear Lie group $G' \subseteq GL(m, \mathbb{C})$. Since ρ is a homomorphism, $\rho = L_{\rho(g)} o \rho o L_{g^{-1}}$, where L_g denotes the left translation by g. Since L_g is differentiable at all points, it is sufficient to show that ρ is differentiable at the identity e. First, note that the exponential maps are local diffeomorphisms (by the inverse function theorem) from $L(G)$ to G. Clearly, the exponential map from $L(G)$ to G is the restriction of the exponential map on $M_n(\mathbb{C})$. Let $\{A_1, A_2, \ldots, A_l\}$ be a basis of $L(G)$. The maps $\phi_i = \rho o \eta_{A_i}$ are continuous homomorphisms from $\mathbb{R}$ to G'. From the above Proposition, there exists $B_i \in L(G')$ for each i such that $\eta_{B_i} = \phi_i = \rho o \eta_{A_i}$, $i = 1, 2, \cdots, l$. This means that $\rho(e^{tA_i}) = e^{tB_i}$. Thus,

$$\rho(e^{t_1A_1}e^{t_2A_2}\cdots e^{t_lA_l}) = e^{t_1B_1}e^{t_2B_2}\cdots e^{t_lB_l}.$$

Consider the map ψ from $\mathbb{R}^l$ to G given by

$$\psi(t_1, t_2, \ldots, t_l) = e^{t_1A_1}e^{t_2A_2}\cdots e^{t_lA_l}.$$

Then $\frac{\partial\psi}{\partial t_1}|_{\overline{0}} = A_1$, $\frac{\partial\psi}{\partial t_2}|_{\overline{0}} = A_2, \cdots, \frac{\partial\psi}{\partial t_l}|_{\overline{0}} = A_l$. Since $\{A_1, A_2, \ldots, A_l\}$ is linearly independent. The Jacobian of ψ at $\overline{0}$ is non-singular. By the inverse function theorem, ψ is a local diffeomorphism around $\overline{0}$. Also note that

$$(t_1, t_2, \ldots t_l) \xrightarrow{\mu} (e^{t_1B_1}, e^{t_2B_2}, \ldots, e^{t_lB_l}) \mapsto e^{t_1B_1}e^{t_2B_2}\cdots e^{t_lB_l}$$

is differentiable. It follows that ρ and $\mu o \psi^{-1}$ are same at e. Since μ and ψ^{-1} are differentiable at the corresponding identities, ρ is differentiable at e. Further, we have the induced differential map ρ' from $L(G)$ to $L(G')$ given by the equation

$$\rho(e^{tA}) = e^{t\rho'(A)}, t \in \mathbb{R}, A \in L(G).$$

Corollary 5.1.32 together with the continuity of ρ implies that ρ' is a Lie algebra homomorphism from $L(G)$ to $L(G')$. ♯

Proposition 5.1.38 *Let G be a connected linear Lie group. Then, the subgroup $< exp(L(G)) >$ generated by the image of exp is G itself.*

Proof By the inverse function theorem $exp(L(G))$ contains a neighborhood of identity. Thus, the subgroup $H = < exp(L(G)) >$ of G contains a neighborhood of identity. If $x \in H$, then xU is a neighborhood of x which is contained in H. This shows that H is an open subgroup G. Thus, $H = G - \bigcup_{a \notin H} aH$ is also a closed subgroup of G. Since G is connected $H = G$. $\sharp$

The following Corollary is immediate.

Corollary 5.1.39 *Any continuous homomorphism from a connected linear Lie group G to a linear Lie group G' is uniquely determined by the homomorphism induced on their Lie algebras.* $\sharp$

Conversely, we wish to see as to when every Lie algebra homomorphism from $L(G)$ to $L(G')$ determines a unique continuous homomorphism from G to G'. Indeed, L induces an equivalence from the category of simply connected Lie groups to the category of Lie algebras. For the purpose, we quickly introduce the concept of covering spaces, covering groups, and state some basic facts. The proofs of these facts can be found in the Algebraic Topology by Spanier.

A continuous surjective map $p : E \to X$ is called a **covering projection** if X has an open cover $\{U_\alpha \mid \alpha \in \Lambda\}$ such that each U_α is evenly covered by p in the sense that $p^{-1}(U_\alpha)$ is disjoint union of open sets so that the restriction of p to each one of them is a homeomorphism on to U_α. Clearly, p is an open map and each fiber $p^{-1}(\{x\})$ is a discrete subspace of E. X is called the base space and E is called the total space of p. The base spaces are assumed to be connected.

Proposition 5.1.40 (Unique path lifting prperty) *Let $p : E \to X$ be a covering projection. Let $x_0 \in X$ and σ be a path in X with initial point x_0. This means that σ is a continuous map from the closed interval $[0, 1]$ to X such that $\sigma(0) = x_0$. Let $e_0 \in E$ such that $p(e_0) = x_0$. Then, there is a unique path $\hat{\sigma}$ in E with initial point e_0 such that $po\hat{\sigma} = \sigma$.* $\sharp$

A space X is called a path connected space if any two points of X can be joined by a path. All the spaces are assumed to be path connected. A path in X is called a loop if its starting point and the end points are same. Let $x_0 \in X$. Let $\Omega(X, x_0)$ denote the set of all loops in X based at x_0. Two members σ and τ in $\Omega(X, x_0)$ are said to be homotopic if there is a continuous map H from $[0, 1] \times [0, 1]$ to X such that $H(s, 0) = \sigma(s), H(s, 1) = \tau(s)$ and $H(0, t) = x_0 = H(1, t)$. This simply means that if we treat $\Omega(X, x_0)$ as a subspace of the space $X^{[0,1]}$ (of all paths in X) with compact open topology, then σ and τ are joined by a path in $\Omega(X, x_0)$. The notation $\sigma \approx \tau$ stands to say that σ is homotopic to τ. The relation $\approx$ is an equivalence relation on $\Omega(X, x_0)$. We denote the quotient set $\Omega(X, x_0)/\approx$ by $\pi_1(X, x_0)$. The equivalence class determined by σ is denoted by $[\sigma]$. Let σ and τ be members of $\Omega(X, x_0)$. Define a map $\sigma \star \tau$ from $[0, 1]$ to X by $\sigma \star \tau(t) = \sigma(2t)$ if $t \in [0, \frac{1}{2}]$ and $\sigma \star \tau(t) = \tau(2t - 1)$ if $t \in [\frac{1}{2}, 1]$. Clearly, $\sigma \star \tau$ belongs to $\Omega(X, x_0)$. If $\sigma \approx \sigma'$

and $\tau \approx \tau'$, then it can be shown that $(\sigma \star \tau) \approx (\sigma' \star \tau')$ Thus, we have a product $\cdot$ in $\pi_1(X, x_0)$ defined by by $[\sigma] \cdot [\tau] = [\sigma \star \tau]$. Further, $\pi_1(X, x_0)$ is a group with respect to this product. The element $[\omega_0]$ is the identity, where ω_0 is the constant loop at x_0. The inverse of $[\sigma]$ is $[\sigma^{-1}]$, where $\sigma^{-1}(t) = \sigma(1-t)$. The group $\pi_1(X, x_0)$ is called the **Fundamental group** or the **Homotopy group** of X based at x_0. If X is path connected, then $\pi_1(X, x_0)$ is independent (up to isomorphism) of the base point. A base point preserving continuous map f from the pointed topological space (X, x_0) to a pointed topological space (Y, y_0) defines a homomorphism $f_\sharp$ from $\pi_1(X, x_0)$ to $\pi_1(Y, y_0)$ by putting $f_\sharp([\sigma]) = [fo\sigma]$. Indeed, π_1 defines a functor from the category of pointed topological spaces to the category of groups.

Let $p : E \longrightarrow X$ be a covering space, where X is a path connected space. Let x_0 be a base point of X and e_0 be a point in E over x_0. Let σ be a loop in X at x_0. Then from unique path lifting property of p, there is a unique path $\hat{\sigma}$ in E with initial point e_0. Further, if τ is another loop in X at x_0 which is homotopic to τ and $\hat{\tau}$ is the lifting of τ (with intial point e_0), then $\hat{\sigma}$ is homotopic to $\hat{\tau}$. The fundamental group $\pi_1(X, x_0)$ acts on the fiber $p^{-1}(x_0)$ as follows. Let $[\sigma] \in \pi_1(X, x_0)$ and $e \in p^{-1}(x_0)$. Put $[\sigma]e = \hat{\sigma}(1)$, where $\hat{\sigma}$ is the unique lifting of σ with initial point e. The action is transitive and the isotropy subgroup $\pi_1(X, x_0)_{e_0}$ of the action at e_0 is $p_\sharp(\pi_1(E, e_0))$. It follows that $p_\sharp$ is injective and $[\pi_1(X, x_0) : p_\sharp(E, e_0)] = | p^{-1}(\{x_0\}) |$. We again state a basic theorem in the theory of covering spaces:

Theorem 5.1.41 *Let Y be a connected and locally path connected space and $f : (Y, y_0) \longrightarrow (X, x_0)$ be a continuous map. Then, f can be lifted to a continuous map $\hat{f}$ from (Y, y_0) to (E, e_0) if and only if $f_\sharp(\pi_1(Y, y_0)) \subseteq p_\sharp(\pi_1(E, e_0))$.* $\sharp$

A connected space X is said to be a simply connected space if $\pi_1(X, x_0)$ is trivial. A connected space X is said to be locally simply connected if every point has a fundamental system of neighborhoods consisting of simply connected subspaces. Thus, every manifold (being locally homeomorphic to $\mathbb{R}^n$ for some n) is locally simply connected space. Since every Lie algebra over $\mathbb{R}$ (being homeomorphic to a vector space over $\mathbb{R}$) is simply connected and every Lie group is locally homeomorphic to its Lie algebra, it follows that every Lie group is locally simply connected. We have the following fundamental result in covering space theory:

Proposition 5.1.42 *Let X be a locally simply connected space. Then, there is a covering space $\hat{p} : \hat{X} \longrightarrow X$ with base X which is universal in the sense that given any covering space $p : E \longrightarrow X$ with base X, there is a unique covering map ϕ from $\hat{X}$ to E such that $po\phi = \hat{p}$.* $\sharp$

Clearly, the covering space $\hat{p} : \hat{X} \longrightarrow X$ described in the above proposition is unique up to covering isomorphism. The space $\hat{X}$ is called the **universal covering space** of X. Every covering space over X is in between $\hat{p} : \hat{X} \longrightarrow X$. There is a Galois like correspondence between the subgroups of $\pi_1(X, x_0)$ and intermediary covering projections. Indeed, the group $\pi_1(X, x_0)$ is the group of covering transformations of $\hat{p} : \hat{X} \longrightarrow X$.

Now, let G be a Lie group and $\hat{p} : \hat{G} \longrightarrow G$ be a covering projection with $\hat{G}$ connected and locally path connected. Let $\hat{e}$ be an element in $p^{-1}(\{e\})$. Then from the lifting properties of covering projection, we get a product on $\hat{G}$ with respect to which $\hat{G}$ is a group with $\hat{p}$ a homomorphism. $\hat{p}$ is a local homeomorphism, $\hat{G}$ is a Lie group, and $\hat{p}$ is a Lie group homomorphism. $\hat{e}$ is the identity. Since $p^{-1}(\{e\}) = Ker\,\hat{p}$ is discrete normal subgroup of $\hat{G}$, it is contained in the $\mathbb{Z}(\hat{G})$. Conversely, if H is a discrete normal subgroup of $\hat{G}$, then $\nu : \hat{G} \longrightarrow \hat{G}/H$ is a covering map. Suppose that $\hat{G}$ is the universal covering group of G with $\hat{p} : \hat{G} \longrightarrow G$ the corresponding covering projection. Then $G \approx \hat{G}/Ker\,\hat{p}$, and $Ker\,\hat{p} = \pi_1(G)$. The Group $\hat{G}$ is called the simply connected form of G. If further, $\hat{G}$ is such that $Z(\hat{G})$ is discrete, then $\hat{G}/Z(\hat{G})$ is a semi-simple Lie group and it is called the adjoint form of G. For example, $SU(2)$ is a simply connected form of $SO(3)$ and $SO(3)$ is the adjoint form of $SU(2)$.

Theorem 5.1.43 *Let $G \subseteq GL(n, \mathbb{C})$ be a simply connected linear Lie group and $G' \subseteq GL(m, \mathbb{C})$ be a Lie group. Then every Lie algebra homomorphism from $L(G)$ to $L(G')$ is the differential of a unique continuous homomorphism from G to G'. In particular, L defines an equivalence from the category of simply connected Lie groups to a category of Lie algebras.*

Proof Consider the linear Lie group

$$G \times G' \subseteq GL(n, \mathbb{C}) \times GL(m, \mathbb{C}) \subseteq GL(n+m, \mathbb{C}).$$

Let τ be a Lie algebra homomorphism from $L(G)$ to $L(G')$. Then $L(G \times G')$ is $L(G) \times L(G')$. Indeed, given any continuous homomorphism ρ from $\mathbb{R}$ to $G \times G'$, there is a uninque pair $(A, B) \in L(G) \times L(G')$ such that $\rho(s) = (e^{sA}, e^{sB})$. The graph g_τ of τ is a Lie sub algebra of $L(G) \times L(G')$. Let H be the subgroup of $G \times G'$ generated by $exp(g_\tau)$. Then H is immersed subgroup of $G \times G'$. The first projection map p_1 from H to G induces an isomorphism from the Lie algebra g_τ to $L(G)$. In other words, p_1 is a locally bijective covering map from $< exp(g_\tau) > \ to\ H$. This shows that $p_1 : H \longrightarrow G$ is a covering map. Since G is simply connected, p_1 is an isomorphism. The map $p_2 p_1^{-1}$ is the required homomorphism. ♯

Theorem 5.1.44 *A complex Torus $S^1 \times S^1 \times \cdots S^1$ is a compact connected abelian Lie group. Conversely, every compact connected abelian Lie group is a complex torus.*

Proof Clearly, a complex Torus $\underbrace{S^1 \times S^1 \times \cdots \times S^1}_{n}$, being isomorphic to the closed subgroup of $GL(n, \mathbb{C})$, is a compact connected abelian Lie group. Conversely, let G be a compact connected abelian Lie group. Since

$$Lim_{n \to \infty}[e^{\frac{A}{n}}, e^{\frac{B}{n}}]^{n^2} = e^{[A,B]},$$

it follows that $e^{[A,B]} = I$ for all $A, B \in L(G)$. Thus, $[A, B] = 0$ for all $A, B \in L(G)$. As such, the exponential map exp is a continuous homomorphism from the additive group of $L(G)$ to G. Since $exp(L(G))$ is a open subgroup of G and G is connected, exp is surjective also. Consequently, $L(G)/Ker(exp)$ is topologically isomorphic to G. Since G is compact, $Ker(exp)$ is a lattice in $L(G)$. This means that $L(G)/Ker(exp)$ is isomorphic to $\mathbb{R}^n/\mathbb{Z}^n \approx \underbrace{S^1 \times S^1 \times \cdots \times S^1}_{n}$ for some n. ♯

Theorem 5.1.45 *Every compact connected complex Lie group is a complex Torus.*

Proof Let G be a compact connected complex Lie group. Consider the adjoint representation $Ad : G \longrightarrow GL(L(G)) \subseteq End(L(G))$ given by

$$Ad(g)(A) = gAg^{-1}.$$

Clearly, Ad is a holomorphic representation. Since G is compact, by the maximum principle, Ad is constant. Thus, $gAg^{-1} = A$ for all $g \in G$ and $A \in L(G)$. Consequently, $exp(L(G))$ lies in the center of G. Since G is connected, $exp(L(G))$ generates G. This shows that G is abelian. The result follows from the above Proposition. ♯

Example 5.1.46 Consider $SU(2)$. $A \in L(SU(2))$ if and only if $exp(A)^{-1} = (exp(A))^\star = exp(A^\star)$ and $Det(exp(A)) = 1$. Hence $L(SU(2)) = \{A \in M_n(\mathbb{C}) \mid A^\star = -A \; and \; TrA = 0\} = su(2)$. Thus,

$$su(2) = \{\begin{bmatrix} ix_1 & x_2 + ix_3 \\ -x_2 + ix_3 & -ix_1 \end{bmatrix} \mid x_1, x_2, x_3 \in \mathbb{R}\} \approx \mathbb{R}^3.$$

Clearly, $su(2)$ has a standard basis $\{X_1, X_2, X_3\}$ over $\mathbb{R}$, where

$$X_1 = \frac{1}{2}\begin{bmatrix} 0 & i \\ i & 0 \end{bmatrix}, X_2 = \frac{1}{2}\begin{bmatrix} 0 & 1 \\ -1 & 0 \end{bmatrix}, X_3 = \frac{1}{2}\begin{bmatrix} i & 0 \\ 0 & -i \end{bmatrix}.$$

The basis is so chosen that $[X_1, X_2] = X_3$, $[X_2, X_3] = X_1$, and $[X_3, X_1] = X_2$. The representation $Ad : SU(2) \longrightarrow GL(su(2)) \approx GL(3, \mathbb{R})$ is given by $Ad(g)(A) = gAg^{-1}$. Clearly, $\| A \| = Det(A) = \| Ad(g)(A) \|$ for all $g \in SU(2)$ and $A \in su(2)$. This means that $Ad(g)$ is an orthogonal transformation for each $g \in SU(2)$. Since $SU(2)$ is simply connected, Ad is a surjective homomorphism from $SU(2)$ to $SO(3)$. Clearly, the kernel of Ad is $\{I, -I\}$. Thus, $SU(2)$ is a double cover of $SO(3)$ and $\pi_1(SO(3)) \approx \mathbb{Z}_2$.

Since $SU(2)$ is simply connected, the differential induces a bijective correspondence $\rho \mapsto \rho'$ from the class of representations of $SU(2)$ to the class of representations of $su(2)$ under which irreducible representations correspond. We have already determined all irreducible representations ρ_n of $SU(2)$ as described in Example 5.1.28. We describe the differentials of these representations. Denote the polynomial X^kY^{n-k} in $\wp_n$ by by ϕ_k. Then $\{\phi_k \mid 0 \leq k \leq n\}$ is a basis of $\wp_n$. Since

$$\frac{d}{dt}(\begin{bmatrix} cos\frac{t}{2} & isin\frac{t}{2} \\ isin\frac{t}{2} & cos\frac{t}{2} \end{bmatrix})|_{t=0} = X_1,$$

$$e^{tX_1} = \begin{bmatrix} cos\frac{t}{2} & isin\frac{t}{2} \\ isin\frac{t}{2} & cos\frac{t}{2} \end{bmatrix}.$$

Similarly,

$$e^{tX_2} = \begin{bmatrix} cos\frac{t}{2} & -sin\frac{t}{2} \\ sin\frac{t}{2} & cos\frac{t}{2} \end{bmatrix},$$

and

$$e^{tX_3} = \begin{bmatrix} e^{i\frac{t}{2}} & 0 \\ 0 & e^{-i\frac{t}{2}} \end{bmatrix}.$$

Thus,

$$\rho_n'(X_1)(\phi_k)(X,Y) = \frac{d}{dt}(\phi_k([X,Y]e^{tX_1}))|_{t=0} = \frac{d}{dt}\left(\left(Xcos\frac{t}{2} + iYsin\frac{t}{2}\right)^k \times \left(iXsin\frac{t}{2} + Ycos\frac{t}{2}\right)^{n-k}\right)|_{t=0} = \frac{i}{2}[k\phi_{k-1} + (n-k)\phi_{k+1}].$$

Similarly,

$$\rho_n'(X_2)(\phi_k)(X,Y) = \frac{1}{2}[k\phi_{k-1} - (n-k)\phi_{k+1}],$$

and

$$\rho_n'(X_3)(\phi_k)(X,Y) = \frac{2k-n}{2}i\phi_k.$$

Note that the eigenvalues of $i\rho_n'(X_3)$ are called weights. Further, $\frac{n}{2}$ is the maximal weight.

Real Forms, and Complexifications

Now, we shall see as to how and why the representations of $SU(2)$ are important in the representation theory. Note that $su(2)$ is not a Lie algebra over $\mathbb{C}$.

Let L be a Lie algebra over $\mathbb{R}$. Consider $L_{\mathbb{C}} = L \otimes_{\mathbb{R}} \mathbb{C}$. This is a vector space over the field $\mathbb{C}$ of complex numbers. It can be thought of as extension of scalars. We have an obvious Lie product on $L_{\mathbb{C}}$ with respect to which it is a Lie algebra over $\mathbb{C}$. $L_{\mathbb{C}}$ is called the **complexification** of L. The association $L \to L_{\mathbb{C}}$ is a functor from the category of real Lie algebras to the category of complex Lie algebras which is adjoint to the forgetful functor from complex Lie algebras to the category of real Lie algebras.

The following Theorem is not hard to see if we realize that the complex Lie group $GL(n,\mathbb{C})$ is the complexification of the compact real Lie group $U(n)$.

Theorem 5.1.47 *Let K be a compact connected real Lie group. Then, the complexification $K_{\mathbb{C}}$ of K is an analytic group such that $L(K_{\mathbb{C}})$ is the complexification of $L(K)$. Further, $\pi_1(K) \approx \pi_1(K_{\mathbb{C}})$. Moreover, all complex representations of K*

can be uniquely extended to a holomorphic representation of $K_{\mathbb{C}}$ and under this correspondence, irreducible representations correspond. ♯

Example 5.1.48 Every complex matrix can be expressed as $A + iB$ where A and B are in $u(n)$. Thus, $u(n)_{\mathbb{C}} = gl(n, \mathbb{C})$. Similarly, $gl(n, \mathbb{R})_{\mathbb{C}} = gl(n, \mathbb{C})$, $su(n)_{\mathbb{C}} = sl(n, \mathbb{C}) = sl(n, \mathbb{R})_{\mathbb{C}}$. $U(n)_{\mathbb{C}} = GL(n, \mathbb{C}) = GL(n, \mathbb{R})_{\mathbb{C}}$, $SU(n, \mathbb{C})_{\mathbb{C}} = SL(n, \mathbb{C})$.

Induced Representations

We introduce the concept of induced representation and Frobenius reciprocity for their use in the subsequent sections. Let H be a closed subgroup of a compact (or more generally a locally compact) group G. Let ρ be a representation of G on V. Then, the restriction of ρ to H denoted by $\rho|_H$ is a representation of H. One may observe, by means of an example, that the restriction of an irreducible representation need not be an irreducible representation(the two-dimensional irreducible representation of S_3 when restricted to A_3 is not irreducible). We wish to describe the adjoint to the restriction. Note that $G/^l H$ carries a G-invariant integral. Let η be a unitary representation of H on W. Let V be the space of continuous functions f (vanishing at infinity in case of locally compact group) from G to W such that $f(hg) = \eta(h)(f(g))$ for all $h \in H$ and $g \in G$. Then, we have a representation η^G of G on V which is given by $\eta^G(g)(f)(x) = f(xg^{-1})$. The representation η^G is called the **induced representation** induced by the representation η of the subgroup H. Here, we shall restrict ourself to finite groups, and in this case the induced representation η^G has another convenient description as follows:

Let W be a left $\mathbb{C}(H)$—module affording the representation η of H. $\mathbb{C}(H)$ is a sub algebra of $\mathbb{C}(G)$, and as such $\mathbb{C}(G)$ is a bi-($\mathbb{C}(G)$, $\mathbb{C}(H)$) module. Hence $V = \mathbb{C}(G) \otimes_{\mathbb{C}(H)} W$ is a left $\mathbb{C}(G)$—module. This gives us a representation of G which we denote by η^G, and call it the **induced** representation of G induced by the representation η of the subgroup H of G. Let S be a left transversal to H in G. Then $\mathbb{C}(G)$ as right $\mathbb{C}(H)$—module can be written as $\oplus\Sigma_{x\in S} x\mathbb{C}(H)$. Thus, V can be written as

$$V = \oplus\Sigma_{x\in S} x \otimes W.$$

Consider an element $x \otimes w$, $w \in W$, $x \in S$ in one of the direct summands of V. Suppose that $gx = yh$, $h \in H$ and $y \in S$. Then $\eta^G(g)(x \otimes w) = g(x \otimes w) = gx \otimes w = yh \otimes w = y \otimes hw = y \otimes w'$, where $w' = hw = \eta(h)(w)$. Clearly, $Dim V = Dim W \cdot [G : H]$. Thus, $deg \eta^G = deg(\eta) \cdot [G : H]$. If $\{w_1, w_2, \cdots, w_r\}$ is a basis of W and $S = \{x_1, x_2, \cdots, x_s\}$. Then $\{x_i \otimes w_j \mid 1 \le i \le s, 1 \le j \le r\}$ is a basis of V. The character χ_{η^G} of G is denoted by χ_η^G, and it is called the **induced character**.

Proposition 5.1.49 *Let H be a subgroup of a finite group G. Let ρ be a representation of H. Then*

$$\chi_\rho^G(g) = \frac{1}{\mid H \mid} \Sigma_{x\in G} \chi'_\rho(xgx^{-1}),$$

where $\chi'_\rho = \chi_\rho$ on H and 0 on $G - H$.

Proof Let ρ be a representation of H on W, and $\{w_1, w_2, \cdots, w_r\}$ be a basis of W. Let $S = \{x_1, x_2, \cdots x_s\}$ be a left transversal to H in G. Let $V = \mathbb{C}(G) \otimes_{\mathbb{C}(H)} W$. Then as observed, $\{x_i \otimes w_j\}$ form a basis of V. Now, the basis element $x_i \otimes w_j$ will contribute in the diagonal entry of $\rho^G(g)$ only if $gx_i = x_i h$ for some $h \in H$, and then $\rho^G(g)(x_i \otimes w_j) = x_i \otimes \rho(h)(w_j)$. Thus, for such a x_i, the sum of the contributions in the diagonal entries of $\rho^G(g)$ corresponding to the set $\{x_i \otimes w_j, |, 1 \leq j \leq r\}$ is $\chi_\rho(x_i^{-1} g x_i)$ to the diagonal entry. This shows that

$$\chi_\rho^G(g) = \Sigma_{i=1}^{s} \chi_\rho'(x_i^{-1} g x_i) = \frac{1}{\mid H \mid} \Sigma_{x \in G} \chi_\rho'(x^{-1} g x).$$

The last equality holds because χ_ρ is a class function. ♯

Theorem 5.1.50 *(Frobenius reciprocity Law) Let H be a subgroup of a finite group G. Let ρ be a representation of H, and η be a representation of G. Then*

$$< \chi_\rho^G, \chi_\eta >_G = < \chi_\rho, \chi_{\eta_H} >_H,$$

where η_H denotes the restriction of η to H, $<, >_G$ denotes the inner product in $\mathbb{C}(G)$, and $<, >_H$ denotes the inner product in $\mathbb{C}(H)$.

Proof We have

$$\begin{aligned}
< \chi_\rho^G, \chi_\eta >_G &= \frac{1}{\mid G \mid} \Sigma_{g \in G} \chi_\rho^G(g) \chi_\eta(g^{-1}) \\
&= \frac{1}{\mid G \mid} \Sigma_{g \in G} \left(\frac{1}{\mid H \mid} \Sigma_{x \in G} \chi_\rho'(x^{-1} g x) \chi_\eta(x^{-1} g^{-1} x) \right) \\
&= \frac{1}{\mid H \mid} \Sigma_{y \in G} \chi_\rho'(y) \chi_\eta(y^{-1}) \\
&= \frac{1}{\mid H \mid} \Sigma_{h \in H} \chi_\rho(h) \chi_{\eta_H}(h^{-1}) = < \chi_\rho, \chi_{\eta_H} >_H .
\end{aligned}$$

♯

In practice, to determine irreducible representations of a group G, we look at the representations of some special type of subgroups, induce it to G, and then decompose it into irreducible representations.

Remark 5.1.51 Observe that the Frobenious reciprocity holds even if we replace characters by the class functions.

Example 5.1.52 Let H be a subgroup of a finite group G. Let S be a left transversal to H in G. Let 1_H denote the trivial representation of H, and $V = \oplus \Sigma_{x \in S} x \otimes \mathbb{C}$ the right vector space over $\mathbb{C}$ with $\tilde{S} = \{x \otimes 1 \mid x \in S\}$ as a basis. Then, the induced representation 1_H^G is the representation of G on V given by $1_H^G(g)(x \otimes 1) = gx \otimes 1 = yk \otimes 1 = y \otimes 1_H(k)(1) = y \otimes 1$, where $gx = yk$, $y \in S$, $k \in H$. The character $\chi_{1_H^G}$ is given by $\chi_{1_H^G}(g) = trace 1_H^G(g) = \mid \{x \in S \mid gx = xk \ for \ some \ k \in H\} \mid = \mid \{x \in S \mid gxH = xH\} \mid$ for all $g \in G$. Using the Frobenious reciprocity law,

$$< \chi_{1_H^G}, \chi_{1_G} >_G = < \chi_{1_H}, \chi_{1_G}/H >_H = < \chi_{1_H}, \chi_{1_H} >_H = 1.$$

It follows that the trivial representation 1_G of G appears once and only once in the representation of 1_H^G as the direct sum of irreducible representations. More explicitly, $1_H^G = 1_G \oplus s_H(G)$, where $s_H(G)$ is the representation of G with no summands as 1_G. We shall call $s_H(G)$ the standard representation of G induced by the subgroup H of G. What is $s_{\{e\}}(G)$? Describe the representation $s_H(G)$. Further,

$$\frac{1}{\mid G \mid}\Sigma_{g\in G}\chi_{1_H^G}(g)\chi_{1_G}(g^{-1}) = 1.$$

In turn,

$$\frac{1}{\mid G \mid}\Sigma_{g\in G} \mid \{x \in S \mid gxH = xH\} \mid = 1.$$

Now, let θ be a left transitive action of G on X. Then θ induces a representation ρ of G on the vector space $\mathbb{C}X$ over $\mathbb{C}$ with X as a basis. If H is the isotropy subgroup of the action at a point $x_1 \in X$, then X can be realized as a left transversal to H in G, and the representation ρ is equivalent to 1_H^G. Thus, in this case,

$$\frac{1}{\mid G \mid}\Sigma_{g\in G} \mid \{x \in S \mid g\theta x = x\} \mid = 1.$$

More generally, let G be a finite group which acts on a finite set X through a left action θ. Let V a vector space over $\mathbb{C}$ with X as a basis. The action θ of G on X determines a representation ρ of G on V. Let $\{X_1, X_2, \ldots, X_r\}$ be the set of distinct orbits of the action. The action of G on X induces transitive actions of G on each X_i. Further, $V = \mathbb{C}X = \mathbb{C}X_1 \oplus \mathbb{C}X_2 \oplus \cdots \oplus \mathbb{C}X_r$, and ρ induces representations ρ_i of G on $\mathbb{C}X_i$ for each i with $\rho = \rho_1 \oplus \rho_2 \oplus \cdots \oplus \rho_r$. Let H_i denote the isotropy subgroup of the action at a point $x_i \in X_i$. Then as observed above, $\rho_i = 1_{H_i}^G$ for each i, and the character $\chi_\rho = \Sigma_i \chi_{\rho_i}$. In turn, using the Frobenius reciprocity,

$$< \chi_\rho, \chi_{1_G} >_G = \Sigma_i < \chi_{1_{H_i}}, \chi_{1_G} >_G = \Sigma_i < \chi_{1_{H_i}}, \chi_{1_{H_i}} >_{H_i} = r.$$

We get

$$\frac{1}{\mid G \mid}\Sigma_{g\in G} \mid \{x \in S \mid g\theta x = x\} \mid = r,$$

where r is the number of orbits of the action. Also note that $1_{\{e\}}^G$ is the regular action ρ_{reg} of G.

Example 5.1.53 Let G be a finite group which acts transitively on a finite set X through a left action θ. Let H denote the isotropy subgroup of the action at a point $x \in X$. Then H also acts on X. Let ρ denote the representation of G associated with the action θ. Then $\rho = 1_H^G$, and the representation of H associated with the induced action of H on X is the restriction $1_H^G/H$. It follows from the discussion in the above example that the number r of H—orbits of the action is given by

$$r = < \chi_{1^G_H/H}, \chi_{1_H} >_H = < \chi_{1^G_H}, \chi_{1^G_H} >_G = \frac{1}{|G|}\Sigma_{g\in G}\chi_\rho(g)\chi_\rho(g^{-1}).$$

Further, $\chi_\rho(g) = trace\rho(g)$ is the number of fixed points of the action of the element g on X, and which is the same as the number of fixed points of g^{-1}. This shows that $\chi_\rho(g) = \chi_\rho(g^{-1})$. Thus,

$$r = \frac{1}{|G|}\Sigma_{g\in G}(\chi_\rho(g))^2 = \frac{1}{|G|}\Sigma_{g\in G}(|\{x \in X \mid g\theta x = x\}|)^2.$$

Let us further assume that G acts doubly transitively on X. Then, the isotropy subgroup H of the action at $x_0 \in X$ acts transitively on $X - \{x_0\}$, and so the number of orbits of the action of H on X is 2. From the above discussion, it follows that

$$\frac{1}{|G|}\Sigma_{g\in G}(|\{x \in X \mid g\theta x = x\}|)^2 = 2.$$

Also $1^G_H = 1_G \oplus s_H(G)$, where 1_G does not appear as a summand in $s_H(G)$. Hence

$$< \chi_{s_H(G)}, \chi_{s_H(G)} >_G = 1.$$

This shows that the standard representation $s_H(G)$ of G is irreducible provided that the action of G on X is transitive as well as doubly transitive (For example, S_n or $A_n, n \geq 4$ acts transitively as well as doubly transitively on a set containing n elements).

Let H and K be subgroups of a group G. A subset $KgH = \{kgh \mid k \in K \text{ and } h \in H\}$ is called a **(K, H) double coset**. The set of all (K, H) double cosets will be denoted by $[K, G, H]$. What are $[\{e\}, G, H]$, $[H, G, \{e\}]$ and $[G, G, H]$? It is easily observed that $[K, G, H]$ is a partition of G. The set of representatives obtained by choosing one and only one member from each (K, H)—double coset is called a **double coset representative system**. For convenience, we choose e to represent double coset KH. Let S be a left transversal to H in G. Then $[K, G, H] = \{KsH \mid s \in S\}$. Further, G and so also K acts on S in a natural manner. It follows that the number of K—orbits of this action is precisely the number $|[K, G, H]|$ of (K, H)—double cosets. Using the arguments in Examples 5.1.52 and 5.1.53, we see that

$$|[K, G, H]| = \frac{1}{|K|}\Sigma_{k\in K}|\{x \in S \mid k\theta x = x\}|.$$

Since $k\theta x = x$ if and only if $k \in x^{-1}Hx \bigcap K$, it follows that

$$|[K, G, H]| = \frac{1}{|K|}\Sigma_{x\in S}|H^{x^{-1}} \bigcap K|.$$

We leave the proof of the following two Propositions as exercises.

Proposition 5.1.54 (Mackey restriction formula) *Let H and K be subgroups of G. Let χ be a character of K and χ_K^G the induced character on G. Let $\chi_K^G|_H$ denote the restriction of χ_K^G to H. Then*

$$\chi_K^G|_H = \sum_{x \in S} (\chi^x|_{H \bigcap K^x})^H,$$

where S is the double coset representative system of the pair (H, K) in G, and χ^x is the character of $K^x = x^{-1}Kx$ given by $\chi^x(x^{-1}kx) = \chi(k)$. ♯

The following Proposition is useful in determining the irreducible components of induced representations.

Proposition 5.1.55 *Let η be a character of H and χ a character of K. Then*

$$< \eta^G, \chi^G > = \sum_{x \in S} < \eta|_{H \bigcap K^x}, \chi^x|_{H \bigcap K^x} > .$$

(note that the last inner product is on $\mathbb{C}(H \bigcap K^x)$). ♯

Exercises

5.1.1 Determine all irreducible representations of D_8.
5.1.2 Determine all irreducible representations of a non abelian group of order p^3.
5.1.3 Determine all irreducible representations of A_4.
5.1.4 Describe the induced representation $1_{S_{n-1}}^{S_n}$.
5.1.5 Prove the last two propositions of the section.

5.2 Representations of S_n, and of $GL(2.q)$

In this section, we describe all irreducible representations of symmetric groups and of linear groups $GL(2, q)$. As already observed, the number of equivalence classes of irreducible representations is the class number of the group. In general, we may not be able to parametrize the irreducible representations in terms of the conjugacy classes of the group. However, for symmetric group, we parametrize the irreducible representations with the set of conjugacy classes and give their explicit constructions.

Every conjugacy class of S_n determines and is determined uniquely by a partition $n = \lambda_1 + \lambda_2 + \cdots + \lambda_r$, where $\lambda_1 \geq \lambda_2 \geq \cdots \geq \lambda_r$. This partition will be denoted by λ. The number of partitions of n is denoted by $p(n)$. The function p , thus obtained, is called the **partition function**. It can be seen that $p(n)$ is the coefficient of t^n in the power series representation of $\prod_{r=1}^{\infty} \frac{1}{1-t^r}$. Evidently, the power series is convergent for $| t |< 1$. The partition function has nice arithmetical properties and has been widely studied. Asymptotically $p(n) \approx \frac{1}{4\sqrt{3}n} e^{\pi\sqrt{\frac{2n}{3}}}$.

Let $\lambda = (\lambda_1, \lambda_2, \cdots, \lambda_r)$ be a partition of n. Thus, $\lambda_1 \geq \lambda_2 \geq \cdots \geq \lambda_r \geq 1$ such that $n = \lambda_1 + \lambda_2 + \cdots + \lambda_r$. To λ, we associate a diagram consisting of r rows of boxes such that the i_{th} row contains λ_i boxes and all the rows are aligned from left. Thus, for example, for the partition $\lambda = (5, 4, 2, 2)$ of 13, the associated diagram is as given below

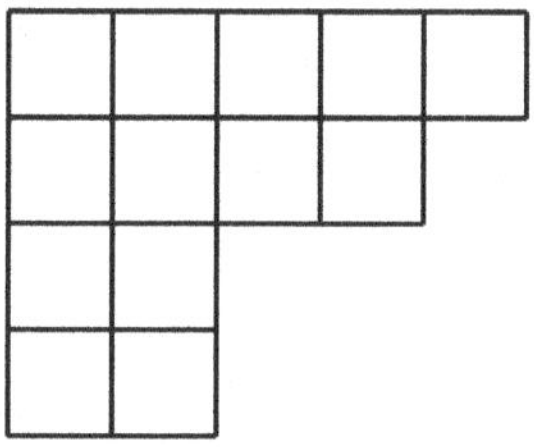

The diagram associated with the partition λ is called the **Young Diagram** of λ. If we interchange the rows and the columns of Young diagram of (λ), then the partition associated with the new diagram is called **conjugate partition** of (λ). Thus, the partition $(\mu) = (4, 4, 2, 2, 1)$ of 13 is conjugate to $(5, 4, 2, 2)$. The partition $(5, 4, 2, 2, 1)$ of 14 is self conjugate. Using the Young diagrams, we shall decompose the regular representation of S_n in to irreducible representations. In a Young diagram associated with a partition λ of n, we fill up the boxes of the diagram by the sequence $1, 2, \ldots, n$ in a sequence starting from the first row to get a tableau which is called the **Young Tableau** of the partition. Thus, the Young Tableau associated with the partition $(5, 4, 2, 2)$ of 13 is

1	2	3	4	5
6	7	8	9	
10	11			
12	13			

Let λ be a partition of n. Let P_λ denote the subgroup $\{p \in S_n \mid p\ fixes\ each\ row\ of\ the\ YoungTableau\ associated\ to\ \lambda\}$ of S_n and Q_λ denote the subgroup $\{p \in S_n \mid p\ fixes\ each\ column\ of\ the\ Young\ Tableau\ associated\ to\ \lambda\}$ of S_n. Consider the elements $a_\lambda = \sum_{p \in P_\lambda} p$ and $b_\lambda = \sum_{p \in Q_\lambda} signp\ p$ of the group algebra $\mathbb{C}(S_n)$. Given any vector space V, $\otimes^n V = \underbrace{V \otimes V \otimes \cdots \otimes V}_{n}$ is a $\mathbb{C}(S_n)$-module by putting $p \cdot (v_1 \otimes v_2 \otimes \cdots \otimes v_n) = v_{p(1)} \otimes v_{p(2)} \otimes \cdots \otimes v_{p(n)}$. This induces a ring homomorphism ρ from $\mathbb{C}(S_n)$ to $End(\otimes^n V)$. Clearly, the image $\rho(a_\lambda)(\otimes^n V)$ is the subspace $Sym^{\lambda_1}(V) \otimes Sym^{\lambda_2}(V) \otimes \cdots \otimes Sym^{\lambda_r}(V)$ of $\otimes^n V$, and the image $\rho(b_\lambda)(\otimes^n V)$ is the subspace $\bigwedge^{\mu_1}(V) \otimes \bigwedge^{\mu_2}(V) \otimes \cdots \otimes \bigwedge^{\mu_t}(V)$ of $\otimes^n V$, where $(\mu_1, \mu_2, \ldots, \mu_t)$ is conjugate to the partition λ. The element $c_\lambda = a_\lambda b_\lambda$ of $\mathbb{C}(S_n)$ is called the **symmetrizer** of the partition λ.

Example 5.2.1 1. Consider $S_2 = \{I, p\}$, where p is the transposition $(1, 2)$. We have two partitions $\lambda = (2)$ and $\mu = (1, 1)$ of 2. They are conjugate to each other. The corresponding Young Tableau are given below

1	2

1
2

Evidently, a_λ is the element $I + p$ of $\mathbb{C}(S_2)$, $b_\lambda = 1, c_\lambda = a_\lambda. c_\lambda^2 = 2c_\lambda = 2(I + p), a_\mu = 1, b_\mu = I - p = c_\mu$, and $c_\mu^2 = 2(I - p) = 2c_\mu$. Also $(a + bp)(I + p) = (a + b)(I + p)$ and $(a + bp)(I - p) = (a - b)(I - p)$. Since $\{I + p, I - p\}$ form a $\mathbb{C}$-basis of $\mathbb{C}(S_2)$, $\mathbb{C}(S_2) = \mathbb{C}(I + p) \oplus \mathbb{C}(I - p)$ is the decomposition of $\mathbb{C}(S_2)$ as direct sum of simple submodules. Since $(I + p)p = I + p$ and $(I - p)p = -(I - p)$, $\mathbb{C}(I + p)$ corresponds to the trivial irreducible representation and $\mathbb{C}(I - p)$ corresponds to the sign representation of S_2

2. Consider S_3. We have three partitions $\lambda = (3),\ \mu = (2, 1),\ and\ \nu = (1, 1, 1)$. The corresponding Young Tableau are as follows:

1	2	3

1	2
3	

1
2
3

Further, $a_\lambda = \sum_{p \in S_3} p,\ b_\lambda = 1,\ c_\lambda = a_\lambda,\ c_\lambda^2 = 6c_\lambda,\ a_\mu = I + (1, 2),\ b_\mu = I - (1, 3),\ c_\mu = a_\mu b_\mu = I - (1, 3) + (1, 2) - (1, 3, 2),\ c_\mu^2 = 3c_\mu,\ a_\nu = 1,\ b_\nu = \sum_{p \in S_3} (sign(p))p,\ c_\nu = b_\nu,\ and\ c_\nu^2 = 6c_\nu.$

The submodule $\mathbb{C}(S_3)c_\lambda = \mathbb{C}c_\lambda$ is the one-dimensional simple submodule of $\mathbb{C}(S_3)$ affording the trivial representation, since $pc_\lambda = c_\lambda$ for all $p \in S_3$. Similarly, $\mathbb{C}(S_3)c_\nu = \mathbb{C}c_\nu$ and $pc_\nu = sign(p)c_\nu$. Hence $\mathbb{C}(S_3)c_\nu$ affords the sign representation of S_3. Finally, it can be observed that the submodule $\mathbb{C}(S_3)c_\mu$ affords the standard irreducible representation ρ_s of S_3 and $\mathbb{C}(S_3)$ as a module over itself is isomorphic to $\mathbb{C}c_\lambda \oplus \mathbb{C}c_\nu \oplus \mathbb{C}(S_3)c_\mu \oplus \mathbb{C}(S_3)c_\mu$.

More generally, we have the following Theorem:

Theorem 5.2.2 *Consider S_n. Let λ be a partition of n. Then $\mathbb{C}(S_n)c_\lambda$ is simple $\mathbb{C}(S_n)$-module affording an irreducible representation ρ_λ of S_n. Conversely, every irreducible representation of S_n is of the form ρ_λ for a unique partition λ of n. $c_\lambda^2 = m_\lambda c_\lambda$, where $m_\lambda = \frac{n!}{deg\rho_\lambda}$.*

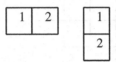

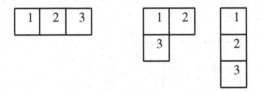

We need a Lemma to prove the Theorem.

Lemma 5.2.3 *(i)* $pc_\lambda sign(q)q = c_\lambda$ *for all* $p \in P_\lambda$ *and* $q \in Q_\lambda$.

(ii) $P_\lambda \bigcap Q_\lambda = \{I\}$. *Further,* $g \in \mathbb{C}(S_n) - P_\lambda Q_\lambda$ *if and only if* $g = \tau g q$ *for some transposition* $\tau \in P_\lambda$ *and* $q \in Q_\lambda$.

(iii) Let $x = \sum_{g \in S_n} \alpha_g g$ *be a member of* $\mathbb{C}(S_n)$. *Then* $pxsign(q)q = x$ *for all* $p \in P_\lambda$ *and* $q \in Q_\lambda$ *if and only if* $x = \alpha c_\lambda$ *for some* $\alpha \in \mathbb{C}$.

(iv) Let $x \in \mathbb{C}(S_n)$. *Then* $c_\lambda x c_\lambda = \alpha_x c_\lambda$ *for some scalar* α_x.

(v) $c_\lambda^2 = m_\lambda c_\lambda$ *for some scalar* m_λ.

(vi) Suppose that $\lambda > \mu$ *under the lexicographic order. Then* $a_\lambda x b_\mu = 0$ *for all* $x \in \mathbb{C}(S_n)$.

(vii) If $\lambda > \mu$, *then* $c_\lambda c_\mu = 0$.

Proof (i) Note that $pa_\lambda = a_\lambda = a_\lambda p$ and $sign(q)qb_\lambda = b_\lambda = b_\lambda sign(q)q$ for all $p \in P_\lambda$ and $q \in Q_\lambda$. Consequently, $pc_\lambda sign(q)q = pa_\lambda b_\lambda sign(q)q = a_\lambda b_\lambda = c_\lambda$ for all partitions λ.

(ii) Since identity is the only permutation which fixes each row and each column of a Young tableau, $P_\lambda \bigcap Q_\lambda = \{I\}$. Next, let $g = \tau g q$, where τ is a transposition in P_λ and $q \in Q_\lambda$. Suppose that $g = pq'$, where $p \in P_\lambda$ and $q' \in Q_\lambda$. Then $pq' = \tau p q' q$. But, then $p^{-1}\tau p = q'qq'^{-1}$ belongs to $P_\lambda \bigcap Q_\lambda = \{I\}$. This is a contradiction, since $\tau \neq I$. This shows that $g \notin P_\lambda Q_\lambda$. Conversely, suppose that $g^{-1}\tau g \notin Q_\lambda$ for any transposition $\tau \in P_\lambda$. Let T denote the Young Tableau associated with the partition λ. If $h \in S_n$, then hT will denote the Tableau obtained by replacing i with $h(i)$ in the Tableau T. Since $g^{-1}\tau g \notin Q_\lambda$ for any transposition $\tau \in P_\lambda$, there is a $p_1 \in P_\lambda$ and a $q_1 \in gQ_\lambda g^{-1}$ such that $p_1 T$ and $q_1 g T$ have same first row. Repeating the argument successively, we see that there is a $p \in P_\lambda$ and $q \in gQ_\lambda g^{-1}$ such that $pT = qgT$. This means that $p = qg$. But, then $g = pg^{-1}q^{-1}g$ belongs to $P_\lambda Q_\lambda$. This proves (ii).

(iii) One way implication follows from (i). Let $x = \sum_{g \in S_n} \alpha_g g$ be a member of $\mathbb{C}(S_n)$ such that $pxsign(q)q = x$ for all $p \in P_\lambda$ and $q \in Q_\lambda$. We show that $x = \alpha c_\lambda$ for some scalar α. Since $p(\sum_{g \in S_n} \alpha_g g)sign(q)q = \sum_{g \in S_n} \alpha_g g$ for all $p \in P_\lambda$ and $q \in Q_\lambda$, $\alpha_{pgq} = sign(q)\alpha_g$ for all $p \in P_\lambda$, $q \in Q_\lambda$, and $g \in S_n$. This means that $\alpha_{pq} = sign(q)\alpha_1$ for all $p \in P_\lambda$ and $q \in Q_\lambda$. Thus, $x = \alpha_1 \sum_{p \in P_\lambda, q \in Q_\lambda} psign(q)q + \sum_{g \notin P_\lambda Q_\lambda} \alpha_g g = \alpha_1 c_\lambda + \sum_{g \notin P_\lambda Q_\lambda} \alpha_g g$. It is sufficient, therefore, to show that $\alpha_g = 0$ for all $g \notin P_\lambda Q_\lambda$. If $g \notin P_\lambda Q_\lambda$, then from (ii), it follows that $g = \tau g q$ for some transposition τ in P_λ and $q \in Q_\lambda$. From the earlier observation, $\alpha_g = -\alpha_g$. This means that $\alpha_g = 0$.

(iv) Evidently, $pc_\lambda x c_\lambda sign(q)q = c_\lambda x c_\lambda$. The result follows from (iii).

(v) Follows from (iv) by putting $x = 1$.

(vi) It is sufficient to show that $a_\lambda g b_\mu = 0$ for all $g \in S_n$. Let T^λ denote the Young tableau associated with λ and T^μ that associated with μ. Then gT^μ is the Tableau which can be used to construct $gb_\mu g^{-1}$. Since $\lambda > \mu$, there is a pair of natural numbers which lie in same row of T^λ and also in the same column of gT^μ. If τ is the transposition determined by the pair of natural numbers, then $a_\lambda \tau = a_\lambda$, $\tau g b_\mu g^{-1} = -gb_\mu g^{-1}$. Hence $a_\lambda g b_\mu g^{-1} = a_\lambda \tau\tau g b_\mu g^{-1} = -a_\lambda g b_\mu g^{-1}$. This means that $a_\lambda g b_\mu g^{-1} = 0$ or equivalently, $a_\lambda g b_\mu = 0$.

(vii) Follows from (vi) by putting $x = b_\lambda a_\mu$. ♯

Proof of Theorem 5.2.2. By Lemma 5.2.3 (iv), $c_\lambda \mathbb{C}(S_n)c_\lambda \subseteq \mathbb{C}c_\lambda$. Clearly, $\mathbb{C}(S_n)c_\lambda$ is a submodule of $C(S_n)$. Let W be a submodule of $\mathbb{C}(S_n)c_\lambda$. Then $c_\lambda W$ (being a subspace consisting of scalar multiples of c_λ) is $\mathbb{C}c_\lambda$ or it is $\{0\}$. Suppose that $c_\lambda W = \mathbb{C}c_\lambda$. Since W is a $\mathbb{C}(S_n)$-module, $\mathbb{C}c_\lambda \subseteq W$ and, in particular, $c_\lambda \in W$. Thus, in this case, $\mathbb{C}(S_n)c_\lambda = W$. Next, suppose that $c_\lambda W$ is $\{0\}$. Then $W \cdot W \subseteq \mathbb{C}(S_n)c_\lambda W = \{0\}$. Since W is a submodule of $\mathbb{C}(S_n)$, there is a $w \in \mathbb{C}(S_n)$ with $w = w^2$ such that $W = \mathbb{C}(S_n)w$. It follows that $W = \{0\}$. Thus, $\mathbb{C}(S_n)c_\lambda$ is a simple submodule of $\mathbb{C}(S_n)$ (note that $\mathbb{C}(S_n)c_\lambda \neq \{0\}$) affording an irreducible representation of S_n.

Now, suppose that $\lambda > \mu$. Since $c_\lambda \mathbb{C}(S_n)c_\lambda = \mathbb{C}c_\lambda \neq \{0\}$ but $c_\lambda \mathbb{C}(S_n)c_\mu = \{0\}$ (Lemma 5.2.3), $\mathbb{C}(S_n)c_\lambda \not\approx \mathbb{C}(S_n)c_\mu$. Since the number of equivalence classes of simple $\mathbb{C}(S_n)$-modules is equal to the number of partitions of n, any simple $\mathbb{C}(S_n)$-module is isomorphic to one and only one member of the set $\{\mathbb{C}(S_n)c_\lambda \mid \lambda\ is\ a\ partition\ of\ n\}$. This determines all irreducible representations of S_n.

Finally, we show that the degree $deg(\rho_\lambda)$ of the irreducible representation ρ_λ afforded by the module $\mathbb{C}(S_n)c_\lambda$ is $\frac{n!}{m_\lambda}$, where m_λ is the scalar given in Lemma 5.2.3(v). Consider the right multiplication r_{c_λ} on $\mathbb{C}(S_n)$ by c_λ. Clearly, it is a $\mathbb{C}(S_n)$-module homomorphism whose image is $\mathbb{C}(S_n)c_\lambda$. By Lemma 5.2.3(v), r_{c_λ} is multiplication by the scalar m_λ on $\mathbb{C}(S_n)c_\lambda$. Thus, $Tr(r_{c_\lambda}) = m_\lambda Dim(\mathbb{C}(S_n)c_\lambda) = m_\lambda deg(\rho_\lambda)$. On the other hand, the coefficient of $g \in S_n$ in $g \cdot c_\lambda$ is 1. This means that $Tr(r_{c_\lambda}) = \mid S_n \mid = n!$. The result follows. ♯

Corollary 5.2.4 *As a* $\mathbb{C}(S_n)$*-module*

$$\mathbb{C}(S_n) \approx \oplus \sum_{\lambda \in P(n)} \frac{n!}{m_\lambda} \mathbb{C}(S_n)c_\lambda,$$

and as a ring

$$\mathbb{C}(S_n) \approx \prod_{\lambda \in P(n)} M_{\frac{n!}{m_\lambda}}(\mathbb{C}),$$

where $P(n)$ *is the set of partitions of* n.

Proof Follows from the above Theorem and the equations (1) and (3) of Sect. 5.1. ♯

Remark 5.2.5 Since $c_\lambda^2 = m_\lambda c_\lambda$ is integral multiple of c_λ, $c_\lambda \in \mathbb{Q}(S_n)$. Thus, all irreducible representations of S_n are realizable over $\mathbb{Q}$.

Frobenius Character Formula

Next, we state the Frobenius character formula for the character χ_λ of the representation ρ_λ associated with the partition λ. Let Σ_n denote the set $\{\alpha = (\alpha_1, \alpha_2, \ldots, \alpha_n) \mid \alpha_i \geq 0\ and\ \sum_{i=1}^{n} i\alpha_i = n\}$. Let C_α denote the conjugacy class of S_n consisting of permutations which are products of α_1 cycles of length 1, α_2 cycles of length 2, $\cdots$, and α_n cycles of length n (all pairwise disjoint). Every conjugacy class of S_n is uniquely represented by C_α for some $\alpha \in \Sigma_n$. Consider the polynomial algebra $\mathbb{C}[X_1, X_2, \ldots, X_n]$ in n variables. Let $P_i(X)$ denote the polynomial $X_1^i + X_2^i + \cdots + X_n^i$ and $\Delta(X)$ denote the polynomial $\prod_{1 \leq i < j \leq n}(X_i - X_j)$. Let $[f(X_1, X_2, \ldots, X_n)]_{(t_1, t_2, \ldots, t_n)}$ denote the coefficient of $X_1^{t_1} X_2^{t_2} \cdots X_n^{t_n}$ in the polynomial $f(X_1, X_2, \ldots, X_n)$. Let $\lambda = (\lambda_1, \lambda_2, \ldots, \lambda_r)$ be a partition of n. Extend it to a sequence $(\lambda_1, \lambda_2, \cdots, \lambda_n)$ by putting $\lambda_i = 0$ for all $i \geq r+1$. Put $l_1^\lambda = \lambda_1 + n - 1, l_2^\lambda = \lambda_2 + n - 2, \ldots, l_n^\lambda = \lambda_n$.

Theorem 5.2.6 (Frobenius Character Formula) $\chi_\lambda(C_\alpha) = [\Delta(X) \prod_{i=1}^n P_i(X)^{\alpha_i})]_{(l_1^\lambda, l_2^\lambda, \ldots, l_n^\lambda)}$. ♯

The proof of the formula involves pain staking computations and we leave it. Let us have some applications and examples.

Consider S_3 and $\lambda = (2, 1)$. So $\lambda_1 = 2$, $\lambda_2 = 1$, and $\lambda_3 = 0$. The representation ρ_λ is the standard representation. We have three conjugacy classes, $\{I\} = C_\alpha$ corresponding to the sequence $\alpha = (3, 0, 0)$, $C_\beta = \{(1, 2), (2, 3), (1, 3)\}$ corresponding to $\beta = (0, 3, 0)$, and $C_\gamma = \{(1, 2, 3), (1, 3, 2)\}$ corresponding to the sequence $\gamma = (0, 0, 3)$. Clearly, $l_1^\lambda = 2 + 3 - 1 = 4$, $l_2^\lambda = 1 + 1 = 2$, $l_3^\lambda = 0$. $\chi_\lambda(C_\alpha)$ is the coefficient of $X_1^4 X_2^2$ in $[(X_1 - X_2)(X_2 - X_3)(X_1 - X_3)(X_1 + X_2 + X_3)^3 (X_1^2 + X_2^2 + X_3^2)^0 (X_1^3 + X_2^3 + X_3^3)^0]$. Thus, $\chi_\lambda(C_\alpha) = 2 = deg(\rho_\lambda)$. $\chi_\lambda(C_\beta)$ is the coefficient of $X_1^4 X_2^2$ in $[(X_1 - X_2)(X_2 - X_3)(X_1 - X_3)(X_1 + X_2 + X_3)^0 (X_1^2 + X_2^2 + X_3^2)^3 (X_1^3 + X_2^3 + X_3^3)^0]$. Thus, $\chi_\lambda(C_\beta) = 0$. Similarly, $\chi_\lambda(C_\gamma) = -1$. Note that this agrees with our earlier computation of the character of the standard representation ρ_s of S_3.

Corollary 5.2.7 *Let λ be a partition of n. Then*

$$deg(\rho_\lambda) = \frac{n!}{l_1^\lambda! l_2^\lambda! \cdots l_n^\lambda!} \prod_{1 \leq i < j \leq r} (l_i^\lambda - l_j^\lambda).$$

Proof Since $deg(\rho_\lambda) = \chi_\lambda(I)$ and the conjugacy class of I corresponds to the sequence $(n, 0, \ldots, 0)$ (I is product of n disjoint cycles of length 1), $deg(\rho_\lambda)$ is the coefficient of $X_1^{l_1^\lambda} X_2^{l_2^\lambda} \cdots X_r^{l_r^\lambda}$ in the polynomial $\prod_{1 \leq i < j \leq r}(X_i - X_j) \cdot (X_1 + X_2 + \cdots + X_n)^n$. Now, $\prod_{1 \leq i < j \leq n}(X_i - X_j)$ is a suitable Vandermonde determinant, and as such,

$$\prod_{1 \leq i < j \leq n} (X_i - X_j) = \sum_{p \in S_n} sign(p) X_1^{(p(n)-1)} X_2^{(p(n-1)-1)} \cdot X_n^{(p(1)-1)}.$$

Further,

$$(X_1 + X_2 + \cdots + X_n)^n = \sum_{r_1 + r_2 + \cdots + r_n = n} \frac{n!}{r_1! r_2! \cdots r_n!} X_1^{r_1} X_2^{r_2} \cdots X_n^{r_n}$$

(use induction on n). Clearly, the coefficient of $X_1^{l_1^\lambda} X_2^{l_2^\lambda} \cdots X_n^{l_n^\lambda}$ in

$$(\sum_{p \in S_n} sign(p) X_1^{(p(n)-1)} X_2^{(p(n-1)-1)} X_n^{(p(1)-1)})(\sum_{r_1+r_2+\cdots+r_n=n} \frac{n!}{r_1!r_2!\cdots r_n!} X_1^{r_1} X_2^{r_2} \cdots X_n^{r_n})$$

is

$$\sum sign(p) \frac{n!}{(l_1^\lambda - p(n) + 1)!(l_2^\lambda - p(n-1) + 1)! \cdots (l_n^\lambda - p(1) + 1)!},$$

where the sum is taken over all permutations $p \in S_n$ for which $l_{n-i+1}^\lambda - p(i) + 1 \geq 0$ for all i. We can rewrite it as

$$\frac{n!}{l_1^\lambda! l_2^\lambda! \cdots l_n^\lambda!} \sum_{p \in S_n | p(i)-1 \leq l_{n-i+1}^\lambda \forall i} sign(p) \prod_{i=1}^{n} l_i^\lambda (l_i^\lambda - 1) \cdots (l_i^\lambda - p(n-i+1) + 2)$$

which, in turn, equals to $\frac{n!}{l_1^\lambda! l_2^\lambda! \cdots l_n^\lambda!} \cdot D$, where D is the determinant of the matrix

$$\begin{bmatrix} 1 & l_1^\lambda & l_1^\lambda(l_1^\lambda - 1) & \cdots & l_1^\lambda(l_1^\lambda - 1) \cdots (l_1^\lambda - n + 1) \\ 1 & l_2^\lambda & l_2^\lambda(l_2^\lambda - 1) & \cdots & l_2^\lambda(l_2^\lambda - 1) \cdots (l_2^\lambda - n + 1) \\ \cdot & \cdot & \cdot & \cdot & \cdot \\ \cdot & \cdot & \cdot & \cdot & \cdot \\ \cdot & \cdot & \cdot & \cdot & \cdot \\ 1 & l_n^\lambda & l_n^\lambda(l_n^\lambda - 1) & \cdots & l_n^\lambda(l_n^\lambda - 1) \cdots (l_n^\lambda - n + 1) \end{bmatrix}.$$

Clearly, D equals to the determinant of a Vandermonde matrix

$$\begin{bmatrix} 1 & l_1^\lambda & (l_1^\lambda)^2 & \cdots & (l_1^\lambda)^n \\ 1 & l_2^\lambda & (l_2^\lambda)^2 & \cdots & (l_2^\lambda)^n \\ \cdot & \cdot & \cdot & \cdot & \cdot \\ \cdot & \cdot & \cdot & \cdot & \cdot \\ \cdot & \cdot & \cdot & \cdot & \cdot \\ 1 & l_n^\lambda & (l_n^\lambda)^2 & \cdots & (l_n^\lambda)^n \end{bmatrix}.$$

This means that $D = \prod_{1 \leq i < j \leq n} (l_i^\lambda - l_j^\lambda)$. Thus, $deg(\rho_\lambda) = \frac{n!}{l_1^\lambda! l_2^\lambda! \cdots l_n^\lambda!} \prod_{1 \leq i < j \leq n} (l_i^\lambda - l_j^\lambda)$. ♯

Now, we wish to describe the irreducible representations of A_n with the knowledge of representations of S_n. More generally, let H be a subgroup of G of index 2. Let η be a representation of H on W. Let $a \in G - H$. Define a representation η^a of H on W by $\eta^a(h) = \eta(aha^{-1})$. Clearly, η is irreducible if and only if η^a is irreducible. Let b be another element of $G - H$. Then $b = ka$ for some $k \in H$ and η^a is equivalent to η^b. For, then $\eta^b(h) = \eta(bhb^{-1}) = \eta(kah(ka)^{-1}) = \eta(k)\eta^a(h)\eta(k)^{-1}$ for all $h \in H$. Thus, η^a is unique up to equivalence. This representation is called the representation conjugate to η.

Further, we have a nontrivial one-dimensional representation τ of G on $\mathbb{C}$ given by $\tau(h) = 1$ for all $h \in H$ and $\tau(g) = -1$ for all $g \notin H$. For a given representations ρ of G on V, we have a representation ρ^τ of G on V which is given by $\rho^\tau(g)(v) = \rho(g)(v)$ for $g \in H$ and $\rho^\tau(g)(v) = -\rho(g)(v)$ for $g \notin H$. Indeed, $\rho^\tau \approx \rho \otimes \tau$. Note that $\rho^\tau|_H = \rho|_H$.

Theorem 5.2.8 *Let ρ be an irreducible representation of G on V and H a subgroup of index 2 in G. Then one and only one of the following hold:*

1. $\rho \not\approx \rho^\tau$ and $\rho|_H$ is irreducible representation of H which is conjugate to itself. Also $(\rho|_H)^G = \rho \oplus \rho^\tau$.

2. $\rho \approx \rho^\tau$, and $\rho|_H = \eta \oplus \eta'$, where η and η' are nonequivalent irreducible representations of H and which are conjugate to each other. Further, $\eta^G \approx \eta'^G \approx \rho$.

Finally, every irreducible representation of H is uniquely obtained in this manner.

Proof Since ρ is irreducible representation of G,

$$1 = <\chi_\rho, \chi_\rho>_G = \frac{1}{|G|}\sum_{g\in G} |\chi_\rho(g)|^2 = \frac{1}{|G|}(\sum_{h\in H} |\chi_\rho(h)|^2 + \sum_{g\notin H} |\chi_\rho(g)|^2).$$

Thus,

$$|G| = 2|H| = \sum_{h\in H} |\chi_\rho(h)|^2 + \sum_{g\notin H} |\chi_\rho(g)|^2 .$$

Since $\frac{1}{|H|}\sum_{h\in H} |\chi_\rho(h)|^2 = <\chi_{\rho|_H}, \chi_{\rho|_H}>_H$ is an integer, $\sum_{h\in H} |\chi_\rho(h)|^2$ is an integral multiple of $|H|$. Hence $\sum_{h\in H} |\chi_\rho(h)|^2 = |H|$ or $2|H|$. Suppose that $\sum_{h\in H} |\chi_\rho(h)|^2 = |H|$. Then $<\chi_{\rho|_H}, \chi_{\rho|_H}>_H = 1$ and hence $\rho|_H$ is irreducible. Further, in this case $\sum_{g\notin H} |\chi_\rho(g)|^2 = |H|$ and so $\rho \not\approx \rho^\tau$. Also $\rho|_H^G = \rho \oplus \rho^\tau$.

Next, suppose that $\sum_{h\in H} |\chi_\rho(h)|^2 = 2|H|$. Then $\sum_{g\notin H} \chi_\rho(g) = 0$. This means that $\rho \approx \rho^\tau$. Further, in this case $\rho|_H = \eta \oplus \eta'$, where η and η' are irreducible and conjugate to each other. Since ρ is irreducible, $\eta \not\approx \eta'$.

The final assertion also follows if we look and analyze the induced representation η^G of an irreducible representation η of H. ♯

Now, we describe the conjugacy classes of H in terms of the conjugacy classes in G.

Proposition 5.2.9 *Let H be a subgroup of G of index 2. Then a conjugacy class C of H is either a conjugacy class of G or else there is a unique conjugacy class $C' \neq C$ of H such that $C \bigcup C'$ is a conjugacy class of G.*

Proof Let $C = \{khk^{-1} \mid k \in H\}$ be a conjugacy class of H determined by $h \in H$. Then $C \subseteq \hat{C} = \{ghg^{-1} \mid g \in G\}$. Suppose that $\hat{C} \neq C$. Then, there is an element $g \notin H$ such that $ghg^{-1} \notin C$. We show that $\hat{C} - C$ is a conjugacy class of H determined by ghg^{-1}. Let $k \in H$. Since $ghg^{-1} \notin C$, $kghg^{-1}k^{-1} \notin C$. Hence the conjugacy class of H determined by ghg^{-1} is contained in $\hat{C} - C$. Let xhx^{-1} be a member of $\hat{C} - C$. Then $x \notin H$. Since $[G : H] = 2$, $x = kg$ for some $k \in H$. But then $xhx^{-1} =$

$kghg^{-1}k^{-1}$ belongs to the conjugacy class of H determined by ghg^{-1}. This shows that $C \bigcup C' = \hat{C}$ is a conjugacy class of G, where C' is a conjugacy class of H. ♯

Definition 5.2.10 A conjugacy class C of H is called a **split conjugacy class** of H if there is another conjugacy class C' of H such that $C \bigcup C'$ is a conjugacy class of G. The pair (C, C') is called a **split pair** of conjugacy classes.

Example 5.2.11 Consider the subgroup A_3 of S_3. The pair $(\{(1, 2, 3)\}, \{(1, 3, 2)\})$ is a split pair of conjugacy classes of A_3. Determine split conjugacy classes of A_4 considered as a subgroup of S_4.

Remark 5.2.12 More generally, the above discussion can be carried out (with a little care) for irreducible representations of a normal subgroup of prime index or even more generally for a normal subgroup of finite index.

The above discussion can be used to determine the irreducible representations of A_n as exercises (Exercises 5.2.8–5.2.13).

Representations of $GL(2, q)$ and $SL(2, q)$

Our next aim is to describe the irreducible representations and characters of $GL(2, q)$, $SL(2, q)$, and of $PSL(2, q)$, where q is a prime power. We need to describe the conjugacy classes of these groups. Recall that Exercises 4.5.1 and 4.5.2 of Algebra 3 describe these conjugacy classes. However, since we are going to use it, we shall give solutions to those exercises.

1. Since $Z(GL(2, q)) = \{aI \mid a \in F_q^\star\}$, there are $q - 1$ singleton conjugacy classes $\{\{aI\} \mid a \in F_q^\star\}$.

2. For $a \in F_q^\star$, consider the matrix

$$A_a = \begin{bmatrix} a & 1 \\ 0 & a \end{bmatrix}.$$

Since similar matrices have same eigenvalues, A_a is conjugate to A_b if and only if $a = b$. Further,

$$\begin{bmatrix} x & y \\ z & t \end{bmatrix} \cdot \begin{bmatrix} a & 1 \\ 0 & a \end{bmatrix} = \begin{bmatrix} a & 1 \\ 0 & a \end{bmatrix} \cdot \begin{bmatrix} x & y \\ z & t \end{bmatrix}$$

if and only if $x = t \in F_q^\star$ and $z = 0$. Thus, the centralizer of A_a contains $q(q - 1)$ elements. Since $\mid GL(2, q) \mid = (q^2 - 1)(q^2 - q)$ elements, The conjugacy class $\overline{A_a}$ determined by A_a contains $q^2 - 1$ elements. There are $q - 1$ such conjugacy classes.

3. Consider

$$A_{a,b} = \begin{bmatrix} a & 0 \\ 0 & b \end{bmatrix},$$

where $a, b \in F_q^\star, a \neq b$. Clearly, $A_{a,b}$ is conjugate to $A_{c,d}$ if and only if $\{a, b\} = \{c, d\}$. Further,

$$\begin{bmatrix} x & y \\ z & t \end{bmatrix} \cdot \begin{bmatrix} a & 0 \\ 0 & b \end{bmatrix} = \begin{bmatrix} a & 0 \\ 0 & b \end{bmatrix} \cdot \begin{bmatrix} x & y \\ z & t \end{bmatrix},$$

$a \neq b$ if and only if $x = t \in F_q^\star$ and $z = 0 = y$.Thus, the centralizer of $A_{a,b}$ is the diagonal subgroup of $GL(2, q)$ containing $(q-1)^2$ elements. Consequently, the conjugacy class $\overline{A_{a,b}}$ contains $\frac{(q^2-1)(q^2-q)}{(q-1)^2} = q(q+1)$ elements. There are $\frac{(q-1)(q-2)}{2}$ such conjugacy classes.

4. Assume that q is odd. Let ξ be a generator of $F_q^\star$. Since q is odd, $X^2 - \xi$ is irreducible in $F_q[X]$. Consequently,

$$B_{a,b} = \begin{bmatrix} a & b\xi \\ b & a \end{bmatrix}$$

is a member of $GL(2, q)$, whenever $a \in F_q^\star$ or $b \in F_q^\star$. Let ζ be a root of $X^2 - \xi$ in the quadratic extension F_{q^2} of F_q. Define a map η from $F_{q^2}^\star$ to $GL(2, q)$ by $\eta(a + b\zeta) = B_{a,b}$. Clearly, η is an injective homomorphism of groups. Note that $B_{a,0}$ is scalar matrix aI. The eigenvalues of $B_{a,b}$ are $a \pm b\zeta$. Thus, $B_{a,b}$ is conjugate to $B_{c,d}$ if and only if $a = c$ and $b = \pm c$. Further assume that $b \neq 0$. Then

$$\begin{bmatrix} x & y \\ z & t \end{bmatrix} \cdot \begin{bmatrix} a & b\xi \\ b & a \end{bmatrix} = \begin{bmatrix} a & b\xi \\ b & a \end{bmatrix} \cdot \begin{bmatrix} x & y \\ z & t \end{bmatrix},$$

if and only if $x = t$ and $y = \xi z$. Thus, the centralizer of $B_{a,b}, b \neq 0$ contains $q(q-1)$ elements. Hence the conjugacy class $\overline{B_{a,b}}$ contains $\frac{(q^2-1)(q^2-q)}{2q(q-1)} = \frac{q^2-1}{2}$ elements. The number of such conjugacy classes is $q^2 - q$.

The number of elements in the union of the conjugacy classes of the types 1, 2, 3, and 4 is $(q-1) + (q-1)(q^2-1) + \frac{(q-1)(q-2)}{2} + \frac{q(q-1)}{2} = (q^2-1)(q^2-q) = \mid GL(2, q) \mid$. Thus, these are all conjugacy classes of $GL(2, q)$. The number of conjugacy classes is $2(q-1) + \frac{(q-1)(q-2)}{2} + \frac{q(q-1)}{2} = q^2 - 1$. Consequently, there are $q^2 - 1$ irreducible representations of $GL(2, q)$. We wish to determine them.

1. For each group homomorphism α from $F_q^\star$ to $\mathbb{C}^\star$, we have a one-dimensional representation ρ_α of $GL(2, q)$ given by $\rho_\alpha(g) = \alpha(Det(g))$. Note that there are $q-1$ homomorphisms from $F_q^\star$ to $\mathbb{C}^\star$ each determined uniquely by a choice of $(q-1)_{th}$ root of unity in $\mathbb{C}$. This gives us $q-1$ irreducible one-dimensional irreducible representations $\{\rho_\alpha \mid \alpha \in Hom(F_q^\star, \mathbb{C}^\star)\}$. The characters χ_{ρ_α} are given by $\chi_{\rho_\alpha}(aI) = \alpha(a)^2 = \chi_{\rho_\alpha}(A_a)$, $\chi_{\rho_\alpha}(A_{a,b}) = \alpha(a)\alpha(b)$, and $\chi_{\rho_\alpha}(B_{a,b}) = \alpha(a^2 - \xi b^2)$.

2. Consider the projective line $P^1 F_q$ over F_q. By the definition $P^1 F_q$ is the quotient set $(F_q \times F_q - \{(0,0)\})/\equiv$, where $(a, b) \equiv (c, d)$ if and only if there is a $\lambda \in F_q^\star$ such that $(\lambda a, \lambda b) = (c, d)$. The equivalence class determined by (a, b) is denoted by $[a, b]$. Thus, $[a, b] = [c, d]$ if and only if $c = \lambda a$ and $d = \lambda b$ for some $\lambda \in F_q^\star$. Clearly, $P^1 F_q = \{[\lambda, 1] \mid \lambda \in F_q\} \bigcup \{[0, 1]\}$ contains $q+1$ elements. $GL(2, q)$ acts transitively on $P^1 F_q$ in obvious manner. Express $P^1 F_q$ as ordered set $\{x_1, x_2, \ldots, x_{q+1}\}$. The action of $GL(2, q)$ on $P^1 F_q$ gives a representation of ρ of $GL(2, q)$ of degree $q+1$ on $\oplus \sum_{i=1}^{q+1} \mathbb{C}x_i$. We have the trivial subrepresentation ρ_1 of ρ on the subspace $\mathbb{C}(x_1 + x_2 + \cdots x_{q+1})$ of $\oplus \sum_{i=1}^{q+1} \mathbb{C}x_i$. Clearly, $\rho = \rho_1 \oplus \rho_s$,

where ρ_s is the subrepresentation of ρ on the subspace $V = \{\sum_{i=1}^{q+1} a_i x_i \mid \sum_{i=1}^{q+1} a_i = 0\}$. Clearly, $\chi_{\rho_s}(aI) = DimV = q$ for all a. We have a basis $S = \{x_i - x_{i+1} \mid 1 \leq i \leq q\}$ of V. It can be easily observed that $\chi_{\rho_s}(A_a) = 0$ for all a, $\chi_{\rho_s}(A_{a,b}) = 1$, and $\chi_{\rho_s}(B_{a,b}) = -1$. Consequently,

$$
\begin{aligned}
< \chi_{\rho_s}, \chi_{\rho_s} > &= \frac{1}{\mid GL(2,q) \mid}[(q-1)\cdot q^2 \\
&+ (q-1)\cdot 0 + \frac{(q-1)(q-2)}{2}(q^2+q)\cdot 1^2 \\
&+ \frac{(q^2-q)}{2}(q^2-q)\cdot(-1)^2] = 1.
\end{aligned}
$$

Thus, ρ_s is irreducible. For each homomorphism α from $F_q^\star$ to $\mathbb{C}^\star$, Consider the representation $\rho_s^\alpha = \rho_s \otimes \rho_\alpha$. Clearly, $\chi_{\rho_s^\alpha} = \chi_{\rho_s}\chi_{\rho_\alpha}$. It can be easily observed that $< \chi_{\rho_s^\alpha}, \chi_{\rho_s^\alpha} > = 1$. This gives $q-1$ irreducible representations $\{\rho_s^\alpha \mid \alpha \in Hom(F_q^\star, \mathbb{C}^\star)\}$ each of degree q. Their characters are described as above.

3. Next, consider the Borel subgroup

$$
B = \{\begin{bmatrix} a & c \\ 0 & b \end{bmatrix} \mid a, b \in F_q^\star, c \in F_q\},
$$

and the unipotent subgroup

$$
U = \{\begin{bmatrix} 1 & c \\ 0 & 1 \end{bmatrix} \mid c \in F_q\},
$$

of $GL(2,q)$. Since B/U is isomorphic to $F_q^\star \times F_q^\star$, for each pair $\alpha, \beta \in Hom(F_q^\star, \mathbb{C}^\star)$, we have a one-dimensional representation $\eta_{\alpha,\beta}$ of B which is given by $\eta_{\alpha,\beta}(\begin{bmatrix} a & c \\ 0 & b \end{bmatrix}) = \alpha(a)\beta(b)$. Let $\rho_{\alpha,\beta}$ denote the induced representation $\eta_{\alpha,\beta}^{GL(2,q)}$. Clearly, $\rho_{\alpha,\beta}$ is of degree $[GL(2,q) : B] = q+1$. Further, using Proposition 5.1.49, we can see that $\chi_{\rho_{\alpha,\beta}}(aI) = (q+1)\alpha(a)\beta(a)$, $\chi_{\rho_{\alpha,\beta}}(A_a) = \alpha(a)\beta(a)$, $\chi_{\rho_{\alpha,\beta}}(A_{a,b}) = \alpha(a)\beta(b) + \alpha(b)\beta(a)$, and $\chi_{\rho_{\alpha,\beta}}(B_{a,b}) = 0$. It follows that $\rho_{\alpha,\beta} = \rho_{\beta,\alpha}$, and $\rho_{\alpha,\alpha} = \rho_s^\alpha \oplus \rho_\alpha$. It also follows that $< \chi_{\rho_{\alpha,\beta}}, \chi_{\rho_{\alpha,\beta}} > = 1$ for $\alpha \neq \beta$. This gives us $\frac{1}{2}(q-1)(q-2)$ irreducible representations $\{\rho_{\alpha,\beta} \mid \alpha \neq \beta\}$ whose characters are given above.

4. Finally, we find the remaining $\frac{1}{2}q(q-1)$ irreducible representations of $GL(2,q)$. We discover these irreducible representations as components of the representations induced by the representations of the subgroup $L = \{B_{a,b} \mid (a,b) \in F_q \times F_q - \{(0,0)\}\}$ of $GL(2,q)$ and the tensor products of earlier representations. Note that L is isomorphic to $F_{q^2}^\star$ under the isomorphism $B_{a,b} \mapsto (a + b\zeta)$. We have $q^2 - 1$ one-dimensional irreducible representations of L. Let μ be a one-dimensional representation of L. Consider the induced representation $\mu^{GL(2,q)}$ of $GL(2,q)$ which is of degree $[GL(2,q) : L] = q^2 - 1$. It can be easily seen that

$\chi_{\mu^{GL(2,q)}}(aI) = q(q-1)\mu(aI = B_{a,0})$, $\chi_{\mu^{GL(2,q)}}(A_a) = 0 = \chi_{\mu^{GL(2,q)}}(A_{a,b})$, and $\chi_{\mu^{GL(2,q)}}(B_{a,b}) = \mu(B_{0,1}) + \mu(B_{0,1})^q$. Since $(aI)^q = aI$ and $B(0,1)^{q^2} = B(0,1)$, it follows that $\chi_{\mu^{GL(2,q)}} = \chi_{(\mu^q)^{GL(2,q)}}$. This way, we get $\frac{1}{2}q(q-1)$ distinct induced representations $\{\mu^{GL(2,q)} \mid \mu^q \neq \mu\}$ from the irreducible representations of L. These representations are not irreducible as $< \chi_{\mu^{GL(2,q)}}, \chi_{\mu^{GL(2,q)}} > = q-1$ for μ with property $\mu^q \neq \mu$. Now, let us look at the tensor product representations $\rho_s^\alpha \otimes \rho_\beta$, $\rho_{\alpha,\beta} \otimes \rho_\gamma$ and $\rho_s \otimes \rho_{\alpha,\beta}$. Since $\rho_s^\alpha \otimes \rho_\beta \approx \rho_s^{\alpha\beta}$ and $\rho_{\alpha,\beta} \otimes \rho_\gamma \approx \rho_{\alpha\gamma,\beta\gamma}$, they do not give any new irreducible representations. However, $\rho_s^{\alpha,\beta} = \rho_s \otimes \rho_{\alpha,\beta}$ and in particular, $\rho_s^{\alpha,1}$ contains new irreducible representations. It can be easily seen that the character $\chi_{\rho_s^{\alpha,1}}$ takes the values $q(q+1)\alpha(a),\ 0,\ \alpha(a)+\alpha(b),\ and\ 0$ on the conjugacy classes $\{aI\}$, $\overline{A_a}$, $\overline{A_{a,b}}$, and $\overline{B_{a,b}}$, respectively. The following identities can be easily checked:

$$< \chi_{\rho_s^{\alpha,1}}, \chi_{\rho_{\alpha,1}} > = 2 \quad \cdots (1).$$

$$< \chi_{\mu^{GL(2,q)}}, \chi_{\rho_{\alpha,1}} > = 1 \quad \cdots (2),$$

where $\mu(B(a,0)) = \alpha(a)$ for all $a \in F_q^\star$.

$$< \chi_{\rho_s^{\alpha,1}}, \chi_{\rho_s^{\alpha,1}} > = q+3 \quad \cdots (3).$$

$$< \chi_{\rho_s^{\alpha,1}}, \chi_{\mu^{GL(2,q)}} > = q \quad \cdots (1),$$

where $\mu(B(a,0)) = \alpha(a)$ for all $a \in F_q^\star$. The above identities ensure that $\rho_{\alpha,1}$ is a subrepresentation of $\rho_s^{\alpha,1}$. Again, since $< \chi_{\rho_s^{\alpha,1}}, \chi_{\mu^{GL(2,q)}} > = q$ and $< \chi_{\mu^{GL(2,q)}}, \chi_{\mu^{GL(2,q)}} > = q-1$, $\mu^{GL(2,q)}$ and $\rho_s^{\alpha,1}$ have many common irreducible representations. This prompts us to look at the virtual (perhaps)chararacter

$$\chi_{\rho_s^{\alpha,1}} - \chi_{\rho_{\alpha,1}} - \chi_{\mu^{GL(2,q)}} = \chi_{\hat{\mu}}.$$

Using the above 4 identities, we see that $\chi_{\hat{\mu}}$ takes the values $(q-1)\alpha(a),\ \alpha(a),\ 0$, and $-(\mu(B_{0,1}) + (\mu(B_{0,1}))^q)$ on the four types of conjugacy classes of $GL(2,q)$ as already described. This means that $< \chi_{\hat{\mu}}, \chi_{\hat{\mu}} > = 1$ and $\chi_{\hat{\mu}}(1) = q-1$. This shows that $\chi_{\hat{\mu}}$ is a genuine irreducible character of $GL(2,q)$ of degree $q-1$. Consequently, we have the set $\{\hat{\mu} \mid \mu \in Hom(L, \mathbb{C}^\star)\ with\ \mu^q \neq \mu\}$ of $\frac{1}{2}q(q-1)$ irreducible representations of $GL(2,q)$.

This completes the description of all irreducible representations of $GL(2,q)$ together with their characters.

Representations of $SL(2,q)$

Let us list the conjugacy classes of $SL(2,q)$.

1. Since q is odd, $Z(SL(2,q)) = \{I, -I\}$. Thus, $\{I\}$ and $\{-I\}$ are two singleton conjugacy classes of $SL(2,q)$.

2. It can be easily seen that the centralizer of the matrix $\begin{bmatrix} 1 & 1 \\ 0 & 1 \end{bmatrix}$ is the group consisting of the matrices of the types $\begin{bmatrix} a & b \\ 0 & a \end{bmatrix}$, where $a^2 = 1$. Thus, the centralizer of $\begin{bmatrix} 1 & 1 \\ 0 & 1 \end{bmatrix}$ contains $2q$ elements. Since $\mid SL(2,q) \mid = q(q^2-1)$, the conjugacy class $\overline{\begin{bmatrix} 1 & 1 \\ 0 & 1 \end{bmatrix}}$ contains $\frac{q^2-1}{2}$ elements.

3. Observe that $\begin{bmatrix} 1 & 1 \\ 0 & 1 \end{bmatrix}$ is not conjugate to $\begin{bmatrix} 1 & \xi \\ 0 & 1 \end{bmatrix}$, and as above the conjugacy class $\overline{\begin{bmatrix} 1 & \xi \\ 0 & 1 \end{bmatrix}}$ also contains $\frac{q^2-1}{2}$ elements.

4. Clearly, $\begin{bmatrix} -1 & 1 \\ 0 & -1 \end{bmatrix}$ is not conjugate to any of the matrices of the type 1, 2, and 3. As above the conjugacy class $\overline{\begin{bmatrix} -1 & 1 \\ 0 & -1 \end{bmatrix}}$ again contains $\frac{q^2-1}{2}$ elements.

5. The conjugacy class $\overline{\begin{bmatrix} -1 & \xi \\ 0 & -1 \end{bmatrix}}$ is different from the previous conjugacy classes, and it also contains $\frac{q^2-1}{2}$ elements.

6. Let $a \neq \pm 1$. Then only diagonal matrices in $SL(2,q)$ commute with $\begin{bmatrix} a & 0 \\ 0 & a^{-1} \end{bmatrix}$. Thus, the conjugacy class $\overline{\begin{bmatrix} a & 0 \\ 0 & a^{-1} \end{bmatrix}}$ again contains $q(q+1)$ elements. There are $\frac{q-3}{2}$ such conjugacy classes.

7. Finally, for $a \neq \pm 1$, $\overline{\begin{bmatrix} a & b\xi \\ b & a^{-1} \end{bmatrix}}$ is a conjugacy class different from the previous conjugacy classes which contains $q(q-1)$ elements. The number of such conjugacy classes is $\frac{q-1}{2}$.

This exhaust all elements of $SL(2,q)$. There are $q+4$ conjugacy classes of $SL(2,q)$ thus obtained.

To determine the $q+4$ irreducible representations of $SL(2,q)$, we first look at the restrictions of the representations of $GL(2,q)$ to $SL(2,q)$.

1. The representations ρ_α restricted to $SL(2,q)$ are all trivial irreducible representation of $SL(2,q)$.

2. Since $\rho_\alpha|_{SL(2,q)}$ is a trivial representation, the restriction $\rho_s^\alpha|_{SL(2,q)}$ of the tensor product representation ρ_s^α is equivalent to the restriction $\rho_s|_{SL(2,q)}$. Looking at the values of the character χ_{ρ_s} on the conjugacy classes of $SL(2,q)$ as described above, we see that

$$\begin{aligned} < \chi_{\rho_s|_{SL(2,q)}}, \chi_{\rho_s|_{SL(2,q)}} > &= \frac{1}{q(q^2-1)} \sum_{h \in SL(2,q)} \mid \chi_{\rho_s|_{SL(2,q)}}(h) \mid^2 \\ &= \frac{1}{q(q^2-1)} [q^2 + q^2 + q(q+1)\frac{q-3}{2} + q(q-1)\frac{q-1}{2}] = 1. \end{aligned}$$

This shows that $\rho_s|_{SL(2,q)}$ is an irreducible representation of $SL(2, q)$ of degree q.

3. Similarly, we can see that $\rho_{\alpha,1}|_{SL(2,q)}$ is irreducible provided that $\alpha^2 \neq 1$. Further, then $\rho_{\alpha,1}|_{SL(2,q)} \approx \rho_{\beta,1}|_{SL(2,q)}$ if and only if $\alpha = \beta$ or $\alpha^{-1} = \beta$. There are $\frac{q-3}{2}$ such irreducible representations which are of degree $q+1$. Again if $\alpha^2 = 1$ and $\alpha \neq 1$, then $\rho_{\alpha,1}$ is direct sum of two nonequivalent irreducible representations which are different from the earlier irreducible representations.

4. Recall the representations $\hat{\mu}$ of $GL(2, q)$, where $\mu \in Hom(L, \mathbb{C}^\star)$ such that $\mu^q \neq \mu$. Let H denote the subgroup $\{B_{a,b} \in L \mid B_{a,b}^{q+1} = I\}$. It can be seen that $\hat{\mu}|_{SL(2,q)}$ is irreducible provided that $\mu^2|_H \neq 1$. Further, $\hat{\mu}|_{SL(2,q)} = \hat{\nu}|_{SL(2,q)}$ if and only if $\mu|_H = \nu|_H$. We have $\frac{q-1}{2}$ such nonequivalent irreducible representations different from the earlier ones. Further, $\hat{\mu}|_{SL(2,q)} = \widehat{\mu^{-1}}|_{SL(2,q)}$. Next, if $\mu^2|_H = 1$ and $\mu|_H \neq 1$, then $\hat{\mu}$ is direct sum of two nonequivalent irreducible representations different from the earlier representations.

This gives all $q+4$ irreducible representations (together with their characters) of $SL(2, q)$.

Exercises

5.2.1 Show that the representation associated with the partition (n) is the trivial representation and the representation associated with $(1, 1, \ldots, 1)$ is the *sign* representation.

5.2.2 Show that the representation associated with the partition $(n-1, 1)$ is the standard representation of S_n. What partition cooresponds to its exterior square?

5.2.3 Using the Frobenius character formula, compute the character table of S_4 and S_5.

5.2.4 Show that $m_\lambda = \frac{l_1^\lambda! l_2^\lambda! \cdots l_n^\lambda!}{\prod_{i<j}(l_i^\lambda - l_j^\lambda)}$.

5.2.5 Hook length of a box in a Young diagram is defined to be the number of boxes which are directly towards the right of the box or it is directly below the box including the box itself. Thus, for example, the hook length of the box containing the number 7 in the Young Tableau of the partition $(5, 4, 2, 2)$ of 13 is 5. Show that $deg(\rho_\lambda)$ is $\frac{n!}{h_\lambda}$, where h_λ is the product of hook lengths of the boxes in the Young diagram of λ.

5.2.6 Find the degree of the representation ρ_λ, where $\lambda = (5, 4, 2, 2)$ is the partition of 13.

5.2.7 Show that the representation afforded by the module $\mathbb{C}(S_n)a_\lambda$ is the representation $1_{S_\lambda}^{S_n}$ induced by the trivial representation of $S_\lambda = S_{\lambda_1} \times S_{\lambda_2} \times \cdots \times S_{\lambda_n}$.

5.2.8 Let $\hat{\lambda}$ be a partition of n which is conjugate to λ. Show that $\rho_{\hat{\lambda}} \approx \rho_{\lambda}^{\tau}$.

5.2.9 Suppose that $\hat{\lambda} \neq \lambda$. Use Theorem 5.2.8. to show that $\rho_\lambda|_{A_n}$ is irreducible representation of A_n.

5.2.10 Suppose that $\hat{\lambda} = \lambda$. Show that $\rho_\lambda|_{A_n}$ is direct sum of two nonequivalent irreducible representations of A_n which are conjugate to each other.

5.2.11 Show that a symmetric Young diagram determines a split pair of conjugacy classes of A_n.
5.2.12 Find the split pair of conjugacy classes of A_5 and A_7.
5.2.13 Determine the irreducible representations and irreducible characters of A_5 and A_6.
5.2.14 Determine the irreducible characters of $PGL(2, q)$ by using the knowledge of the irreducible characters of $GL(2, q)$. Using the fact $PSL(2, q)$ is a subgroup of $PGL(2, q)$, find the irreducible characters of $PSL(2, q)$.

5.3 Steinberg Characters

This section is devoted to introduce the Steinberg character as a virtual character in the character ring of a Chevalley group and establish that it is, indeed, a genuine irreducible character. We also evaluate it on different type of elements.

Let $G(V, F_q)$ be a Chevalley group associated with (L, V, F_q), where L is a simple Lie algebra over $\mathbb{C}$ with a root system Φ. Fix a basis Δ of Φ. For simplicity of notation, we shall denote $G(V, F_q)$ by G, $B(V, F_q)$ by B, $U(V, F_q)$ by U, $H(V, F_q)$ by H, and $N(V, F_q)$ by N. Recall (Definition 4.3.44) that a principal parabolic subgroup of G is a subgroup $G_\pi = \bigcup_{\sigma \in W_\pi} B\sigma B = BW_\pi B = BNB$ of G, where $\pi \subseteq \Delta$, W_π is a parabolic subgroup of $W(\Phi)$ generated by $\{\sigma_\alpha \mid \alpha \in \pi\}$, and $N_\pi = \nu^{-1}(W_\pi)$, ν being the quotient map $N \mapsto N/H \approx W(\Phi)$.

For each subset π of Δ, let D_π denote the subset $\{\sigma \in W(\Phi) \mid \sigma(\pi) \subseteq \Phi^+\}$. It can be seen that D_π is a coset representative system of $W(\Phi)/W_\pi$ (termed as distiguished coset representative system). Note that $1 \in D_\pi$. For each $\sigma \in D_\pi$, select a unique member $n_\sigma \in \nu^{-1}(\sigma)$ with $n_1 = 1$. The set $N_\pi = \{n_\sigma \mid \sigma \in D_\pi\}$ is a coset representative system of G_π in G. Next, Let π and π' be subsets of Δ. Then a double coset $W_\pi \sigma W_{\pi'}$ has a unique element of smallest length. Let $D_{\pi,\pi'}$ denote the subset $D_\pi^{-1} \bigcap D_{\pi'}$. Then it can be shown that $D_{\pi,\pi'}$ is a double coset representative system of W_π, $W_{\pi'}$ in $W(\Phi)$. Indeed, the unique element of $W_\pi \sigma W_{\pi'}$ of smallest length belongs to $D_{\pi,\pi'}$. $D_{\pi,\pi'}$ is called the distinguished double coset representative system of W_π and $W_{\pi'}$ in $W(\Phi)$. Again, let $N_{\pi,\pi'}$ be the set obtained by selecting one and only one member from $\nu^{-1}(\sigma)$ for each $\sigma \in D_{\pi,\pi'}$. Clearly, $\mid N_{\pi,\pi'} \mid = \mid D_{\pi,\pi'} \mid$, and $N_{\pi,\pi'}$ is a double coset representative system of G_π, $G_{\pi'}$ in G. Further, if $\sigma \in D_{\pi,\pi'}$, then $W_\pi \bigcap \sigma W_{\pi'} \sigma^{-1} = W_{\pi''}$, where $\pi'' = \pi \bigcap \sigma(\pi')$.

The virtual character St_G of G defined by

$$St_G = \sum_{\pi \subseteq \Delta} (-1)^{|\pi|} 1_{G_\pi}^G$$

is called the **Steinberg Character** of G. We shall show that St_G is, indeed, a genuine irreducible character. For the purpose, we shall study the corresponding virtual character

$$st_{W(\Phi)} = \sum_{\pi \subseteq \Delta} (-1)^{|\pi|} 1_{W_\pi}^{W^\Phi}$$

of $W(\Phi)$.

For each subset π of Δ, let C_π denote the subset $\{v \in E \mid < \alpha, v >= 0 \; for \; all \; \alpha \in \pi \; and \; < \alpha, v >> 0 \; for \; all \; \alpha \notin \pi\}$ of E. Since the $W(\Phi)$ acts simply transitively on the set of all Weyl chambers (Theorem 2.2.21), the following Proposition is immediate.

Proposition 5.3.1 *Let $\sigma \in W(\Phi)$. Then, the following are equivalent:*
(i) $\sigma(C_\pi) = C_\pi$.
(ii) $\sigma(v) = v$ for all $v \in C_\pi$.
(iii) $\sigma \in W_\pi$. ♯

Proposition 5.3.2 *Let V be a subspace of E. Let Σ_V be a simplicial complex consisting of simplexes of a Coxeter complex contained in V. Let $n_i^{\Sigma_V}$ be the number of simplexes in Σ_V which are of dimension i. Then*

$$\sum_i (-1)^i n_i^{\Sigma_V} = (-1)^{DimV}.$$

Proof The proof is by the induction on the number n_{Σ_V} of hyperplanes defining the complex Σ_V. Suppose that $n_{\Sigma_V} = 1$. Let P be the hyperplane defining Σ_V. We have three simplexes P_+, P_-, and P defining Σ_V. Clearly, $Dim(P_+) = DimV = Dim(P_-)$, and $DimP = DimV - 1$. Thus, we have two simplexes of dimension equal to $DimV$, and one simplex of dimension equal to $DimV - 1$. Consequently,

$$\sum_i (-1)^i n_i^{\hat{\Sigma}_V} = 2(-1)^{DimV} + (-1)^{DimV-1} = (-1)^{DimV)}.$$

The result follows for Σ_V. Assume the result for Σ_V with $n_{\Sigma_V} = m$. One adds an extra hyperplane P in Σ_V to obtain a complex $\hat{\Sigma}_V$ with $n_{\hat{\Sigma}_V} = m + 1$. Each simplex of Σ_V of dimension i is replaced by three simplexes , two of dimension i, and one of dimension $i - 1$. Thus,

$$\sum_i (-1)^i n_i^{\hat{\Sigma}_V} = \sum_i (-1)^i n_i^{\Sigma_V} = (-1)^{DimV)}.$$

The result follows. ♯

Proposition 5.3.3 $st_{W(\Phi)}(\sigma) = (-1)^{l(\sigma)} = Det\sigma.$

Proof Since $Det\sigma_\alpha = -1$ for all $\alpha \in \Phi$, it follows that $Det(\sigma) = (-1)^{l(\sigma)}$. It remains to show that $st_{W(\Phi)}(\sigma) = Det(\sigma)$. By Proposition 5.1.49,

$$\chi_{1_{W_\pi}^{W(\Phi)}}(\sigma) = \frac{1}{\mid W_\pi \mid} \sum_{\tau \in W(\Phi), \tau^{-1}\sigma\tau \in W_\pi} 1 \cdots (1).$$

Now, $\tau^{-1}\sigma\tau \in W_\pi$ if and only if $\sigma(\tau(C_\pi)) = \tau(C_\pi)$ (Proposition 5.3.1). Thus,

$$\chi_{1_{W_\pi}^{W(\Phi)}}(\sigma) = \frac{1}{\mid W_\pi \mid} \mid W_\pi \mid r_\pi(\sigma) = r_\pi(\sigma) \quad \cdots (2),$$

where $r_\pi(\sigma)$ denotes the number of simplexes $\tau(C_\pi)$ which are fixed by σ. In other words, $r_\pi(\sigma)$ is the number of simplexes contained in the subspace $V = \{v \in E \mid \sigma(v) = v\}$ of E. This determines a complex Σ_π contained in V. Let t_i be the number of simplexes of Σ_π of dimension i. From Proposition 5.3.2,

$$\sum_i (-1)^i t_i = (-1)^{DimV} \quad \cdots (3).$$

It is clear from definition of C_π that $DimC_\pi = l - \mid \pi \mid$. Thus, $t_i = \sum_{|\pi|=l-i} r_\pi(\sigma)$. substituting the value of t_i in the above equation,

$$(-1)^l \sum_{\pi \subseteq \Delta} (-1)^{|\pi|} r_\pi(\sigma) = (-1)^{DimV} \quad \cdots (4),$$

or equivalently,

$$\sum_{\pi \subseteq \Delta} (-1)^{|\pi|} r_\pi(\sigma) = (-1)^{l-DimV}.$$

From the Eq. 2,

$$st_{W(\Phi)} = \sum_{\pi \subseteq \Delta} (-1)^{|\pi|} r_\pi(\sigma) = (-1)^{l-DimV} \quad \cdots (5).$$

Finally, consider $Det(\sigma)$. Since σ is an orthogonal transformation, the eigenvalues of σ are 1, -1, and $e^{i\theta}$. Further, if $e^{i\theta}$ is an eigenvalue , then $e^{-i\theta}$ is also an eigenvalue. Again the multiplicity of the eigenvalue 1 is $DimV$. Thus, $Det(\sigma) = (-1)^{l-DimV} = st_{W(\Phi)}(\sigma)$. ♯

Proposition 5.3.4 $< St_G, St_G > = 1$.

Proof Since inner product is bi-linear on the character ring,
$< St_G, St_G >$
$= \sum_{\pi \subseteq \Delta} \sum_{\pi' \subseteq \Delta} (-1)^{|\pi|}(-1)^{|\pi'|} < 1_{G_\pi}^G, 1_{G_{\pi'}}^G >$
$= \sum_{\pi \subseteq \Delta} \sum_{\pi' \subseteq \Delta} (-1)^{|\pi|}(-1)^{|\pi'|} \sum_{n \in N_{\pi,\pi'}} < 1_{G_\pi \bigcap G_{\pi'}^n}, 1_{G_\pi \bigcap G_{\pi'}^n}^n >$ (by Proposition 5.1.55),
where $N_{\pi,\pi'}$ is a double coset representative system of G_π and $G_{\pi'}$ in G. Thus,

$$< St_G, St_G > = \sum_{\pi \subseteq \Delta} \sum_{\pi' \subseteq \Delta} (-1)^{|\pi|}(-1)^{|\pi'|} \mid N_{\pi,\pi'} \mid .$$

Clearly, $\mid N_{\pi,\pi'} \mid = \mid D_{\pi,\pi'} \mid$, where $D_{\pi,\pi'}$ is distinguished double coset representative system of W_π and $W_{\pi'}$ in $W(\Phi)$. Thus,
$< St_G,\ St_G >$
$= \sum_{\pi\subseteq\Delta}\sum_{\pi'\subseteq\Delta}(-1)^{|\pi|}(-1)^{|\pi'|} \mid W(\Phi)_{\pi,\pi'} \mid$
$= < st_{W(\Phi)},\ st_{W(\Phi)} >$
$= \frac{1}{|W(\Phi)|}\sum_{\sigma\in W(\Phi)} st_{W(\Phi)}(\sigma)^2$
$= \frac{1}{|W(\Phi)|}\sum_{\sigma\in W(\Phi)} Det(\sigma)^2$
$= 1.$ ♯

Corollary 5.3.5 *St_G is an irreducible character or $-St_G$ is an irreducible character.*

Proof Since St_G is in the character ring of G, $St_G = n_1\chi_1 + n_2\chi_2 + \cdots + n_r\chi_r$, where χ_i are pairwise nonequivalent irreducible characters of G and $n_i \in \mathbb{Z}$. Since $< St_G,\ St_G > = 1$, $n_i^2 = 1$ for a unique i while the rest of n_j are 0. This means that $St_G = \pm\chi_i$ for some i. ♯

Proposition 5.3.6 $< 1_B^G,\ St_G > = 1.$

Proof $< 1_B^G,\ St_G >$
$= \sum_{\pi\subseteq\Delta}(-1)^{|\pi|} < 1_B^G,\ 1_{G_\pi}^G >$
$= \sum_{\pi\subseteq\Delta}(-1)^{|\pi|} \mid N_{\Delta,\pi} \mid$ (by Proposition 5.1.55),
$= \sum_{\pi\subseteq\Delta}(-1)^{|\pi|} \mid W_{\Delta,\pi} \mid$
$= \sum_{\pi\subseteq\Delta}(-1)^{|\pi|} < 1_1^{W(\Phi)},\ 1_{W_\pi}^{W(\Phi)} >$
$= < 1_1^{W(\Phi)}, st_{W(\Phi)} >$
$= < 1,\ 1 >$ (using the Frobenius reciprocity)
$= 1.$ ♯

The following corollary is immediate.

Corollary 5.3.7 *St_G is an irreducible character which appears in 1_B^G with multiplicity 1.* ♯

Corollary 5.3.8 *Let $x \in G$ such that x does not belong to any proper parabolic subgroup. Then $St_G(x) = (-1)^l$.*

Proof For $\pi \neq \Delta$, no conjugate of x will lie in G_π. Consequently,

$$St_G(x) = \sum_{\pi\subseteq\Delta}(-1)^{|\pi|}1_{G_\pi}^G(x) = (-1)^{|\Delta|}1_{G_\Delta}^G(x) = (-1)^l 1_G^G(x) = (-1)^l.$$

♯

The parabolic subgroup G_π has the unipotent radical (largest normal unipotent subgroup) $U_\pi = U\bigcap(\sigma_0^\pi)^{-1}U\sigma_0^\pi$, where σ_0^π is the unique element of W_π which takes π to Φ_π^- (Φ_π denotes the root system $\Phi\bigcap < \pi >$ in the subspace $< \pi >$ of E). If L_π denotes the subgroup generated by $H\bigcup\{X_\alpha^V(F_q) \mid \alpha \in \Phi_\pi\}$, then G_π is the semidirect product $U_\pi \succ L_\pi$. This decomposition is called the **Levi decomposition** of the parabolic subgroup G_π. The subgroup L_π is called the **Levi subgroup** of G_π. Further, L_π admits a Tits system (B_π, N_π), where $B_\pi = U_{\sigma_0^\pi} H = \prod_{\alpha \in \Phi_\pi^+} X_\alpha^V(F_q)$ and $N_\pi = \nu^{-1}(W_\pi)$. The parabolic subgroups of L_π are precisely, the subgroups of the form $G_{\pi'} \bigcap L_\pi$, where $\pi' \subseteq \pi$. Correspondingly, we have the Steinberg character St_{L_π} of L_π which is given by

$$St_{L_\pi} = \sum_{\pi' \subseteq \pi} (-1)^{|\pi'|} 1_{G_{\pi'} \bigcap L_\pi}^{L_\pi}.$$

Proposition 5.3.9 $St_G|_{G_\pi} = St_{L_\pi}^{G_\pi}$. ♯

Before proving the Proposition, we obtain some interesting and useful consequences as Corollaries:

Corollary 5.3.10 *Let g be an element of G_π such that $x^{-1}gx \notin G_{\pi'}$ for all $\pi' \subset \pi$ and $x \in G$. Then, the following hold:*
(i) $St_G(g) = 0$ whenever g is not conjugate in G_π to any element of L_π.
(ii) $St_G(g) = (-1)^{|\pi|} \mid C_{U_\pi}(g) \mid$ whenever $g \in L_\pi$.

Proof (i) Since $g \in G_\pi$ and it is not conjugate in G_π to any element of L_π, it follows from the above Proposition that

$$St_G(g) = St_G|_{G_\pi}(g) = st_{L_\pi}^{G_\pi}(g) = \frac{1}{\mid L_\pi \mid} \sum_{x \in G_\pi \text{ and } x^{-1}gx \in L_\pi} st_{L_\pi}(x^{-1}gx) = 0.$$

(ii) Suppose that $g \in L_\pi$. Again by the above Proposition,

$$St_G(g) = St_G|_{G_\pi}(g) = st_{L_\pi}^{G_\pi}(g) = \frac{1}{\mid L_\pi \mid} \sum_{x \in G_\pi \text{ and } x^{-1}gx \in L_\pi} st_{L_\pi}(x^{-1}gx).$$

Let $x = uv \in G_\pi = L_\pi U_\pi$, where $u \in L_\pi$ and $v \in U_\pi$. Then $u^{-1}gu \in L_\pi$. Suppose that $x^{-1}gx = v^{-1}u^{-1}guv$ belongs to L_π. Then $[u^{-1}gu, v] = u^{-1}g^{-1}uv^{-1}u^{-1}guv$ belongs to U_π, since U_π is normal in G_π and $v \in U_\pi$. Also $[u^{-1}gu, v]$ belongs to L_π. Since $L_\pi \bigcap U_\pi = \{1\}$, $v \in C_{U_\pi}(u^{-1}gu)$. Conversely, if $v \in C_{U_\pi}(u^{-1}gu)$, then $x^{-1}gx \in L_\pi$. Consequently, using Proposition 5.3.9, we obtain

$$\begin{aligned} St_G(g) = st_{L_\pi}^{G_\pi}(g) &= \frac{1}{\mid L_\pi \mid} \sum_{u \in L_\pi} \sum_{v \in C_{U_\pi}(u^{-1}gu)} st_{L_\pi}(u^{-1}gu) \\ &= \frac{1}{\mid L_\pi \mid} \sum_{u \in L_\pi} \mid C_{U_\pi}(u^{-1}gu) \mid st_{L_\pi}(g) \\ &= \mid C_{U_\pi}(g) \mid st_{L_\pi}(g) = (-1)^{|\pi|} \mid C_{U_\pi}(g) \mid \text{ (Corollary 5.3.8).} \end{aligned}$$

♯

Taking π to be empty set, and using the part (ii) of the above Proposition, we obtain the following Corollary:

Corollary 5.3.11 $St_G(1) = (-1)^{|\emptyset|} \mid C_U(1) \mid = \mid U \mid$. ♯

Thus, the degree of the Steinberg irreducible character is $\mid U \mid$ and

$$\sum_{\pi \subseteq \Delta} (-1)^{|\pi|}[G : G_\pi] = \sum_{\pi \subseteq \Delta} (-1)^{|\pi|} 1_{G_\pi}^G(1) = St_G(1) = \mid U \mid .$$

Corollary 5.3.12 *Let $g \in G_\pi$. Suppose that for any proper subset π' of π, g is not conjugate in G to any element of $G_{\pi'}$. Then*
(i) $\mid C_{U_\pi}(g) \mid = \mid C_G(g) \mid_p$, and
(ii) $St_G(g) = (-1)^{|\pi|} \mid C_G(g) \mid_p$,
where p is the prime with $q = p^r$ for some r and n_p denote highest power of p dividing n.

Proof (i) From Corollary 5.1.23,

$$\frac{St_G(g) \mid G \mid}{St_G(1) \mid C_G(g) \mid} = (-1)^{|\pi|} \frac{\mid C_{U_\pi}(g) \mid\mid G \mid}{\mid U \mid\mid C_G(g) \mid} = (-1)^{|\pi|} \frac{[G : U]}{[C_G(g) : C_{U_\pi}(g)]}$$

is an algebraic integer. Since a rational algebraic integer is necessarily an integer, it follows that $[C_G(g) : C_{U_\pi}(g)]$ divides $[G : U] = \mid G \mid_{p'}$. Consequently, $[C_G(g) : C_{U_\pi}(g)]$ is co-prime to p. Since $\mid C_{U_\pi}(g) \mid$ is a power of p, $\mid C_G(g) \mid_p = \mid C_{U_\pi}(g) \mid$. This proves (i).

The part (ii) follows from Corollary 5.3.10(ii). ♯

Corollary 5.3.13 *If g does not belong to any proper parabolic subgroup of G, then $\mid C_G(g) \mid$ is co-prime to p. In particular, $\mid g \mid$ is also co-prime to p.*

Proof Under the hypothesis of the Corollary, $g \in G_\Delta$ and no conjugate of g belongs to G_π for $\pi \subset \Delta$. Since $U_\Delta = \{1\}$, it follows from the above Corollary that $\mid C_G(g) \mid_p = 1$. Thus, $\mid C_G(g) \mid$ is co-prime to p. Since $< g > \subseteq C_G(g)$, $\mid g \mid$ is also co-prime to p. ♯

Corollary 5.3.14 *Let $g \in G_\pi - \bigcup_{\pi' \subset \pi}(\bigcup_{x \in G} x^{-1}G_{\pi'}x)$. Then g is conjugate to an element of L_π if and only if p is co-prime to the order of g.*

Proof Suppose that g is conjugate to an element of L_π. Since a proper parabolic subgroup of L_π is conjugate in L_π to $L_\pi \bigcap G_{\pi'}$ for some proper subset π' of π, g is not conjugate to any proper parabolic subgroup of L_π. It follows from the above Corollary that a conjugate of g has its order co-prime to p. Thus, order of g is co-prime to p.

Conversely, suppose that the order of g is co-prime to p. Since U_π is normal subgroup of G_π, $< U_\pi, \{g\} > = U_\pi < g >$. Also, since $\mid U_\pi \mid$ is a power of p,

$U_\pi \bigcap < g > = \{1\}$. Thus, $< g >$ and $U_\pi < g > \bigcap L_\pi$ are complements to U_π in $U_\pi < g >$. By the Schur Zasenhauss Lemma (Theorem 10.1.33, Algebra 2), $< g >$ is conjugate to $U_\pi < g > \bigcap L_\pi$. This shows that g is conjugate in G_π to an element of L_π. ♯

Corollary 5.3.15 *Let $g \in G$. Then, the following hold:*
(i) If the order of g is a multiple of p, then $St_G(g) = 0$.
(ii) If the order of g is co-prime to p, then $St_G(g) = \pm \mid C_G(g) \mid$. Indeed, if g is conjugate to an element of L_π but for $\pi' \subset \pi$ it is not conjugate to $L_{\pi'}$, then $St_G(g) = (-1)^{|\pi|} \mid C_G(g) \mid_p$.

Proof (i) Let g be an element of G. Then, there is a subset π of Δ such that g is conjugate to an element of G_π but for $\pi' \subset \pi$, it is not conjugate to any element in $G_{\pi'}$ (if worst comes, we may take $\pi = \emptyset$). If the order of g is a multiple of p, then by Corollary 5.3.14, then g is not conjugate to any element of L_π. From Corollary 5.3.10, $St_G(g) = 0$.
(ii) If order of g is co-prime to p, then by Corollary 5.3.12, $St_G(g) = (-1)^{|\pi|} \mid C_G(g) \mid_p$. ♯

The proof of Proposition 5.3.9 is pain staking. As such the reader may skip the proof of the Proposition and the following Lemmas needed to prove the Proposition.

Lemma 5.3.16 $\sigma = \sigma_0^\pi \sigma_0$ *is the unique element of $W(\Phi)$ satisfying the conditions: (i) $\sigma^{-1} \in D_\pi$, and (ii) $\Phi^+ \bigcap \sigma(\Delta) \subseteq \pi$.*

Proof Since $(\sigma_0^\pi)^{-1}(\pi) \subseteq \Phi^-$, $\sigma^{-1}(\pi) = \sigma_0^{-1}(\sigma_0^\pi)^{-1}(\pi) \subseteq \Phi^+$. Thus, $\sigma^{-1} \in D_\pi$. Next, consider $\sigma(\Delta) = \sigma_0^\pi\sigma_0(\Delta)$. Since $\sigma_0(\Delta) \subseteq \Phi^-$ and $(\sigma_0^\pi(\Phi^-) \bigcap \Phi^+) \subseteq \pi$, it follows that $\Phi^+ \bigcap \sigma(\Delta) \subseteq \pi$.

Conversely, let τ be an element of $W(\Phi)$ such that $\tau^{-1} \in D_\pi$ and $\Phi^+ \bigcap \tau(\Delta) \subseteq \pi$. Let $\alpha \in \Delta$. If $\tau(\alpha) \in \Phi^+$, then $\tau(\alpha) \in \pi$. This means that $\tau(\alpha) \in \pi$ or $\tau(\alpha) \in \Phi^-$. If $\tau(\alpha) \in \pi$, then $\sigma_0^\pi\tau(\alpha) \in \Phi^-$ (from the definition of σ_0^π). Suppose that $\tau(\alpha) \in \Phi^-$. If $\sigma_0^\pi\tau(\alpha) \in \Phi^+$, then $-\tau(\alpha) \in \Phi_\pi^+$ and so $-\alpha \in \tau^{-1}(\Phi_\pi^+)$. This is a contradiction, since $\tau^{-1}(\Phi_\pi^+) \subseteq \Phi^+$. This shows that if $\tau(\alpha) \in \Phi^-$, then $\sigma_0^\pi\tau(\alpha) \in \Phi^-$. It follows that $\sigma_0^\pi\tau(\alpha) \in \Phi^-$ for each $\alpha \in \Delta$. Thus, $\sigma_0^\pi\tau = \sigma_0$ or equivalently, $\tau = \sigma_0^\pi\sigma_0$. ♯

Lemma 5.3.17 *For all $\pi' \subseteq \pi$,*

$$(B \bigcap n_0^\pi n_0 B(n_0^\pi n_0)^{-1}) N_{\pi'} (B \bigcap n_0^\pi n_0 B(n_0^\pi n_0)^{-1}) = L_\pi \bigcap G_{\pi'},$$

where $\nu(n_0^\pi) = \sigma_0^\pi$, $\nu(n_0) = \sigma_0$, ν being the natural quotient map from N to $N/H \approx W(\Phi)$. In particular, the standard parabolic subgroups of L_π are of the form $L_\pi \bigcap G_{\pi'}$.

Proof Clearly,

$$
\begin{aligned}
(B\bigcap n_0^{\pi}n_0B(n_0^{\pi}n_0)^{-1}) &= (B\bigcap (n_0n_0^{\pi})^{-1}B(n_0n_0^{\pi}))\\
&= (UH\bigcap (n_0n_0^{\pi})^{-1}U(n_0n_0^{\pi})H = U_{\sigma_0^{\pi}}H = B_{\pi}.
\end{aligned}
$$

Since (B_{π}, N_{π}) is a Tits pair for L_{π},

$$
\begin{aligned}
&(B\bigcap n_0^{\pi}n_0B(n_0^{\pi}n_0)^{-1})N_{\pi'}(B\bigcap n_0^{\pi}n_0B(n_0^{\pi}n_0)^{-1})\\
&= B_{\pi}N_{\pi'}B_{\pi} = B_{\pi}N_{\pi}B_{\pi}\bigcap BN_{\pi'}B = L_{\pi}\bigcap G_{\pi'}.
\end{aligned}
$$

♯

Lemma 5.3.18 *Let* $n \in N_{\pi,\pi'}$ *and* $\sigma = \nu(n)$. *Then*

$$
G_{\pi}\bigcap nG_{\pi'}n^{-1} = (B\bigcap nBn^{-1})N_{\pi\bigcap\sigma(\pi')}(B\bigcap nBn^{-1}) = G_{\pi\bigcap\sigma(\pi')}.
$$

Proof Let $n_1 \in N_{\pi}, n_2 \in N_{\pi'}, \sigma_1 = \nu(n_1), \sigma_2 = \nu(n_2)$. Since $\sigma = \nu(n)$ belongs to $D_{\pi,\pi'}$, by the definition of $D_{\pi,\pi'}$, $l(\sigma^{-1}\sigma_1) = l(\sigma^{-1}) + l(\sigma_1)$ and $l(\sigma_2\sigma^{-1}) = l(\sigma_2) + l(\sigma^{-1})$. It follows that

$$
B = (B\bigcap nBn^{-1})(B\bigcap n_1Bn_1^{-1}) = (B\bigcap n_2^{-1}Bn_2)(B\bigcap n^{-1}Bn).
$$

In turn,

$$
\begin{aligned}
&(Bn_1B)\bigcap nBn_2Bn^{-1}\\
&= n((n^{-1}Bn_1B)\bigcap (Bn_2Bn^{-1}))\\
&= n((n^{-1}(B\bigcap nBn^{-1}))(B\bigcap n_1Bn_1^{-1})n_1B\\
&\bigcap (Bn_2(B\bigcap n_2^{-1}Bn_2)(B\bigcap n^{-1}Bn)n^{-1}))\\
&= n(((B\bigcap n^{-1}Bn)n^{-1}n_1B)\\
&\bigcap (Bn_2n^{-1}(B\bigcap nBn^{-1}))).
\end{aligned}
$$

Observe that $BnB = Bn'B$ if and only if $\nu(n) = \nu(n')$ and the intersection $BnB\bigcap Bn'B$ is empty set otherwise. Thus, if $(Bn_1B)\bigcap nBn_2Bn^{-1}$ is nonempty, then $\sigma^{-1}\sigma_1 = \sigma_2\sigma^{-1}$ or equivalently, σ_1 belongs to $W_{\pi}\bigcap\sigma W_{\pi'}\sigma^{-1} = W_{\pi\bigcap\sigma(\pi')}$. Further, then

$$
(Bn_1B)\bigcap nBn_2Bn^{-1} = (B\bigcap nBn^{-1})n_1(B\bigcap nBn^{-1}).
$$

This proves the Lemma. ♯

Proof of Proposition 5.3.9 $St_G|_{G_\pi} = (\sum_{\pi' \subseteq \Delta}(-1)^{|\pi'|}1^G_{G_{\pi'}})|_{G_\pi}$
$= \sum_{\pi' \subseteq \Delta}(-1)^{|\pi'|}(1^G_{G_{\pi'}})|_{G_\pi}$ (restriction is linear)
$= \sum_{\pi' \subseteq \Delta}(-1)^{|\pi'|}\sum_{n \in N_{\pi,\pi'}}(1^{G_\pi}_{G_\pi \bigcap nG_{\pi'}n^{-1}})$ (by the Mackey restriction formula), where $N_{\pi,\pi'}$ is the distinguished double coset representative system of G_π and $G_{\pi'}$ in G. Next, by Lemma 5.3.18,

$$G_\pi \bigcap nG_{\pi'}n^{-1} = (B \bigcap nBn^{-1})N_{\pi \bigcap \sigma(\pi')}(B \bigcap nBn^{-1}) = G_\pi \bigcap G_{\sigma(\pi')}.$$

For simplification, denote $(B \bigcap nBn^{-1})N_{\pi \bigcap \sigma(\pi')}(B \bigcap nBn^{-1})$ by $A^\sigma_{\pi'}$. Then,
$St_G|_{G_\pi} = \sum_{\pi' \subseteq \Delta}(-1)^{|\pi'|}\sum_{n \in N_{\pi,\pi'}} 1^{G_\pi}_{A^\sigma_{\pi'}}$ $(\nu(n) = \sigma)$
$= \sum_{\pi' \subseteq \Delta}(-1)^{|\pi'|}\sum_{\sigma \in D_{\pi,\pi'}} 1^{G_\pi}_{A^\sigma_{\pi'}}$
$= \sum_{\pi' \subseteq \Delta}(-1)^{|\pi'|}\sum_{\sigma \in D^{-1}_\pi, \pi' \subseteq (\sigma^{-1}(\Phi^+) \bigcap \Delta)} 1^{G_\pi}_{A^\sigma_{\pi'}}$, since $D_{\pi,\pi'} = D^{-1}_\pi \bigcap D_{\pi'}$
$= \sum_{\sigma \in D^{-1}_\pi}\sum_{\pi' \subseteq \sigma^{-1}(\Phi^+) \bigcap \Delta}(-1)^{|\pi'|}1^{G_\pi}_{A^\sigma_{\pi'}} \quad \cdots (\star)$.

Now, $\pi' = (\pi' \bigcap \sigma^{-1}(\pi)) \bigcup (\pi' \bigcap (\Phi - \sigma^{-1}(\pi))$. Clearly,

$$\pi \bigcap \sigma(\pi') = \pi \bigcap \sigma(\pi' \bigcap \sigma^{-1}(\pi)).$$

Hence from $(\star)$, we obtain
$St_G|_{G_\pi}$
$= \sum_{\sigma \in D^{-1}_\pi}\sum_{(\pi' \bigcap \sigma^{-1}(\pi)) \subseteq ((\sigma^{-1}(\Phi^+) \bigcap \Delta) \bigcap \sigma^{-1}(\pi))}(-1)^{|\pi' \bigcap \sigma^{-1}(\pi)|}$
$\sum_{(\pi' \bigcap (\Phi - \sigma^{-1}(\pi))) \subseteq ((\sigma^{-1}(\Phi^+) \bigcap \Delta) \bigcap (\Phi - \sigma^{-1}(\pi)))}(-1)^{|\Phi - \sigma^{-1}(\pi)|}1^{G_\pi}_{A^\sigma_{\pi'}} \quad \cdots (\star\star)$.

Again, if $(\sigma^{-1}(\Phi^+) \bigcap \Delta) \bigcap (\Phi - \sigma^{-1}(\pi)) \neq \emptyset$, then

$$\sum_{(\pi' \bigcap (\Phi - \sigma^{-1}(\pi))) \subseteq (\sigma^{-1}(\Phi^+) \bigcap \Delta) \bigcap (\Phi - \sigma^{-1}(\pi))} (-1)^{|(\pi' \bigcap (\Phi - \sigma^{-1}(\pi)))|} = 0.$$

Thus,

$$St_G|_{G_\pi} = \sum_{\sigma \in D^{-1}_\pi, (\sigma^{-1}(\Phi^+) \bigcap \Delta) \subseteq \sigma^{-1}(\pi)} \sum_{\pi' \bigcap \sigma^{-1}(\pi)} (-1)^{|\pi' \bigcap \sigma^{-1}(\pi)|}1^{G_\pi}_{A^\sigma_{\pi'}}.$$

By Lemma 5.3.16, $\tau = \sigma^\pi_0\sigma_0$ is the only element of $W(\Phi)$ such that $\tau \in D^{-1}_\pi$ and $(\tau^{-1}(\Phi^+) \bigcap \Delta) \subseteq \tau^{-1}(\pi)$. Further, for such $\tau = \sigma^\pi_0\sigma_0$, $(\tau^{-1}(\Phi^+) \bigcap \Delta) = -\sigma_0(\Delta)$. Hence

$$St_G|_{G_\pi} = \sum_{(\pi' \bigcap \tau^{-1}(\pi)) \subseteq -\sigma_0(\Delta)} (-1)^{|\pi' \bigcap \tau^{-1}(\pi)|}1^{G_\pi}_{A^\tau_{\pi' \bigcap \tau^{-1}(\pi)}}.$$

Put $\pi'' = \tau(\pi' \bigcap \tau^{-1}(\pi))$. Then $\pi'' \subseteq \pi$, and we see that

$$St_G|_{G_\pi} = \sum_{\pi'' \subseteq \pi} (-1)^{|\pi''|} 1_{A^\tau_{\pi''}}^{G_\pi}.$$

By Lemma 5.3.17, $A^\tau_{\pi''} = L_\pi \bigcap G_{\pi''}$. Hence

$$St_G|_{G_\pi} = \sum_{\pi'' \subseteq \pi} (-1)^{|\pi''|} 1_{L_\pi \bigcap G_{\pi''}}^{G_\pi}.$$

In turn,

$$St_G|_{G_\pi} = \sum_{\pi'' \subseteq \pi} (-1)^{|\pi''|} (1_{L_\pi \bigcap G_{\pi''}}^{L_\pi})^{G_\pi},$$

or

$$St_G|_{G_\pi} = (\sum_{\pi'' \subseteq \pi} (-1)^{|\pi''|} 1_{L_\pi \bigcap G_{\pi''}}^{L_\pi})^{G_\pi} = st_{L_\pi}^{G_\pi}.$$

♯

Finally, it is natural to ask about the Steinberg representation which affords the Steinberg character described above. The basic idea of Steinberg was to use the alternating character of the Weyl group $W(\Phi)$ to construct a left ideal of the group algebra $\mathbb{C}(G)$ (indeed of $\mathbb{C}G(V, K)$) which describes the Steinberg representation. Let t denote the section of the natural surjective homomorphism ν from N to $W(\Phi)$ with $t_1 = 1$. Thus, $\nu(t_\sigma) = \sigma$. Consider the element e of $\mathbb{C}(G)$ given by

$$e = (\frac{| U |}{| G |} \sum_{b \in B} b) \sum_{\sigma \in W(\Phi)} (-1)^{Det \sigma} t_\sigma,$$

and the left module $\mathbb{C}(G)e$ over $\mathbb{C}(G)$. The representation thus obtained is the **Steinberg representation** which affords the Steinberg character described earlier. Indeed, Curtis ("The Steinberg Character of a finite group with (B, N) pair", J. Algebra (4) 1966) in 1966 discovered the alternating sum formula for the Steinberg character.

Further the alternating sum formula for the Steinberg character prompted Solomon to describe Steinberg module as homology module of Coxeter complex and the character as Euler Characteristic.

Exercises

5.3.1 Describe the Steinberg character and the Steinberg representation of $SL(2, q)$, $GL(2, q)$, and also of $PSL(2, q)$.

5.3.2 Describe the Steinberg character and the Steinberg representation of B_2, G_2, and also of F_4

5.4 Principal and Discrete Series Representations

In this section, we shall decompose the set of irreducible representations of a Chevalley group in to different equivalence classes and study them. In its spirit, most of these are due to Harish-Chandra.

Proposition 5.4.1 *Let H be a normal subgroup of a finite group G. Let χ be a character of G. Then, the map $T_{G/H}(\chi)$ from G to $\mathbb{C}$ defined by*

$$T_{G/H}(\chi)(g) = \frac{1}{\mid H \mid}\sum_{h\in H}\chi(hg)$$

is also a character of G. More generally, if χ is a generalized character of G, then $T_{G/H}(\chi)$ is also a generalized character of G.

Proof It is sufficient to prove the first statement. Let V be a $\mathbb{C}(G)$-module affording the character χ. Consider the element

$$e = \frac{1}{\mid H \mid}\sum_{h\in H} h$$

of $\mathbb{C}(G)$. Clearly, e is idempotent ($e^2 = e$) and $geg^{-1} = e$. Thus, $geV = geg^{-1}gV = egV = eV$ for all $g \in G$. This means that eV is a $\mathbb{C}(G)$-submodule of V. Since e is idempotent, $V = eV \oplus (1-e)V$. Further, $eV = \{v \in V \mid hv = v \ for \ all \ h \in H\} = V^H$. Let μ denote the character afforded by the submodule eV. Then

$$\mu(g) = \chi(ge) = \frac{1}{\mid H \mid}\sum_{h\in H}\chi(gh) = \frac{1}{\mid H \mid}\sum_{h\in H}\chi(ghg^{-1}g) = \frac{1}{\mid H \mid}\sum_{k\in H}\chi(kg).$$

This shows that $\mu = T_{G/H}(\chi)$. ♯

Definition 5.4.2 The character $T_{G/H}(\chi)$ of G is called the **truncation** of the character χ by H. If G is a semi-direct product $H \succ K$, then $T_{G/H}(\chi)$ is also denoted by $T_K(\chi)$.

For simplicity, from now onward, we shall denote $G(V, K)$ by G and $G(V, K)^F = G(V, F_q)$ by G^F. Let χ be an irreducible character of G^F. Let G_π be a principal parabolic subgroup of G^F, where $\pi \subseteq \Delta$. Let $G_\pi = U_\pi L_\pi$ be the standard Levi decomposition of G_π, where U_π is the unipotent radical of G_π and L_π is the corresponding standard Levi subgroup.

Definition 5.4.3 An irreducible character χ of G^F is called a **Cuspidal Character** of G^F if $T_{G_\pi/U_\pi}(\chi) = 0$ for all proper principal parabolic subgroups G_π, $\pi \subset \Delta$ (observe that the notation $A \subset B$ means that A is a proper subset of B).

Since any proper parabolic subgroup P of G^F is similar to proper principal parabolic subgroup, χ is cuspidal if and only if $T_{P/U}(\chi) = 0$ for all proper parabolic subgroup. If G is a torus or a Borel subgroup, then all the irreducible representations will be termed as cuspidal representations.

Proposition 5.4.4 *Let χ be an irreducible character of G^F. Then, the following conditions are equivalent:*

(i) χ is cuspidal.

(ii) $< \chi, 1_{U_\pi}^{G^F} > = 0$ for all proper subset π of Δ.

(iii) $< \chi|_{U_\pi}, 1 >_{U_\pi} = 0$ for all proper subset π of Δ.

(iv) The trivial representation 1 of U_π does not appear in the direct sum decomposition of $\rho|_{U_\pi}$ as irreducible representations.

(v) $\sum_{u\in U_\pi} \chi(ug) = 0$ for all $g \in G^F$ and $\pi \subset \Delta$.

(vi) $\sum_{u\in U_\pi} \chi(gu) = 0$ for all $g \in G^F$ and $\pi \subset \Delta$.

Proof $((i) \Longrightarrow (ii))$: Assume (i). Then $T_{G_\pi/U_\pi}(\chi)(g) = 0$ for all $g \in G^F$ and $\pi \subset \Delta$. In particular,

$$T_{G_\pi/U_\pi}(\chi)(1) = \frac{1}{\mid U_\pi \mid} \sum_{u\in U_\pi} \chi(u) = 0$$

for all $\pi \subset \Delta$. This means that $< \chi|_{U_\pi}, 1 >_{U_\pi} = 0$ for all $\pi \subset \Delta$. By the Frobenius reciprocity $< \chi, 1_{U_\pi}^{G_F} > = 0$.

$((ii) \Longrightarrow (iii))$: Immediate from the Frobenius reciprocity.

$((iii) \Longrightarrow (iv))$: Immediate.

$((iv) \Longrightarrow (v))$: Assume (iv). Ler ρ be a representation of G^F affording the character χ and η be an irreducible component of $\rho|_{U_\pi}$. Then from the Schur orthogonality relation,

$$< \eta_{ij}, (1_{U_\pi})_{11} >_{U_\pi} = 0$$

for all i, j. Consequently,

$$\sum_{u\in U_\pi} \eta_{ij}(u) = 0$$

for all i, j. Hence

$$\sum_{u\in U_\pi} \eta(u) = 0$$

for all irreducible components of$\rho|_{U_\pi}$. It follows that

$$\sum_{u\in U_\pi} \rho(u) = 0.$$

Thus,

$$0 = (\sum_{u\in U_\pi} \rho(u))\rho(g) = (\sum_{u\in U_\pi} \rho(ug))$$

for all $g \in G^F$. Hence

$$\sum_{u\in U_\pi} \chi(ug) = \sum_{u\in U_\pi} Tr(\rho(ug)) = Tr((\sum_{u\in U_\pi} \rho(ug))) = 0$$

for all $g \in G^F$.

$((v) \Longrightarrow (vi))$: Since

$$\sum_{u\in U_\pi} \chi(ug) = \sum_{u\in U_\pi} \chi(gg^{-1}ug) = \sum_{v\in U_\pi} \chi(gv),$$

(v) and (vi) are equivalent.

$((vi) \Longrightarrow (i))$: Evident from the definition of cuspidal character. ♯

Now, let G_π be a principal parabolic subgroup of G^F and $G_\pi = U_\pi L_\pi$ the standard Levi decomposition. If ρ is an irreducible representation of L_π, then we have a corresponding irreducible representation $\hat{\rho}$ of G_π which is given by $\hat{\rho}(ul) = \rho(l)$, where $u \in U_\pi$ and $l \in L_\pi$. Recall (Sect. 5.3) that L_π is also a group with (B, N) pair. Thus, we can talk of parabolic subgroups of L_π and also the notion of cuspidal characters of L_π as in case of G^F.

Theorem 5.4.5 *Let χ be an irreducible character of G^F. Then, there is a subset π of Δ and an irreducible cuspidal character μ of L_π such that*

$$< \chi, \hat{\mu}^{G^F} >\neq 0,$$

where $\hat{\mu}$ is the representation $\mu o \nu$ of G_π, ν being the natural projection of G_π on to L_π.

Proof Consider the set $S = \{\pi \subseteq \Delta \mid < \chi, 1_{U_\pi}^{G^F} > = < \chi_{U_\pi}, 1_{U_\pi} >\neq 0\}$. Clearly, $U_\Delta = \{1\}$ and hence $< \chi|_{U_\Delta}, 1_{U_\Delta} >\neq 0$. Hence $\Delta \in S$ and so $S \neq \emptyset$. Let π be a minimal member of S. Let V be a simple G^F-module affording the character χ. Consider the subspace $V^{U_\pi} = \{v \in V \mid uv = v \; for \; all \; u \in U_\pi\}$ of fixed points of U_π. Since $< \chi|_{U_\pi}, 1_{U_\pi} >\neq 0$, $V^{U_\pi} \neq \{0\}$. Let $g \in G_\pi$ and $v \in V^{U_\pi}$. Then for all $u \in U_\pi$, $ugv = gg^{-1}ugv = gv$, since $g^{-1}ug \in U_\pi$. Thus V^{U_π} is a G_π-module. In particular, V^{U_π} is a L_π-module. Let μ be the character of L_π afforded by this module. Suppose that $\mu = \sum_{i=1}^r \mu_i$ be the representation of μ as sum of irreducible characters. Let $\hat{\mu}$ and $\hat{\mu}_i$ denote the corresponding characters of G_π. Indeed, the G_π-module V^π affords the character $\hat{\mu}$ and $\hat{\mu} = \sum_{i=1}^r \hat{\mu}_i$. Note that $\hat{\mu}_i$ are also irreducible characters of G_π. Since V_π is a G_π submodule of V, $\chi|_{G_\pi} = \hat{\mu}$ and each $\hat{\mu}_i$ is irreducible component of $\chi|_{G_\pi}$. Consequently, by the Frobenius reciprocity,

$$< \hat{\mu_i}^{G^F}, \chi >_{G^F} = < \hat{\mu}_i, \chi|_{G_\pi} >\neq 0.$$

This means that χ appears as an irreducible component of the induced character $\hat{\mu_i}^{G^F}$ for each i. It suffices to show that each μ_i is a cuspidal character of L_π. Suppose contrary. Suppose that μ_i is not cuspidal character of L_π for some i. Recall that a principal parabolic subgroup of L_π is of the form $G_{\pi'} \bigcap L_\pi$, where $\pi' \subseteq \pi$. Further, the unipotent radical of $G_{\pi'} \bigcap L_\pi$ is $U_{\pi'} \bigcap L_\pi$ and $L_{\pi'}$ is the Levi part. More explicitly, $G_{\pi'} \bigcap L_\pi = (U_{\pi'} \bigcap L_\pi)L_{\pi'}$. Since μ_i is not cuspidal, using Proposition 5.4.4, we see that there is a proper subset π' of π such that

$$< \mu_i|_{U_{\pi'} \bigcap L_\pi}, 1_{U_{\pi'} \bigcap L_\pi} > \neq 0.$$

This means that there is a nonzero element $v \in V^{U_\pi}$ such that $uv = v$ for all $u \in U_{\pi'} \bigcap L_\pi$. But already $uv = v$ for all $u \in U_\pi$. Since $U_{\pi'} = U_\pi(U_{\pi'} \bigcap L_\pi)$, $uv = v$ for all $u \in U_{\pi'}$. Thus, $< \chi|_{U_{\pi'}}, 1_{U_{\pi'}} > \neq 0$. Hence $\pi' \in S$. This is a contradiction to the the minimality of π. This shows that μ_i is cuspidal for each i. ♯

We say that a subset π of Δ is an associate to a subset π' of Δ if there is an element $\sigma \in W(\Phi)$ such that $\sigma(\pi) = \pi'$. Clearly, the relation "is associate to" is an equivalence relation on the power set $\wp(\Delta)$ of Δ. Let Σ be a set obtained by selecting one and only one member from each equivalence class. Let $\pi \in \Sigma$. Let $W_{(\pi)}$ denote the subgroup $\{\sigma \in W(\Phi) \mid \sigma(\pi) = \pi\}$ of $W(\Phi)$. Let $Ch(\pi)$ denote the set of irreducible cuspidal character of L_π. If $\mu \in Ch(\pi)$ and $\sigma \in W_{(\pi)}$. It can be easily observed that $\mu^\sigma \in Ch(\pi)$, where $\mu^\sigma(l) = n_\sigma l n_\sigma^{-1}$. Thus, $W_{(\pi)}$ acts on $Ch(\pi)$. Let $\hat{Ch}(\pi)$ be a set obtained by selecting one and only one member from each orbit of the action of $W_{(\pi)}$ on $Ch(\pi)$. Let X_π denote the set $\{\chi \mid \chi\ is\ an\ irreducible\ component\ of \hat{\mu}^{G^F}\ for\ some\ \mu \in \hat{Ch}(\pi)\}$.

Theorem 5.4.6 *An irreducible representation of G^F belongs to one and only one X_π, $\pi \in \Sigma$.* ♯

The proof of the above Theorem is a little involved and we leave it. The reader may refer to "Finite Groups of Lie Types, Conjugacy Classes and Characters" by Carter for a proof.

An irreducible character is said to lie in π**-series** if it belongs to X_π. The members of $X_\emptyset$ are called **Principal Series** irreducible characters and the members of X_Δ are called **Discrete Series** irreducible characters. Clearly, the parabolic subgroup $G_\emptyset = B^F$ and the Levi decomposition is $B^F = U^F T^F$. Thus, the principal series representations of G^F are irreducible components of $\Theta_{B^F}^{G^F}$, where Θ is an irreducible character of T^F and Θ_{B^F} is given by $\Theta_{B^F}(ut) = \Theta(t)$, $u \in U^F$, $t \in T^F$. Next, the parabolic subgroup $G_\Delta = G^F$, $U_\Delta = 1$ and $L_\Delta = G^F$. As such the discrete series representations of G^F are precisely the cuspidal representations of G^F.

The following Proposition follows from Proposition 5.1.55 and the Bruhat decomposition of G^F:

Proposition 5.4.7 *It Θ and λ are irreducible characters of $T^F \subset B^F$, then*

$$< \Theta_{B^F}^{G^F}, \lambda_{B^F}^{G^F} > = | \{\sigma \in W(\Phi) \mid \Theta^\sigma = \lambda\} |,$$

where as usual $\Theta^\sigma(t) = \Theta(n_\sigma t n_\sigma^{-1}$. In particular, if Θ and λ are not conjugate in the above sense, then

$$< \Theta_{B^F}^{G^F}, \lambda_{B^F}^{G^F} > = 0,$$

and $\Theta_{B^F}^{G^F}$ *is irreducible if and only if* Θ *is regular in the sense that* $\Theta^\sigma = \Theta$ *if and only if* σ *is the identity element.* ♮

Exercises

5.4.1 Describe the series representations of $GL(2, q)$, $SL(2, q)$, $PSL(2, q)$, $A_2(q)$, and $B_2(q)$.

5.5 Deligne–Lusztig Generalized Characters

In this section, we introduce Deligne–Lusztig generalized characters and study them briefly. Deligne and Lusztig introduced these generalized characters with the help of l-adic cohomology (see Sect. 4.5 of Algebra 4) and settled the conjecture of Macdonald which asserts that the irreducible characters of a Chevalley group can be parametrized in terms of the pairs (T, Θ), where T is a F-stable maximal Torus and Θ is a character of T^F.

We list some of the important properties of the l-adic cohomology functors with compact support for their use in the discussions involving Deligne–Lusztig characters (see Algebra 3, Theorem 4.5.7).

P_1. $H_c^m(-, \overline{\mathbb{Q}_{(l)}})$ defines a contravariant functor from the category of schemes with finite morphisms to the category of vector spaces over the algebraic closure $\overline{\mathbb{Q}_{(l)}}$ of the field $\mathbb{Q}_{(l)}$ of l-adic numbers. In particular, we have a representation ρ of $Aut(X)$ on the space $H_c^m(X, \overline{\mathbb{Q}_{(l)}})$ given by $\rho(\sigma) = (H_c^m(\sigma))^t$(note that an automorphism of a scheme is a finite morphism). Further, $H_c^m(X, \overline{\mathbb{Q}_{(l)}})$ may be nonzero only when $0 \leq m \leq 2DimX$.

P_2. If X is an affine space A_K^n, then $H_c^{2n}(X, \overline{\mathbb{Q}_{(l)}}) \approx \overline{\mathbb{Q}_{(l)}}$ and $H_c^m(X, \overline{\mathbb{Q}_{(l)}}) = \{0\}$ for $m \neq 2n$.

P_3. Let p be a prime different from l and $q = p^r$. Let A be a set of polynomials in $F_q[X_1, X_2, \ldots, X_n]$ defined over the Galois field F_q. Let K be a field extension of F_q, and $Y(K) = V(A)$ denote the variety determined by A over K. The map F from $Y(K)$ to $Y(K)$ given by $F(\alpha_1, \alpha_2, \cdots, \alpha_n) = (\alpha_1^q, \alpha_2^q, \ldots, \alpha_n^q)$ is called the **Standard Frobenius map**. Evidently, $Y(K)^F = Y(F_q)$. More generally, an F_q structure on a scheme (X, ϑ_X) is a scheme (X_0, ϑ_{X_0}) over F_q together with an isomorphism from (X, ϑ_X) to the fiber product $X_0 \times_{Spec\ F_q} Spec\ K$ of the schemes (X_0, ϑ_{X_0}) and $(Spec\ K, \vartheta^K)$ over $(Spec\ F_q, \vartheta^{F_q})$. Let F_0 denote the morphism $(I_{X_0}, F_0^\sharp)$ from (X_0, ϑ_{X_0}) to itself given by $(F_0^\sharp)_U(a) = a^q$ for all $a \in \vartheta_{X_0}(U)$ and for all open subsets U of X_0. The morphism $F_0 \times 1$ from $X_0 \times_{Spec\ F_q} Spec\ K$ to itself induces an morphism F from (X, ϑ_X) to itself. This morphism is called the F_q- Frobenius map on (X, ϑ_X).

Grothendieck Trace Formula. Let l be a prime different from p. Then

$$| X^F | = \sum_{r=0}^{2n} (-1)^r trace(F, H_c^r(X, \overline{\mathbb{Q}_{(l)}}),$$

where $DimX = n$ and F is the F_q- Frobenius map described above. Evidently, F^m is the F_{q^m}- Frobenius map. Observe that this was an essential requirement (the hypothesis(iv) of Theorem 4.5.2, of Algebra 3.) to establish a part of Weil conjectures.

P_4. If g is an automorphism of X of finite order, then the Lefschetz number $L(g, X)$ given by

$$L(g, X) = \sum_{m=0}^{2DimX} (-1)^m Trace(g, H_c^m(X, \overline{\mathbb{Q}_{(l)}}))$$

is an integer which is independent of l.

P_5. Let X and Y be algebraic varieties and f be a morphism from X to Y such that each fiber $f^{-1}(\{y\})$ is an affine variety isomorphic to A_K^n. Let g and g' be automorphisms of X and Y, respectively, which are of finite orders. Suppose that $fog = g'of$. Then $L(g, X) = L(g', Y)$.

P_6. Let $\{X_i \mid 1 \leq i \leq n\}$ be a pairwise disjoint finite family of locally closed subsets of X such that $X = \bigcup_{i=1}^n X_i$. Let g be an automorphism of X of finite order such that $g(X_i) = X_i$ for each i. Then

$$L(g, X) = \sum_{i=1}^{n} L(g, X_i).$$

Further, suppose that $\bigcup_{i=1}^m X_i$ is closed for each $m \leq n$. Let G be a finite group of automorphisms of X such that each X_i is invariant under G. Let $H_c^r(X_i, \overline{\mathbb{Q}_{(l)}})_\Theta$ be a subspace of $H_c^r(X_i, \overline{\mathbb{Q}_{(l)}})$ affording an irreducible character Θ of G. Suppose that $H_c^r(X_i, \overline{\mathbb{Q}_{(l)}})_\Theta = 0$ for all r and i. Then $H_c^r(X, \overline{\mathbb{Q}_{(l)}})_\Theta = 0$ for all r.

P_7. Let $\{X_i \mid 1 \leq i \leq n\}$ be a pairwise disjoint finite family of closed subsets of X such that $X = \bigcup_{i=1}^n X_i$. Let G be an automorphism group of X of finite order such that for each pair i, j, there is a $g \in G$ such that $g(X_i) = X_j$. Let $H = \{g \in G \mid g(X_1) = X_1\}$ be the subgroup of G. Then, the generalized character $g \mapsto L(g, X)$ of G is induced by the generalized character $h \mapsto L(h, X_1)$ of H.

P_8. Let X be an affine variety and G be a finite group of automorphisms of X. Then

$$H_c^m(X/G, \overline{\mathbb{Q}_{(l)}}) \approx H_c^m(X, \overline{\mathbb{Q}_{(l)}})^G,$$

where $H_c^m(X, \overline{\mathbb{Q}_{(l)}})^G$ is the subspace of the fixed points of the action of G on $H_c^m(X, \overline{\mathbb{Q}_{(l)}})$. Further, let f be a G-equivariant automorphism of X of finite order and g be an automorphism of X/G such that $\nu o f = g o \nu$, where ν is the quotient map from X to X/G. Then

$$L(g, X/G) = \frac{1}{| G |} \sum_{x \in G} L(fox, X).$$

P_9. If X and Y are algebraic varieties, then we have the Kunneth formula

$$H_c^m(X \times Y, \overline{\mathbb{Q}_{(l)}}) = \oplus \sum_{r+s=m} (H_c^r(X, \overline{\mathbb{Q}_{(l)}}) \otimes H_c^r(X, \overline{\mathbb{Q}_{(l)}})).$$

Further,

$$L(g \times g', X \times Y) = L(g, X)L(g', Y),$$

where g and g' are automorphisms of X and Y, respectively, which are of finite orders.

P_{10}. Let g be an automorphism of X of finite order. Suppose that $g = su = us$, where order of s is co-prime to p and order of u is a power of p, p being a prime. Let X^s denote the fixed point set of s. Then

$$L(g, X) = L(u, X^s).$$

In particular, if $X^s = \emptyset$, then $L(g, X) = 0$.

P_{11}. If X is finite and g is an automorphism of X, then $L(g, X) = | X^g |$.

P_{12}. If G is a connected algebraic group which acts as a group of automorphisms of X, then the induced action of G on $H_c^m(X, \overline{\mathbb{Q}_{(l)}})$ is the trivial action. ♯

Consider $G(V, K)$, and $G(V, F_q) = G(V, K)^F$, where K is an algebraically closed field of characteristic p, $q = p^m$ for some m and F the Frobenius map $a \mapsto a^q$. For simplicity in the following discussions, we shall denote $G(V, K)$ by G and $G(V, F_q) = G(V, K)^F$ by G^F.

F-stable Maximal Tori and F-stable Borel subgroups

Proposition 5.5.1 *Let Ω^F denote the set of all pairs (B, T) such that B is a F-stable Borel subgroup of G containing the F-stable maximal torus T. Then Ω^F is nonempty and G^F acts transitively on Ω^F by inner conjugation. Further, Ω^F is a finite set.*

Proof We first show that $\Omega^F \neq \emptyset$. Let T be a maximal torus (not necessarily F-stable) contained in a Borel subgroup B. Then $F(B)$ is also a Borel subgroup of G. Since any two Borel subgroups of G are conjugate, $F(B) = aBa^{-1}$ for some $a \in G$. By the Lang–Steinberg Theorem, there is an element $b \in G$ such that $a = F(b)^{-1}b$. Then $\hat{B} = bBb^{-1}$ is a Borel subgroup of G and $F(\hat{B}) = F(b)F(B)F(b)^{-1} = ba^{-1}F(B)(ba^{-1})^{-1} = ba^{-1}aBa^{-1}ab = bBb^{-1} = \hat{B}$. Thus, $\hat{B}$ is F-stable Borel subgroup. $\hat{B}$ is also connected algebraic group and bTb^{-1} is a maximal torus of G contained in $\hat{B}$. Again, since $\hat{B}$ is F-stable, $F(bTb^{-1})$ is maximal torus contained in $\hat{B}$. Hence as in the previous case, there is a $u \in \hat{B}$ such that $\hat{T} = ubTb^{-1}u^{-1}$ is F-stable. This shows that $(\hat{B}, \hat{T})$ belongs to Ω^F.

Next, if $(B, T) \in \Omega^F$ and $g \in G^F$, then it is clear that (gBg^{-1}, gTg^{-1}) belongs to Ω^F. This means that G^F acts on Ω^F by inner conjugation. Let (B, T) and (B', T') be members of Ω^F. Since any two Borel subgroups of G are conjugate in G, there is a $g \in G$ such that $gBg^{-1} = B'$. Again, since B and B' are F-stable, $gBg^{-1} = F(g)F(B)F(g)^{-1} = F(g)BF(g)^{-1}$. Thus, $g^{-1}F(g)B = Bg^{-1}F(g)$.

Since $N_G(B) = B$ (B being a maximal solvable subgroup), $g^{-1}F(g) \in B$. From the surjectivity of the Lang map (Lang–Steinberg Theorem) applied on B, there is an element $b \in B$ such that $g^{-1}F(g) = b^{-1}F(b)$. But, then $F(gb^{-1}) = gb^{-1} \in G^F$. Clearly, $gb^{-1}B(gb^{-1})^{-1} = B'$. Also $T_0 = gb^{-1}T(gb^{-1})^{-1}$ is a maximal F-stable subgroup of G contained in B'. Since T' is also a F-stable maximal subgroup contained in B', there is an element $u \in B'$ such that $uT_0u^{-1} = T'$. Consequently,

$$F(u)T_0F(u)^{-1} = F(u)F(T_0)F(u)^{-1} = F(T') = T' = uT_0u^{-1}.$$

This means that $u^{-1}F(u)T_0 = T_0u^{-1}F(u)$. It follows that $u^{-1}F(u) \in N_G(T_0)$. Since B' is F-stable and $u \in B'$, $u^{-1}F(u) \in B' \bigcap N_G(T_0)$. Applying the Lang–Steinberg Theorem to $B' \bigcap N_G(T_0)$ we get an element $v \in B' \bigcap N_G(T_0)$ such that $u^{-1}F(u) = v^{-1}F(v)$. Clearly, $z = uv^{-1} \in G^F \bigcap B'$. Thus , $x = zgb^{-1} \in G^F$ is such that $xBx^{-1} = B'$ and also $xTx^{-1} = T'$. The last assertion follows, since G^F is finite. ♯

Corollary 5.5.2 *G^F acts transitively on the set of all F-stable Borel subgroups of G through inner conjugation.* ♯

Definition 5.5.3 A maximal F-stable torus T is called a **F- split maximal torus** if it is contained in a F-stable Borel subgroup. All other F-stable maximal tori are called non-split maximal tori. If T is a F-split maximal torus of G, then T^F is called a **split Maximal Torus** of G^F. If T is a non-split F-stable maximal torus of G, then T^F is called a non-split maximal torus of G^F.

The following Corollary is immediate from Proposition 5.5.1.

Corollary 5.5.4 *All split maximal tori of G^F are conjugate in G^F.* ♯

Proposition 5.5.1 ensures the existence of F-split maximal tori. Our next aim is to discuss non-F-split maximal tori.

Let T be a F-stable maximal (not necessarily split) torus of G. Let (B_0, T_0) be a member of Ω^F (see Proposition 5.5.1). Since any two maximal tori of G are conjugatw, we have an element $a \in G$ such that $T_0 = aTa^{-1}$. Since T_0 is F-stable,

$$\begin{aligned} T_0 &= aTa^{-1} = F(T_0) = F(a)F(T)F(a)^{-1} \\ &= F(a)TF(a)^{-1} = F(a)a^{-1}T_0aF(a)^{-1}. \end{aligned}$$

This means that $aF(a)^{-1} \in N_G(T_0)$. Now,

$$T^F = \{a^{-1}ta \mid t \in T_0 \; and \; F(a)^{-1}F(t)F(a) = a^{-1}ta\}.$$

Thus,

$$aT^Fa^{-1} = \{t \in T_0 \mid aF(a)^{-1}F(t)F(a)a^{-1} = t\}.$$

This means that T^F is isomorphic (conjugate) to the subgroup of T_0 consisting of the fixed points of the automorphism $\eta = i_{aF(a)^{-1}} o F$ (i_x denotes the inner automorphism determined by x). Thus, every F-stable subgroup of $T = a^{-1}T_0a$ determines an element $aF(a)^{-1} \in N(T_0)$ which, in turn, determines a unique elememrnt σ_T of the Weyl group. Conversely, given an element σ of the Weyl group, we have an element $n_\sigma \in N_G(T_0)$. representing σ. By the Lang–Steinberg there is an element there is an element a in G such that $aF(a)^{-1} = n_\sigma$. Then

$$F(a^{-1}T_0a) = F(a)^{-1}T_0F(a) = a^{-1}aF(a)^{-1}T_0F(a)a^{-1}a = a^{-1}T_0a.$$

This means that $T = a^{-1}T_0a$ is F-stable and T^F is isomorphic to the subgroup of T_0 consisting of the fixed points of $i_{n_\sigma} oF$. The F-stable subgroup $T = a^{-1}T_0a$ (T^F) is said to be obtained from split F-stable maximal torus T_0 (T_0^F) by twisting with σ. We shall denote σ by σ_T, where $T = a^{-1}T_0a$.

Example 5.5.5 Consider $G = GL(2, K)$, where K is an algebraically closed field of characteristic $p \neq 0$. Let F be the Frobenius given by $F([a_{ij}]) = [a_{ij}^q]$, where $q = p^r$. Then $G^F = GL(2, q)$. The pair $(B_0, T_0) \in \Omega^F$, where B_0 is the subgroup of $GL(2, K)$ consisting of upper triangular matrices and T_0 is the subgroup of diagonal matrices. Observe that the matrix $\begin{bmatrix} 0 & 1 \\ 1 & 0 \end{bmatrix}$ belongs to $N_G(T_0)$. Indeed, $N_G(T_0) = T_0 \bigcup \begin{bmatrix} 0 & 1 \\ 1 & 0 \end{bmatrix} T_0$, and $W(T_0) = N_G(T_0)/T_0$ is of order 2. Let $A = \begin{bmatrix} a & b \\ c & d \end{bmatrix}$ be a member of $GL(2, K)$ such that $AF(A)^{-1} = \begin{bmatrix} 0 & 1 \\ 1 & 0 \end{bmatrix}$. Then

$$\begin{bmatrix} a & b \\ c & d \end{bmatrix} = \begin{bmatrix} 0 & 1 \\ 1 & 0 \end{bmatrix} \begin{bmatrix} a^q & b^q \\ c^q & d^q \end{bmatrix}.$$

This means that $a = c^q,\ c = a^q,\ b = d^q, \ and\ d = b^q$. Consequently, $a^{q^2} = a \neq 0$, $b^{q^2} = b \neq 0$, and $A = \begin{bmatrix} a & b \\ a^q & b^q \end{bmatrix}$, where $a, b \in F_{q^2}^\star$ and $ab^q - ba^q \neq 0$. For such A, $T = A^{-1}T_0A$ is non-split maximal torus of $GL(2, K)$. Further, then T^F is conjugate to the subgroup of T_0 consisting of the matrices of of the form $\begin{bmatrix} a & 0 \\ 0 & a^q \end{bmatrix}$, where $a \in F_{q^2}^\star$. Thus, T^F is a cyclic group of order $q^2 - 1$.

Similarly, non-split maximal torus in $SL(2, q)$ is isomorphic to the subgroup of T_0 consisting of the matrices of the type $\begin{bmatrix} a & 0 \\ 0 & a^q \end{bmatrix}$, where $a \in F_{q^2}^\star$ and $a^{q+1} = 1$. This is a cyclic group of order $q + 1$.

Let T be a F-stable maximal Torus of G and B a Borel subgroup containing T (note that B need not be F-stable). Then $B = UT$, where U is the unipotent radical of B. Recall the Lang map L from G to G which is given by $L(g) = g^{-1}F(g)$. Thus, $G^F = L^{-1}(1)$. Put $\hat{X} = L^{-1}(U)$. Since L is a morphism of varities, $\hat{X}$ is an affine algebraic variety. Recall the l-adic cohomology functors $H_c^r(-, \overline{\mathbb{Q}_l})$ (with compact support) from the category of algebraic varieties to the category of vector spaces over $\overline{\mathbb{Q}_l}$. For each $g \in G^F$, we have the automorphism $x \mapsto gx$ of $\hat{X}$ which, in turn, induces an automorphism $v \mapsto gv$ of the vector space $H_c^r(\hat{X}, \overline{\mathbb{Q}_l})$ for each r. This defines a left G^F-module structure on $H_c^r(\hat{X}, \overline{\mathbb{Q}_l})$, and also a bi-$(G^F, T^F)$-module structure on $H_c^r(\hat{X}, \overline{\mathbb{Q}_l})$. Denote the corresponding representation of $G^F \times T^F$ by ρ_r. Let Θ be an irreducible character of T^F. Since $\Theta(a)$ is an algebraic integer for each $a \in T^F$ and $\overline{\mathbb{Q}_l}$ is an algebraically closed field containing $\mathbb{Q}$, $\Theta \in Hom(T^F, \mathbb{C}^\star) = Hom(T^F, \overline{\mathbb{Q}_l})$. Let $H_c^r(\hat{X}, \overline{\mathbb{Q}_l})_\Theta$ denote the subspace $\{v \in H_c^r(\hat{X}, \overline{\mathbb{Q}_l}) \mid va = \Theta(a)v \ \forall a \in T^F\}$ of $H_c^r(\hat{X}, \overline{\mathbb{Q}_l})$. Further, if $g \in G^F$ and $v \in H_c^r(\hat{X}, \overline{\mathbb{Q}_l})_\Theta$, then

$$(gv)a = g(va) = g\Theta(a)v = \Theta(a)(gv)$$

and so $gv \in H_c^r(\hat{X}, \overline{\mathbb{Q}_l})_\Theta$. Thus, $H_c^r(\hat{X}, \overline{\mathbb{Q}_l})_\Theta$ is a left G^F-module. Let η_r denote the corresponding representation of G^F. Since $Tr(\eta_r(g))$ is an algebraic integer, we have a map $R^B_{T,\Theta}$ from G^F to $\overline{\mathbb{Q}_l} \bigcap \mathbb{C}$ defined by

$$R^B_{T,\Theta}(g) = \sum_{r=0}^{2m} (-1)^r Tr(\eta_r(g)) \quad \cdots \ (1),$$

where $m = Dim\hat{X}$. This generalized character is called a **Deligne–Lusztig Character** of G^F. We shall see soon that $R^B_{T,\Theta}$ does not depend on the choice of the Borel subgroup B containing T.

To describe a convenient formula for $R^B_{T,\Theta}$, we need the following two Lemmas:

Lemma 5.5.6 *Let $\Theta : A \longrightarrow K^\star$ be an irreducible character of a finite abelian group A, where K is an algebraically closed field of characteristic 0. Let V be a finite-dimensional vector space over K which is a right A-module. Consider the element*

$$e = \frac{1}{\mid A \mid} \sum_{a \in A} \Theta(a^{-1})a$$

of the group algebra $K(A)$. Then e is idempotent in the sense that $e^2 = e$ and $Ve = \{v \in V \mid va = \Theta(a)v\}$.

Proof $e^2 = (\frac{1}{|A|} \sum_{a \in A} \Theta(a^{-1})a)(\frac{1}{|A|} \sum_{a \in A} \Theta(a^{-1})a)$
$= \frac{1}{|A|^2} \sum_{a \in A} \sum_{b \in A} \Theta((ab)^{-1})ab$
$= \frac{1}{|A|^2} \mid A \mid \sum_{c \in A} \Theta(c^{-1})c$
$= \frac{1}{|A|} \sum_{c \in A} \Theta(c^{-1})c = e.$

Next, we denote $\{v \in V \mid va = \Theta(a)v\}$ by V_Θ. We need to show that $V_\Theta = Ve$. Let $v \in V_\Theta$. Then $va = \Theta(a)v$ and so

$$
\begin{aligned}
ve &= v\frac{1}{\mid A\mid}\sum_{a\in A}\Theta(a^{-1})a = \frac{1}{\mid A\mid}\sum_{a\in A}\Theta(a^{-1})va \\
&= \frac{1}{\mid A\mid}\sum_{a\in A}\Theta(a^{-1})\Theta(a)v = \frac{1}{\mid A\mid}\sum_{a\in A}\Theta(1)v = v.
\end{aligned}
$$

This shows that $V_\Theta \subseteq Ve$. Let $v \in Ve$. Then $v = we$ for some $w \in V$. In turn,

$$
va = wea = w(\frac{1}{\mid A\mid}\sum_{b\in A}\Theta(b^{-1})b)a.
$$

Putting $b = a^{-1}c$, we obtain that

$$
va = \Theta(a)w\frac{1}{\mid A\mid}\sum_{c\in A}\Theta(c^{-1})c = \Theta(a)we = \Theta(a)v.
$$

This means that $v \in V_\Theta$ and so $Ve \subseteq V_\Theta$. ♯

Lemma 5.5.7 *Let V be a finite-dimensional vector space over an algebraically closed field K of characteristic 0. Suppose that V is a bi- (G, A)-module, where G is a finite group and A is a finite abelian group. Let ρ denote the corresponding representation of $G \times A$ on V. Let Θ be an irreducible character of A. Then V_Θ is a G- module. Further, if we denote the cooresponding representation of G by η, then*

$$
Tr(\rho(g, e)) = Tr(\eta(g))
$$

for all $g \in G$.

Proof Clearly, V is a $G \times A$-module, where the module structure is given by $(g, a)v = gva$. Let $v = we$ belongs to $Ve = V_\Theta$. Then $gv = gwe \in Ve = V_\Theta$. Thus Ve is a G-module. Further, $(g, e)v = gwee = gwe = gv$. Thus, Ve is a (g, e)-subspace. Also

$$
(g, e)(v(1 - e)) = gv - gv = 0.
$$

Since $V = Ve \oplus V(1 - e)$, it follows that $Tr(\rho(g, e)) = Tr(\eta(g))$. ♯

Theorem 5.5.8 *Let $g \in G^F$. Let T, Θ, and B be as above. Then*

$$
R^B_{T,\Theta}(g) = \frac{1}{\mid T^F\mid}\sum_{t\in T^F}\Theta(t^{-1})L(\chi(g, t), \hat{X}),
$$

where $L(\chi(g, t), \hat{X})$ is the Lefschetz number of the automorphism $\chi(g, t)$ of $\hat{X}$ given by $\chi(g, t)(x) = gxt$.

Proof Recall that the Lefschetz number $L(\chi, X)$ of an automorphism χ of X is given by

$$L(\chi, X) = \sum_{r=0}^{2DimX} (-1)^r Tr(H_c^r(\chi)).$$

Now, by the definition

$$R_{T,\Theta}^B(g) = \sum_{r=0}^{2m} (-1)^r Tr(\eta_r(g)),$$

where η_r is the representation of G^F as given before. From Lemma 5.5.7, $Tr(\eta_r(g)) = Tr(\rho_r(g, e))$, where ρ_r is the representation of $G^F \times T^F$ as described earlier. Thus,

$$R_{T,\Theta}^B(g) = \sum_{r=0}^{2m} (-1)^r Tr(\rho_r(g, e)),$$

where

$$e = \frac{1}{\mid T^F \mid} \sum_{t \in T^F} \Theta(t^{-1})t.$$

Consequently,
$R_{T,\Theta}^B(g)$
$= \sum_{r=0}^{2m} Tr(\rho_r(g, \frac{1}{|T^F|} \sum_{t \in T^F} \Theta(t^{-1})t))$
$= \frac{1}{|T^F|} \sum_{t \in T^F} \Theta(t^{-1}) \sum_{r=0}^{2m} (-1)^r Tr(\rho_r(g, t))$
$= \frac{1}{|T^F|} \sum_{t \in T^F} \Theta(t^{-1}) \sum_{r=0}^{2Dim\hat{X}} (-1)^r Tr(H^r(\chi(g, t))$
$= \frac{1}{|T^F|} \sum_{t \in T^F} \Theta(t^{-1}) L(\chi(g, t), \hat{X})$ (by the definition). $\sharp$

Theorem 5.5.9 *Let T be a maximally split F-stable maximal torus of G which is contained in F-stable Borel subgroup B of G. Let Θ be an irreducible character of T^F and Θ_{B^F} the one-dimensional representation $B^F \longrightarrow B^F/U^F \approx T^F \stackrel{\Theta}{\to} \mathbb{C}^\star \bigcap \overline{\mathbb{Q}_l}^\star$ of B^F. Then $R_{T,\Theta}^B = \Theta_{B^F}^{G^F}$.*

Proof Let T be a maximally split F-stable maximal torus of G which is contained in F-stable Borel subgroup B of G. Fix the member $(B, T) \in \Omega^F$ (refer to Proposition 5.5.1.). We shall denote the set of all F-stable Borel subgroups of G by $\hat{\Omega}^F$. Thus, the first projection is a surjective G^F equivariant map from Ω^F to $\hat{\Omega}^F$. Let $g \in \hat{X} = L^{-1}(U)$. Then $L(g) = g^{-1}F(g) \in U \subseteq B$ and hence

$$F(gBg^{-1}) = F(g)BF(g)^{-1} = gg^{-1}F(g)BF(g)^{-1}gg^{-1} = gBg^{-1}.$$

This means that gBg^{-1} belongs to $\hat{\Omega}^F$. Consequently, we get a map η from $\hat{X}$ to $\hat{\Omega}^F$ which is given by $\eta(g) = gBg^{-1}$. Since any element of $\hat{\Omega}^F$ is of the form gBg^{-1} for some $g \in G^F = L^{-1}(\{1\}) \subseteq L^{-1}(U) = \hat{X}$ (Corollary 5.5.2), η is surjective.

Now, G^F acts on $\hat{X}$ through left multiplication and it also acts on $\hat{\Omega}^F$ through inner conjugation. Since

$$\eta(gx) = gxB(gx)^{-1} = gxBx^{-1}g^{-1} = g\eta(x),$$

η is an equivariant map. We write $x \approx y$ if $\eta(x) = \eta(y)$. Clearly, $\approx$ is an equivalence relation on $\hat{X}$. Since η is equivariant and G^F acts transitively on $\hat{\Omega}^F$, it also acts transitively on the finite set $\hat{X}/\approx = \{\overline{1} = \overline{x_1}, \overline{x_2}, \dots, \overline{x_r}\}$ and the action is given by $g\overline{x} = \overline{gx}$. Evidently, $\overline{1} = \hat{X} \bigcap B$ and $\overline{x} = \eta^{-1}(\eta(x))$ is a closed subset of $\hat{X}$ for all $x \in \hat{X}$ (note that $\eta(1) = B$). Further, if $t \in T^F$ and $x \in \hat{X}$, then $\eta(xt) = \eta(x)$. Thus, $\overline{xt} = \overline{x}$ for each $x \in \hat{X}$ and $t \in T^F$. Consequently, the action of T^F on $\hat{X}$ through right multiplication induces action on each equivalence class $\overline{x}$ in $\hat{X}/\approx$. Now, from Theorem 5.5.8,

$$R^B_{T,\Theta}(g) = \frac{1}{\mid T^F \mid}\sum_{t \in T^F} \Theta(t^{-1})L(\chi(g,t), \hat{X}),$$

where $L(\chi(g,t), \hat{X})$ is the Lefschetz number of the automorphism $\chi(g,t)$ of $\hat{X}$ given by $\chi(g,t)(x) = gxt$. Clearly, the stabilizer of $\overline{1}$ is $G^F \bigcap B = B^F$. We have a generalized character μ of $B^F \times T^F$ on $\overline{1}$ which is given by

$$\mu(b,t) = L(\chi(b,t), \overline{1}).$$

The generalized character $(g,t) \mapsto L(\chi(g,t), \hat{X})$ of $G^F \times T^F$ on $\hat{X}$ is the induced character $\mu^{G^F \times T^F}$, thanks to the property P_7 of the l-adic cohomology. Consequently, the generalized character $R^B_{T,\Theta}$ is induced by the generalized character

$$b \mapsto \frac{1}{\mid T^F \mid}\sum_{t \in T^F} \Theta(t^{-1})L(\chi(b,t), \overline{1}).$$

Next, we show that $\overline{1} = \hat{X} \bigcap B = T^F U$: Let $b = tu$ be a member of B, where $t \in T$ and $u \in U$. Then $L(b) = u^{-1}t^{-1}F(t)F(u)$. Thus, $b \in \hat{X}$ if and only if $L(b) = u^{-1}t^{-1}F(t)F(u)$ belongs to U. In other words, $b \in \hat{X}$ if and only if $t^{-1}F(t) \in U \bigcap T = \{1\}$. It follows that $\overline{1} = T^F U$. Consider the natural quotient map ν from $\overline{1} = T^F U$ to $T^F U/U \approx T^F/(T^F \bigcap U) \approx B^F/U^F$. Clearly, ν is a surjective morphism of varieties. Further, each fiber tU of ν is an affine variety isomorphic to $U \approx K^N$, where $N = DimU$. It follows from the property P_5 of the l-adic cohomology that

$$L(\chi(b,t), \overline{1}) = L(\chi(b,t), T^F U) = L(\nu o \chi(b,t), T^F U/U) = L(\chi(b,t), B^F/U^F).$$

Next, from the property P_{11},

$$L(\chi(b,t),\overline{1}) = L(\chi(b,t), B^F/U^F) = |(B^F/U^F)^{\chi(b,t)}|,$$

where $(B^F/U^F)^{\chi(b,t)}$ denote the set of fixed points of $\chi(b,t)$ on B^F/U^F. Let sU^F be a member of B^F/U^F, where $s \in T^F$. Now, sU^F belongs to $(B^F/U^F)^{\chi(b,t)}$ if and only if $bsU^Ft = sU^F$ or equivalently $bU^F = t^{-1}U^F$. Thus, if $b \notin t^{-1}U^F$, then $|(B^F/U^F)^{\chi(b,t)}| = 0$. Suppose that $b \in t^{-1}U^F$ ($tb \in U^F$). Then $|(B^F/U^F)^{\chi(b,t)}| = |T^F|$, since for each $b \in B^F$, there is a unique $t \in T^F$ such that $b \in t^{-1}U^F$. It follows that $R^B_{T,\Theta}$ is induced by the character $\hat{\mu}$ of B^F which is given by $\hat{\mu}(b) = \Theta(t^{-1})$, where t is the unique element of T such that $b \in t^{-1}U^F$. Evidently, $\hat{\mu} = \Theta_{B^F}$. This proves that $R^B_{T,\Theta} = \Theta^{G^F}_{B^F}$. ♯

Next, suppose that T is F-stable maximal torus which is not maximally split. Then T is not contained in any F-stable Borel subgroup of G. We describe $R^B_{T,\Theta}(g)$ in terms of the semi-simple and unipotent parts of g. More explicitly, we have the following Proposition:

Proposition 5.5.10 *Let $g \in G^F$. Suppose that $g = su$, where s is the semi-simple part and u is the unipotent part of g. Then s and u belong to G^F. Let $\hat{X}^{s,t}$ denote the subvariety $\{x \in \hat{X} \mid sxt = x\}$. Then $u\hat{X}^{s,t} \subseteq \hat{X}^{s,t}$ and*

$$R^B_{T,\Theta}(g) = \frac{1}{|T^F|}\sum_{t\in T^F}\Theta(t^{-1})L(l_u, \hat{X}^{s,t}),$$

where l_u is the left multiplication by u on $\hat{X}^{s,t}$.

Proof Note that $F(s)$ is semi-simple and $F(u)$ is unipotent. Since $su = g = F(g) = F(s)F(u)$ and a Jordan decomposition is unique, it follows that $F(s) = s$ and $F(u) = u$. Hence $s, u \in G^F$. Next, let $x \in \hat{X}^{s,t}$. Then $sxt = x$. But then $suxt = usxt = ux$. This means that $ux \in \hat{X}^{s,t}$. Now, by Theorem 5.5.8,

$$R^B_{T,\Theta}(g) = R^B_{T,\Theta}(su) = \frac{1}{|T^F|}\sum_{t\in T^F}\Theta(t^{-1})L(\chi(su,t), \hat{X}).$$

Since s is semi-simple, the order of s is co-prime to p (if $a^q = a$, then order of a divides $q-1 = p^m - 1$) and the order of u is a power of p. It follows from the property P_{10} of the l-adic cohomology that $L(\chi(su,t),\hat{X}) = L(l_u, \hat{X}^{s,t})$. This shows that

$$R^B_{T,\Theta}(g) = \frac{1}{|T^F|}\sum_{t\in T^F}\Theta(t^{-1})L(l_u, \hat{X}^{s,t}).$$

♯

Corollary 5.5.11 *Let $u \in G^F$ be a unipotent element. Let T be a F-stable maximal torus of G and Θ be any irreducible character of T^F. Then*

$$R^B_{T,\Theta}(u) = \frac{1}{|T^F|}L(u,\hat{X}) = R_{T,1}(u)$$

is an integer which is independent of Θ.

Proof Since $\hat{X}^{1,t} = \{x \in \hat{X} \mid xt = x\}$ is empty set if $t \neq 1$ and it is $\hat{X}$ if $t = 1$, it follows from the above Proposition that

$$R^B_{T,\Theta}(u) = \frac{1}{\mid T^F \mid}\sum_{t \in T^F} \Theta(t^{-1})L(l_u, \hat{X}^{1,t}) = \frac{1}{\mid T^F \mid}L(u, \hat{X}).$$

for all irreducible character Θ. By the property P_4 of the l-adic cohomology , $L(u, \hat{X})$ is an integer. Thus, $R^B_{T,\Theta}(u)$ is a rational number. Since a generalized character is an algebraic integer, and a rational algebraic integer is an integer, $R^B_{T,\Theta}(u)$ is an integer. ♯

Let $g \in G^F$ and $g = su$, where s is semi-simple and u is unipotent. Then $s \in G^F$ and also $u \in G^F$. Suppose that G is connected and simply connected. Then it is a fact that the centralizer $C_G(s)$ of s is connected and reductive. If $xs = sx$, then $F(x)s = F(x)F(s) = F(xs) = F(sx) = F(s)F(x) = sF(x)$. This means that $C_G(s)$ is also F- stable. Further, $x^{-1}Tx$ is contained in $C_G(s)$ whenever $x \in G^F$. Clearly, u is unipotent element of $C_G(s)$. However, in general $C_G(s)$ need not be connected. We state the following more general formula for $R^B_{T,\Theta}$ in terms of semi-simple and unipotent parts.

Theorem 5.5.12 *Let* $g \in G^F$ *and* $g = su$ *the Jordan decomposition, where* s *is semi-simple and* u *is unipotent. Then*

$$R^B_{T,\Theta}(g) = \frac{1}{\mid (C_G(s)^0) \mid}\sum_{x \in G^F \text{ and } xsx^{-1} \in T^F} \Theta(xsx^{-1})R^{B\bigcap(C_G(s))^0}_{x^{-1}Tx,1}(u),$$

where $C_G(s)^0$ *denote the component of identity of* $C_G(s)$.*(Note that* $B\bigcap(C_G(s))^0$ *is a* F*-stable Borel subgroup of* $(C_G(s))^0$.*).* ♯

Observe that Corollary 5.5.11, is a particular case of the above Theorem. We denote $R^B_{T,\Theta}(u)$ by $Q^G_T(u)$ and the function Q^G_T defined on the unipotent elements by $Q^G_T(u) = R^B_{T,1}(u)$ is called the **Green function**. This function was explored by J.A.Green in 1955 to study representations of $GL(n, q)$.

Our next aim is to state a formula for the inner product of Deligne–Lusztig generalized characters. Let T and T' be two F-stable maximal tori of G. Let $N(T, T')$ denote the subset $\{g \in G \mid gTg^{-1} = T'\}$ of G. Clearly, if $g \in N(T, T')$, then $gt \in N(T, T')$ for all $t \in T$. Thus, $N(T, T')$ is a union of a set of right cosets of T in G. Let $W(T, T')$ denote the set $\{Tg \mid g \in N(T, T')\}$. In particular, $W(T, T) = W(T)$ is the isotropy subgroup at T of the inner conjugation action of G on the set of all maximal tori of G. Since T and T' are F-stable, $gTg^{-1} = T'$ implies that $F(g)TF(g)^{-1} = T'$. Thus $N(T, T')$ is F-stable and each right coset of $N(T, T')$ mod (T) is F-stable. In particular, F acts on $W(T, T')$ and we can talk of $W(T, T')^F$. Let Θ and Θ' be irreducible characters of T^F and T'^F, respectively. Let $w = T^Fn \in W(T, T')^F$, where

$n \in N(T, T')^F$. Define a character $(\Theta')^w$ of T^F by putting $(\Theta')^w(t) = \Theta'(n^{-1}tn)$. We state the following formula for inner product of Deligne–Lusztig genralized characters. We leave the proof as it is lengthy and involved. The reader may refer to "Finite Groups of Lie Types: Conjugacy classes and Complex Characters" by R. W. Carter.

Theorem 5.5.13 $< R^B_{T,\Theta},\ R^{B'}_{T',\Theta'} > = |\ \{w \in W(T, T')^F \mid (\Theta')^w = \Theta\}\ |$. ♯

Corollary 5.5.14 *Suppose that* Θ *is in general position in the sense that* $(\Theta)^w = \Theta$ *if and only if* w *is identity elemement of* $W(T)^F$. *Then*

$$< R^B_{T,\Theta},\ R^B_{T,\Theta} > = 1,$$

and in particular, $R^B_{T,\Theta}$ *or* $-R^B_{T,\Theta}$ *is an irreducible character.* ♯

Indeed, if T is obtained by twisting with σ_T from the split F-stable maximal torus T_0, and Θ is in general position, then $Det\sigma_T R^B_{T,\Theta}$ is irreducible.

Corollary 5.5.15 *Suppose that the* F*-stable subgroups* T *and* T' *are not* G^F *conjugate. Then*

$$< R^B_{T,\Theta},\ R^{B'}_{T',\Theta} > = 0.$$

♯

Corollary 5.5.16 $R^B_{T,\Theta}$ *is independent of the Borel subgroup* B *containing* T.

Proof Le B and B' be Borel subgroups of G containing the F-stable maximal torus T. From Theorem 5.5.13,

$$< R^B_{T,\Theta},\ R^B_{T,\Theta} >=< R^B_{T,\Theta},\ R^{B'}_{T,\Theta} >=< R^{B'}_{T,\Theta},\ R^{B'}_{T,\Theta} > .$$

Consequently,

$$< R^B_{T,\Theta} - R^{B'}_{T,\Theta},\ R^B_{T,\Theta} - R^{B'}_{T,\Theta} > = 0.$$

Thus, $R^B_{T,\Theta} = R^{B'}_{T,\Theta}$. ♯

We can unambiguously write $R_{T,\Theta}$ for $R^B_{T,\Theta}$. The following Proposition expresses the regular character χ_{reg} of G^F in terms of Deligne–Lusztig characters.

Proposition 5.5.17 $\chi_{reg} = \frac{1}{|G^F|_p} \sum_{T \in M_F} \sum_{\Theta \in \hat{T^F}} Det\sigma_T R_{T,\Theta}$, *where* M_F *is the set of* F*-stable maximal tori.* ♯

Corollary 5.5.18 *For every irreducible character* χ *of* G^F, *there is a pair* (T, Θ) *such that*

$$< R_{T,\Theta},\ \chi > \neq 0.$$

♯

Exercises

5.5.1 Describe Deligne–Lusztig characters of $GL(2, q)$, $SL(2, q)$, and verify the results of this section.

5.5.2 Describe all F -stable subgroups of $GL(3, K)$ and find the orders of T^F for F-stable subgroups.

Bibliography

1. Artin et al., *SGA 4 Lecture Notes in Mathematics*, 269, 270, 305, Springer, (1972–73)
2. A. Borel, *Linear Algebraic Groups*, Benjamin, New York, 1969.
3. N. Bourbaki, *Lie Groups and Lie algebras*, Springer, 1989.
4. R. W. Carter, *Simple Groups of Lie Type*, John Wiley, London, 1972.
5. R. W. Carter, *Finite Groups of Lie Type, Conjugacy classes and Complex characters*, John Wiley, 1985.
6. C. Chevalley, *Sur certains groupes simples*, Tohoku Math. J, 7(1955), (14–66).
7. C. Chevalley, *Invariants of finite groups generated by reflections*, Amer. J. Math. 77. 778–782, 1955.
8. C. W. Curtis, *Chevalley Groups and related Topics*, Finite Simple Groups, edited by Powel and Higman, Academic press, 1971.
9. Curtis and Reiner, *Representation theory of finite groups and associative algebras*, AMS, New ed, 2006.
10. M. L. Curtis, *Matrix Groups*, Springer- Verlag, 1984.
11. P. Deligne, *La conjecture Weil I*, Publ. Math. IHES, 43, 203–307,1974.
12. P. Deligne, *La conjecture Weil II*, Publ. Math. IHES, 52, 137–252, 1980.
13. P. Deligne, *SGA* $4\frac{1}{2}$, *Cohomology étale*, Springer, Lecture Notes in Mathematics, 569, 1977.
14. P. Deligne and G. Lusztig, *Representations of Reductive Groups over finite fields*, Ann. of Math., 103, 1976.
15. J. Dieudonne, *La geometrie des groupes classiques*, Springer, 1963.
16. J. A. Green, *The characters of finite general linear groups*, TAMS, 80, (402–447), 1955.
17. Harish-Chandra, *Some application of universal enveloping algebra of a semi-simple algebra* TAMS, 70, 28–96, 1951.
18. J. E. Humphreys, *Introduction to Lie Algebras and Representation Theory*, GTM, Springer, 1972.
19. N. Jacobson, *Lie Algebras*, Dover Publication, 1979.
20. B. Kostant, *A formula for the multiplicity of a weight*, TAMS, 93, 53–73, 1959.
21. B. Kostant, *Groups over* $\mathbb{Z}$, *Algebraic Groups and Discontinuous Subgroups*, Proc. Symp. Pure Math. IX, Providence: Amer. Math Soc, 1966.
22. G. Lusztig, *Representations of finite Chevalley Groups*, CBMS Regional Conference Series in Mathematics, AMS, 39, 1977.

R. Lal, *Algebra 4*, Infosys Science Foundation Series,
https://doi.org/10.1007/978-981-16-0475-1

23. J. P.Serre, *Linear Representations of finite groups*, GTM, Springer, 1996.
24. J. P.Serre, *Lie Algebras and Lie Groups*, Benjamin, New York, 1965.
25. J. P.Serre, *Complex Semisimple Lie Algebras*, Springer Monographs in Mathematics, 2001.
26. R. Steinberg, *Lectures on Chevalley Groups*, Yale University, 1967.
27. Bhama Srinivasan, *Representations of Finite Chevalley Groups*, Lecture notes in Mathematics, Springer, 764, 1979.
28. M. Suzuki, *Group Theory I and II*, Springer, 1980.

Index

R. Lal, *Algebra 4*, Infosys Science Foundation Series,
https://doi.org/10.1007/978-981-16-0475-1

Printed in the United States
by Baker & Taylor Publisher Services